Guide to
the Vascular Flora
of Illinois

Robert H. Mohlenbrock

Southern Illinois University Press
Carbondale and Edwardsville

Feffer & Simons, Inc.
London and Amsterdam

Library of Congress Cataloging in Publication Data

Mohlenbrock, Robert H
  Guide to the vascular flora of Illinois.

  Includes index.
  1. Botany—Illinois. I. Title.
QK157.M63   581.9'773   75-22414
ISBN 0-8093-0704-9
ISBN 0-8093-0756-1 pbk.

Contents

Preface

THIS book is intended to be a field guide to the vascular flora of Illinois. It provides keys to 3,047 taxa of ferns, gymnosperms, and flowering plants in the state. For each taxon there is given a common name if one is in general usage in Illinois, followed by an indication of flowering time for angiospermous plants. A statement of habitat is provided for each taxon, as well as a general comment on its distribution in Illinois. For plants considered rare and therefore known from only a very few counties, those counties are usually enumerated.

The sequence of families follows that found in the eighth edition of Asa Gray's *Manual of Botany* (Fernald, 1950), except that the Phrymaceae is placed next to the Verbenaceae in this work. I am aware that there is criticism to the Englerian sequence of families found in Gray's *Manual,* and that consideration of data derived from cytological, biochemical, and anatomical studies indicates different affinities for many of our families. Nonetheless, because systems devised by Cronquist (1968), Thorne (1968), and others are unfamiliar to the majority of field botanists, both professional and amateur, the more familiar Englerian system is employed in this work.

The nomenclature used has been derived after consultation with many floras, monographs, and revisions. Where the nomenclature adopted in this work is at variance with that found in the eighth edition of Gray's *Manual of Botany* (Fernald, 1950), the third edition of the Nathaniel Lord Britton and Addison Brown *Illustrated Flora of the Northern United States, Canada, and the British Possessions* . . . (Gleason, 1952), and the third edition of *Flora of Illinois* (Jones, 1963), the nomenclature used in these three works is indicated, followed by the abbreviations F, G, and J, respectively.

Included in this manual are all taxa known to be native in Illinois either at present or in the past, and all alien vascular plants for which there is evidence that they maintain themselves year after year without cultivation. No taxon is included in this work unless it has been verified by the writer or by Mr. Floyd Swink who is responsible for many recent discoveries in the Chicago area.

Illinois extends for nearly 380 miles north to south, more than one-third the distance from Canada to the Gulf of Mexico. Its maximum width is about 200 miles. The total number of square miles is just less than 58,000. The highest point of elevation is at Charles Mound in the extreme northwestern corner of the state, with an elevation above mean sea level of 1,241 feet. Average elevation for the state is 600 feet above sea level.

The French botanist André Michaux was apparently the first plant collector to visit Illinois when he crossed into the territory from Vincennes in 1795, traversed westward to Kaskaskia, and then southward and out of the state. In this short visit, Michaux collected fifty-two plants which were new to science. In 1818, Constantine Rafinesque botanized in the southeastern corner of the state.

The first flora for the state was prepared by Dr. S. B. Mead, a physician from Augusta, Illinois, who published, in 1846, a list of 900 species of plants known to occur in Illinois. In 1857, Increase Lapham wrote *A catalogue of the plants of the state of Illinois* in which he recorded 1,127 plants for the state. By 1876, H. N. Patterson was able to list 1,542 species of plants from Illinois.

For nearly three quarters of a century, no updated accounting of the plants of the entire state was available. In 1945, 1950, and 1963, G. N. Jones published the three editions of *Flora of Illinois* in which he listed 2,125, 2,200, and 2,400 species, respectively. Since I began work on the forty-volume "Illustrated Flora of Illinois" in 1960, more than 500 taxa of vascular plants have been added to the Illinois flora. This *Guide to the Vascular Flora of Illinois* accounts for 2,699 species, 265 lesser taxa, and 83 hybrids in Illinois. Continued study and exploration are adding several dozen taxa yearly to the Illinois flora.

I am indebted to Dr. Dan K. Evans who contributed *Carex* to this work, and to Dr. Richard P. Wunderlin who wrote the Compositae. Dr. Robert Haynes of Ohio State University has offered suggestions in *Potamogeton,* and Dr. James Richardson, Dr. Robert A. DeFilipps, and Dr. Kenneth L. Weik have helped with *Sagittaria, Juncus,* and the Lemnaceae, respectively.

I am grateful to the Illinois Nature Preserves Commission who has kindly let me use *The Natural Divisions of Illinois* in its entirety, and to Mr. John Schwegman, who was the principal author.

Mr. Floyd Swink of the Morton Arboretum has been a constant

source of inspiration for this book. Without his expertise, much would be lacking from this work.

Lastly I give much thanks to my wife, Beverly, who typed the several difficult drafts of the manuscript.

Robert H. Mohlenbrock

Southern Illinois University
August 1974

County Map of Illinois

Guide to the Vascular Flora of Illinois

1	Wisconsin Driftless Division
2	Rock River Hill Country Division
	a Freeport Section
	b Oregon Section
3	Northeastern Morainal Division
	a Morainal Section
	b Lake Michigan Dunes Section
	c Chicago Lake Plain Section
	d Winnebago Drift Section
4	Grand Prairie Division
	a Grand Prairie Section
	b Springfield Section
	c Western Section
	d Green River Lowland Section
	e Kankakee Sand Area Section
5	Upper Mississippi River and Illinois River Bottomlands Division
	a Illinois River Section
	b Mississippi River Section
6	Illinois River and Mississippi River Sand Areas Division
	a Illinois River Section
	b Mississippi River Section
7	Western Forest-Prairie Division
	a Galesburg Section
	b Carlinville Section
8	Middle Mississippi Border Division
	a Glaciated Section
	b Driftless Section
9	Southern Till Plain Division
	a Effingham Plain Section
	b Mt Vernon Hill Country Section
10	Wabash Border Division
	a Bottomlands Section
	b Southern Uplands Section
	c Vermilion River Section
11	Ozark Division
	a Northern Section
	b Central Section
	c Southern Section
12	Lower Mississippi River Bottomlands Division
	a Northern Section
	b Southern Section
13	Shawnee Hills Division
	a Greater Shawnee Hills Section
	b Lesser Shawnee Hills Section
14	Coastal Plain Division
	a Cretaceous Hills Section
	b Bottomlands Section

The Natural Divisions of Illinois

Courtesy of the Illinois Nature Preserves Commission

The Natural Divisions of Illinois

1. Introduction

The Natural Divisions of Illinois [1] is a classification of the natural environments and biotic communities of Illinois based on physiography, flora, and fauna. The general approach was first applied to Illinois by Vestal,[2] who developed a preliminary vegetation map of the state based mainly on topography, soil type, and natural vegetation.

The natural divisions of Illinois have been derived through an expansion of Vestal's approach. Factors considered in delimiting the 14 natural divisions are topography, soils, bedrock, glacial history, and the distribution of plants and animals. The divisions are divided into 33 sections based on differences that are of the same nature as those used to separate the divisions but deemed less significant. The role of each factor in developing this system is discussed in the text.

TOPOGRAPHY

Glacial history has played an important role in shaping Illinois's topography. Except for relatively small driftless areas in northwestern, west-central, and extreme southern Illinois, the topography can be characterized as a more or less dissected plain of glacial till. Rugged topography is characteristic of the driftless areas and of much of the till plain along the major river valleys. Other areas of significant relief within the till plain are the youthful moraines of northeastern Illinois and areas of thin drift and rolling topography in the Rock River valley and southeastern Illinois. The remainder of

---

[1] Used by permission of the Illinois Nature Preserves Commission. Written primarily by John Schwegman.

[2] Vestal, A. G. 1931. A Preliminary Vegetation Map of Illinois. Transactions of the Illinois State Academy of Science 23:204–17.

this till plain is an arc of older, eroded Illinoian till south and west of the younger, less dissected Wisconsinan till.

Topography has influenced the biota of Illinois by controlling the diversity of habitats. In the glaciated regions, forest was restricted mainly to the rugged topography of stream valleys and moraines while prairie occupied most of the level uplands and some of the broad floodplains. In general, the more rugged the topography the greater the diversity of habitats. Youthful topography, such as the Wisconsinan till plain, is poorly drained, resulting in an abundance of aquatic habitats. Drainage has been improved by ditching in most prairie regions of the till plain.

SOILS

The principal soil features influencing the recognition of divisions and sections are sand areas, soils derived from glacial till, deep loess soils, and claypan soils. The description of soil conditions within the divisions is only generalized. Alluvial soils occur in every division, but they are not usually mentioned in the text.

Some of the most important features of soils are parent material, texture, and degree of development. The general terms "heavy soils," "light soils," and "sandy soils" refer to texture. Heavy soils are predominantly clay; moisture generally moves slowly through such soils and aeration is poor. They often provide a more rigorous environment for plant growth. Light soils contain a mixture of clay, silt, and sand, are well aerated, and readily yield moisture to plants. Sandy soils are well aerated and have poor moisture-holding ability. They provide a severe environment for plants because of their droughtiness and susceptibility to wind erosion.

The parent material for most Illinois soils is loess, a wind-deposited silt. Loess deposits are deepest near the major river valleys and become thinner and of finer texture with increased distance. This mantle of loess has reduced the diversity of soil types in much of the state. Deep loess soils are generally less weathered, and more fertile than those derived from thin loess, especially in the southern third of Illinois.

Soils derived primarily from glacial till are restricted to northeastern Illinois, where geographic position has prevented deep accumulations of loess. The texture of these soils ranges from gravel to clay. The diversity of soil parent material is partly responsible for the great variety of environments and varied biota of this region.

Sand derived from glacial outwash is another important parent material that occurs as terraces along major streams throughout Illinois. In some instances the soils developed from sand support biota distinctive enough to warrant recognition in the natural divisions. Local sand areas are not listed because they are generally not considered to be of statewide significance.

Alluvial soils tend to be lighter in central and northern Illinois than in southern Illinois. They frequently support distinctive biota.

Peat and bedrock are parent materials that give rise to soils which often support unusual plants and animals. Although limited in area, they have contributed to the recognition of some divisions and sections.

The degree of soil development is a function of time and climate. Older soils tend to be more strongly developed and have many of their soluble minerals and clay particles leached from the surface layers. The same is true of soils developed in areas of higher precipitation, as in southern Illinois. A strongly developed soil may have a claypan formed by the accumulation of clay particles in the subsoil. Claypan restricts penetration by roots and movement of ground water, thus sometimes determining the dominant vegetation over wide areas. For example, the strongly developed claypan soils of south-central Illinois determine to a great extent the distribution of post oak flatwoods. These soils are a primary factor in the recognition of the Southern Till Plain Division.

Younger soils are derived from relatively recent deposits of till, loess, or alluvium, or exposure of bedrock. They are not strongly developed, are relatively rich in minerals, and lack claypans.

BEDROCK

Bedrock differences are often reflected in the topography, especially in driftless areas. Bedrock also controls the plant life to some extent through the influence it has on the thin soils developed from it. Bedrock type can determine the presence of caves and other habitats.

The Ozark Division shows the effect of bedrock differences. The Northern Section (Monroe County) and the Southern Section (Union and Alexander counties) are both parts of the Salem Plateau of the Ozarks and were presumably uplifted at about the same time. The Southern Section is underlain by cherty limestone which is resistant to erosion and has produced very rugged topography with surface drainage. Steep cherty slopes in this section support acid-soil plants such as shortleaf pine, azalea, and farkleberry. On the other hand, the Northern Section is underlain by relatively pure limestone that is less resistant to erosion and groundwater solution, producing gentler topography with some internal drainage through sinkholes and solution cavities. Natural sinkhole ponds are common. Most of the acid-soil plants are lacking, and the flora includes a distinctive assemblage of Ozark "limestone glade" species.

NATURAL VEGETATION

The dominant species and some of the most frequent associates in the principal plant communities of each natural division are listed in the text, but no attempt is made to characterize the total vegetation of each division. Generally the descriptions pertain to natural vegetation in Illinois about the time of settlement. At present this vegetation has been completely destroyed by man over vast sections of the state.

The striking contrast between prairie and forest vegetation that existed at the time of settlement is of little consequence to the recognition of natural divisions, as both forest and prairie or prairie-like communities occur in every division. However, the relative abundance of the prairie community was important in delimiting some divisions.

At the time of the settlement of Illinois there were considerable areas termed "barrens" by land surveyors. Apparently the barrens developed into forest or were cleared for agriculture soon after settlement. In any event, they were destroyed before being thoroughly described. They probably occurred in all the natural divisions, but they are not included in the vegetation accounts because their species composition is unknown.

FLORA

An analysis of the distribution of vascular plants in relation to physiographic provinces, bedrock types, soil types, topography, and drainage systems reveals a considerable number of distinctive species apparently restricted to certain provinces or natural features. These restricted distributions, especially for such dominant plants as trees, were important in delimiting the natural divisions. Many of these distinctive plants are listed in the descriptions of the divisions.

FAUNA

The distribution of birds in Illinois is less adequately documented than that of other classes of vertebrates. Nor does their distribution correlate as well with the natural divisions recognized here, being more heavily influenced by climate and by artificial and transitory conditions. For these reasons, distinctive species of birds are mentioned in the descriptions of only a few natural divisions.

Examination of the distribution of non-avian vertebrates in Illinois reveals many correlations with physiographic features, vegetation, watersheds, and soil types. The present work indicates considerable correlation between the natural divisions and patterns of fish distribution, and to a lesser extent with those of mammals.

The vertebrates that are considered distinctive of certain divisions and species that have influenced the location of divisional boundaries are mentioned in the descriptions of the divisions. There is no attempt to characterize the fauna of each division.

No attempt has been made to examine the distribution of invertebrates, or their relation to the natural divisions. The plains scorpion is mentioned as a distinctive feature of the Ozark Division, but this is the only exception.

CLIMATE

Climate has played a role in the delineation of some divisions and sections by its effects on the distribution of plants and animals, especially in extreme northern Illinois and southern Illinois. However, there is little direct correspondence between climatic regions and the natural divisions.

Illinois has a continental climate with hot humid summers and cold winters. This climate varies considerably from north to south in temperature and precipitation. The northern counties have lower mean annual temperatures, shorter growing seasons, and less precipitation than the southern counties.

The mean annual temperature ranges from about 47° F. in the north to about 59° F. in the south. The mean January temperature ranges from 22° F. in the north to 36° F. in the south, and the mean July temperature ranges from 73° F. in the north to 80° F. in the south. The average length of the growing season ranges from less than 160 days in northwestern Illinois to more than 200 days in the southern tip of the state.

Average annual precipitation increases from about 32 inches in the north to 47 inches in the south. Although the southern counties receive the most rainfall, much of this occurs in winter. Precipitation during the growing season is more nearly uniform throughout the state.

2. Wisconsin Driftless Division

The Wisconsin Driftless Division is part of an area extending from northwestern Illinois into Wisconsin, Iowa, and Minnesota that apparently escaped Pleistocene glaciation. This division is one of the

most maturely developed land surfaces in Illinois and is characterized by rugged terrain that originally was mostly forested. It has the
coldest climate in the state. It contains several distinctive plants of
northern affinity and some species that may represent relics of the
pre-ice-age flora. The division contains lead deposits.

BEDROCK

The Wisconsin Driftless Division is a maturely dissected upland of
Ordovician and Silurian limestone, dolomite, and shale. Bedrock
crops out along the major watercourses. Prominent "mounds"
capped with the more resistant dolomite are common. A mineralized zone containing deposits of lead and zinc is an important feature. Caves are known in the dolomite.

TOPOGRAPHY

The topography of the Wisconsin Driftless Division is one of rolling hills and great relief, particularly along interior stream canyons.
High erosional remnants (including Charles Mound, the highest
point in Illinois, with an elevation of 1,257 feet), are prominent features. There are loess-capped bluffs and palisades along the Mississippi River valley and ravines and bluffs throughout the division.

SOILS

The soils of this division have developed from loess or, on
steeper slopes, from loess on bedrock. The loess soils are derived
from thick deposits and are weakly to moderately developed. The
soils on bedrock are thin to moderately thick and well drained.

PLANT COMMUNITIES

FOREST   The original vegetation of the Wisconsin Driftless Division was predominantly upland hardwood forest dominated by
black oak (*Quercus velutina*) and white oak (*Quercus alba*) on dry
sites and by sugar maple (*Acer saccharum*), basswood (*Tilia americana*), and red oak (*Quercus rubra*) on mesic sites. Floodplain
forests dominated by silver maple (*Acer saccharinum*), American
elm (*Ulmus americana*), and green ash (*Fraxinus pennsylvanica* var.
subintegerrima) occupy alluvial soils of the stream valleys. Cliffs
and cool, shaded slopes of this division often support white pine
(*Pinus strobus*), Canada yew (*Taxus canadensis*), and white birch
(*Betula papyrifera*).

PRAIRIE   Dry prairie on the rolling uplands contained such species of the northern Great Plains as the plains buttercup (*Ranunculus rhomboideus*), pasque flower (*Anemone patens*), June grass (*Koeleria macrantha*), and Wilcox's panic grass (*Panicum wilcoxianum*), along with the dominant little bluestem (*Schizachyrium scoparium*) and side-oats grama (*Bouteloua curtipendula*). Areas of mesic and wet prairie were infrequent. Frink's Prairie and Jules Prairie were two extensive prairies on the uplands. Loess hill prairies dominated by little bluestem and side-oats grama occur on the steep southwest-facing bluffs above the Mississippi River floodplain.

AQUATIC HABITATS

The main aquatic habitats of the Wisconsin Driftless Division are creeks and rivers. Springs are local, and intermittent streams are characteristic of the ravines.

SPECIAL FEATURES

The Wisconsin Driftless Division contains several distinctive plants considered relicts of preglacial or interglacial floras, such as jeweled shooting star (*Dodecatheon amethystinum*), sullivantia (*Sullivantia renifolia*), and cliff goldenrod (*Solidago sciaphila*). Cool, shaded ravines, cliffs, and river bluffs provided habitat for relict populations of some distinctive northern plants including woodland white violet (*Viola incognita*), bird's-eye primrose (*Primula mistassinica*), American stickseed (*Hackelia americana*), and moschatel (*Adoxa moschatellina*). Some of these distinctive plants are restricted to river bluffs of the Mississippi River valley, and others grow only in interior stream canyons.

Principal Natural Features

FOREST: Dry upland, mesic upland, floodplain.

PRAIRIE: Dry, mesic, wet.

LOESS HILL PRAIRIE

BEDROCK: Outcrops of dolomite and shale, zinc and lead deposits, caves.

TOPOGRAPHY: Mississippi River bluffs, interior stream canyons, ravines, ridges and "mounds," level to rolling upland.

AQUATIC HABITATS: Creeks, rivers.

SPECIAL FEATURES: Northern and preglacial or interglacial plants.

3. Rock River Hill Country Division

The Rock River Hill Country Division is a region of rolling topography that is drained by the Rock River. It has a thin mantle of glacial till. Prairie formerly occupied the larger expanses of level uplands, but forest was equally abundant along the water courses and in the more dissected uplands. Several distinctive plant species occur in this division. The Freeport and Oregon sections are recognized on the basis of bedrock types and resultant floral differences.

GLACIAL HISTORY

The Rock River Hill Country Division is thinly mantled with glacial drift from the Illinoian and early Wisconsinan stages of Pleistocene glaciation. The Pecatonica lobe of the Altonian substage of the Wisconsinan glacial stage extended westward in this division into Stephenson and Ogle counties.

BEDROCK

The bedrock is primarily Ordovician and Silurian dolomite and limestone. Outcrops of the dolomite occur throughout the division, particularly along the streams. St. Peter sandstone underlies the Oregon Sections and crops out frequently. There are caves in the dolomite near the Mississippi River.

TOPOGRAPHY

The topography varies from level to rolling, with river valleys, and "dells" or bluffs along streams throughout the division. The Oregon Section is very rough, with bluffs, ridges, and ravines in the sandstone. The Pecatonica River meanders through a broad plain formed by sediments of a glacial lake.

SOILS

The soils have developed primarily from moderately thick loess and thin loess on bedrock. Soils developed from glacial outwash occur in the major river valleys. Small areas of soils developed from glacial till occur throughout the division.

PLANT COMMUNITIES

FOREST The forests are similar to those of the Wisconsin Driftless Division. They occurred on slopes and areas protected from prairie fires. The dry upland forests are dominated by black oak, white oak, bur oak (*Quercus macrocarpa*), and wild black cherry (*Prunus serotina*). Mesic sites support forests dominated by sugar maple, basswood, slippery elm (*Ulmus rubra*), and red oak. White pine, Canada yew, and yellow birch (*Betula lutea*) are occasionally found on cool, shaded bluffs. The floodplain forests are particularly well developed along the Pecatonica River and are dominated by silver maple, black willow (*Salix nigra*), cottonwood (*Populus deltoides*), American elm, and ashes (*Fraxinus* spp.). The Oregon Section is heavily forested with black oak, white oak, and bur oak.

PRAIRIE   Prairies occupied much of the level to rolling uplands. The largest was known as Shannon Prairie. The dry upland prairies contained the floral elements of the northern Great Plains listed for the Wisconsin Driftless Division. Mesic prairies were the most common type and contained the species typical of the Grand Prairie, the dominant grasses being big bluestem (*Andropogon gerardii*), Indian grass (*Sorghastrum nutans*), and prairie dropseed (*Sporobolus heterolepis*). Wet prairies contained cord grass (*Spartina pectinata*), bluejoint grass (*Calamagrostis canadensis*), and big bluestem.

MARSH   Marshes in poorly drained parts of the prairies and along the streams and river floodplains are dominated by cattail (*Typha* sp.), bulrushes (*Scirpus* spp.), sedges (Cyperaceae), and occasionally common reed (*Phragmites australis*).

AQUATIC HABITATS

The major aquatic habitats of the Rock River Hill Country Division are rivers and creeks. Meander scar sloughs characterized the Pecatonica River floodplain. Springs and seepage areas occur locally, especially in the sandstone areas of the Oregon Section.

FREEPORT SECTION   The Freeport Section includes most of the Rock River Hill Country Division and is characterized by rolling hills and dolomite and limestone bedrock.

Principal Natural Features

FOREST: Dry upland, mesic upland, floodplain.
PRAIRIE: Dry, mesic, wet.
MARSH
BEDROCK: Outcrops of dolomite and limestone caves.
TOPOGRAPHY: Stream canyons and bluffs, floodplain, rolling uplands, meander scars.
AQUATIC HABITATS: Rivers, creeks, sloughs.
SPECIAL FEATURE: Northern relict plants.

OREGON SECTION   The Oregon Section is distinguished from the rest of the division by its sandstone bedrock and the unique northern plants associated with it. These distinctive northern relicts include ground pine (*Lycopodium dendroideum*), rusty woodsia (*Woodsia ilvensis*), oak fern (*Gymnocarpium dryopteris*), and American wintergreen (*Pyrola americana*).

Principal Natural Features

FOREST: Dry upland, mesic upland, floodplain.
PRAIRIE: Dry, mesic, wet.
BEDROCK: Sandstone outcrops.
TOPOGRAPHY: Rolling uplands, floodplain, ravines, bluffs.
AQUATIC HABITATS: Rivers, creeks.
SPECIAL FEATURE: Northern relict plants.

4. Northeastern Morainal Division

The Northeastern Morainal Division is the region of most recent glaciation in Illinois. Glacial landforms are common features and are responsible for the rough topography over most of the area. Lake-bed deposits and beach sands are also frequent features. Unlike most of Illinois, the soils of this division are derived from glacial drift rather than loess. Drainage is poorly developed, and many natural lakes are found. This division contains distinctive northern and eastern floral elements including the bog community. Several species of animals are known in Illinois only from this area.

The sections are recognized because of differences in topography, soil, glacial history, flora, and fauna.

BEDROCK

The bedrock is primarily Ordovician and Silurian limestone and dolomite with some shale. The bedrock is deeply buried by glacial drift but limestone crops out along some of the streams.

GLACIAL HISTORY

The Northeastern Morainal Division is covered with deep glacial drift from the Altonian and Woodfordian substages of the Wisconsinan glacial stage. Moraines, kames, eskers, and other glacial landforms occur throughout the division.

TOPOGRAPHY

Moraines and morainic systems are dominant topographic features and account for the rough, hilly, and rolling terrain of most of the division. There are outwash plains at the fronts of major terminal moraines, such as the Marengo Ridge. The Chicago lake plain and ancient beach ridges are prominent features of the Chicago area and were formed by sediments of glacial Lake Chicago. Bluffs were formed along Lake Michigan north of Chicago during highwater stages of glacial Lake Chicago.

Sand dunes are present along Lake Michigan north of Waukegan and east of the Sugar River in Winnebago County. The Lake Michigan dunes are well developed and are associated with the beach area. This is the only division to have a natural beach-and-dunes association. Ridges and swales occur in the sand area north of Waukegan and in the Chicago lake plain.

SOILS

The soils are derived primarily from glacial drift, lake bed sediments, beach deposits, and peat. They range from very poorly drained to well drained on the uplands. They are diverse in texture, ranging from gravel and sand to silty clay loams. The many different soils are responsible for the diversity of plant communities found in this division.

PLANT COMMUNITIES

FOREST   Bur oak and white oak dominate the dry upland forests on moraines and other glacial landforms of this division. Many of the forests were typical oak openings but have been heavily grazed or have rapidly grown up to closed, dense forests due to the effective exclusion of fire. The mesic upland forests were dominated by sugar maple, basswood, red oak, and white ash (*Fraxinus americana*). Notable northern shrubs of these mesic forests are highbush cranberry (*Viburnum trilobum*) and red elder (*Sambucus pubens*). Beech (*Fagus grandifolia*) occurred in a few ravines along Lake Michigan. The dominant species of the floodplain forests are silver maple, green ash, and American elm. Some of the poorly drained upland forests are dominated by swamp white oak (*Quercus bicolor*). Black oak and Hill's oak (*Quercus ellipsoidalis*) occur on the sandy soils of the dunes and ridges, forming savanna-like or scrub-oak forests. Jack pine (*Pinus banksiana*) and white pine once grew on sandy ridges in the Chicago lake plain. Tamarack (*Larix laricina*) occurs in some poorly drained depressions of the Valparaiso morainic system. Plants essentially restricted in Illinois to the forests of the Northeastern Morainal Division include maple-leaf viburnum (*Viburnum acerifolium*), round-lobed hepatica (*Hepatica nobilis* var. *obtusa*), wood reed (*Cinna latifolia*), large-leaved aster (*Aster macrophyllus*), moccasin flower (*Cypripedium acaule*), and purple trillium (*Trillium erectum*).

PRAIRIE The presettlement vegetation of the Northeastern Morainal Divison was about 60 percent prairie. Dry prairie on the gravel moraines and eroded bluffs was dominated by little bluestem and side-oats grama, and contained several species typical of the western plains. The mesic prairie and wet prairie were dominated by prairie dropseed, big bluestem, Indian grass, switch grass (*Panicum virgatum*), cord grass, and bluejoint grass, and contained many characteristic forbs. Extensive areas of sand prairie occurred in the sand areas of the Chicago lake plain, east of the Sugar River in Winnebago County, and in the Lake Michigan Dunes Section. Some distinctive plants of sandy prairie and sandy open woods are sweet-fern (*Comptonia peregrina*), speckled alder (*Alnus rugosa*), yellow fringed orchid (*Habenaria ciliaris*), fringed gentian (*Gentiana crinita*), and small fringed gentian (*Gentiana procera*). Colic-root (*Aletris farinosa*), hardhack (*Spiraea tomentosa*), and lupine (*Lupinus perennis*) occur in sand prairies near Lake Michigan.

FEN A fen is a type of wet prairie with an alkaline water source. Fens are associated with calcareous springs and seeps and also occur in swales and on low ground near lakes in areas of calcareous ground water. They warrant recognition because of their distinctive species composition. Some of the notable plants are small, white lady's slipper (*Cypripedium candidum*), grass-of-Parnassus (*Parnassia glauca*), meadow spikemoss (*Selaginella apoda*), Ohio goldenrod (*Solidago ohioensis*), Kalm's lobelia (*Lobelia kalmii*), shrubby cinquefoil (*Potentialla fruticosa*), low calamint (*Satureja arkansana*), and white camass (*Zigadenus glaucus*).

MARSH Marshes are conspicuous plant communities of the Northeastern Morainal Division, common because of the poorly drained soils. Marshes are generally dominated by cattails and bulrushes, with common reed locally abundant.

SEDGE MEADOW Many of the marshes of this division grade into an open community where sedges dominate instead of cattails and bulrushes, forming sedge meadows. These are often associated with wet-shrub communities dominated by dogwoods (*Cornus* spp.), willows (*Salix* spp.) and, infrequently, speckled alder (*Alnus rugosa*).

BOG True bogs are found in Illinois only in the Northeastern

Morainal Division. These have formed in poorly drained depressions in the Valparaiso morainic system and contain many distinctive plants such as pitcher plant (*Sarracenia purpurea*), sundew (*Drosera* sp.), cranberry (*Vaccinium macrocarpon*), leatherleaf (*Chamaedaphne calyculata*), poison sumac (*Rhus vernix*), winterberry (*Ilex verticillata*), and dwarf birch (*Betula pumila*). All stages of bog succession—young, mature, and old—are represented.

AQUATIC HABITATS

The Northeastern Morainal Division is poorly drained and has many aquatic habitats. It contains all of Illinois's glacial lakes and is the only division on Lake Michigan. The glacial lakes are generally of two types: those with a peat base, and those with a sand or marl base. The biota of each type is different. Some species of water marigold (*Bidens beckii*) and two species of bladderwort (*Utricularia* spp.) are restricted to the glacial lakes of this division.

DISTINCTIVE FAUNA

The pugnose shiner, blackchin shiner, and banded killifish are known in Illinois only from the glacial lakes of this division. The alewife, American smelt, lake chub, and ninespine stickleback are fishes that occur in Illinois only along the shore of Lake Michigan. Lake trout and lake whitefish are species restricted to the deeper waters of Lake Michigan. The common tern breeds in Illinois in only a few locations near Lake Michigan, and the piping plover did so until fairly recently. Breeding of the golden-winged warbler in Illinois is restricted to this division, where it has been known to hybridize with the blue-winged warbler. Breeding of the Nashville warbler in Illinois is reported only from this division. Except for casual occurrences, the large population of oldsquaws that winters on Illinois waters is restricted to Lake Michigan; and much the same is true for wintering white-winged, common, and surf scoters. The occurrence in Illinois of several species of migrant waterbirds such as the parasitic, pomarine, and long-tailed jaegers, the knot, and Bonaparte's and little gull is limited almost entirely to Lake Michigan or its shores. Other animals restricted to this division are the spotted turtle, blue-spotted salamander, and pigmy shrew.

MORAINAL SECTION The Morainal Section of the Northeastern Morainal Division encompasses the moraines and morainic systems of the late advances of the Woodfordian substage of Wisconsinan glaciation. This section contains most of Illinois's glacial lakes as well as its true bogs. Glacial landforms are well represented.

Principal Natural Features

FOREST: Dry upland, mesic upland, wet upland, floodplain tamarack swamp.

PRAIRIE: Dry, mesic, wet.

FEN

MARSH

SEDGE MEADOW

BOG: Youthful, mature, old.

GLACIAL LANDFORMS: Moraines, kames, eskers, drumlins kettleholes.

TOPOGRAPHY: Rolling upland, ravines, lake bluffs.

AQUATIC HABITATS: Rivers, creeks, glacial lakes, sloughs.

LAKE MICHIGAN DUNES SECTION   This section is recognized because of the unique flora of its dunes and beaches. This flora includes some sand-binding plants that differ from those of the inland sand deposits, such as beach grass (*Ammophila breviligulata*), creeping juniper (*Juniperus horizontalis*), and bearberry (*Arctostaphylos uva-ursi*). Plant communities range from sand prairie to marsh, to scrub-oak forest. The continuing vegetational succession from shifting sand to stabilized sand results in a wide variety of plant associations.

Principal Natural Features

FOREST: Scrub-oak.

SAND PRAIRIE: Dry, mesic, and wet.

FEN

MARSH

TOPOGRAPHY: Beach, ridges and swales, dunes.

AQUATIC HABITATS: Creeks, Lake Michigan.

CHICAGO LAKE PLAIN SECTION   The Chicago lake plain is a flat, poorly drained area of lake-bed sediments deposited by glacial Lake Chicago. It is recognized because of its special topography and physiographic history. Long ridges of shore-deposited sand are conspicuous topographic features. A few natural lakes occur near Calumet City. The original vegetation was mostly prairie and marsh, with scrub-oak forests on sandy ridges. Black gum (*Nyssa sylvatica*) and sassafras (*Sassafras albidum*) are found in some of the wet forests. Thismia (*Thismia americana*), one of the most unusual plants of the American flora, occurred only in the Chicago Lake Plain Section, near Lake Calumet, but is now considered extinct because this area has been converted to indusry.

Principal Natural Features

FOREST: Scrub-oak, mesic upland, floodplain.
PRAIRIE: Dry, mesic, wet.
FEN
MARSH
TOPOGRAPHY: Lake plain, ridges and swales.
AQUATIC HABITATS: Lakes, creeks, Lake Michigan.

WINNEBAGO DRIFT SECTION This section encompasses the Winnebago Formation of Altonian drift. This early Wisconsinan drift is better drained than that of the Morainal Section, having fewer marshes and no glacial lakes. The original vegetation was predominantly prairie, with oak openings, dry upland forest, and well-developed floodplain forests. Glacial outwash is extensive along many of the creeks and rivers. Wet prairie and marsh with large sedge meadows or "prairie bogs" occur in the sand area along Coon Creek. Dunes have formed along the east bank of the Sugar River and support sand prairie vegetation and dry upland forests of black oak and Hill's oak.

Distinctive features of this section are the extensive gravel hill prairies which once extended along the eroded east bluffs of the Rock River valley into Wisconsin. These prairies contained many western elements, including pasque flower, plains buttercup, and prairie smoke (*Geum triflorum*).

Principal Natural Features

FOREST: Dry upland, mesic upland, floodplain.
PRAIRIE: Dry, mesic, wet.
SAND PRAIRIE: Dry, mesic, wet.
MARSH
SEDGE MEADOW
BEDROCK: Outcrops of limestone and sandstone.
TOPOGRAPHY: Dunes, outwash plains, river terraces, meander scars.
AQUATIC HABITATS: Rivers, creeks, sloughs.

5. Grand Prairie Division

The Grand Prairie Division is a vast plain formerly occupied primarily by tall-grass prairie. The soils were developed from recently deposited loess, lake-bed sediments, and outwash and are generally very fertile. Natural drainage was poor, resulting in many marshes and prairie potholes. Forest bordered the rivers and there were occasional groves on moraines and other prominent glacial landforms. The sections of this division are differentiated on the basis of soils, topography, and glacial history.

At one time bison grazed the prairies and waterfowl in great numbers occupied the marshes and potholes. The steel plow brought about the rapid destruction of the vast Illinois prairies. Ditches and tile lines drained almost all of the marshes and potholes. The bison were gone by 1814. The abundant waterfowl were displaced. The giant Canada goose was extirpated as a breeding bird, and other characteristic species disappeared or became scarce. The prairie, once seemingly limitless, is now one of the rarest plant communities in Illinois, with only pitifully small and often degraded patches remaining.

GLACIAL HISTORY

The Grand Prairie is a rather level, poorly drained plain of glacial drift from the Illinoian and Wisconsinan stages of Pleistocene glaciation. Repeated advances and retreats of the Wisconsinan glaciers created a series of moraines and morainic systems of which the Shelbyville and Bloomington morainic systems are conspicuous.

BEDROCK

The bedrock, deeply buried by glacial drift, crops out only along the larger rivers. Major outcrops of sandstone are found near Ottawa along the Illinois and Fox rivers. Dolomite crops out along the Kankakee River west of Kankakee.

TOPOGRAPHY

The topography of the Grand Prairie Division is generally level to rolling, with the major stream valleys and the extensive systems of moraines providing the greatest relief. Large flat expanses of lake bed deposits are found in Lasalle, Kendall, Will, Grundy, Livingston, Ford, Iroquois, Kankakee, and Douglas counties. Extensive out wash plains and sand dunes are found in Kankakee and Iroquois counties and in the valleys of the Green River and lower Rock River. The major rivers have well-developed floodplains, and in many areas there are ravines in the bluffs.

SOILS

The soils are relatively young and high in organic content, having developed from a thin to moderately thick layer of loess, glacial drift, or lake-bed sediments. Soils developed from sand, muck, and peat exist in the Kankakee Sand Area and Green River Lowland sections. Deep loess occurs along the Illinois and the lower Sangamon rivers.

PLANT COMMUNITIES

FOREST   The forests of the Grand Prairie Division are generally associated with the stream valleys and moraines. On dry sites, the forests are dominated by white oak, black oak, and shagbark hickory (*Carya ovata*), with shingle oak (*Quercus imbricaria*) and bur oak frequent associates. On mesic sites these species are replaced by basswood, sugar maple, slippery elm, American elm, hackberry (*Celtis occidentalis*), red oak, and white ash. Black walnut (*Juglans nigra*), bitternut hickory (*Carya cordiformis*) and, in the northern part, bigtooth aspen (*Populus grandidentata*) are common. The floodplain forests are of the silver maple-American elm-ash type. The prairie groves were influenced by recurrent fires and are generally of two types: one dominated by bur oak and the other dominated by American elm and hackberry. Sandy soils in the Green River Lowland and Kankakee Sand Area sections support scrub forests of black oak.

PRAIRIE   The vast prairies of Illinois were once one of its most remarkable features. They contained several hundred species of grasses and forbs and were interspersed by numerous marshes and prairie potholes. At the time of settlement, wet and mesic prairie were the most widespread plant communities of the Grand Prairie Division. Mesic prairie was dominated by big bluestem, Indian grass, prairie dropseed, switch grass, and little bluestem and contained many characteristic prairie plants such as leadplant (*Amorpha canescens*), compass plant (*Silphium laciniatum*), prairie dock (*Silphium terebinthinaceum*), and rattlesnake master (*Eryngium yuccifolium*). The wet sites were dominated by cord grass, sedges, and bluejoint grass and supported such species as ironweed (*Vernonia fasciculata*), boneset (*Eupatorium perfoliatum*), swamp milkweed (*Asclepias incarnata*), and water hemlock (*Cicuta macu-lata*).

Dry upland prairie occurs mainly on steep slopes along the Illi-nois River, on loess bluffs along the lower Sangamon River, and on gravel moraines and kames. The dominant species of the dry prairies are little bluestem and side-oats grama. They commonly contain such forbs as scurf pea (*Psoralea tenuiflora*), pale beard tongue (*Penstemon pallidus*), false boneset (*Kuhnia eupatorioides*), cylindrical blazing star (*Liatris cylindracea*), and fringed puccoor (*Lithospermum incisum*).

Sand prairie occurs in the Kankakee Sand Area and Green River Lowland sections. Little bluestem, fall witch-grass (*Leptoloma cog natum*), and sand dropseed (*Sporobolus cryptandrus*) are the im-portant grasses, and goat's rue (*Tephrosia virginiana*) and spotted monarda (*Monarda punctata*) are common forbs. June grass and porcupine grass (*Stipa spartea*) are commom. The sand prairies contain several species of Great Plains affinity such as western ragweed (*Ambrosia psilostachya*), prickly-pear cactus (*Opuntia ra finesquii*), poppy mallow (*Callirhoe triangulata*), hairy grama (*Bou teloua hirsuta*), western sunflower (*Helianthus occidentalis*), silky aster (*Aster sericeus*), and flax-leaved aster (*Aster linariifolius*).

MARSH The formerly common marshes and prairie potholes of the Grand Prairie Division were dominated by bulrushes, sedges bur-reeds (*Sparganium* spp.), cattails, and common reed and con-tained many species of aquatic and semiaquatic plants such as ar rowhead (*Sagittaria* spp.), water plantain (*Alisma subcordatum*), pondweed (*Potamogeton* spp.), pickerelweed (*Pontederia cordata*), beggar-ticks (*Bidens* spp.), and water crowfoot (*Ranunculus* spp.).

AQUATIC HABITATS

The aquatic habitats of the Grand Prairie Division are rivers creeks, and prairie potholes.

DISTINCTIVE FAUNA

Distinctive animals of the Grand Prairie Division are Blanding's turtle, western smooth green snake, western fox snake, eastern plains garter snake, Kirtland's water snake, northern lined snake Franklin's ground squirrel, and thirteen-lined ground squirrel. None of these animals are completely restricted to this division, but all are most abundant here. Several amphibians and reptiles common outside the limits of Wisconsinan glaciation are conspicuously ab sent from the Grand Prairie.

GRAND PRAIRIE SECTION   The Grand Prairie Section encompasses the area outside the Northeastern Morainal Division that was covered by the Woodfordian substage of the Wisconsinan stage of Pleistocene glaciation, excluding the outwash and sand areas. The Shelbyville and Bloomington morainic systems form the boundaries of this section. Mesic black-soil prairie, marshes, and prairie potholes in the young, poorly-drained drift are characterstic. Glacial landforms are common. The Kankakee mallow (*Iliamna remota*) and Chase aster (*Aster schreberi*) are endemic to Illinois and restricted to small areas of the Grand Prairie Section. The Kankakee mallow is found only on an island in the Kankakee River, and Chase aster is known only from three counties near Peoria.

Principal Natural Features

FOREST: Floodplain, mesic upland, dry upland.
PRAIRIE GROVE
PRAIRIE: Wet, mesic, dry.
MARSH
BEDROCK: Sandstone and dolomite outcrops.
TOPOGRAPHY: Level to rolling upland, floodplain, ravines, river bluffs, lake plains, glacial landforms.
AQUATIC HABITATS: Prairie potholes, rivers, creeks.
SPECIAL FEATURES: Endemic plants.

SPRINGFIELD SECTION   The Springfield Section is part of the Illinoian drift, and the drainage system is better developed than in the younger Wisconsinan drift of the Grand Prairie Section. This section was mostly prairie in presettlement times. Deep loess deposits that support dry hill prairie occur along the lower Sangamon River. Large tracts of floodplain forest exist in the valley of the lower Sangamon River and its tributaries.

Principle Natural Features

FOREST: Floodplain, mesic upland, dry upland.
PRAIRIE: Wet, mesic, dry.
LOESS HILL PRAIRIE
MARSH
BEDROCK: Outcrops.
TOPOGRAPHY: Level to rolling upland, floodplain, ravines, river bluffs.
AQUATIC HABITATS: River, creeks.

WESTERN SECTION   The Western Section was predominantly prairie in presettlement times and therefore is in the Grand Prairie Division. It is part of the older, dissected Illinoian drift.

Principal Natural Features

FOREST: Floodplain, mesic upland, dry upland.
PRAIRIE: Wet, mesic, dry.
MARSH
BEDROCK: Outcrops.
TOPOGRAPHY: Level to rolling upland, floodplain, ravines.
AQUATIC HABITATS: Creeks.

GREEN RIVER LOWLAND SECTION   The broad valley of the Green River and lower Rock River was formed by glacial melt waters. Much glacial outwash was deposited, and sand flats and dunes developed. This section had extensive marshes and wet prairies. It has scrub-oak forests on the dry sandy ridges and floodplain forests along the rivers. Sand prairie occupied the sand flats and dunes. Most of this section has been disturbed by grazing, drainage, and cultivation. There are some active dunes where the stabilizing cover has been removed.

Principal Natural Features.

FOREST: Scrub-oak, floodplain.
PRAIRIE: Wet, mesic.
SAND PRAIRIE: Wet, mesic, dry.
MARSH
TOPOGRAPHY: Outwash plain, dunes.
AQUATIC HABITATS: Rivers.

KANKAKEE SAND AREA SECTION   The sand of the Kankakee Sand Area Section was deposited by the Kankakee Flood during late Wisconsinan glaciation. Sand prairie and marsh were the predominant vegetation of this section before the land was drained for cultivation. Primrose violet (*Viola primulifolia*) and Carey's smartweed (*Polygonum careyi*) are restricted in Illinois to this section. Scrub-oak forests occur on drier sites. The clear, well-vegetated sand-bottomed streams contain such unusual fishes as the weed shiner, iron color shiner, and least darter.

Principal Natural Features

FOREST: Scrub-oak.
SAND PRAIRIE: Wet, mesic, dry.
MARSH
TOPOGRAPHY: Outwash plain, dunes.
AQUATIC HABITATS: Creeks.

6. Upper Mississippi River and Illinois River Bottomlands Division

The Upper Mississippi River and Illinois River Bottomlands Division encompasses the rivers and floodplains of the Mississippi River above its confluence with the Missouri River and the bottomlands and associated backwater lakes of the Illinois River and its major tributaries south of LaSalle. It does not include the major sand deposits, which are in a separate division. Much of the division was originally forested but prairie and marsh also occurred. The more sluggish nature of the Illinois River and its distinctive backwater lakes distinguishes the Illinois River Section from the Upper Mississippi River Section.

BEDROCK

The bedrock of the two river valleys is deeply covered by alluvial deposits.

TOPOGRAPHY

The bottomlands of the Mississippi River and the Illinois River are characterized by broad floodplains and gravel terraces formed by glacial flood waters.

SOILS

The soils are from recent alluvium and glacial outwash. They are poorly drained, alkaline to slightly acidic, and vary from sandy to clayey. In general they are lighter than the alluvial soils of the Lower Mississippi River Bottomlands Division.

PLANT COMMUNITIES

FOREST The bottomland forests are generally dominated by silver maple, American elm, and green ash. Pin oak (*Quercus palus-*

tris) is the most important oak; pecan (*Carya illinoensis*), bur oak, sycamore (*Platanus occidentalis*), honey locust (*Gleditsia tria-canthos*), hickories (*Carya* spp.), and black walnut are frequent. Black willow and river birch (*Betula nigra*) are common in the Mississippi River Section. A few southern lowland species, including water locust (*Gleditsia aquatica*), overcup oak (*Quercus lyrata*), sugarberry (*Celtis laevigata*), deciduous holly (*Ilex decidua*), and swamp privet (*Forestiera acuminata*) range into the southern part of this division.

PRAIRIE   In presettlement times mesic prairie and wet prairie occurred in the broad bottomlands. The species composition of these prairies was similar to that of the prairies of the Grand Prairie Division.

MARSH   Marshes containing the typical marsh species are important features throughout both sections.

SPRING BOGS   Spring-fed bogs with peat deposits are found on terraces along the Illinois River. These are unique to the Illinois River and the species composition differs from that of the bogs of the Northeastern Morainal Division. Distinctive plants include black ash (*Fraxinus nigra*), willows, poison sumac, and skunk cabbage (*Symplocarpus foetidus*).

AQUATIC HABITATS

Oxbow lakes occur in both the Illinois River valley and the Mississippi River valley. Backwater lakes are distinctive of the Illinois River valley, and the Illinois River is more sluggish than the Mississippi River. Springs are common in gravel terraces along the Illinois River. The bottomland forests are subject to prolonged periods of flooding.

DISTINCTIVE FAUNA

The fish faunas of the Illinois River and the upper Mississippi River are similar although they differ somewhat from that of the silt-laden Mississippi River south of its confluence with the Missouri River.

ILLINOIS RIVER SECTION   The Illinois River Section is distinguished from the Mississippi River Section by its distinctive backwater lakes and differences in forest vegetation. The spring bogs along the river bluffs are a special feature.

Principal Natural Features

FOREST: Bottomland.
PRAIRIE: Wet, mesic.
MARSH
SPRING BOGS
TOPOGRAPHY: River floodplain, river terraces.
AQUATIC HABITATS: Backwater lakes, oxbow lakes, rivers.

MISSISSIPPI RIVER SECTION   This section is composed of several distinct bottomlands along the Mississippi River, from Wisconsin to Calhoun County. Most of the prairies of this section have been drained for agriculture. Forests are still found along the river inside the levees and on the river islands.

Principal Natural Features

FOREST: Bottomland.
PRAIRIE: Wet, mesic.
MARSH
TOPOGRAPHY: River floodplain, river terraces.
AQUATIC HABITATS: Oxbow lakes, rivers.

7. Illinois River and Mississippi River Sand Areas Division

The Illinois River and Mississippi River Sand Areas Division encompasses the sand areas and dunes in the bottomlands of the Illinois and Mississippi rivers and includes the "perched dunes" atop the bluffs near Hanover in JoDaviess County. Scrub-oak forest and dry sand prairie are the natural vegetation of this division. Several plant species found here are more typical of the short-grass prairies to the west of Illinois. Several "relict" western amphibians and reptiles are known only from these sand areas. The two sections are distinguished because of differences in flora and fauna.

TOPOGRAPHY

The topography is generally one of level to rolling plains of sand deposited by glacial melt waters and blown into widespread areas east of the rivers. In many areas the sand has migrated onto the bluffs and uplands east of the river terraces. In places, dunes 20 to

40 feet high have formed and blowouts are common in unstabilized sand.

SOILS

The soils are derived from sand and sandy material. Other soils in depressions surrounded by sand are also in this division. The soils are generally droughty and subject to wind erosion. Low areas are generally wet.

PLANT COMMUNITIES

FOREST   The forests of this division are limited to scrubby stands dominated by black oak in the Mississippi River Section and black and blackjack oaks (*Quercus marilandica*) in the Illinois River Section.

PRAIRIE   Sand prairie, composed of such species as little bluestem, June grass, Indian grass, and porcupine grass, is the major community of this division. The sand prairie habitats range from dry to wet and include such plants as goat's rue, spotted monarda, prickly-pear cactus, tubercled three-awned grass (*Aristida tuberculosa*), poppy mallow, and fall witch-grass. The dry sites have western floral elements, including sand love-grass (*Eragrostis trichodes*), hairy grama and, rarely, bladderpod (*Lesquerella ludoviciana*), and Patterson's bindweed (*Stylisma pickeringii*). The mesic and wet sites have prairie vegetation similar to that of the Grand Prairie Division. There are also various plant associations related to unstabilized sand. Long-leaved calamovilfa (*Calamovilfa longifolia*) is one of the principal sand binders.

MARSH   Marsh occurs in low, poorly drained areas.

DISTINCTIVE FAUNA

Distinctive animals are the bullsnake, plains hognose snake, Illinois mud turtle, and Illinois chorus frog. The Illinois chorus frog is restricted to the Illinois River Section. The white-tailed jackrabbit is found in Illinois only in the northern part of the Mississippi River Section. The lark sparrow breeds most commonly in Illinois in the sandy habitats of this division, and a northern outlier population of the summer tanager breeds in forests of the Illinois River Section.

ILLINOIS RIVER SECTION   The Illinois River Section is distinguished from the Mississippi River Section on the basis of floral and faunal differences.

Principal Natural Features

FOREST: Scrub-oak.
SAND PRAIRIE: Dry, mesic, wet.
MARSH
TOPOGRAPHY: Dunes, blowouts, level to rolling plain.

MISSISSIPPI RIVER SECTION   The Mississippi River Section has floral elements absent from the Illinois River Section, including rock spikemoss (*Selaginella rupestris*), rough-seeded rock-pink (*Talinum rugospermum*), and beach-heath (*Hudsonia tomentosa*). Rock spikemoss and beach-heath form large mats that stabilize blowouts.

Principal Natural Features

FOREST: Scrub-oak.
SAND PRAIRIE: Dry, mesic, wet.
MARSH
TOPOGRAPHY: Dunes, blowouts, level to rolling plain.

8. Western Forest-Prairie Division

The Western Forest-Prairie Division is a strongly dissected glacial till plain of Illinoian and Kansan age. At the time of settlement, forest was the predominant vegetation, but there was also considerable prairie on the level uplands. The prairie soils were developed from loess and are fertile. The two sections are geographically separated by the Illinois River valley and also have some faunal differences.

GLACIAL HISTORY

Most of the bedrock is covered by glacial drift from the Illinoian stage of Pleistocene glaciation. There is an area of older Kansan drift in the western parts of Pike and Adams counties.

BEDROCK

Pennsylvanian and Mississippian bedrocks of limestone, sandstone, shale, and coal crop out frequently along the major streams.

TOPOGRAPHY

The till plain is strongly dissected, with many ravines in the level to rolling uplands. Floodplains are developed along the major streams.

SOILS

Most of the soils are fairly young, having developed from four to five feet of loess. The prairie soils are high in organic matter and are similar to those of the Grand Prairie Division. The forest soils are acidic and low in organic matter. Relatively small areas of droughty, fine-textured soils have developed in till on some steep slopes.

PLANT COMMUNITIES

FOREST   The upland forests consist of an oak-hickory association with black oak, white oak, and several species of hickory as the dominants. Scattered sites of fine-textured soils support a post oak-blackjack oak community. This forest community also occurred on the margins of the prairies, perhaps because of fires. The mesic forests contain white oak, red oak, basswood, sugar maple, and slippery elm. The floodplain forests are dominated by silver maple, American elm, ashes, and box elder (*Acer negundo*).

PRAIRIE   In presettlement times large prairies existed on the uplands of this division. Carthage Prairie, Hancock Prairie, and Bushnell Prairie were in the Galesburg Section; and String Prairie and Brown's Prairie were in the Carlinville Section. The prairie vegetation was similar to that of the Grand Prairie Division, but wet prairie was less frequent.

MARSH   Poorly drained areas with marsh vegetation are less frequent than in the Grand Prairie Division but are similar in composition.

AQUATIC HABITATS

The aquatic habitats of this well-drained division consist mainly of rivers and creeks.

DISTINCTIVE FAUNA

There is some continuity of animal life between the Western Forest-Prairie Division and the Southern Till Plain Division. The five-lined skink, the broad-headed skink, and the ornate box turtle are species that occur throughout the Southern Till Plain Division and range into this division but are absent from the adjoining Grand Prairie Division.

GALESBURG SECTION The Galesburg Section is the area north of the broad Illinois River valley and is·distinguished from the Carlinville Section because of its separate location. There were about equal amounts of forest and prairie in this section at the time of settlement, with the forests primarily in the well-dissected areas along tributaries of the Illinois River.

Principal Natural Features

FOREST: Dry upland, mesic upland, floodplain.
PRAIRIE: Dry, mesic, wet.
MARSH
BEDROCK: Outcrops.
TOPOGRAPHY: Level to rolling uplands, ravines, floodplain.
AQUATIC HABITATS: Rivers, creeks.

CARLINVILLE SECTION The Carlinville Section is the area of well-dissected land southeast of the Illinois River valley. The original vegetation of this section was mostly forest, with only about 12 percent of the area prairie.

Principal Natural Features

FOREST: Dry upland, mesic upland, floodplain.
PRAIRIE: Dry, mesic, wet.
MARSH
BEDROCK: Outcrops.
TOPOGRAPHY: Level to rolling uplands, ravines, floodplain.
AQUATIC HABITATS: Creeks.

9. Middle Mississippi Border Division

The Middle Mississippi Border Division consists of a relatively narrow band of river bluffs and rugged terrain bordering the Mississippi River floodplain from Rock Island County to St. Clair County and the lower Illinois River floodplain. Forest is the predominant

natural vegetation but hill prairies are common on the west-facing
bluffs. The soils were generally developed from very deep loess.
Limestone cliffs are common features. This division is best distin
guished from the river bluffs to the north and south of it by the ab
sence of certain floral and faunal elements. The Driftless Section
was never glaciated and is distinguished from the remainder of the
division.

BEDROCK

The bedrock of the Middle Mississippi Border Division consists of
limestone and sandstone, with dolomite associated with the sand
stone in Calhoun County. Outcrops and cliffs of limestone are com
mon along the bluffs, and sandstone outcrops occur in Rock Island
and Calhoun counties. Cretaceous gravels and clays occur in Pike
and Adams counties. There are caves throughout the division in the
limestone and dolomite but they are most abundant in unglaciated
Calhoun County.

GLACIAL HISTORY

Most of the division was glaciated during the Illinoian stage of
Pleistocene glaciation. An area of older Kansan drift is in western
Hancock, Adams, and Pike counties. Calhoun County and parts of
Pike and Adams counties apparently escaped Pleistocene glacia
tion.

TOPOGRAPHY

The Middle Mississippi Border Division is greatly dissected, par
ticularly along the major streams where there are bluffs and ra
vines. Sinkhole plains are most common in the southern part of the
division. The Driftless Section is higher and has more rugged to
pography than the Glaciated Section.

SOILS

Most of the soils on the uplands have developed from deep, well
drained loess. Isolated areas of heavy soils are also present, espe
cially in the southern part of the division.

PLANT COMMUNITIES

FOREST The vegetation of the Middle Mississippi Border Divi
sion is mostly mesic and dry forests associated with the dissected

uplands. The forests of the dry sites are dominated by black oak and white oak. Sugar maple, basswood, red oak, hackberry, slippery elm, and black walnut are major components of the forests on mesic sites. Floodplain forests along the creeks contain silver maple, hickories, cottonwood, and sycamore. Post oak (*Quercus stellata*) is common on the heavy soils and near ridge tops.

PRAIRIE   Prairies of the Middle Mississippi Border Division are limited to the steep slopes and ridges of deep loess atop the river bluffs. The prairies are dominated by little bluestem and side-oats grama, with purple prairie clover (*Petalostemum purpureum*) and flowering spurge (*Euphorbia corollata*) among the most frequent forbs. Scurf pea, a distinctive western plant, is also common. Stickleaf (*Mentzelia oligosperma*), another western plant, reaches its northeastern limits on the exposed limestone ledges of this division and of the Northern Section of the Ozark Division.

AQUATIC HABITATS

There are creeks throughout the division and sinkhole ponds in the Driftless Section.

DISTINCTIVE FAUNA

The dark-sided salamander and western worm snake are restricted in Illinois to the Middle Mississippi Border Division. Forested glens of this division serve as major nighttime roosting places for wintering bald eagles.

GLACIATED SECTION   The topography of this section has been modified by the Illinoian and Kansan stages of Pleistocene glaciation. Limestone underlies most of the Glaciated Section and frequently forms cliffs along the river bluffs.

Principal Natural Features

FOREST: Dry upland, mesic upland, floodplain.
LOESS HILL PRAIRIE
BEDROCK: Limestone outcrops, caves.
TOPOGRAPHY: River bluffs, ravines, floodplain, sinkhole plain.
AQUATIC HABITATS: Creeks.
SPECIAL FEATURE: Roosting areas for wintering bald eagles.

DRIFTLESS SECTION   The Driftless Section apparently escaped Pleistocene glaciation. Its topography is rougher than that of the

Glaciated Section, and it has many sinkholes and sinkhole ponds. Except for the jeweled shooting star, the Driftless Section is not known to harbor preglacial relict plants.

Principal Natural Features

FOREST: Dry upland, mesic upland, floodplain.
LOESS HILL PRAIRIE
BEDROCK: Limestone and sandstone outcrops, caves.
TOPOGRAPHY: River bluffs, ravines, floodplain, sinkhole plain.
AQUATIC HABITATS: Creeks, sinkhole ponds.
SPECIAL FEATURE: Roosting areas for wintering bald eagles.

10. Southern Till Plain Division

The Southern Till Plain Division encompasses most of the area of dissected Illinoian glacial till plain south of the Shelbyville Moraine and the Sangamon River and Macoupin Creek watersheds. Both forest and prairie were present at the time of settlement. The soils are relatively poor because of their high clay content and the frequent occurrence of a "claypan" subsoil. Post oak flatwood forest is characteristic of the division. The two sections are distinguished because of topographic differences.

BEDROCK

The bedrock of the Southern Till Plain Division consists of sandstone, limestone, coal, and shale, which commonly crop out in the eastern and southeastern parts of the division. Bedrock lies near the surface in the Mt. Vernon Hill Country Section.

GLACIAL HISTORY

The Illinoian stage of Pleistocene glaciation reached the southernmost limit of North American continental glaciation just beyond the limits of this division. The Southern Till Plain Division is entirely covered by Illinoian till. Glacial landforms are common only in the northwestern part of the division.

TOPOGRAPHY

The glacial till of the Southern Till Plain Division becomes thinner from north to south. The bedrock of the Mt. Vernon Hill Country

Section is near the surface, accounting for its hilly and rolling topography. The Effingham Plain Section is a nearly level to dissected till plain. There are broad floodplains along the major
streams and there are ravines in the bluffs along the stream valleys.

SOILS

The soils on the uplands are light colored and strongly developed, with poor internal drainage. They have developed from thin
loess and till under both forest and prairie vegetation. Fragipan and
claypan layers are characteristic of the upland soils. Some of the
prairie soils have a high sodium content and are known locally as
"alkaline slicks."

PLANT COMMUNITIES

FOREST   The level, poorly drained, heavy soils on the uplands of
the Southern Till Plain Division support a characteristic flatwoods
forest of post oak, swamp white oak, blackjack oak, and pin oak.
Forests on the uplands, where drainage is slightly improved, include black oak, shingle oak, mockernut hickory (*Carya tomentosa*), and shagbark hickory in the post oak community. The forests
on the slopes along stream valleys are dominated by white oak,
shingle oak, and black oak on the drier southern and western exposures, with hickories, white ash, basswood, sugar maple, wild
black cherry, slippery elm, and black walnut with the oaks on the
more mesic sites. Forests in the broad floodplains of the Kaskaskia
and Big Muddy rivers are dominated by silver maple, willows, sycamore, and American elm near the rivers, with pin oak, white oak,
hickories, ashes, hackberry, and honey locust on the heavier soils
farther from the rivers. Pin oak occasionally grows in nearly pure
stands over large areas of the floodplain. The floodplain forests of
the smaller streams have a higher percentage of oaks than the
floodplain forests of central and northern Illinois. Pin oak and shingle oak are dominant, with white oak, red oak, hickories, black walnut, river birch, and cottonwood occasional associates. Shumard
oak (*Quercus shumardii*) and sweet gum (*Liquidambar styraciflua*)
grow in the floodplain of the Big Muddy River but are not generally
abundant.

PRAIRIE   At the time of settlement about 40 percent of the
uplands of the Southern Till Plain Division supported prairie vegetation. Most of the prairie was of the mesic tall-grass type characteristic of the Grand Prairie Division. Twelve Mile Prairie and Look-
ing Glass Prairie were two large expanses of prairie. Mesic prairies
extended along the west side of the division almost to the limit of

glaciation but were rare in the southeastern part. Wet prairie was not common but did occur in parts of the Kaskaskia River floodplain. It is not known whether the alkaline slicks of this division supported a unique prairie flora.

MARSH   Marshes were associated with the stream floodplains of this division.

AQUATIC HABITATS

The aquatic habitats of the Southern Till Plain Division consist of rivers, creeks, and oxbow lakes.

DISTINCTIVE FAUNA

The northern crayfish frog, northern fence lizard, ground skink, five-lined skink, and broad-headed skink are common in the Southern Till Plain Division but are rare or absent from the Grand Prairie Division.

EFFINGHAM PLAIN SECTION   The Effingham Plain Section is a relatively flat plain drained by the Kaskaskia River. It originally was mostly prairie. Post oak flatwoods are characteristic of the uplands of this section.

A few flocks of the greater prairie chicken remain within this section. Sanctuaries are being established to maintain the population by providing nesting habitat.

Principal Natural Features

FOREST: Upland flatwoods, dry upland, mesic upland, and floodplain.

PRAIRIE: Wet, mesic, dry.

MARSH

BEDROCK: Outcrops.

TOPOGRAPHY: Level to rolling upland, dissected till plain ravines, floodplain.

AQUATIC HABITATS: Creeks, rivers, oxbow lakes.

SPECIAL FEATURES: Greater prairie chicken population, alkaline slicks.

MT. VERNON HILL COUNTRY SECTION   The Mt. Vernon Hill Country Section is distinguished from the Effingham Plain Section by its rolling, hilly topography. In presettlement times upland forests covered most of this section. The striped shiner and the stoneroller, two fishes of nearly statewide distribution, are absent from most of the Mt. Vernon Hill Country Section. The broad bottomlands of the major rivers that drain the eastern part of this section are considered to be part of the Wabash Border Division.

Principal Natural Features

FOREST: Upland flatwoods, dry upland, mesic upland, floodplain.

PRAIRIE: Wet, mesic, dry.

MARSH

BEDROCK: Outcrops

TOPOGRAPHY: Rolling till plain, dissected till plain, ravines, floodplain.

AQUATIC HABITATS: Creeks, rivers, oxbow lakes.

11. Wabash Border Division

The Wabash Border Division includes the bottomlands of the Wabash River and its major tributaries, the loess-covered uplands bordering the Wabash River, and the forests of the Vermilion River, Little Vermilion River, and Crab Apple Creek. This is a region of lowland oak forests containing beech, tuliptree (*Liriodendron tulipifera*), and other trees typical of the forests to the east of Illinois. The Wabash River drainage contains several distinctive fishes. The sections are distinguished by differences in topography, glacial history, and flora and fauna.

BEDROCK

The bedrock of the Wabash Border Division consists of limestone, sandstone, coal, and shale. Small outcrops of bedrock occur along some of the streams throughout the division. Bedrock outcrops along the river bluffs are few and never form the towering cliffs characteristic of the western border of Illinois.

GLACIAL HISTORY

All but a small area in the southern part of the Wabash Border Division was subject to Pleistocene glaciation. The Vermilion River dissects younger Wisconsinan drift, whereas the remainder of the division is the older Illinoian till plain. There are areas of outwash sand along the lower Wabash River.

TOPOGRAPHY

The topography of the Wabash Border Division is relatively gentle. The Southern Uplands Section is a low, eroded till plain, with bluffs above the bottomlands. The Vermilion River Section has more rugged topography resulting from erosion of its streams into Wisconsinan drift. The rivers have broad floodplains formed by glacial lakes, which include terrace deposits and many meander scars.

SOILS

The soils in the Bottomlands Section range from floodplain soils to terrace soils. The soils of the Southern Uplands Section have developed from moderately deep loess deposits. Soils of the Vermilion River Section are derived from thin loess over loamy till. Despite the diversity of soils, this division is united by the continuity of its forest vegetation and its fish and amphibian faunas.

PLANT COMMUNITIES

FOREST   The forests of the bottomlands are developed either on Recent alluvium or on Pleistocene terraces, which are generally better drained. Bottomland forests in the floodplains of the major rivers are dominated by pin oak, overcup oak, swamp white oak, swamp chestnut oak (*Quercus michauxii*), bur oak, cherrybark oak (*Quercus pagodaefolia*), and Shumard oak. Commonly associated with the oaks are sweetgum, hackberry, American elm, kingnut hickory (*Carya laciniosa*), silver maple, and pecan. The forests on the terraces contain a mixture of the tree species found on the floodplains and those found on the uplands. Shumard oak, bur oak, and sweetgum are the commonest, with pin oak and swamp white oak in poorly drained areas. The best-drained terraces contain shagbark hickory and tuliptree. Other characteristic trees are kingnut hickory, Kentucky coffeetree (*Gymnocladus dioica*), and hackberry. Along the rivers black willows, cottonwood, sycamore, and silver maple predominate. Sloughs near the Wabash River and Saline River contain bald cypress (*Taxodium distichum*) and some other southern swamp species.

The mesic forests of the Southern Uplands Section and the Vermilion River Section are dominated by white oak, red oak, and sugar maple and frequently contain beech and tuliptree. Forests on drier sites are dominated by black oak and hickories. Forests in the floodplains of the Southern Uplands Section and Vermilion River Section contain silver maple, cottonwood, willows, sycamore, and American elm.

PRAIRIE   At the time of settlement some of the Southern Uplands Section supported mesic prairie, while mesic and wet prairie occurred in the Bottomlands Section. The prairies of this division were similar to those of the Grand Prairie Division.

MARSH   The poorly drained bottomlands contain large areas of marsh associated with the sloughs and meander scars. These are dominated by cord grass and river bulrush (*Scirpus fluviatilis*).

AQUATIC HABITATS

The poorly drained Bottomlands Section of the Wabash Border Division contains many aquatic habitats, including rivers, oxbow lakes, and sloughs. The Southern Uplands and Vermilion River sections are better drained, and the aquatic habitats consist mainly of creeks and rivers.

DISTINCTIVE FAUNA

Fishes limited in Illinois to the Wabash Border Division are the river chub, river redhorse, mountain madtom, and greenside darter. The northern bigeye chub and bluebreast darter are known in Illinois only from the Vermilion River system, and the harlequin darter is kown only from the Embarras River. The red-backed salamander is essentially restricted in Illinois to this division.

BOTTOMLANDS SECTION   The Bottomlands Section of the Wabash Border Division encompasses the bottomland forests, sloughs, marshes, and oxbow lakes in the floodplains of the Wabash River, the Ohio River, and their major tributaries. Bottomland forests are the predominant vegetation with wet prairie and marsh associated with the sloughs.

Principal Natural Features

FOREST: Floodplain, terrace, swamp.
PRAIRIE: Mesic, wet.
MARSH
TOPOGRAPHY: River floodplain, terrace deposits, meander scars.
AQUATIC HABITATS: Oxbow lakes, sloughs, rivers.

SOUTHERN UPLANDS SECTION   The Southern Uplands Section contains the dry and mesic upland forests on the deep loess bluffs along the Wabash River. The upland forests of white oak, sugar maple, beech, and sweetgum are the predominant plant community. Some sandstone ravines support an unusual combination of plant species that includes some relict northern species.

Principal Natural Features

FOREST: Dry upland, mesic upland, floodplain.
PRAIRIE: Mesic, dry.
BEDROCK: Outcrops.
TOPOGRAPHY: Dissected till plain, river bluffs, ravines.
AQUATIC HABITATS: Creeks.
SPECIAL FEATURE: Relict northern plants.

VERMILION RIVER SECTION   The Vermilion River Section is characterized by rugged topography and the beech-maple forests in the ravines along the Vermilion River and its tributaries. The beech-maple forests represent an important climax deciduous forest type of the northeastern United States, which is found in Illinois only in the extreme eastern and southern portions.

Principal Natural Features

FOREST: Dry upland, mesic upland, floodplain.
PRAIRIE: Dry, mesic, wet.
TOPOGRAPHY: Dissected till plain, ravines, floodplain.
AQUATIC HABITATS: Rivers, creeks.

12. Ozark Division

The Ozark Division consists of the Illinois part of the Salem Plateau of the Ozark uplift from northern Monroe County southward and includes the glaciated sandstone ravines in Randolph County. The area is mostly forested, but many hill prairies occur in the Northern Section. The division contains many Ozarkian, southern, and southwestern plants and animals that are rare or absent elsewhere in Illinois. The sections are based on differences in bedrock, topography, flora, and fauna.

BEDROCK

Most of the Ozark Division is part of the Salem Plateau of the Ozark uplift. The northern part of the division is underlain by relatively pure limestone, while the southern part is underlain by cherty limestone that is more resistant to erosion. Sandstone underlies the Central Section of the division. Bedrock crops out in all sections. Caves and sinkholes are numerous in the limestone of the Northern Section and less so in the Southern Section.

GLACIAL HISTORY

Part of the Northern Section and all of the Central Section of the Ozark Division were glaciated during the Illinoian stage of Pleistocene glaciation. The Southern Section is driftless.

TOPOGRAPHY

The topography of the Ozark Division is that of a maturely dissected plateau with steep bluffs along the Mississippi River. There are ravines and stream canyons throughout the division, especially in the sandstone of the Central Section. The Northern Section has a well-developed sinkhole plain topography.

SOILS

Most of the soils of the Ozark Division are derived from deep loess, but thin soils occur over the bedrock outcrops along the river bluffs and in interior ravines. Some of the soils of the Southern Section are derived from bedrock and are acidic.

PLANT COMMUNITIES

FOREST   At the time of settlement the Ozark Division was almost entirely forested. The forests of the Northern section consist in part of red oak, sugar maple, basswood, and Ohio buckeye (*Aesculus glabra*) on the mesic sites, with white oak, black oak, and hickories on the ridge tops. Beech and tuliptree occur in the mesic forests of the Central Section. The forests of the Southern Section contain a rich assemblage of tree species including cucumbertree (*Magnolia acuminata*), blackgum, butternut (*Juglans cinerea*), black walnut, and bitternut hickory in addition to red oak, sugar maple, basswood, white oak, black oak, Ohio buckeye, beech, and tuliptree. Yellow-wood (*Cladrastis lutea*) occurs in the Southern Section in Alexander County. A mixed association of white oak, red oak, sycamore, American elm, river birch, wild black cherry, and cottonwood is found along the stream floodplains throughout the division. Stands of shortleaf pine (*Pinus echinata*) are found in the

Southern Section and as far north as Piney Creek in Randolph County.

PRAIRIE Loess hill prairies are common on the river bluffs in the Northern Section of the Ozark Division but are rare in the Southern Section and absent from the Central Section. These prairies have a similar species composition to that of loess hill prairies of the Middle Mississippi Border Division, with little blue-stem and side-oats grama dominating; but they have several species of plants that are restricted to this division in Illinois.

AQUATIC HABITATS

The Ozark Division has few aquatic habitats. Ponds occur in some of the sinkholes. Springs occur at cave entrances and at the bases of some of the bluffs. Creeks are the commonest aquatic feature of the division.

DISTINCTIVE FAUNA

The plains scorpion, eastern narrow-mouthed toad, and eastern coachwhip are restricted in Illinois to the Northern Section of the Ozark Division, while the spring cavefish, northern blacktail shiner, and scarlet snake are known only from the Southern Section. The northern flat-headed snake occurs in both the Northern and Southern sections but is found nowhere else in the State.

NORTHERN SECTION The Northern Section is distinguished from the other sections of the Ozark Division by its limestone bedrock, numerous caves and sinkholes, unique plant and animal species, and forest composition. Plant species unique to this section are the reticulate-seeded spurge (*Euphorbia spathulata*), stiff bedstraw (*Galium virgatum*), Missouri black-eyed susan (*Rudbeckia fulgida* var. *missouriensis*), and small heliotrope (*Heliotropium tenellum*). These plants grow in hill prairies or on exposed limestone ledges.

Principal Natural Features

FOREST: Dry upland, mesic upland, floodplain.
LOESS HILL PRAIRIE
BEDROCK: Sinkholes, caves, limestone outcrops.
TOPOGRAPHY: Sinkhole plain, ravines, river bluffs, floodplain.
AQUATIC HABITATS: Sinkhole ponds, creeks, springs.
SPECIAL FEATURES: Ozark "limestone glade" plants, distinctive reptiles and amphibians.

CENTRAL SECTION   The Central Section of the Ozark Division is distinguished because of its sandstone bedrock, forest composition, and distinctive flora. Distinctive plants of this section include Harvey's buttercup (*Ranunculus harveyi*), large-flowered rock-pink (*Talinum calycinum*), and Bradley's spleenwort (*Asplenium bradleyi*). These species apparently entered Illinois from the Missouri Ozarks after Illinoian glaciation.

Principal Natural Features

FOREST: Dry upland, mesic upland, floodplain.
BEDROCK: Sandstone outcrops.
TOPOGRAPHY: Hills, ravines, stream canyons, floodplain.
AQUATIC HABITATS: Creeks.
SPECIAL FEATURE: Ozarkian floral element.

SOUTHERN SECTION   The Southern Section of the Ozark Division is distinguished by its bedrock, topography, glacial history, unique fauna, forest composition, and distinctive southern and Ozarkian flora. Black spleenwort (*Asplenium resiliens*), shortleaf pine, azalea (*Rhododendron prinophyllum*), and big-leaf snowbell-bush (*Styrax grandifolia*) are part of the distinctive floral element.

Principal Natural Features

FOREST: Dry upland, mesic upland, floodplain.
LOESS HILL PRAIRIE
BEDROCK: Outcrops.
TOPOGRAPHY: Steep ravines, river bluffs, floodplain.
AQUATIC HABITATS: Creeks, springs, sinkhole ponds.
SPECIAL FEATURES: Ozarkian floral element, spring cavefish.

13. Lower Mississippi River Bottomlands Division

The Lower Mississippi River Bottomlands Division includes the Mississippi River and its floodplain from Alton to the Thebes Gorge. The Mississippi River is muddy here due to the silt load brought in by the Missouri River. Its fish fauna contains a distinctive assemblage of silt-tolerant plains species.

The Northern Section (the American Bottom) originally contained prairies, marshes, and forest. The Southern Section was densely forested. The forests of this division contain a greater number of tree species than the forests of the Upper Mississippi River, including some southern lowland species.

TOPOGRAPHY

The broad bottomlands of the Lower Mississippi River Bottomlands Division were formed by glacial flood waters. Since the retreat of the glaciers the river has meandered through this broad floodplain, and many meander scars and oxbow lakes remain.

SOILS

The soils are generally fine textured, with areas of both sandy, well drained soils and clay soils with poor internal drainage. The soils of this division have developed from alluvium.

PLANT COMMUNITIES

FOREST   Except for areas of wet prairie and marsh in the Northern Section, the division was entirely forested in presettlement times. The bottomland forests on the light soils include silver maple, ashes, American elm, honey locust, sugarberry, and pecan. Beech, basswood, and red buckeye (*Aesculus discolor*) grew as associated species on the loamy soils in the Southern Section; but most of these soils, being better drained and fertile, have been cleared for agriculture. Bottomland forests on heavy soils of the Southern Section are dominated by pin oak, overcup oak, Shumard oak, and cherrybark oak in association with kingnut hickory, sugarberry, and sweetgum. Bottomland swamps in the Southern Section are dominated by pumpkin ash (*Fraxinus tomentosa*), swamp cottonwood (*Populus heterophylla*), Drummond's red maple (*Acer rubrum* var. *drummondii*), and water locust. Bald cypress grows in the Southern Section as far north as the southern edge of Union County. Tupelo (*Nyssa aquatica*) and some other coastal plain species are absent.

PRAIRIE   There were relatively large areas of wet and mesic prairie in the Northern Section of the Lower Mississippi River Bottomlands Division, but most have been drained for agriculture.

MARSH   The Northern Section contained large marshes dominated by river bulrush, cattail, lotus (*Nelumbo lutea*), and pickerelweed. Some of the marshes remain, even though much of the area is organized into drainage districts.

AQUATIC HABITATS

The silt-laden Mississippi River below the Missouri River provides aquatic habitats somewhat different from those of the upper Mississippi River. Oxbow lakes and sloughs are common features of the bottomlands. The springfed swamps of northwestern Union County provide a unique habitat for several species of fish.

DISTINCTIVE FAUNA

The herpetofauna of the Lower Mississippi River Bottomlands Division is similar to that of the bottomlands of the Coastal Plain Division. It includes the western cottonmouth, green water snake, green treefrog, western bird-voiced treefrog, and mole salamander. The fish fauna of this division includes several species not found elsewhere in Illinois. The bantam sunfish is found only in the springfed swamps in northwestern Union Country; and the Alabama shad, plains minnow, sturgeon chub, flathead chub, and sicklefin chub are found in the Mississippi River. The banded pigmy sunfish is known only from the springfed swamps of this division and from the bottomland swamps of the Coastal Plain Division.

NORTHERN SECTION   The Northern Section is distinguished by its forest composition, the presence of wet prairies and marshes, and the absence of the coastal plain trees of the Southern Section. The bottomlands of this section near St. Louis are called the "American Bottoms."

Principal Natural Features

FOREST: Bottomland.
PRAIRIE: Wet, mesic.
MARSH
TOPOGRAPHY: River floodplain, meander scars.
AQUATIC HABITATS: Oxbow lakes, Mississippi River.

SOUTHERN SECTION   The bottomland forests of the Southern Section contain a greater number of tree species, including some bottomland swamp species typical of the coastal plain. The composition of the forests in this section varies with the soils.

Principal Natural Features

FOREST: Bottomland on heavy soils, bottomland on light soils, bottomland swamp.

TOPOGRAPHY: River floodplain, meander scars.
AQUATIC HABITATS: Oxbow lakes, Mississippi River.
SPECIAL FEATURE: Springfed swamps.

14. Shawnee Hills Division

The Shawnee Hills extend across the southern tip of the state from Fountain Bluff on the Mississippi River to the Shawneetown Hills near the mouth of the Wabash River. This unglaciated hill country is characterized by a high east-west escarpment of sandstone cliffs forming the Greater Shawnee Hills and a series of lower hills underlain by limestone and sandstone known as the Lesser Shawnee Hills. Originally this division was mostly forested and considerable forest remains to the present time. There are a number of distinctive plant species restricted to this division of Illinois.

BEDROCK

The Greater Shawnee Hills form a band along the northern edge of the division and consist of massive Pennsylvanian sandstone strata that dip northward toward the Illinois Basin. The range of hills averages 10 miles wide and borders the Lesser Shawnee Hills to the south. The Lesser Shawnee Hills are underlain by Mississippian limestone and sandstone, and sinkholes and caves are locally common features. Mineralized faults containing fluorspar and zinc, silver, and other metals exist in the eastern part of the Shawnee Hills Division. Iron deposits are found in Hardin County. There is a dome containing an igneous rock core in western Hardin County, and outcrops of igneous rock occur in the Lesser Shawnee Hills Section.

TOPOGRAPHY

The topography of the Shawnee Hills Division is very rugged, with many bluffs and ravines. The north slopes of the Greater Shawnee Hills Section are relatively gentle, but the south slopes consist of many escarpments, cliffs, and overhanging bluffs. Streams have eroded canyons in the sandstone. The Lesser Shawnee Hills average about 200 feet lower than the Greater Shawnee Hills. The Lesser Shawnee Hills have local areas of sinkhole topography.

SOILS

The soils are derived mainly from loess. Narrow bands of moderately developed deep loess soils occur along the Mississippi River in Jackson County and along the Ohio River in eastern Hardin County; however, most of the soils are derived from thinner loess and are strongly developed. Claypan and fragipan layers are frequent.

PLANT COMMUNITIES

FOREST   At the time of settlement most of the Shawnee Hills Division supported forest, and considerable land remains timbered. Most of the upland forests are dominated by white oak, black oak, and shagbark hickory with post oak, blackjack oak, scarlet oak (*Quercus coccinea*), and pignut hickory (*Carya glabra*) on dry sites. Deep mesic ravines have a forest community of red oak, beech, tuliptree, bitternut hickory, sugar maple, and white ash with black walnut, butternut, Ohio buckeye, and basswood occasional. The floodplain forests also contain sycamore, Kentucky coffeetree, sugarberry, and honey locust.

PRAIRIE   The Lesser Shawnee Hills Section contains limestone glades which support a dry prairie vegetation like that of hill prairies in western Illinois, but with the addition of such southern plants as wild blue sage (*Salvia pitcheri*) and heart-leaved tragia (*Tragia cordata*).

AQUATIC HABITATS

The Shawnee Hills Division has numerous clear rocky streams and creeks. Sinkhole ponds are found in the Lesser Shawnee Hills Section.

DISTINCTIVE FAUNA

The streams support several distinctive fishes including black-spotted topminnow, spottail darter, and stripetail darter. The latter species is restricted to extreme southeastern Illinois.

GREATER SHAWNEE HILLS SECTION   The Greater Shawnee Hills Section is distinguished by its sandstone bedrock, topography, and distinctive plants. Filmy fern (*Trichomanes boschianum*), Virginia saxifrage (*Saxifraga virginiensis*), small-flowered rock-pink (*Talinum parviflorum*), thread-leaved evening primrose (*Oenothera linifolia*), synandra (*Synandra hispidula*), French's shooting star

(*Dodecatheon frenchii*), and small-flowered alumroot (*Heuchera parviflora*) are some of the distinctive plants of this section. French's shooting star and small-flowered alumroot are abundant in their restricted habitats and may have persisted since preglacial times. Deep ravines and sandstone ledges along the larger streams support relict northern plants such as clubmosses (*Lycopodium* spp.), sphagnum (*Sphagnum* sp.), and barren strawberry (*Waldsteinia fragarioides*). Except for synandra, these distinctive plants are absent from a seemingly suitable habitat in the glaciated area of southwestern Williamson County.

Principal Natural Features

FOREST: Dry upland, mesic ravine, floodplain.
BEDROCK: Outcrops, faults.
TOPOGRAPHY: Bluffs, ravines, stream canyons, floodplain, overhanging cliffs.
AQUATIC HABITATS: Creeks.
SPECIAL FEATURE: Northern relict plants, preglacial relict plants.

LESSER SHAWNEE HILLS SECTION   The Lesser Shawnee Hills Section is distinguished by its limestone bedrock and sinkhole topography. The fluorspar deposits near Cave in Rock and Rosiclare in Hardin County are world famous. Caves are common features of the limestone bluffs. Distinctive plants of this section are wild mock-orange (*Philadelphus pubescens*) and great chickweed (*Stellaria pubera*).

Principal Natural Features

FOREST: Dry upland, mesic ravine, floodplain.
LIMESTONE GLADES
BEDROCK: Sinkholes, caves, sandstone outcrops, limestone outcrops, igneous outcrops, mineralized faults.
TOPOGRAPHY: Sinkhole plain, interior bluffs, river bluffs, ravines, floodplain.
AQUATIC HABITATS: Sinkhole ponds, creeks.
SPECIAL FEATURE: Fluorspar deposits.

15. Coastal Plain Division

The Coastal Plain Division is a region of swampy forested bottomlands and low clay and gravel hills that is the northernmost extension of the Gulf Coastal Plain Province of North America. Bald cypress-tupelo gum swamps are a unique feature of this division in Illinois as are many southern animals and plants found within them. The division encompasses the bottomlands of the Cache, Ohio, and Mississippi rivers and hills capped by Cretaceous and Tertiary sand, gravel, and clay. It has a relatively mild climate, the warmest in the state. The two sections distinguish between upland and lowland environments.

GLACIAL HISTORY

The Coastal Plain Divison was never subjected to Pleistocene glaciation, but it has been influenced by glacial floodwaters. Glacial Lake Cache was formed in the valleys of the Cache and Ohio rivers during a late stage of glaciation. Pleistocene deposits form the terraces in the bottomlands of the rivers.

BEDROCK

The Cretaceous Hills Section is composed of unconsolidated sediments of Cretaceous and Tertiary sands, gravels, and clays. The bedrock of the Bottomlands Section is deeply buried by alluvium.

TOPOGRAPHY

The Bottomlands Section of the Coastal Plain Division consists of the broad floodplain at the confluence of the Ohio and Mississippi rivers, the broad floodplain of the Cache River, and the terraces and meander scars in the floodplains. The topography of the Cretaceous Hills Section is steep to rolling.

SOILS

The soils of the uplands are derived from relatively thin loess and in a few places from gravel. Some areas in the eastern part of the Cretaceous Hills Section have gravel exposures. The soils of the Bottomlands Section range from Recent alluvium to older terrace soils. The terrace soils tend to be hardpan clays; but areas of sand occur, especially in the western part of the section. The alluvial soils are generally heavy except near the rivers.

PLANT COMMUNITIES

FOREST   The presettlement vegetation of the Coastal Plain Division was mostly forest. The upland forests of the Cretaceous Hills Section are similar to those of the Shawnee Hills with black oak, white oak, red oak, cherrybark oak, blackgum, tuliptree, shagbark hickory, and pignut hickory the common trees. In the extreme western end of these hills beech and cucumbertree are also found. A large stand of native chestnuts (*Castanea dentata*) formerly grew near Olmsted in Pulaski County. Mesic ravines along the Ohio River contain southeastern floral elements that include the silverbell (*Halesia carolina*).

The bottomland forests consist in part of Shumard oak, cherrybark oak, swamp white oak, swamp chestnut oak, pin oak, overcup oak, kingnut hickory, shagbark hickory, bitternut hickory, ashes, sweetgum, blackgum, honey locust, sugarberry, pecan, wild black cherry, and catalpa (*Catalpa speciosa*). Beech, tuliptree, and cucumbertree grow on the better-drained bottomland soils. Pin oak is the commonest tree on the heavier terrace soils, along with post oak and willow oak (*Quercus phellos*). Silver maple and American elm grow along the streams.

The bottomland swamps of this division contain an association of bald cypress, tupelo gum, swamp cottonwood, Drummond's red maple, water locust, pumpkin ash, and overcup oak. Water hickory (*Carya aquatica*) and planer-tree (*Planera aquatica*) are occasional on better-drained soils.

PRAIRIE   The Cretaceous Hills Section contained small dry prairies on the uplands of the eastern part and mesic prairies in some of the broad creek bottoms. The few persisting remnants indicate that little bluestem was the dominant species of the dry upland prairies and that big bluestem, Indian grass, and gama grass (*Tripsacum dactyloides*) were the dominant species of the mesic prairies. Prairie dropseed and cord grass are absent from the prairies of the Cretaceous Hills Section.

SOUTHERN SEEP SPRING BOG   Seep springs in the eastern end of the Cretaceous Hills Section are very acidic. They are generally dominated by sedges, royal fern (*Osmunda regalis*), lady fern (*Athyrium filix-femina*), and cinnamon fern (*Osmunda cinnamomea*) and may contain large areas of sphagnum moss. These seep springs contain a distinctive southeastern flora which includes netted chain fern (*Woodwardia areolata*), screw-stem (*Bartonia paniculata*), and incomperta sedge (*Carex incomperta*).

AQUATIC HABITATS

The aquatic habitats of the bottomlands of the Coastal Plain Division include rivers, creeks, oxbow lakes, and sloughs. The Cretaceous Hills have creeks and seep springs.

DISTINCTIVE FAUNA

The herpetofauna of the swampy lowlands of the Coastal Plain Division is similar to that of the Lower Mississippi River Bottomlands Division and includes the western cottonmouth, green water snake, green treefrog, western bird-voiced treefrog, and mole salamander. The range of the dusky salamander in Illinois is essentially restricted to the Creataceous Hills Section.

CRETACEOUS HILLS SECTION The Cretaceous Hills Section encompasses the rolling hills of unconsolidated Cretaceous and Tertiary sands, gravels, and clays. Found in these hills are fossil beds from the Cretaceous period.

Principal Natural Features

FOREST: Dry upland, mesic upland.
PRAIRIE: Dry, mesic.
SOUTHERN SEEP SPRING BOG
BEDROCK: Cretaceous and Tertiary sands, gravels and clays, Cretaceous fossil beds.
TOPOGRAPHY: Steep to rolling hills.
AQUATIC HABITATS: Creeks, seep springs.

BOTTOMLANDS SECTION The Bottomlands Section encompasses the bottomland forests, oxbow lakes, sloughs, and rivers of the Coastal Plain Division. This section includes the remannts of the once vast bald cypress and tupelo gum swamps along the rivers.

Principal Natural Features

FOREST: Floodplain, terrace, swamp.
TOPOGRAPHY: Terrace ridges, floodplain, meander scars.
AQUATIC HABITATS: Backwater swamps, sloughs, oxbow lakes, creeks, rivers.

General Key to Groups of Illinois Vascular Plants

I. Plants Ferns, Quillworts, Clubmosses, or Horsetails, not Producing Seeds

1. Leaves grass-like; corm at base of leaves mostly bilobed .... 4. Isoetaceae, p. 72
1. Leaves not grass-like; corm absent 2
2. Stem conspicuously jointed and with longitudinal ridges .... 1. Equisetaceae, p. 69
2. Stem not conspicuously jointed and without longitudinal ridges 3

3. Plants true aquatics, either rooted or floating 4
3. Plants terrestrial, although occasionally in boggy situations 5
4. Plants rooted; blades shaped like a 4-leaf clover; blades and sporocarps
with long stalks .... 9. Marsileaceae, p. 82
4. Plants floating; blades very small, bilobed; blades and sporocarps sessile ....10. Salviniaceae, p. 82
5. Leaves less than 2 cm long, often scale-like 6
5. Leaves more than 2 cm long 7
6. Creeping plants with erect branches; leaves without a ligule ....
.... 2. Lycopodiaceae, p. 70
6. Creeping or tufted plants without erect branches; leaves with a ligule
near the axil ....3. Selaginellaceae, p. 71
7. Sporangia borne on a branch arising from the base of the sterile portion
of the blade .... 5. Ophioglossaceae, p. 72
7. Sporangia borne on the back of or on the margin of green leaves, or on
wholly fertile leaves or pinnae arising directly from the rhizome, or with
fertile fronds terminated by a fruiting panicle 8
8. Blades delicate, translucent, only a single layer of cells thick; sporangia
borne at the base of a bristle-like projection from the margin of the
blade .... 7. Hymenophyllaceae p. 74
8. Blades more firm, opaque, several layers of cells thick; sporangia not associated with a marginal bristle 9
9. Fertile and sterile portions of leaves essentially alike in form, except for
a possible size differentiation ....8. Polypodiaceae, p. 74
9. Fertile and sterile portions of leaves greatly dissimilar in form 10
10. Sporangia naked and solitary (although borne in loose panicles or clusters) on modified pinnae, these borne at the apex of the leaf, or centrally, or the whole leaf fertile .... 6. Osmundaceae, p. 73
10. Sporangia covered by an indusium or, if not, then borne in well-defined
sori ....8. Polypodiaceae, p. 74

II. Plants Cacti or Pitcher-like or Completely Non-green

1. Plants green 2
1. Plants non-green 3
2. Plants cacti; leaves not pitcher-like .... 110. Cactaceae, p. 323
2. Plants not cacti; leaves modified into elongated pitchers ....
.... 73. Sarraceniaceae, p. 257
3. Plants twining or creeping .....*Cuscuta* in 133. Convolvulaceae, p. 356
3. Plants erect 4
4. Stamens 1 or 3; ovary inferior 5
4. Stamens 4, 8, or 10; ovary superior 6
5. Plants up to 1.5 cm tall; stamens 3 *Thismia* in 37. Burmanniaceae, p. 189
5. Plants well over 1.5 cm tall; stamen 1 .... 38. Orchidaceae, p. 189
6. Flowers actinomorphic; stamens 8 or 10 ....122. Ericaceae, p. 340
6. Flowers zygomorphic; stamens 4 .... 144. Orobanchaceae, p. 393

III. Plants Climbing or Twining

1. Leaves simple 2
1. Leaves compound 25

2. Leaves alternate 3
2. Leaves opposite or whorled 18
3. Plants with tendrils 4
3. Tendrils absent 8
4. Leaves entire, although sometimes lobed 5
4. Leaves toothed 6
5. Leaves unlobed, or with a pair of basal lobes; stems often with bristles or prickles; perianth 6-parted, undifferentiated 34. Smilacaceae, p. 185
5. Leaves 3-lobed; stems neither bristly nor prickly; sepals and petals differentiated, each 4- to 5-parted ....108. Passifloraceae, p. 322
6. Petals united to about the middle; stamens 3 153. Cucurbitaceae, p. 406
6. Petals free to the base; stamens 4–5 7
7. Flowers borne in racemes; petals falling away as flower opens ....
.... 100. Vitaceae, p. 309
7. Flowers solitary; petals persistent during flowering ....
.... 108. Passifloraceae, p. 322
8. Leaves entire, although sometimes lobed 9
8. Leaves toothed and sometimes lobed as well 16
9. Stems bearing ocreae at the nodes .... 51. Polygonaceae, p. 212
9. Stems not bearing ocreae at the nodes 10
10. Calyx united, elongated, S-shaped, with 3 lobes at the apex ....
.... 50. Aristolochiaceae, p. 212
10. Calyx free or united, not long and S-shaped, usually with more than 3 parts 11
11. Flowers unisexual 12
11. Flowers bisexual 13
12. Ovary superior; fruit a drupe .... 65. Menispermaceae, p. 240
12. Ovary inferior; fruit a 3-angled capsule .... 35. Diocoreaceae, p. 186
13. Ovary inferior; aerial roots abundant . *Hedera* in 119. Araliaceae, p. 331
13. Ovary superior; aerial roots sparse or absent 14
14. Corolla absent; stamens 7–10; fruit a 3-angled achene ....
.... *Brunnichia* in 51. Polygonaceae, p. 212
14. Corolla 5-parted; stamens 5; fruit a berry or capsule 15
15. Corolla deeply 5-lobed; fruit a berry .... 140. Solanaceae, p. 379
15. Corolla shallowly 5-lobed; fruit a capsule . 133. Convolvulaceae, p. 355
16. Fruit a capsule; seeds arillate .... 93. Celastraceae, p. 304
16. Fruit a berry or drupe; seeds not arillate 17
17. Fruit a berry; leaves coarsely serrate, often lobed; tendrils present ....
.... 100. Vitaceae, p. 309
17. Fruit a drupe; leaves serrulate, unlobed; tendrils absent ....
.... *Berchemia* in 99. Rhamnaceae, p. 308
18. Leaves serrate or lobed 19
18. Leaves entire 20
19. None of the leaves lobed; leaves evergreen; aerial roots present; seeds arillate .... 93. Celastraceae, p. 304
19. Some of the leaves lobed; leaves not evergreen; aerial roots absent; seeds without arils ....*Humulus* in 46. Moraceae, p. 210
20. Latex present 21
20. Latex absent 22
21. Fruits a pair of follicles; filaments free from each other ....
.... 131. Apocynaceae, p. 351
21. Fruit a single follicle; filaments united into a tube ....
.... 132. Asclepiadaceae, p. 352
22. Flowers in a loose head, sharing a common receptacle; fruit an achene
.... *Mikania* in 155. Compositae, p. 436

22. Flowers not in a head and not sharing the same receptacle; fruit a capsule or berry or follicle 2

23. Perianth 6-parted; stamens 3 or 6; fruit a 3-angled capsule ....
.... 35. Dioscoreaceae, p. 18

23. Perianth composed of a 5-parted calyx and a 5-parted corolla; stamens 5 fruit a berry or follicle 2

24. Flowers actinomorphic; fruit a follicle *Vinca* in 131. Apocynaceae, p. 35

24. Flowers zygomorphic; fruit a berry *Lonicera* in 149. Caprifoliaceae, p. 40

25. Leaves opposite 2

25. Leaves alternate 2

26. Leaves bifoliolate ....*Bignonia* in 142. Bignoniaceae, p. 39.

26. Leaves with 3 or more leaflets 2

27. Leaves trifoliolate ....*Clematis* in 63. Ranunculaceae, p. 23

27. Leaves with 5 or more leaflets 2

28. Corolla trumpet-shaped, orange or red; sepals 5; stamens 5 ....
.... *Campsis* in 142. Bignoniaceae, p. 39

28. Corolla absent; sepals 4; stamens numerous ....
.... *Clematis* in 63. Ranunculaceae, p. 23

29. Leaves trifoliolate 3

29. Leaves with 2, 4, or 5 or more leaflets 3

30. Flowers zygomorphic; stamens 10 ....80. Leguminosae, p. 27

30. Flowers actinomorphic; stamens 5 3

31. Petals free; aerial roots present; plants poison to the touch ....
.... 91. Anacardiaceae, p. 30

31. Petals united at the base; aerial roots absent; plants not poison to the touch .... 140. Solanaceae, p. 37

32. Leaves with an even number of leaflets ....80. Leguminosae, p. 27

32. Leaves with an odd number of leaflets 33

33. Leaves palmately compound .... 100. Vitaceae, p. 30

33. Leaves pinnately compound 3

34. Leaves once pinnate 35

34. Leaves bi- or tri-pinnate or -ternate 36

35. Flowers zygomorphic; stamens 10 ....80. Leguminosae, p. 27

35. Flowers actinomorphic; stamens 5 ....153. Cucurbitaceae, p. 40

36. Flowers actinomorphic; petals 5; stamens 5; fruit a berry ....
.... *Ampelopsis* in 100. Vitaceae, p. 30

36. Flowers zygomorphic; petals 4; stamens 6 or 8; fruit a capsule 3

37. Stamens 8; corolla not subcordate; fruit bladdery ....
.... *Cardiospermum* in 97. Sapindaceae, p. 307

37. Stamens 6; corolla subcordate; fruit not bladdery ....
.... *Adlumia* in 69. Papaveraceae, p. 244

IV. Plants Woody; Leaves Less Than 3 mm Wide, Usually Needle-like (Scale-like in Some Genera of Cupressaceae)

1. Some or all the leaves scale-like, overlapping, up to 0.4 mm long .....
.... 14. Cupressaceae, p. 84

1. Some or all the leaves needle-like or awl-shaped 2

2. Needle-like leaves borne in fascicles of 2 or more, at least 3.5 cm long or longer ....12. Pinaceae, p. 82

2. Needle-like leaves borne singly, up to 2 cm long 3

3. All or most of the needle-like leaves more than 1.2 cm long, usually flat

tened or with a keel; seeds either borne in a fleshy scarlet cup or in a
woody cone 4
. None of the needle-like leaves more than 1.2 cm long, usually awl-
shaped; seeds borne in a fleshy purplish "berry" ....
.... 14. Cupressaceae, p. 84
. Shrub with evergreen leaves; seeds borne in a fleshy scarlet cup ......
.... 11. Taxaceae, p. 82
. Tree with deciduous leaves; seeds borne in a woody cone with shield-
shaped scales ....13. Taxodiaceae, p. 83

V. Plants Woody; Leaves Compound

. Leaves opposite 2
. Leaves alternate 7
'. Leaves trifoliolate 3
. Leaves with 5 or more leaflets 4
. Leaflets serrulate; flowers perfect; fruits bladdery ....
.... 94. Staphyleaceae, p. 305
. Leaflets coarsely and sparsely toothed or entire; flowers unisexual; fruits
a cluster of samaras ....95. Aceraceae, p. 305
. Leaves palmately compound .... 96. Hippocastanaceae, p. 306
. Leaves pinnately compound 5
. Flowers perfect; petals 5; stamens 5; fruit fleshy ....
.... Sambucus in 149. Caprifoliaceae, p. 400
. Flowers unisexual; petals 4 or absent; stamens 2–12, rarely 5; fruit a
samara or pair of samaras 6
. Fruit a samara; petals 4; stamens 2 .. Fraxinus in 127. Oleaceae, p. 347
. Fruit a pair of samaras; petals absent; stamens 3 or more ....
.... 95. Aceraceae, p. 305
. Leaves, or most of them, bi- or tri-pinnately compound 8
'. Leaves once compound 10
. Trunk and leafstalks coarsely prickly; fruit a cluster of drupes ....
.... 119. Araliaceae, p. 331
. Trunk and leafstalks without prickles, although the trunk sometimes
with stout spines; fruit a legume or bladdery 9
. Leaflets entire or semelate; trunk often bearing stout spines; fruit an
elongated legume ....80. Leguminosae, p. 277
. Leaflets coarsely dentate; trunk spineless; fruit bladdery, 3-parted .....
.... Koelreuteria in 97. Sapindaceae, p. 307
. Leaves trifoliolate 11
. Leaves with 5 or more leaflets 14
. Plants prickly .... 79. Rosaceae, p. 262
. Plants not prickly 12
. Flowers zygomorphic; stamens 10 ....80. Leguminosae, p. 277
. Flowers actinomorphic; stamens 3–5 13
. Fruit an orbicular samara; ovary 2-locular; stigmas 2 85. Rutaceae, p. 296
. Fruit a drupe; ovary 1-locular; stigmas 3 .... 91. Anacardiaceae, p. 303
. Plants with thorns, spines, bristles, or prickles 15
. Plants without thorns, spines, bristles, or prickles 17
. Flowers zygomorphic; stamens 10; fruit a legume ....
.... 80. Leguminosae, p. 277
. Flowers actinomorphic; stamens usually 4–5 or more than 10 16

16. Flowers perfect; stamens numerous; flowers often 2 cm broad or broader .... 79. Rosaceae, p. 26
16. Flowers unisexual; stamens 4–5; flowers less than 1 cm broad .... 85. Rutaceae, p. 29
17. Corolla absent; all flowers unisexual; staminate flowers in catkins ... . 42. Juglandaceae, p. 20
17. Corolla present; some or all flowers perfect; flowers not in catkins 1
18. Flowers zygomorphic; fruit a legume ....80. Leguminosae, p. 27
18. Flowers actinomorphic; fruit not a legume ....
19. Leaflets mostly 5 (rarely 7) .... *Potentilla* in 79. Rosaceae, p. 26
19. Leaflets 7–31 2
20. Ovary superior; stamens 2, 3, 5, or 10 2
20. Ovary inferior; stamens numerous ..... *Sorbus* in 79. Rosaceae, p. 26
21. Ovary simple; stamens 5; fruit a berry .... 91. Anacardiaceae, p. 30
21. Ovary 2- to 5-lobed; stamens 2, 3, or 10; fruit a samara .... 86. Simaroubaceae, p. 29

VI. Plants Woody; Leaves Simple, Alternate

1. Some or all the leaves lobed ....
1. None of the leaves lobed 1
2. Latex present .... 46. Moraceae, p. 20
2. Latex absent ....
3. Flowers without petals ....
3. Flowers with petals ....
4. Leaves densely coated with white felt on the lower surface; pistillate flowers without a calyx .... *Populus* in 40. Salicaceae, p. 19
4. Leaves without a white felt on the lower surface; pistillate flowers with calyx ....
5. Leaves star-shaped, 5- to 7-lobed; twigs often winged .... *Liquidambar* in 77. Hamamelidaceae, p. 26
5. Leaves not star-shaped; twigs unwinged ....
6. Twigs green; leaves with 2 or 3 lobes; fruit a drupe 68. Lauraceae, p. 24
6. Twigs brown or tan; leaves usually with more than 3 lobes; fruit a nut
7. Pistillate flowers in a bur-like head; leaves strongly aromatic; fruit shiny nut subtended by 8 linear bracts ....41. Myricaceae, p. 20
7. Pistillate flowers not in a bur-like head; leaves not aromatic; fruit a acorn .... *Quercus* in 44. Fagaceae, p. 20
8. Flowers unisexual, borne in globose heads ..... 78. Platanaceae, p. 26
8. Flowers perfect, arranged in various inflorescences ....
9. Ovary or ovaries superior 1
9. Ovary inferior 1
10. Stamens united into a long tube .... 103. Malvaceae, p. 31
10. Stamens free from each other 1
11. Stamens inserted on the receptacle; sepals 3, usually colored like the petals .... 66. Magnoliaceae, p. 24
11. Stamens inserted on the calyx; sepals usually 5, often green .... 79. Rosaceae, p. 26
12. Styles 2 per carpel .... 76. Saxifragaceae, p. 25
12. Style 1 per carpel .... 79. Rosaceae, p. 26
13. Leaves toothed 1
13. Leaves entie 3

35. Stamens united into a tube; anthers dehiscing lengthwise ....
.... 126. Styracaceae, p. 34
35. Stamens free; anthers dehiscing by pores at the tip 122. Ericaceae, p. 34
36. Corolla absent 3
36. Corolla present 4
37. Flowers unisexual 3
37. Flowers perfect 4
38. Latex present; plants often spiny; fruit globose, greenish, 4 cm or mor
in diameter .... *Maclura* in 46. Moraceae, p. 20
38. Latex absent; plants not spiny; fruit various 3
39. Pistillate flowers in aments .... 40. Salicaceae, p. 19
39. Pistillate flowers not in aments 4
40. Leaves asymmetrical at the base; trunk warty ....
.... *Celtis* in 45. Ulmaceae, p. 20
40. Leaves symmetrical at the base; trunk not warty 4
41. Ovary inferior; leaves not aromatic when crushed 4
41. Ovary superior; leaves aromatic when crushed ...68. Lauraceae, p. 24
42. Fruit an acorn; styles 3 .... 44. Fagaceae, p. 20
42. Fruit a drupe; style 1 .... 114. Nyssaceae, p. 32
43. Stamens 4; leaves with peltate scales ....
.... *Elaeagnus* in 112. Elaeagnaceae, p. 32
43. Stamens 8; leaves without peltate scales ...111. Thymelaeaceae, p. 32
44. Petals free 4
44. Petals united 5
45. Flowers unisexual .... 114. Nyssaceae, p. 32
45. Flowers perfect 4
46. Flowers pea-shaped; leaves heart-shaped; fruit a legume ....
.... *Cercis* in 80. Leguminosae, p. 27
46. Flowers not pea-shaped; leaves not heart-shaped; fruit no a legume 4
47. Ovary superior 4
47. Ovary inferior 5
48. Plants prickly; sepals 6 .... 64. Berberidaceae, p. 24
48. Plants not prickly; sepals mostly 3 or 5 4
49. Stamens dehiscing by terminal pores; sepals 5; stamens 5 ....
.... 122. Ericaceae, p. 34
49. Stamens dehiscing lengthwise; sepals 3; stamens numerous 5
50. Petals greenish-yellow; fruit cone-like .... 66. Magnoliaceae, p. 24
50. Petals maroon; fruit a thickened, elongated, banana-like berry ......
.... 67. Annonaceae, p. 24
51. Stamens 4; fruit a drupe .... 121. Cornaceae, p. 33
51. Stamens 10; fruit dry .... 126. Styracaceae, p. 34
52. Stamens usually 8 or 16; fruit a large, orange berry; calyx 4-parted ...
.... 125. Ebenaceae, p. 34
52. Stamens 5 or 10; fruit various; calyx 5-parted 5
53. Fertile stamens 8 or 10; ovary inferior 5
53. Fertile stamens 5; ovary superior 5
54. Anthers opening by terminal pores; fruit fleshy, several-seeded ......
.... 122. Ericaceae, p. 34
54. Anthers opening by longitudinal slits; fruit dry, 1-seeded ....
.... 126. Styracaceae, p. 34
55. Plants usually thorny; 5 sterile stamens present; anthers dehiscing
lengthwise .... 124. Sapotaceae, p. 34
55. Plants thornless; sterile stamens absent; anthers dehiscing by termina
pores .... 122. Ericaceae, p. 34

VII. Plants Woody; Leaves Simple, Opposite or Whorled

1. Plants parasitic, growing on the branches of trees ....
.... 49. Loranthaceae, p. 211
1. Plants not parasitic 2
2. Some or all the leaves whorled 3
2. Leaves opposite 5
3. Leaves heart-shaped, most or all of them at least 6 cm broad ....
.... *Catalpa* in 142. Bignoniaceae, p. 392
3. Leaves not heart-shaped, all of them less than 6 cm broad 4
4. Flowers white, borne in dense, globose heads; ovary inferior; corolla 4-toothed; stipules present *Cephealanthus* in 148. Rubiaceae, p. 397
4. Flowers magenta, borne in axillary clusters; ovary superior; petals 5, free; stipules absent ....*Decodon* in 113. Lythraceae, p. 324
5. Flowers borne in catkins; leaves more than four times longer than broad
.... 40. Salicaceae, p. 195
5. Flowers not borne in catkins; leaves less than four times longer than broad 6
6. Leaves palmately lobed 7
6. Leaves unlobed, or rarely pinnately lobed 8
7. Fruit a pair of samaras; ovary superior ....95. Aceraceae, p. 305
7. Fruit a drupe; ovary inferior ....149. Caprifoliaceae, p. 400
8. Leaves toothed 9
8. Leaves entire 12
9. Petals free 10
9. Petals united, or absent 11
0. Stamens 4 or 5; seeds enclosed in a red aril .. 93. Celastraceae, p. 304
0. Stamens 8–40; seeds not arillate .... 76. Saxifragaceae, p. 258
1. Stamens 2; ovary superior ....127. Oleaceae, p. 347
1. Stamens 4–5; ovary inferior ....149. Caprifoliaceae, p. 400
2. Leaves heart-shaped, at least 6 cm long and 6 cm broad ....
.... *Paulownia* in 141. Scrophulariaceae, p. 383
2. Leaves not heart-shaped, never 6 cm broad and 6 cm long 13
3. Ovary superior; stamens 2 or numerous 14
3. Ovary inferior; stamens 4, 8, or 10–40 15
4. Petals united; stamens 2 ....127. Oleaceae, p. 347
4. Petals free; stamens numerous ....104. Hypericaceae, p. 314
5. Petals free or absent; stamens 4, 8, or 10–40 16
5. Petals united; stamens usually 5 ....149. Caprifoliaceae, p. 400
6. Flowers dioecious; stamens 8; petals absent ....
.... *Shepherdia* in 112. Elaeagnaceae, p. 323
6. Flowers perfect; stamens 4 or 10–40; petals present 17
7. Petals usually 5; stamens 10–40; fruit a capsule 76. Saxifragaceae, p. 258
7. Petals 4; stamens 4; fruit a drupe .... 121. Cornaceae, p. 339

VIII. Plants Monocotyledonous, Herbaceous, Usually with Elongated, Parallel-veined Leaves, Flower Parts Usually in 3s, and a Single Cotyledon

1. Plants climbing or twining (if erect, then usually with a few weak tendrils from the upper axils); leaves net-veined; flowers unisexual 2

1. Plants erect or floating in water (tendrils never present); leaves mostl
   parallel-veined; flowers bisexual or unisexual ....
2. Inflorescence umbellate; ovary superior; fruit a berry ....
   .... 34. Smilacaceae, p. 18!
2. Inflorescence glomerulate or paniculate; ovary inferior; fruit a capsul
   .... 35. Dioscoreaceae, p. 18(
3. Plants with one or two whorls of leaves ....
3. Plants with leaves alternate, opposite, basal, or none ....
4. Flowers radially symmetrical; ovary superior; stamens 6 ....
4. Flowers bilaterally symmetrical; ovary inferior; stamen 1 ....
   .... *Isotria* in 38. Orchidaceae, p. 18!
5. Plants never over 50 cm tall; flowers usually borne singly ....
   .... *Trillium* and *Medeola* in 33. Liliaceae, p. 17(
5. Plants more than 50 cm tall; flowers usually more than one ....
   .... *Lilium* in 33. Liliaceae, p. 17(
6. Flowers crowded together on a spadix, often subtended by a spathe
   .... 27. Araceae, p. 16:
6. Flowers not crowded on a spadix (in *Ruppia,* two flowers are borne on ;
   spadix-like structure) ....
7. Plants thalloid, floating in water ....28. Lemnaceae, p. 16&
7. Plants with roots, stems, and leaves, aquatic or terrestrial ....
8. Perianth absent, or reduced to very minute scales (lodicules) or bristle:
8. Perianth present, composed of either calyx or corolla or both (plant:
   with the perianth reduced to minute scales or bristles should be sough
   under the first 8) .... 1!
9. Each flower subtended by one or more sterile scales; plants generall
   not true aquatics .... 1(
9. Flowers not subtended by individual scales; plants mostly aquatics . 1°
10. Leaves 2-ranked; sheaths usually open; stems usually hollow, with soli(
    nodes, often terete; anthers attached above the base 25. Poaceae, p. 9:
10. Leaves 3-ranked (when present); sheaths closed; stems solid, with sof
    nodes, often 3-angled; anthers attached at the base ....
    .... 26. Cyperaceae, p. 13&
11. Plants erect; inflorescence terminal, spicate, thick; leaves very long, lin
    ear, strap-shaped .... 15. Typhaceae, p. 8·
11. Plants not erect; free-floating or sometimes rooted in bottom mud; in
    florescence axillary or terminal and slenderly spicate; leaves not a:
    above 1:
12. Leaves alternate; stamens 2 or 4; inflorescence spicate and usually ter
    minal, or with flowers borne 2 per spadix 1:
12. Leaves opposite; stamen 1; inflorescence not spicate, axillary ...... 1·
13. Stamens 4; flowers in a spike or head; fruit sessile, appearing as ar
    achene upon drying .... 20. Potamogetonaceae, p. 8(
13. Stamens 2; flowers on a short spadix, concealed within the leaf sheath
    fruit stipitate, drupe-like .... 17. Ruppiaceae, p. 8!
14. Carpel one; fruit beakless ....19. Najadaceae, p. 8(
14. Carpels 2–4; fruit beaked ....18. Zannichelliaceae, p. 8(
15. Pistils simple, more than one, separate or slightly coherent at base . 1(
15. Pistil one, compound .... 1&
16. Calyx and corolla differentiated (in color and texture) ....
16. Calyx and corolla undifferentiated (i.e., similar in color and texture) ..
    .... 21. Juncaginaceae, p. 8!
17. Inflorescence umbellate; pistils 6, coherent at base; fruit a follicle ....
    .... 23. Butomaceae, p. 9:

17. Inflorescence not umbellate; pistils 10 or more, free to base; fruit an achene ....22. Alismaceae, p. 90
18. Ovary superior 19
18. Ovary inferior 28
19. Calyx and corolla differentiated (in color and texture) 20
19. Calyx and corolla undifferentiated (i.e., similar in color and texture)   21
20. Flowers crowded together in a dense head; leaves basal ....
.... 29. Xyridaceae, p. 170
20. Flowers borne in cymes or umbels; leaves caulinz ....
.... 30. Commelinaceae, p. 170
21. Flowers unisexual 22
21. Flowers bisexual 24
22. Leaves net-veined; flowers in umbels .... 34. Smilacaceae, p.185
22. Leaves parallel-veined; flowers in globose clusters, racemes, or panicles
.... 23
23. Perianth small, greenish; flowers aggregated in dense globose clusters; stamens 5 .... 16. Sparganiaceae, p. 85
23. Perianth usually conspicuous, greenish, yellowish, white, or bronze-purple; flowers in racemes or panicles; stamens 6 . 33. Liliaceae, p. 176
24. Perianth scarious .... 32. Juncaceae, p. 172
24. Perianth petaloid 25
25. Stamens 3 ....31. Pontederiaceae, p. 172
25. Stamens 6 (or 4) 26
26. Stamens of different sizes ....31. Pontederiaceae, p. 172
26. Stamens all alike 27
27. Leaves evergreen, rigid; stems woody .... *Yucca* in 33. Liliaceae, p. 176
27. Leaves deciduous, mostly not rigid; stems herbaceous ....
.... 33. Liliaceae, p. 176
28. Plants growing in water 29
28. Plants growing on land 31
29. Leaves whorled ....24. Hydrocharitaceae, p. 92
29. Leaves basal, or cauline and alternate 30
30. Stamens 2, or 6–12, never 3; flowers unisexual; styles not petaloid ....
.... 24. Hydrocharitaceae, p. 92
30. Stamens 3; flowers bisexual; styles petaloid ....36. Iridaceae, p. 186
31. Flowers bilaterally symmetrical; stamens 1 or 2 38. Orchidaceae, p. 189
31. Flowers radially symmetrical or nearly so; stamens 3 or 6 32
32. Stamens 3; styles sometimes petaloid ....36. Iridaceae, p. 186
32. Stamens 6; styles not petaloid 33
33. Leaves reduced to scales; plants lacking chlorophyll, at most 4 cm tall
.... 37. Burmanniaceae, p. 189
33. Leaves blade-bearing; plants with chlorophyll, well over 4 cm tall
.... 33. Liliaceae, p. 176

IX. Plants True Aquatics, Spending Their Entire Life In the Water; Dicotyledons

1. Some or all of the leaves deeply divided 2
1. None of the leaves deeply divided 10
2. Flowers crowded together in a solitary head, each flower sharing a common receptacle ....*Bidens* in 155. Compositae, p. 409
2. Flowers not crowded, each with its own receptacle 3
3. Leaves bearing small bladders; flowers zygomorphic ....
.... 145. Lentibulariaceae, p. 394

3. Leaves not bearing small bladders; flowers actinomorphic 4
4. Some or all the flowers unisexual 5
4. Some or all the flowers perfect 6
5. Calyx (or involucre) 8- to 12-cleft; stamens 10–20; ovary unlobed ......
.... 59. Ceratophyllaceae, p. 231
5. Calyx 4-cleft; stamens 4–8; ovary 2- to 4- lobed 117. Haloragidaceae, p. 330
6. Pistils 2 to several, free, or pistil 1 but deeply 2- to 4-lobed 7
6. Pistil one, unlobed 9
7. Sepals 3; petals 3; stamens 3–6 ....62. Cabombaceae, p. 232
7. Sepals 4–5; petals 0, 4, or 5; stamens 3, 4, 8, or numerous 8
8. Sepals 5; petals 5; stamens numerous ....
.... *Ranunculus* in 63. Ranunculaceae, p. 232
8. Sepals 4; petals 0 or 4; stamens 3, 4, or 8 ..117. Haloragidaceae, p. 330
9. Sepals 5; petals 5, united at base; stamens 5 ....
.... *Hottonia* in 123. Primulaceae, p. 344
9. Sepals 4; petals 4, free; stamens 6 ..*Neobeckia* in 71. Cruciferae, p. 245
10. Leaves with conspicuous sheathing stipules (ocreae) at the base ......
.... 51. Polygonaceae, p. 212
10. Leaves without ocreae 11
11. Flowers unisexual; perianth absent; stamen 1 89. Callitrichaceae, p. 302
11. Flowers perfect; perianth present; stamens 4 or more 12
12. Leaves peltate 13
12. Leaves not peltate 14
13. Sepals 3–4; petals 3–4; stamens 12–18 ....
.... *Brasenia* in 62. Cabombaceae, p. 232
13. Sepals and petals indistinguishable, together totalling more than 8 segments; stamens more than 20 .... 61. Nelumbonaceae, p. 231
14. Leaves opposite; stamens 4 15
14. Leaves alternate; stamens 5 or more 16
15. Corolla absent .... *Peplis* in 113. Lythraceae, p. 324
15. Corolla present ....*Bacopa* in 141. Scrophulariaceae, p. 383
16. Petals united; stamens 5 ...*Nymphoides* in 130. Menyanthaceae, p. 350
16. Petals free; stamens 8 or more 17
17. Petals 4–6; stamens 8–12 ....*Jussiaea* in 116. Onagraceae, p. 326
17. Petals more than 6; stamens more than 12 ..60. Nymphaeaceae, p. 231

X. Dicotyledonous Herbs; Leaves Opposite or Whorled

1. Stems or leaves or both fleshy 2
1. Plants not fleshy 4
2. Leaves reduced to minute scales ....
.... *Salicornia* in 52. Chenopodiaceae, p. 217
2. Leaves fleshy, not scale-like 3
3. Flowers yellow or white; sepals 4–5 . *Sedum* in 75. Crassulaceae, p. 257
3. Flowers rose; sepals 2 .... *Talinum* in 57. Portulacaceae, p. 224
4. Flowers crowded in a dense head, each sharing the same receptacle ..
.... 155. Compositae, p. 409
4. Each flower with its own receptacle 5
5. Leaves, or some of them, compound 6
5. Leaves simple 13
6. Leaves ternately divided 7
6. Leaves pinnately or palmately compound 8

7. Flowers greenish or yellowish; ovary 1, inferior . 150. Adoxaceae, p. 404
7. Flowers white or pinkish; ovaries 4–15, superior ....
.... *Anemonella* in 63. Ranunculaceae, p. 232
8. Leaves palmately compound 9
8. Leaves pinnately compound 10
9. Petals 5; stamens 5; leaflets serrate ....*Panax* in 119. Araliaceae, p. 331
9. Petals 4; stamens 6; leaflets usually coarsely toothed ....
.... *Dentaria* in 71. Cruciferae, p. 245
10. Petals united; stamens 3 or 4 11
10. Petals free; stamens 5 12
11. Calyx campanulate; stamens 4 *Seymeria* in 141. Scrophulariaceae, p. 383
11. Calyx reduced to bristles; stamens 3 .... 151. Valerianaceae, p. 405
12. Erect perennials; leaves twice-pinnate; carpels 2 ....
.... *Polytaenia* in 120. Umbelliferae, p. 331
12. Spreading or prostrate annuals; leaves once-pinnate; carpels 4–5 or
8–12 ....84. Zygophyllaceae, p. 296
13. Ovaries more than 1 ....*Anemone* in 63. Ranunculaceae, p. 232
13. Ovary one 14
14. Corolla absent .... : 15
14. Corolla present 25
15. Flowers unisexual 16
15. Flowers perfect 19
16. Calyx absent 17
16. Calyx present 18
17. Ovary 4-lobed; flowers axillary .... 89. Callitrichaceae, p. 302
17. Ovary 3-lobed; flowers borne from involucres 88. Euphorbiaceae, p. 298
18. Stamens 3–4; calyx lobes 3–4 .... 47. Urticaceae, p. 210
18. Stamens 5; calyx lobes 5 .... *Iresine* in 53. Amaranthaceae, p. 221
19. Leaves linear; stamen 1 ....18. Hippuridaceae, p. 330
19. Leaves broader; stamens more than 1 20
20. Calyx adnate to ovary 21
20. Calyx free from ovary 22
21. Sepals 3, maroon; stamens 12 ....50. Aristolochiaceae, p. 212
21. Sepals 4, greenish; stamens 4 .... *Ludwigia* in 116. Onagraceae, p. 326
22. Flowers subtended by a colored involucre ...54. Nyctaginaceae, p. 223
22. Flowers not subtended by an involucre 23
23. Sepals 5; stamens 5 or 10 24
23. Sepals 4; stamens 4 or 8 ....113. Lythraceae, p. 324
24. Stamens attached to the calyx; leaves whorled ...56. Aizoaceae, p. 224
24. Stamens attached to the receptacle; leaves opposite ....
.... 58. Caryophyllaceae, p. 225
25. Petals free 26
25. Petals united 43
26. Flowers unisexual; ovary 3-lobed .... 88. Euphorbiaceae, p. 298
26. Flowers perfect; ovary rarely 3-lobed 27
27. Anthers opening by terminal pores ....115. Melastomaceae, p. 325
27. Anthers splitting lengthwise 28
28. Ovary superior 29
28. Ovary inferior 41
29. Leaves large, umbrella-like, peltate, borne in pairs, or solitary ....
.... *Podophyllum* in 64. Berberidaceae, p. 240
29. Leaves not umbrella-like and peltate 30
30. Leaves palmately lobed or dissected 31
30. Leaves entire or toothed 32

31. Petals fimbriate at the tip; ovary not deeply lobed ....
.... *Mitella* in 76. Saxifragaceae, p. 258
31. Petals not fimbriate at the tip; ovary deeply lobed 83. Geraniaceae, p. 295
32. Leaves toothed .... *Bergia* in 105. Elatinaceae, p. 317
32. Leaves entire 33
33. Sepals 2–3 34
33. Sepals 4–5 36
34. Leaves in whorls; stamens 6 .... 33. Liliaceae, p. 176
34. Leaves opposite; stamens 2, 3, or 5 35
35. Petals 5; stamens 5; leaves a single pair .... 57. Portulacaceae, p. 224
35. Petals 2–3; stamens 2–3; leaves several pairs .. 105. Elatinaceae, p. 317
36. Petals 4–5; flowers actinomorphic 37
36. Petals 3; flowers usually zygomorphic 40
37. Plants without white latex 38
37. Plants with white latex .... 88. Euphorbiaceae, p. 298
38. Ovules and seeds attached to a central column; stamens all free .....
.... 58. Caryophyllaceae, p. 225
38. Ovules and seeds not attached to a central column; stamens usually
united, at least at the base 39
39. Stamens numerous ....104. Hypericaceae, p. 314
39. Stamens 5 .... 81. Linaceae, p. 294
40. Stamens usually 8; plants glabrous or nearly so 87. Polygalaceae, p. 297
40. Stamens 5, 10, or 15; plants pubescent .... 106. Cistaceae, p. 317
41. Flowers subtended by 4 large, white bracts .... 121. Cornaceae, p. 339
41. Flowers not subtended by 4 large, white bracts 42
42. Calyx tube attached to ovary ....116. Onagraceae, p. 326
42. Calyx tube free from ovary ....113. Lythraceae, p. 324
43. Ovary superior 44
43. Ovary inferior 70
44. Stamens and corolla lobes equal in number 45
44. Stamens and corolla lobes different in number 52
45. Ovaries 2, or ovary 1 but deeply 4-lobed 46
45. Ovary 1, unlobed 48
46. Ovaries 2; latex often present 47
46. Ovary 1, deeply 4-lobed; latex absent .... 139. Labiatae, p. 367
47. Stamens free; fruit a pair of follicles ....131. Apocynaceae, p. 351
47. Stamens united into a tube; follicle solitary 132. Asclepiadaceae, p. 352
48. Leaves lobed .... *Ellisia* in 135. Hydrophyllaceae, p. 359
48. Leaves unlobed 49
49. Ovary 1-locular 50
49. Ovary with more than 1 locule 51
50. Stamens opposite the lobes of the corolla .....123. Primulaceae, p. 344
50. Stamens alternate with the lobes of the corolla 129. Gentianaceae, p. 348
51. Flowers in terminal cymes or axillary; corolla not scarious; leaves connected by a stipular line .... 128. Loganiaceae, p. 348
51. Flowers in long-peduncled heads; corolla scarious; leaves without stipules ....147. Plantaginaceae, p. 396
52. Corolla lobes 3; stamens usually 8 ....87. Polygalaceae, p. 297
52. Corolla lobes 5; stamens 2 or 4 53
53. Leaves entire 54
53. Leaves toothed 64
54. Corolla more or less actinomorphic 55
54. Corolla zygomorphic 59
55. Calyx with lobes of unequal size 56

XI. Herbs; Leaves Alternate or Basal, At Least Some of Them Compound

1. Flowers crowded together in heads, each sharing the same receptacle (various clovers may be sought here, but the flowers do not share the same receptacle) .... 155. Compositae, p. 409

1. Flowers not sharing the same receptacle 2
2. Some or all the leaves trifoliolate 3
2. Leaves not trifoliolate 9
3. Flowers papilionaceous; fruit a legume or loment ....
.... 80. Leguminosae, p. 277
3. Flowers not papilionaceous; fruit neither a legume nor loment 4
4. Flowers borne on a spadix, surrounded by a spathe 27. Araceae, p. 167
4. Spadix and spathe absent 5
5. Petals free 6
5. Petals united .... *Menyanthes* in 130. Menyanthaceae, p. 350
6. Ovary superior 7
6. Ovary inferior .... 120. Umbelliferae, p. 331
7. Calyx and corolla each 4-parted .... 70. Capparidaceae, p. 245
7. Calyx and corolla each 5-parted 8
8. Styles 5-parted at summit, united below into a beak; carpels becoming
separate at maturity, beaked .... 82. Oxalidaceae, p. 294
8. Not as above .... 79. Rosaceae, p. 262
9. Flowers borne on a spadix, surrounded by a spathe 27. Araceae, p. 167
9. Spadix and spathe absent 10
10. Flowers zygomorphic 11
10. Flowers actinomorphic 14
11. Petals 4; stamens 6 .... 69. Papaveraceae, p. 242
11. Petals 5; stamens 5 or 10 or more 12
12. Flowers in an umbel .... 120. Umbelliferae, p. 331
12. Flowers not in an umbel 13
13. Stamens 10 or more; fruit a legume; flowers without a spur ....
.... 80. Leguminosae, p. 277
13. Stamens numerous; fruit a follicle; flower spurred ....
.... *Delphinium* in 63. Ranunculaceae, p. 232
14. Flowers unisexual 15
14. Flowers perfect 16
15. Ovaries 4–15; petals absent; calyx free from the ovaries ....
.... *Thalictrum* in 63. Ranunculaceae, p. 232
15. Ovary 1; petals present; calyx adnate to the ovary ....
.... 153. Cucurbitaceae, p. 406
16. Petals absent; stamens 3 .. *Proserpinaca* in 117. Haloragidaceae, p. 330
16. Petals present; stamens 4 or more 17
17. Petals free 18
17. Petals united 25
18. Ovary or ovaries superior 19
18. Ovary inferior 24
19. Stamens attached to the receptacle 20
19. Stamens attached to the perianth 22
20. Ovaries several; stamens usually 5 or 10 or more ....
.... 63. Ranunculaceae, p. 232
20. Ovary solitary; stamens usually 6 21
21. Leaves palmately compound, stipulate ...... 70. Capparidaceae, p. 245
21. Leaves pinnately compound, without stipules ....71. Cruciferae, p. 245
22. Sepals 5; petals 5; stamens 5 or more 23
22. Sepals 3; petals 3; stamens 6 .... 90. Limnanthaceae, p. 303
23. Carpels sharply pointed at base; sepals awned ....
.... *Erodium* in 83. Geraniaceae, p. 295
23. Carpels not sharply pointed at base; sepals not awned ....
.... 79. Rosaceae, p. 262

24. Fruit a fleshy berry; styles usually 5 .... 119. Araliaceae, p. 331
24. Fruit a pair of dry carpels; styles 2 .... 120. Umbelliferae, p. 331
25. Styles 2; ovary 1-locular ....135. Hydrophyllaceae, p. 359
25. Style 1 and bilobed, or 3; ovary 2- or 3-locular 26
26. Ovary 3-locular; styles usually 3 ....134. Polemoniaceae, p. 358
26. Ovary 2-locular; style bilobed .... 140. Solanaceae, p. 379

XII. Dicotyledonous Herbs; Leaves Alternate or Basal, Simple

1. Flowers crowded together in a head, each sharing a common receptacle
.... 155. Compositae, p. 409
1. Flowers not sharing a common receptacle 2
2. Leaves all basal 3
2. At least one or more leaves borne along the stem 12
3. Petals present; stamens rarely 12 4
3. Petals absent; stamens 12 ....*Asarum* in 50. Aristolochiaceae, p. 212
4. Petals free 5
4. Petals united 10
5. Carpels separate, spirally arranged on an elongated receptacle ....
.... *Myosurus* in 63. Ranunculaceae, p. 232
5. Carpels united, not spirally arranged on an elongated receptacle 6
6. Ovary inferior or partly so .... 76. Saxifragaceae, p. 258
6. Ovary superior 7
7. Leaves covered with glandular hairs ....74. Droseraceae, p. 257
7. Leaves not glandular 8
8. Sepals 4–5; petals 4–5; stamens 5–6; rhizomes without red sap 9
8. Sepals 2; petals 8–12; stamens numerous; rhizomes with red sap ......
.... *Sanguinaria* in 69. Papaveraceae, p. 242
9. Flowers actinomorphic; petals 4; stamens 6 ......71. Cruciferae, p. 245
9. Flowers zygomorphic; petals 5; stamens 5 ...... 107. Violaceae, p. 319
10. Corolla 4-parted, scarious; stamens 4 .... 147. Violaceae, p. 319
10. Corolla 5-parted; stamens 5 or 10 11
11. Stamens 5; anthers dehiscing lengthwise .....123. Primulaceae, p. 344
11. Stamens 10; anthers dehiscing by terminal pores 122. Ericaceae, p. 340
12. Corolla absent 13
12. Corolla present 34
13. Flowers unisexual 14
13. Flowers perfect 23
14. Leaves deeply palmately lobed 15
14. Leaves not deeply palmately lobed 16
15. Stamens 5; fruit an achene; leaves not peltate .... 46. Moraceae, p. 209
15. Stamens numerous; fruit a spiny capsule; leaves peltate ....
.... *Ricinus* in 88. Euphorbiaceae, p. 298
16. Leaves with sheaths (ocreae) at the nodes ... 51. Polygonaceae, p. 212
16. Leaves without ocreae 17
17. Calyx absent .... 88. Euphorbiaceae, p. 298
17. Calyx present 18
18. Staminate flowers not in spikes or heads 19
18. Staminate flowers in spikes or heads 20
19. Stamens 5; plants with stinging hairs. *Laportea* in 47. Urticaceae, p. 210

19. Stamens 3; plants without stinging hairs .... *Phyllanthus* in 88. Euphorbiaceae, p. 298
20. Calyx 4-parted; stamens 4 .... 47. Urticaceae, p. 210
20. Calyx not 4-parted; stamens not 4 21
21. Flowers in glomerules 22
21. Flowers not in glomerules .... 88. Euphorbiaceae, p. 298
22. Sepals scarious; bracts scarious .... 52. Amaranthaceae, p. 221
22. Sepals not scarious; bracts absent .... 52. Chenopodiaceae, p. 217
23. Leaves (or some of them) pinnately or palmately lobed 24
23. Leaves not lobed 26
24. Carpels several, free .... 63. Ranunculaceae, p. 232
24. Carpels united 25
25. Flowers in racemes; stamens 2 or 6 .... 71. Cruciferae, p. 245
25. Flowers in glomerules or axillary clusters or panicles; stamens usually 5 .... 52. Chenopodiaceae, p. 217
26. Ovary superior 27
26. Ovary inferior 32
27. Leaves with sheaths (ocreae) at the base ..... 51. Polygonaceae, p. 212
27. Leaves without ocreae 28
28. Calyx absent .... 39. Saururaceae, p. 195
28. Calyx present 29
29. Inflorescences developing opposite each of the upper leaves; carpels 10, united in a ring .... 55. Phytolaccaceae, p. 224
29. Inflorescences terminal or axillary; carpels not united in a ring 30
30. Stamens 10; sepals united ..... *Penthorum* in 75. Crassulaceae, p. 257
30. Stamens 5 or fewer; sepals free 31
31. Sepals scarious; bracts scarious .... 53. Amaranthaceae, p. 221
31. Sepals not scarious; bracts absent .... 52. Chenopodiaceae, p. 217
32. Calyx 3-lobed; stamens 6; flowers zygomorphic .... 50. Aristolochiaceae, p. 212
32. Calyx 4- or 5-lobed; stamens 4–5; flowers actinomorphic 33
33. Stamens 4; calyx 4-lobed; stigma 4-lobed ..... 116. Onagraceae, p. 326
33. Stamens 5; calyx 5-lobed; stigma not 4-lobed .. 48. Santalaceae, p. 211
34. Petals free 35
34. Petals united 57
35. Flowers unisexual 36
35. Flowers perfect 37
36. Styles 8–10; plants at least 1.5 m tall . *Napaea* in 103. Malvaceae, p. 311
36. Styles 2–3; plants less than 1.5 m tall .... 88. Euphorbiaceae, p. 298
37. Anthers dehiscing by small lids near the tip ..64. Berberidaceae, p. 240
37. Anthers dehiscing longitudinally 38
38. Ovary superior 39
38. Ovary inferior 55
39. Plants with fleshy leaves and stems 40
39. Plants not fleshy 41
40. Sepals 2; stamens 5 .... 57. Portulacaceae, p. 224
40. Sepals 4–5; stamens 8 or 10 .... 75. Crassulaceae, p. 257
41. Leaf umbrella-shaped, peltate *Podophyllum* in 64. Berberidaceae, p. 240
41. Leaf not umbrella-shaped nor peltate 42
42. Latex absent 43
42. Latex present .... 88. Euphoribiaceae, p. 298
43. Stamens united into a tube .... 102. Sterculiaceae, p. 311
43. Stamens free from each other 44
44. Stamens attached to the receptacle 45

Descriptive Flora

1. EQUISETACEAE

1. *Equisetum* [*Tourn.*] *L. Horsetail, Scouring Rush*

1. Aerial stems soft and flexible, dimorphic, the sterile ones profusely branched 2
1. Aerial stems hard and rigid (flexible in *E.* X *nelsonii*), monomorphic, usually unbranched (except *E. palustre, E.* X *litorale,* and *E. fluviatile*) 3
2. Ridges of stem spinulose; teeth of sheath with white margins ....
.... 1. *E. pratense*
2. Ridges of stem not spinulose; teeth of sheath without white margins ..
.... 2. *E. arvense*
3. Ridges of stem 3–6; teeth of sheaths 3; central cavity of stem absent ..
.... 3. *E. scirpoides*
3. Ridges of stem usually more than 6; teeth of at least some sheaths more than 3; central cavity of stem present 4
4. Central cavity up to two-thirds the diameter of the stem 5
4. Central cavity more than two-thirds the diameter of the stem 9
5. Stems regularly unbranched above the base 6
5. Stems regularly branched above the base 8
6. Ridges of stem angled, with tubercles in two rows; teeth of sheath with a central groove throughout; stems firm, perennial 7
6. Ridges of stem rounded, with tubercles in one row; teeth of sheath with a central groove only at tip; stems flexible, annual ......6. *E.* X *nelsonii*
7. Ridges of stem strongly grooved, 4- to 10-angled; teeth of sheaths falling off early; uppermost sheath (including deciduous teeth) rarely more than 5 mm long .... 4. *E. variegatum*
7. Ridges of stem slightly grooved, 10- to 17-angled; teeth of sheaths persistent; uppermost sheath (including teeth) nearly always more than 5 mm long ....5. *E.* X *trachyodon*
8. Sheath of first joint of branches with 5–10 teeth; stem with outer row of cavities about same size as central cavity .... 7. *E. palustre*
8. Sheath of first joint of branches with 3–4 teeth; stem with outer row of cavities considerably smaller than central cavity .... 8. *E.* X *litorale*
9. Teeth of sheaths early deciduous, basally connate 10
9. Teeth of sheaths persistent, free or basally connate 12
10. All sheaths gray ....9. *E. hyemale*
10. Upper sheaths green, the lower sometimes gray 11
11. Stems smooth; central cavity three-fourths the diameter of the stem; cone rounded or apiculate at summit .... 10. *E. laevigatum*

11. Stems slightly rough; central cavity four-fifths the diameter of the stem
cone apiculate at summit ....11. *E.* X *ferrissii*
12. Teeth of sheaths free; sheaths brownish or green; stems often
branched; cones not apiculate .... 12. *E. fluviatile*
12. Teeth of sheaths basally connate; sheaths gray; stems usually unbranched; cones apiculate .... 9. *E. hyemale*

1. E. pratense Ehrh. Meadow Horsetail. May–July. North-facing
wooded slope, rare; JoDaviess Co.
2. E. arvense L. Common Horsetail. April–June. Railroad embankments, roadsides, fields, shores; in every co.
3. E. scirpoides Michx. Dwarf Scouring Rush. Aug.–Sept. Moist
shaded woodlands, rare; Lake and McHenry cos.
4. E. variegatum Schleich. Variegated Scouring Rush. July–Oct.
Lake shores and stream banks, margins of swamps, rare; Lake, Cook
Carroll, Warren, Coles, and DeWitt cos.
5. E. X trachyodon A. Br. Horsetail. June–Sept. Lake shores, very
rare; Cook and Mason cos. Considered to be a hybrid between *E.*
variegatum and *E. hyemale. E. variegatum* var. *jesupi* A. A. Eaton—F
G.
6. E. X nelsonii (A. A. Eaton) Schaffner. Horsetail. June–Sept.
Sandy shores, rare; Cook Co. Considered to be a hybrid between *E.*
variegatum and *E. laevigatum. E. variegatum* var. *nelsonii* A. A.
Eaton—F, G.
7. E. palustre L. Marsh Horsetail. June–Sept. Banks of rivers and
streams, rare; Tazewell and Kankakee cos.
8. E. X litorale Kuhl. Horsetail. June–Sept. Shores and banks, rare;
Lee Co. Considered to be a hybrid between *E. fluviatile* and *E. ar-*
vense.
9. E. hyemale L. var. affine (Engelm.) A. A. Eaton. Scouring Rush.
May–Aug. Shores, banks, roadsides; in every co. *E. hyemale* L.—J.
(Including *E. hiemale* var. *pseudohiemale* [Farw.] Morton and *E. hie-*
male var. *elatum* [Engelm.] Morton.)
10. E. laevigatum A. Br. Smooth Scouring Rush. May–July. Open,
moist, sandy areas; common in the n. half, occasional in the s. half. *E.*
hyemale var. *intermedium* A. A. Eaton—F. (Including *E. kansanum*
Schaffn.)
11. E. X ferrissii Clute. Intermediate Scouring Rush. May–Aug.
Banks and shores; scattered throughout the n. half of the state. Considered to be a hybrid between *E. hyemale* and *E. laevigatum.*
12. E. fluviatile L. Water Horsetail. May–Aug. In water of lakes,
streams, and swamps; occasional in the upper half of the state.

2. LYCOPODIACEAE

1. *Lycopodium* [Dill.] *L. Clubmoss*

1. Sporangia axillary along the stem, the leaves subtending the sporangia
appearing no different from ordinary leaves; gemmae frequently present
in the axils of upper leaves 2
1. Sporangia aggregated in terminal cones, the leaves subtending the

sporangia much modified from ordinary leaves (except in *L. inundatum*); gemmae absent 3

2. Leaves linear or lanceolate, broadest near base, entire; stomates present on both surfaces (with 10 X magnification); spores usually over 35μ in diameter ....
.... 1. *L. porophilum*

2. Leaves oblanceolate, broadest near middle, serrulate or rarely entire; stomates present only on lower surfaces (with 10 X magnification); spores usually less than 35μ in diameter .... 2. *L. lucidulum*

3. Leaves subtending sporangia similar to ordinary leaves, green; sterile stems prostrate 4

3. Leaves subtending sporangia much shorter than ordinary leaves, yellow; sterile stems erect or ascending 5

4. Leaves denticulate, about 1 mm broad .... 3. *L. adpressum*

4. Leaves not denticulate, up to 0.7 mm broad .... 4. *L. inundatum*

5. Cones solitary, 5–7 mm thick; leaves 6- to 8-ranked . 5. *L. dendroideum*

5. Cones 2–4, 3–6 mm thick; leaves 4-ranked 6

6. Peduncles 3–10 cm long; leaves with appressed tips, 1–2 mm apart; sporophylls without scarious margins, ovate .... 6. *L. flabelliforme*

6. Peduncles 12–18 cm long; leaves with spreading tips, about 4 mm apart; sporophylls with scarious margins, orbicular .... 7. *L. X habereri*

1. L. porophilum Lloyd & Underw. Cliff Clubmoss. June–Sept. Moist, shaded sandstone cliffs; three n. and three se. cos. *L. selago* var. *patens* (Beauv.) Desv.—G.

2. L. lucidulum Michx. Shining Clubmoss. June–Sept. Leaves serrulate in the upper half. Shaded sandstone cliffs and ravines, rare; scattered in Ill., but apparently absent in the w. and cent. cos.

var. tryonii Mohlenbrock. June–Sept. Leaves entire. Shaded sandstone cliff, rare; Jackson Co. *L. lucidulum* var. *occidentale* (Clute) L. R. Wilson—F, G.

3. L. adpressum Lloyd & Underw. Bog Clubmoss. July–Sept. Wet ground, very rare; Winnebago Co. *L. inundatum* var. *bigelovii* Tuckerm.—F, G.

4. L. inundatum L. Bog Clubmoss. July–Sept. Bogs, rare; Cook Co.

5. L. dendroideum Michx. July–Sept. Ground Pine. Woodlands, in wet, sandy detritus of St. Peter sandstone, very rare; Ogle Co. *L. obscurum* L. var. *dendroideum* (Michx.) D. C. Eaton—F, G.

6. L. flabelliforme (Fern.) Blanch. Ground Pine. July–Sept. Sandstone ledges in Pope and Ogle cos.; adventive under conifers in Cook, DuPage, and several s. Ill. cos. *L. complanatum* var. *flabelliforme* Fern.—F, G.

7. L. X habereri House. Hybrid between *L. flabelliforme* and *L. tristachyum*. Cook Co. (in 1878).

3. SELAGINELLACEAE

1. Selaginella Beauv. Spikemoss

1. Stems weak, herbaceous; leaves 4-ranked, flaccid, obtuse to acute .....
.... 1. *S. apoda*

1. Stems wiry, evergreen; leaves spirally arranged, stiff, subulate-tipped ..
.... 2. *S. rupestris*

1. S. apoda (L.) Fern. Small Spikemoss. June–Oct. Moist shaded areas, grassy margins of streams; local throughout.
2. S. rupestris (L.) Spring. Rock Spikemoss. June–Sept. Dry rocky areas, mostly of sandstone, local; nw. Ill. and Pope and Union cos.

4. ISOETACEAE

1. Isoetes L. Quillwort

1. Sporangium punctate or striate; megaspores tuberculate; microspores spinulose or papillose .... 1. *I. melanopoda*
1. Sporangium neither punctate nor striate; megaspores reticulate, with narrow ridges; microspores smooth or minutely roughened ....
.... 2. *I. engelmannii*

1. I. melanopoda Gay & Dur. Black Quillwort. May–Sept. Shallow water of ponds and ditches; local throughout. (Including f. *pallida* [Engelm.] Fern.)
2. I. engelmannii A. Br. Engelmann's Quillwort. May–Oct. Shallow water of ponds and ditches; St. Clair Co.

5. OPHIOGLOSSACEAE

1. Leaves pinnately compound or simple and deeply lobed, the venation free; sporangia stipitate .... 1. *Botrychium*
1. Leaves simple, entire, the venation net-like; sporangia sessile ....
.... 2. *Ophioglossum*

1. Botrychium Sw. Grape Fern

1. Sterile blades long- or short-petiolate below long or short stalk of fruiting panicle, evergreen (often turning bronze in winter), coriaceous, subcoriaceous, or membranaceous 2
1. Sterile blades sessile below long stalk of fruiting panicle ....
.... 6. *B. virginianum*
2. Sterile blades ternately compound, on long petioles arising near base of plant 3
2. Sterile blades pinnate or pinnatifid, on long or short petioles arising midway or higher above base of plant 5
3. Ultimate segments of the leaf obtuse or rounded; blade fleshy or coriaceous, not turning bronze in winter .... 1. *B. multifidum*
3. Ultimate segments of the leaf acute or subacute; blade subcoriaceous or membranaceous, turning bronze in winter (except *B. biternatum*) ... 4
4. Pinnules entire, crenulate, or lobed; blade subcoriaceous, turning bronze in winter .... 2. *B. dissectum*

4. Pinnules sharply serrate; blade membranaceous, remaining green in winter .... 3. *B. biternatum*
5. Sterile blade petiolate .... 4. *B. simplex*
5. Sterile blade short-petiolate or nearly sessile ... 5. *B. matricariaefolium*

1. B. multifidum (Gmel.) Rupr. ssp. silaifolium (Presl) Clausen. Northern Grape Fern. Aug.–Sept. Rich woodlands, rare; Boone, Lake, McHenry, Ogle, Will, and Winnebago cos. *B. multifidum* (Gmel.) Rupr.—J; *B. multifidum* var. *intermedium* (D. C. Eaton) Farw.—F, G.

2. B. dissectum Spreng. Cut-leaved Grape Fern. Sept.–Nov. Blade of leaf very finely dissected. Open oak-hickory woodlands; scattered throughout the state.

var. obliquum (Muhl.) Clute. Grape Fern. Sept.–Nov. Blade of leaf shallowly divided. Open oak-hickory woodlands and pastures; throughout the state. *B. dissectum* f. *obliquum* (Muhl.) Fern.—F; *B. obliquum* Muhl.—J.

3. B. biternatum (Sav.) Underw. Grape Fern. Sept.–Nov. Open oak-hickory woodlands, rare; Jackson Co. *B. dissectum* var. *tenuifolium* (Underw.) Farw.—F, G.

4. B. simplex E. Hitchc. Grape Fern. May–June. Grassy meadow, very rare; Winnebago Co.

5. B. matricariaefolium A. Br. June–July. Grassy meadow, very rare; Winnebago Co.

6. B. virginianum (L.) Sw. Rattlesnake Fern. June–July. Dry or moist woodlands; in every co.

2. *Ophioglossum* [*Tourn.*] L. Adder's-tongue

1. Leaf rounded or subacute at apex, the veins forming principle areoles without secondary areoles .... 1. *O. vulgatum*
1. Leaf acute, apiculate, the veins forming principle areoles surrounding secondary areoles .... 2. *O. engelmannii*

1. O. vulgatum L. var. pycnostichum Fern. Adder's-tongue Fern. April–June. Leaf deep green, shiny; spores 37–42μ in diameter. Moist woods and shaded sandstone ledges; scattered in the n. and cent. cos., occasional in the extreme s. *O. vulgatum* L.—G, J.

var. pseudopodum (Blake) Farw. Adder's-tongue Fern. April–June. Leaf pale green, dull; spores 50–54μ in diameter. Low, rich woods, not common. *O. vulgatum* L.—G, J.

2.O. engelmannii Prantl. Adder's-tongue Fern. May–July. Limestone ledges, rare; Hardin, Jersey, Johnson, Pope, and Randolph cos.

6. OSMUNDACEAE

1. *Osmunda* [*Tourn.*] L. Royal Fern

1. Leaves bipinnate, the pinnules serrulate; sporangia borne on upper half of leaf .... 1. *O. regalis*
1. Leaves always pinnate-pinnatifid, the pinnules entire; sporangia borne on a separate fertile leaf or centrally on the leaf 2

2. Sporangia borne on a separate fertile leaf .... 2. *O. cinnamomea*
2. Sporangia produced centrally between sterile pinnae   3. *O. claytoniana*

1. O. regalis L. var. spectabilis (Willd.) Gray. Royal Fern. April–June
Swamps, woods, or, in some of the s. cos., moist, sandstone ledges
throughout the state. *O. regalis* L.—J.
2. O. cinnamomea L. Cinnamon Fern. March–June. Swamps and
swampy woods or, rarely, on moist, sandstone ledges; in the n. one-
third of Ill.; also Pope, Johnson, and Massac cos.
3. O. claytoniana L. Interrupted Fern. April–July. Moist, low woodlands or, rarely, in moist depressions along sandstone ledges; occasional in the n. half of Ill., rare in the s. half.

7. HYMENOPHYLLACEAE

1. *Trichomanes L.   Filmy Fern*

1. T. boschianum Sturm. Filmy Fern. June–Sept. Beneath moist,
overhanging sandstone cliffs, rare; Johnson and Pope cos.

8. POLYPODIACEAE

1. Sori on the leaf margin, covered by a recurved outgrowth of the blade   2
1. Sori not marginal or, if so, not covered by leaf margin 7
2. Sori distinct, borne on the veins or vein tips 3
2. Sori continuous 5
3. Leaves distinctly hairy or glandular; petiole unbranched 4
3. Leaves glabrous; petiole branched at the apex .... 2. *Adiantum*
4. Blades glandular-pubescent; rhizome long-creeping, pubescent ....
.... 1. *Dennstaedtia*
4. Blades eglandular; rhizome compact, scaly .... 6. *Cheilanthes*
5. Rhizome without scales; leaves ternately pinnate; coarse ferns at least 50
cm tall, mostly in woods or in fields .... 3. *Pteridium*
5. Rhizome scaly; leaves regularly pinnate; ferns rarely over 35 cm tall,
mostly on rocks 6
6. Stipe green, weak .... 4. *Cryptogramma*
6. Stipe dark brown or purple-brown, wiry .... 5. *Pellaea*
7. Indusium absent 8
7. Indusium present, although sometimes so deeply cleft as to be difficult
to observe, and sometimes soon deciduous 10
8. Blades once-pinnatifid; leaves less than 40 cm long, usually less than 25
cm long .... 7. *Polypodium*
8. Blades bipinnatifid or pinnately divided; leaves over 40 cm long 9
9. Rachis unwinged; leaves ternate .... 11. *Gymnocarpium*
9. Rachis winged (except sometimes between the lowest pinnae); leaves
triangular, but not ternate .... 12. *Thelypteris*

10. Leaves strongly dimorphic (i.e., with the fertile quite different from the sterile) 11
10. Leaves all alike (In *Polystichum,* the upper fertile pinnae are strongly modified; in one *Asplenium,* the leaves are slightly dimorphic.) 13
11. Fertile leaves bipinnate, the sterile ones pinnatifid .... 9. *Onoclea*
11. Fertile leaves once-pinnate, the sterile ones once-pinnate, pinnate-pinnatifid, or pinnatifid 12
12. Veins free .... 10. *Matteuccia*
12. Veins forming a network .... 14. *Woodwardia*
13. Indusium attached at the center or the side of the sorus 14
13. Indusium attached beneath the sporangia (in *Cystopteris,* the indusium may sometimes appear to be lateral) 19
14. Indusium attached at the center of the sorus 15
14. Indusium attached laterally 17
15. Indusium peltate; fertile pinnae apical and modified; leaf once-pinnate .... 8. *Polystichum*
15. Indusium reniform; fertile pinnae identical to sterile pinnae; leaf pinnate-pinnatifid to tripinnate 16
16. Rhizome slender, cord-like; leaves deciduous; petiole without scales, or with scales at most only 5 mm long .... 12. *Thelypteris*
16. Rhizome stout; leaves more or less evergreen; petiole scaly (at least near base), the scales over 5 mm long .... 13. *Dryopteris*
17. Sori parallel to the main vein, appearing to be in a chain .... 14. *Woodwardia*
17. Sori parallel to the lateral veins 18
18. Veins reaching the margin; blades deciduous .... 15. *Athyrium*
18. Veins not reaching the margin; blades usually evergreen 16. *Asplenium*
19. Indusium separating into shreds; petioles rather densely chaffy .... 17. *Woodsia*
19. Indusium hood-like, not separating into shreds; petioles sparsely chaffy, or glabrous .... 18. *Cystopteris*

1. Dennstaedtia Bernh. Hay-scented Fern

1. D. punctilobula (Michx.) Moore. Hay-scented Fern. July–Oct. Moist, shaded sandstone ravines, rare; Johnson and Pope cos; reported from Wabash Co.

2. Adiantum L. Maidenhair Fern

1. A. pedatum [Tourn.] L. Maidenhair Fern. June–Sept. Moist, shaded woodlands; in every co.

3. Pteridium Gled. Bracken Fern

1. P. aquilinum (L.) Kuhn var. latiusculum (Desv.) Underw. Bracken Fern. June–Sept. Margins of ultimate leaf segments hairy, the terminal segments 5–8 mm broad. Open woodlands and fields; common throughout the state, except in the s. cos. where it is occasional. *P. latiusculum* (Desv.) Hieron.—J.

var. pseudocaudatum (Clute) Heller. Bracken Fern. June–Sept. Margins of ultimate leaf segments glabrous or nearly so, the terminal segments to 4.5 mm broad. Open woodlands and fields, rare; Hardin Co.

4. Cryptogramma R. Br.   Cliffbrake Fern

1. C. stelleri (S. G. Gmel.) Prantl. Slender Cliffbrake. June–Sept.
Deep, shaded, limestone ravines, uncommon; restricted to the n. one-
fourth of Ill.

5. Pellaea Link   Cliffbrake Fern

1. Petiole and rachis pubescent nearly throughout; fertile pinnules much
narrower than the sterile ones .... 1. *P. atropurpurea*
1. Petiole and rachis glabrous or sparsely pubescent; fertile and sterile pinnae similar in shape .... 2. *P. glabella*

1. P. atropurpurea (L.) Link. Purple Cliffbrake. June–Sept. On limestone cliffs, frequently under very dry conditions; in the cos. bordering the Mississippi River in the s. half of Ill.; in the cos. bordering the
Ohio River; also Johnson and Ogle cos.
2. P. glabella Mett. Smooth Cliffbrake. June–Sept. Limestone outcroppings; local in the n. cos., extending southward along the Mississippi and Ohio rivers. *P. atropurpurea* var. *bushii* Mack.—G.

6. Cheilanthes Sw.   Lip Fern

1. Leaves densely brown-woolly beneath, to 15 cm long, tripinnate, with
7–12 (–15) pairs of pinnae; on limestone .... 1. *C. feei*
1. Leaves white-villous beneath, some usually well over 15 cm long, bipinnate-pinnatifid, with 12–20 pairs of pinnae; chiefly on sandstone
.... 2. *C. lanosa*

1. C. feei Moore. Baby Lip Fern. June–Sept. Dry, exposed limestone
cliffs; in cos. bordering the Mississippi River and the Ohio River, in
the s. half of Ill.; also JoDaviess and Carroll cos.
2. C. lanosa (Michx.) D. C. Eaton. Hairy Lip Fern. June–Sept. Dry,
exposed cliffs, chiefly of sandstone, local; confined to the s. one-third
of Ill. and Ogle Co. *C. vestita* (Spreng.) Sw.—F.

7. Polypodium [Tourn.] L.   Polypody

1. Blades and petioles scaleless, appearing smooth, the segments minutely
toothed .... 1 *P. vulgare*
1. Lower surface of blades and petioles scaly, appearing pustular, the segments entire .... 2. *P. polypodioides*

1. P. vulgare L. var. virginianum (L.) Eaton. Common Polypody.
June–Oct. Sandstone rock, usually under some shade; occasional in
the s. one-third of Ill., local and rare in the n. one-third, absent from
the cent. one-third. *P. virginianum* L.—F, J; *P. vulgare* L.—G.
2. P. polypodioides (L.) Watt var. michauxianum Weatherby. Gray
Polypody. June–Oct. Sandstone rocks, tree trunks, and branches, not
common; restricted to the s. one-fourth of the state. *P. polypodioides*
(L.) Watt—G, J.

8. *Polystichum* Roth   *Christmas Fern*

1. P. acrostichoides (Michx.) Schott. Christmas Fern. June–Oct. Woods, particularly in rocky soil; in every co. in the s. one-half of Ill., scattered in the n. cos.

9. *Onoclea* L.   *Sensitive Fern*

1. O. sensibilis L. Sensitive Fern. June–Oct. Moist woodlands or low, open ground, common; in nearly every co. (Including f. *obtusilobata* [Schkuhr] Gilbert.)

10. *Matteuccia* Todaro   *Ostrich Fern*

1. M. struthiopteris (L.) Todaro. Ostrich Fern. July–Oct. Rich, moist woodlands; restricted to the n. one-third of the state, usually in the w. cos. *Pteretis pensylvanica* (Willd.) Fern.—F.

11. *Gymnocarpium* Newm.   *Oak Fern*

1. G. dryopteris (L.) Newm. Oak Fern. June–Sept. Deep, moist woodlands, very rare; Ogle and St. Clair cos. *Dryopteris disjuncta* (Ledeb.) C. V. Morton—F, J.

12. *Thelypteris* Schmidel   *Beech Fern*

1. Rachis winged throughout, or only the basal pair of pinnae stalked; indusium none 2
1. Rachis not winged, the pinnae separate from it nearly to apex of blade; indusium cordate 3
2. Rachis winged only above the two basal pinna pairs, the wings not extending to the lowest pair of pinnae; blades pinnate-pinnatifid; rachis rather densely brown-scaly ....1. *T. phegopteris*
2. Rachis winged throughout; blades bipinnatifid; rachis sparsely white-scaly ....
.... 2. *T. hexagonoptera*
3. Lowest pinnae strongly reduced ....3. *T. noveboracensis*
3. Lowest pinnae only slightly reduced or not at all .... 4. *T. palustris*

1. T. phegopteris (L.) Slosson. Long Beech Fern. June–Aug. Moist, shaded woodlands in sandstone areas, local; Cass, Henderson, Jackson, LaSalle, Menard, Ogle, and Pulaski cos. *Dryopteris phegopteris* (L.) Christens.—F, J.
2. T. hexagonoptera (Michx.) Weatherby. Broad Beech Fern. June–Sept. Rich woodlands; scattered throughout the state. *Dryopteris hexagonoptera* (Michx.) Christens.—F, J.
3. T. noveboracensis (L.) Nieuwl. New York Fern. June–Sept. Moist, rarely dry, woodlands, rare; Kane, Kankakee, Monroe, Pope, and Wabash cos. *Dryopteris noveboracensis* (L.) Gray—F, J.
4. T. palustris Schott var. pubescens (Laws.) Fern. Marsh Fern. June–Oct. Marshy ground and in swamps; rather common in the n., rare in the s. *Dryopteris thelypteris* (L.) Gray—F, J.

13. *Dryopteris Adans.* Shield Fern

1. Teeth of leaves bearing short spiny tips; blades pinnate-pinnatifid to tripinnate 2
1. Teeth of leaves without spiny tips; blades pinnate-pinnatifid to bipinnate 6
2. Indusium eglandular 3
2. Indusium glandular 4
3. Leaves uniform, persisting through the winter .... 1. *D. carthusiana*
3. Outer leaves sterile, smaller, spreading, persisting nearly through the winter; inner leaves fertile, larger, erect, not persisting through the winter .... 5. *D. cristata*
4. Leaf broadest near middle; spores abortive ....2. *D.* X *boottii*
4. Leaf broadest near base; spores normal or abortive 5
5. Pinnae at right angles to rachis, abruptly narrowed to the long tip; rhizome suberect; spores normal .... 3. *D. intermedia*
5. Pinnae ascending from the rachis, gradually narrowed to the tip; rhizome short-creeping; spores abortive .... 4. *D.* X *triploidea*
6. Leaves membranaceous or subcoriaceous; sori not marginal 7
6. Leaves coriaceous; sori marginal .... 10. *D. marginalis*
7. Sori borne midway between mid-nerve and margin of leaf segment .. 8
7. Sori borne near midvein of leaf segment 10
8. Leaves dimorphic, the outer sterile ones smaller and more persistent; pinnae at most about 8 cm long .... 5. *D. cristata*
8. Leaves monomorphic; some or all the pinnae over 8 cm long 9
9. Leaves with 10–15 pairs of pinnae .... 6. *D.* X *clintoniana*
9. Leaves with 25–35 pairs of pinnae .... 9. *D.* X *neo-wherryi*
10. Scales at base of stipe dark brown to black; fronds usually more than half as broad as long .... 7. *D. goldiana*
10. Scales at base of stipe pale brown; fronds usually less than half as broad as long .... 8. *D. celsa*

1. *D. carthusiana* (Villars) H. P. Fuchs. Spinulose Woodfern. June–Aug. Moist, rocky woodlands, rarely in dry situations; occasional throughout the state, except for the s. cent. cos. *D. spinulosa* (O. F. Muell.) Watt.—F, J; *D. austriaca* (Jacq.) Woynar var. *spinulosa* (Muell.) Fiori—G.

2. *D.* X *boottii* (Tuckerm.) Underw. Boott's Woodfern. June–Aug. Moist, rocky woodland, very rare; LaSalle Co. Considered a hybrid between *D. intermedia* and *D. cristata*.

3. *D. intermedia* (*Muhl.*) *Gray.* Common Woodfern. June–Aug. Moist, rocky ravines and woodlands; occasional or rare in the n. half and the s. tip of Ill. *D. spinulosa* (O. F. Muell.) Watt var. *intermedia* (Muhl.) Underw.—F; *D. austriaca* (Jacq.) Woynar var. *intermedia* (Muhl.) Morton—G.

4. *D.* X *triploidea* Wherry. Woodfern. June–Sept. Moist, rocky woodland, very rare; Lake Co. Considered to be a hybrid between *D. intermedia* and *D. carthusiana*. *D. spinulosa* (O. F. Muell.) Watt—F; *D. austriaca* (Jacq.) Woynar—G.

5. *D. cristata* (L.) Gray. Crested Fern. June–Sept. Low, moist woodlands; occasional in the n. one-third of the state; also Pope Co.

6. *D.* X *clintoniana* (D. C. Eaton) Dowell. Clinton's Woodfern. June–Sept. Moist, rich woodland, very rare; St. Clair Co. Considered to be a hybrid between *D. cristata* and *D. goldiana*. *D. cristata* (L.) Gray var. *clintoniana* (D. C. Eaton) Underw.—F.

7. D. goldiana (Hook.) Gray. Goldie's Fern. June–Sept. Moist, shaded, often rocky woodlands, not common; scattered through the state.

8. D. celsa (Wm. Palmer) Small. Log Fern. June–Aug. Along railroad near a swamp, very rare; Johnson Co. *D. goldiana* (Hook.) Gray ssp. *celsa* Wm. Palmer—G.

9. D. X neo-wherryi W. H. Wagner. June–Aug. Rich, rocky woods, very rare; Pope Co. Considered to be a hybrid between *D. goldiana* and *D. marginalis.*

10. D. marginalis (L.) Gray. Marginal Fern. June–Oct. Rocky woodlands; common to occasional throughout the state.

14. *Woodwardia* Sm. Chain Fern

1. Leaves monomorphic; sterile blades once-pinnate ...... 1. *W. virginica*
1. Leaves dimorphic; sterile blades pinnatifid .... 2. *W. areolata*

1. W. virginica (L.) Sm. Chain Fern. July–Sept. Bogs, rare; Lake Co.
2. W. areolata (L.) Moore. Netted Chain Fern. July–Oct. Springy marsh in lowland woods, edge of sandstone cliffs, rare; Johnson and Pope cos.

15. *Athyrium* Roth Lady Fern

1. Blades once-pinnate; pinnae entire .... 1. *A. pycnocarpon*
1. Blades pinnate-pinnatifid to bi- (tri-) pinnate; pinnae pinnatifid 2
2. Blades pinnate-pinnatifid; indusium light brown (at maturity), firm, parallel to the veins; sori straight or nearly so .... 2. *A. thelypterioides*
2. Blades at least bipinnate; indusium dark brown (at maturity), membranaceous, many crossing the veins; sori more or less curved ....
.... 3. *A. filix-femina*

1. A. pycnocarpon (Spreng.) Tidestrom. Narrow-leaved Spleenwort. Aug.–Sept. Moist, shaded woodlands; occasional throughout the state, although rare in the extreme n. cos.

2. A. thelypterioides (Michx.) Desv. Silvery Spleenwort. July–Sept. Rich woodlands, occasional to common.

3. A. filix-femina (L.) Roth var. rubellum Gilb. Lady Fern. June–Sept. Fourth or fifth pair of pinnae (from the base) the largest; rachis glandular; indusium eglandular. Moist, open woodlands and borders of swamps; rather common throughout the state. *A. angustum* (Willd.) Presl—J; *A. filix-femina* var. *michauxii* (Spreng.) Farw.—F, G.

var. asplenioides (Michx.) Farw. Southern Lady Fern. June–Sept. Second or third pair of pinnae (from the base) the largest; rachis eglandular; indusium glandular. Moist, rocky woodlands, occasional in the s. cos.

16. *Asplenium* L. Spleenwort

1. Leaves simple, unlobed (rarely a single pair of lobe-like auricles at base); veins forming a network .... 1. *A. rhizophyllum*
1. Leaves simple and pinnatifid, or pinnate to bipinnate-pinnatifid; veins free 2
2. Rachis green throughout 3
2. Rachis partly or entirely brown, or black 6

3. Petiole green throughout .... 2. *A. ruta-muraria*
3. Petiole brown at base, sometimes throughout 4
4. Blades entirely pinnatifid, or with merely the lowest pair of pinnae distinct; spores normal .... 3. *A. pinnatifidum*
4. Blades pinnatifid above, pinnate below for at least two pairs of pinnae; spores abortive 5
5. Petiole brown throughout ....4. *A.* X *gravesii*
5. Petiole brown at base, green above ....5. *A.* X *trudellii*
6. Rachis brown below (usually about half its length), green above. 7
6. Rachis brown or black throughout, or at least for three-fourths the length 9
7. Blades once-pinnate below, pinnatifid above; spores abortive 8
7. Blades bipinnate to bipinnate-pinnatifid; spores normal .. 8. *A. bradleyi*
8. Pinnae approximate to overlapping; base of most of the pinnae truncate; outline of frond oblong-lanceolate .... 6. *A.* X *kentuckiense*
8. Pinnae separated to remote; base of most of the pinnae cuneate; outline of frond linear .... 7. *A. trichomanes* X *A. pinnatifidum*
9. Rachis brown for three-fourths its length, green above; blade pinnatifid for half its length, the tip narrowly caudate .... 9. *A.* X *ebenoides*
9. Rachis brown or black throughout; blade pinnatifid only at apex, the tip not caudate 10
10. Pinnae not auriculate .... 10. *A. trichomanes*
10. Pinnae auriculate at base of upper margin 11
11. Pinnae pairs opposite (sometimes alternate); fertile and sterile leaves similar; auricle of pinna not overlapping rachis .... 11. *A. resiliens*
11. Pinnae pairs alternate; fertile leaves more erect and taller than sterile leaves; auricle of pinna overlapping rachis .... 12. *A. platyneuron*

1. A. rhizophyllum L. Walking Fern. May–Sept. Rocky woodlands, either on sandstone or limestone; scattered throughout the state. *Camptosorus rhizophyllus* (L.) Link—F, G, J.

2. A. ruta-muraria L. Wall-rue Spleenwort. May–Sept. Crevices of limestone cliffs, very rare; known only from "s. Ill.," collected in the middle 1800s. *A. cryptolepis* Fern.—F.

3. A. pinnatifidum Nutt. Pinnatifid Spleenwort. May–Sept. Crevices of sandstone cliffs; occasional in the s. tip of Ill.; also Cumberland, Fulton, and Wabash cos.

4. A. X gravesii Maxon. Graves' Spleenwort. May–Sept. Crevices of sandstone cliffs, very rare; Union and Williamson cos. Considered to be a hybrid between *A. pinnatifidum* and *A. bradleyi.*

5. A. X trudellii Wherry. Trudell's Spleenwort. May–Sept. Crevices of sandstone cliffs; Jackson Co. Considered to be a hybrid between *A. pinnatifidum* and *A. montanum.*

6. A. X kentuckiense McCoy. Kentucky Spleenwort. May–Sept. Crevices of cherty cliffs; Union Co. Considered to be a hybrid between *A. pinnatifidum* and *A. platyneuron.*

7. A. trichomanes X A. pinnatifidum. Wagner's Spleenwort. May–Sept. Crevices of cherty cliffs; Union Co. Considered to be a hybrid between *A. trichomanes* and *A. pinnatifidum.*

8. A. bradleyi D. C. Eaton. Bradley's Spleenwort. June–Sept. Crevices of sandstone or cherty cliffs, rare; Jackson, Randolph, and Union cos.

9. A. X ebenoides R. R. Scott. Scott's Spleenwort. May–Sept. Rocky, sandstone woodlands, rare; Alexander, Jackson, Pope, and

Union cos. Considered to be a hybrid between *A. rhizophyllum* and *A. platyneuron. Asplenosorus* X *ebenoides* (R. R. Scott) Wherry—F, G.

10. A. trichomanes L. Maidenhair Spleenwort. May–Sept. Crevices of shaded cliffs, both limestone and sandstone; occasional in the s. one-fourth of the state; also Cumberland Co.

11. A. resiliens Kunze. Black Spleenwort. May–Sept. Crevices of limestone cliffs, rare; Jackson and Union cos.

12. A. platyneuron (L.) Oakes. Ebony Spleenwort. May–Sept. Dry or moist woodlands; common in the s. half of Ill., becoming less abundant northward. (Including var. *incisum* [Howe] Robinson.)

17. Woodsia R. Br. Woodsia

1. Blades bipinnate-pinnatifid; leaves generally over 25 cm long; petiole stramineous, unjointed; scales of rhizome few .... 1. *W. obtusa*
1. Blades pinnate-pinnatifid; leaves generally less than 20 cm long; petiole brown, jointed below the middle; scales of rhizome numerous ....
.... 2. *W. ilvensis*

1. W. obtusa (Spreng.) Torr. Common Woodsia. May–Oct. Dry or moist rocky woods, occasionally on very exposed sandstone ledges; occasional to common, except in the e. cent. cos.

2. W. ilvensis (L.) R. Br. Rusty Woodsia. June–Oct. Sandstone cliff, very rare; Ogle Co.

18. Cystopteris Bernh. Fragile Fern

1. All veins of ultimate leaf segments running to the sinuses 1. *C. bulbifera*
1. Some or all the veins of ultimate leaf segments running to the teeth . 2
2. All veins of ultimate leaf segments running to the teeth ... 2. *C. fragilis*
2. Some veins of ultimate leaf segments running to the teeth, some running to the sinuses 3
3. Indusium acute and toothed at the apex, eglandular ...... 2. *C. fragilis*
3. Indusium truncate and entire at the apex, more or less glandular ......
.... 3. *C.* X *tennesseensis*

1. C. bulbifera (L.) Bernh. Bladder Fern. June–Sept. Mostly limestone cliffs and woodlands; occasional in the cos. along the Illinois and Mississippi rivers, otherwise rare.

2. C. fragilis (L.) Bernh. Fragile Fern. June–Sept. All veins of ultimate leaf segments running to the teeth; rhizome short, scaly; petiole brown for half its length. Moist woodlands; Union Co.

var. protrusa Weatherby. Fragile Fern. June–Sept. All veins of ultimate leaf segments running to the teeth; rhizome long-creeping, densely hairy; petiole stramineous, except at the base. Moist woodland; common in every co.

var. mackayi Laws. Fragile Fern. June–Sept. Moist woodlands; Jackson and LaSalle cos.

3. C. X tennesseensis Shaver. Tennessee Fragile Fern. June–Sept. Moist woodlands, rare; Champaign and Will cos.

9. MARSILEACEAE

1. Marsilea L.   Waterclover

1. M. quadrifolia L. Waterclover. June–Dec. Introduced into ponds or lakes; scattered and uncommon.

10. SALVINIACEAE

1. Azolla Lam.   Mosquito Fern

1. Plants less than 1 cm in diameter; leaves about 0.5 mm long; glochidia without cross-walls .... 1. *A. caroliniana*
1. Plants 1 cm in diameter or larger; leaves about 0.7 mm long or longer; glochidia with cross-walls .... 2. *A. mexicana*

    1. A. caroliniana Willd. Mosquito Fern. July–Oct. Standing water, very rare; St. Clair Co.
    2. A. mexicana Presl. Mosquito Fern. July–Oct. Ponds and standing water, local; scattered throughout.

11. TAXACEAE

1. Taxus L.   Yew

1. T. canadensis Marsh. Canada Yew. Wooded hillsides, rare; confined to the n. one-fourth of Ill.

12. PINACEAE

1. Leaves in fascicles of 2–5; plants evergreen .... 1. *Pinus*
1. Leaves in fascicles of 20 or more; plants deciduous .... 2. *Larix*

1. Pinus L.   Pine

1. Leaves 5 in a fascicle .... 1. *P. strobus*
1. Leaves 2–3 in a fascicle 2
2. Leaves, or some of them, 3 in a fascicle 3
2. Leaves, or some of them, 2 in a fascicle 6
3. Leaves slender, at most 1.5 mm broad .... 2. *P. echinata*
3. Leaves stout, 1.5–3.0 mm broad 4
4. Most of the leaves more than 7 cm long; scales of cone with a spine up to 3 mm long 5

4. Most of the leaves less than 7 cm long; scales of cone with a spine 5–6 mm long .... 5. *P. pungens*
5. Leaves up to 15 cm long; cone almost as broad as long ... 3. *P. rigida*
5. Most or all the leaves over 15 cm long; cone longer than broad .... 4. *P. taeda*
6. Leaves up to 7 cm long 7
6. Most of the leaves 7 cm long or longer 9
7. Leaves 2–4 cm long; cones shiny .... 6. *P. banksiana*
7. Some of the leaves over 4 cm long; cones not shiny 8
8. Scales of cone without a spine .... 7. *P. sylvestris*
8. Scales of cone with a spine 5–6 mm long .... 5. *P. pungens*
9. None of the leaves over 17 cm long, dark green 10
9. Some of the leaves 17–25 cm long, light green .... 4. *P. taeda*
10. Scales of cone without a spiny tip .... 8. *P. resinosa*
10. Scales of cone with a minute spiny tip .... 2. *P. echinata*

1. P. strobus L. White Pine. Woods; local in the n. one-fourth of Ill.

2. P. echinata Mill. Shortleaf Pine. Rocky soil, rare in the native condition, where it is known from Randolph and Union cos.; widely planted in plantations.

3. P. rigida Mill. Pitch Pine. Native to the e. of Ill.; occasionally encountered as an introduction in the state.

4. P. taeda L. Loblolly Pine. Native primarily s. of Ill.; frequently planted in Ill.

5. P. pungens Lamb. Table Mountain Pine. Native e. and s. of Ill.; rarely encountered as an introduction in Ill.

6. P. banksiana Lamb. Jack Pine. Sandy soil, rare; Cook, Lake, and Ogle cos.

7. P. sylvestris L. Scotch Pine. Native of Europe; occasionally encountered as an introduction in Ill.

8. P. resinosa Ait. Red Pine. Dry woods, rare; LaSalle Co.

2. *Larix Mill. Larch*

1. Scales of cones glabrous, 10–20 in number; most of the leaves less than 2.5 cm long .... 1. *L. laricina*
1. Scales of cones pubescent, at least 30 in number; most of the leaves at least 2.5 cm long .... 2. *L. decidua*

1. L. laricina (DuRoi) K. Koch. American Larch. Tamarack. Bogs, rare; Cook, Lake, and McHenry cos.

2. L. decidua Mill. European Larch. Escaped from cultivation in a few ne. cos.

13. TAXODIACEAE

1. *Taxodium Rich. Bald Cypress*

1. T. distichum (L.) Rich. Bald Cypress. Swamps; extreme s. Ill., n. along the Wabash River to Lawrence Co. (Including f. *confusum* Palmer & Steyerm.)

14. CUPRESSACEAE

1. Branches flattened; leaves all scale-like; fruit a woody cone ... 1. *Thuja*
1. Branches not flattened; leaves all subulate or some scale-like, others subulate; fruit a fleshy berry-like cone .... 2. *Juniperus*

1. Thuja L.   Arbor Vitae

1. T. occidentalis L. Arbor Vitae. White Cedar. Cliffs, bluffs, and bogs, rare; Cook, Kane, Lake, and LaSalle cos.

2. Juniperus L.   Juniper

1. All leaves subulate, borne in whorls of three .... 1. *J. communis*
1. Some leaves subulate, others scale-like, borne opposite 2
2. Prostrate, trailing shrub; fruits more than 6 mm in diameter, 3- to 5-seeded .... 2. *J. horizontalis*
2. Upright tree; fruits 4–6 mm in diameter, 1- to 2-seeded . 3. *J. virginiana*

1. J. communis L. Common Juniper. Plants erect. Sandy soil, rare; Lake Co. *J. canadensis* Burgsdorf—J.
var. depressa Pursh. Ground Juniper. Plants decumbent. Sand dunes, rare; Cook and Lake cos. *J. canadensis* Burgsdorf—J.
2. J. horizontalis Moench. Trailing Juniper. Sand dunes, rare; Cook and Lake cos.
3. J. virginiana L. Red Cedar. Woods, cliffs, and fields; common throughout Ill. (Including var. *crebra* Fern. & Grisc.)

15. TYPHACEAE

1. Typha L.   Cat-tail

1. Pollen grains borne in groups of 4; stigma spatulate; leaves flat, usually 8 or more per plant; staminate and pistillate portions of the inflorescence usually contiguous, but not always .... 1. *T. latifolia*
1. Pollen grains borne singly; stigma linear; leaves somewhat convex, usually less than 8 per plant; staminate and pistillate portions of the inflorescence separated .... 2. *T. angustifolia*

1. T. latifolia L. Common Cat-tail. June–Oct. Low areas, common; in every co.
2. T. angustifolia L. Narrow-leaved Cat-tail. June–Oct. Wet situations, occasional; scattered throughout the state.

16. SPARGANIACEAE

1. Sparganium L. Bur-reed

1. Plants floating or suberect; stems to 30 cm long; staminate head 1; pistillate heads about 1 cm in diameter, the lowest short-pedunculate; beak of achene 0.5–1.5 mm long .... 1. *S. minimum*
1. Plants erect; stems (5–) 30–120 cm tall; staminate heads 3–25; pistillate heads 1.5–3.5 cm in diameter, all sessile; beak of achene 2–6 mm long. 2
2. Central axis bearing 1–4 pistillate heads; achene stipitate, fusiform, tapering to the summit; stigma 1 3
2. Central axis bearing only staminate heads; achene not stipitate, obpyramidal, truncate at the summit; stigmas 2 .... 5. *S. eurycarpum*
3. At least one of the pistillate heads borne above the subtending bract (supra-axillary); inflorescence simple; achene usually greenish-brown, even at maturity .... 2. *S. chlorocarpum*
3. All heads axillary in the subtending bract; inflorescence usually with 1 or more branches; achene pale or dark brown 4
4. Leaves stiff, strongly keeled; branches of inflorescence without any pistillate heads; pistillate heads 2.5–3.5 cm in diameter; body of achene 5–7 mm long, shining, pale brown .... 3. *S. androcladum*
4. Leaves soft, usually not strongly keeled; branches of inflorescence with 1–3 pistillate heads; pistillate heads 1.5–2.5 cm in diameter; body of achene 3–5 mm long, dull, dark brown .... 4. *S. americanum*

1. S. minimum (Hartm.) Fries. Least Bur-reed. July–Aug. Shallow water; McHenry Co.

2. S. chlorocarpum Rydb. Green-fruited Bur-reed. June–Aug. Shallow water; Cook, DeKalb, Lake, Lee, and Union cos.

3. S. androcladum (Engelm.) Morong. Bur-reed. June–July. Shallow water; scattered throughout the state, but not common.

4. S. americanum Nutt. Bur-reed. June–July. Shallow water; restricted to the northern one-fifth of the state.

5. S. eurycarpum Engelm. Bur-reed. June–Aug. Shallow water; occasional throughout the state, but less common in the extreme cos.

17. RUPPIACEAE

1. Ruppia L. Ditch Grass

1. R. maritima L. var. rostrata Agardh. Ditch Grass. July–Oct. Saline waters, rare; Henry, Lake, and Vermilion cos.

18. ZANNICHELLIACEAE

1. *Zannichellia* L.   Horned Pondweed

1. Z. palustris L. Horned Pondweed. June–Oct. Ditches, spring-fed streams, and ponds; occasional in the n. one-half of the state; also Jackson Co.

19. NAJADACEAE

1. *Najas* L.   Naiad

1. Leaves seemingly entire or serrulate; plants monoecious; achene 1.5–3.5 mm long 2
1. Leaves coarsely toothed; plants dioecious; achene 4.0–7.5 mm long....
.... 5. *N. marina*
2. Leaves linear, 0.5–1.4 mm wide, with slightly dilated bases; marginal spinules microscopic 3
2. Leaves filiform, 0.1–0.3 mm wide, with abruptly dilated bases; marginal spinules macroscopic, although minute 4
3. Achene lustrous, with 30–50 rows of obscure, minute, often square areolae; style and 2 stigmas 0.8–1.6 mm long .... 1 *N. flexilis*
3. Achene dull, with 16–24 rows of distinct, large, hexagonal or rectangular areolae; style and 2–3 stigmas less than 1 mm long   2. *N. guadalupensis*
4. Achene with roughened appearance, with 22–40 rows of longer than broad, rectangular areolae; style and 2–3 stigmas 0.8–1.2 mm long .....
.... 3. *N. gracillima*
4. Achene with ribbed appearance, with 10–18 rows of broader than long, rectangular areolae; style and 2 stigmas 1.0–1.4 mm long .. 4. *N. minor*

1. N. flexilis (Willd.) Rostk. & Schmidt. Naiad. July–Sept. Shallow water; occasional throughout the state.

2. N. guadalupensis (Spreng.) Magnus. Naiad. July–Oct. Along shores of ponds and shallow lakes; occasional to uncommon throughout the state.

3. N. gracillima (A. Br.) Magnus. Naiad. July–Sept. Pools, ponds, and lakes with sandy and muddy substrata, rare; Ford, Jackson, and Williamson cos.

4. N. minor All. Naiad. July–Oct. Shallow waters along lake shores; occasional to common in the s. two-thirds of the state, absent elsewhere.

5. N. marina L. Naiad. July–Oct. In water, rare; Lake Co.

20. POTAMOGETONACEAE

1. *Potamogeton* L.   Pondweed

1. Leaves septate, 0.3–0.5 (–2.0) mm broad; peduncles flexible ....
.... 1. *P. pectinatus*

1. Leaves not septate, some or all over 0.5 mm broad; peduncles rigid . 2
2. Leaves uniform 3
2. Leaves of two kinds, the floating usually shorter and broader than the submersed 20
3. Stipules adnate to leaf-base; leaves auriculate at base .. 2. *P. robbinsii*
3. Stipules free from leaf-base; leaves rounded, cordate, or tapering at base, never auriculate 4
4. Leaves sharply serrulate; fruits 5–6 mm long .... 3. *P. crispus*
4. Leaves entire or minutely denticulate; fruits 1.6–5.0 mm long 5
5. Submersed leaves 5–75 mm broad 6
5. Submersed leaves up to 5 mm broad 13
6. Stems compressed; fruits 3-keeled .... 14. *P. epihydrus*
6. Stems terete; fruits rounded on the sides, or with only one prominent keel 7
7. Submersed leaves rounded or cordate at the base, some of them partly clasping the stem 8
7. Submersed leaves tapering to the base, not clasping the stem 9
8. Leaves entire, boat-shaped at the tip; fruits 4–5 mm long, dorsally keeled; stems zigzag .... 4. *P. praelongus*
8. Leaves minutely denticulate, not boat-shaped at the tip; fruits 2–3.5 mm long, not keeled; stems not zigzag .... 5. *P. richardsonii*
9. Leaves entire (sometimes crisped in *P. pulcher*) 10
9. Leaves minutely denticulate 11
10. Fruiting spike 4 cm long or longer; middle and upper submersed leaves broadly lanceolate to ovate, never crisped; fruit tapering to base ....
.... 15. *P. amplifolius*
10. Fruiting spike 2.0–3.5 cm long; middle and upper submersed leaves linear-lanceolate to lanceolate, sometimes crisped; fruit rounded at base .... 16. *P. pulcher*
11. Fruits (including beak) 3.5–4.3 mm long .... 17 *P. nodosus*
11. Fruits (including beak) 1.7–3.5 mm long 12
12. Primary stems up to 1 mm thick; leaves more than 5 times longer than broad .... 20. *P. gramineus*
12. Primary stems usually more than 1 mm thick; leaves less than 5 times longer than broad .... 18. *P. illinoensis*
13. Submersed leaves with 15–35 nerves; fruits quadrate, 4–5 mm long ....
.... 6. *P. zosteriformis*
13. Submersed leaves with (1–) 3–7 (–11) nerves; fruits obovoid to ovoid to nearly orbicular, 1,5–3.5 mm long 14
14. Spikes with 6 or more whorls of flowers 15
14. Spikes with 1–5 whorls of flowers 17
15. Leaves minutely denticulate .... 20. *P. gramineus*
15. Leaves entire 16
16. Spikes uniform; stipules free from base of leaf; fruits 3–5 mm long ....
.... 12. *P. natans*
16. Spikes of two or more kinds; stipules adnate to base of leaf; fruits 1.0–1.5 mm long .... 13. *P. diversifolius*
17. Spikes subcapitate, 2–5 mm long; fruits keeled; sepaloid connectives 0.5–1.0 mm long; leaves glandless at base .... 7. *P. foliosus*
17. Spikes elongate, more or less interrupted, 6–15 mm long; fruits rounded on back; sepaloid connectives 1.2–2.5 mm long; leaves biglandular at base (sometimes glands absent in *P. pusillus*) 18
18. Leaves thin, translucent .... 8. *P. friesii*
18. Leaves firm, opaque 19
19. Stipules chartaceous, white, becoming fibrous .... 9. *P. strictifolius*

19. Stipules membranous, olive, not becoming fibrous ..... 10. *P. pusillus*
20. Leaves entire (submersed leaves sometimes crisped in *P. pulcher*) . 21
20. Leaves minutely denticulate 26
21. Submersed leaves up to 2 mm broad, 1- to 5-nerved 22
21. Submersed leaves 5–75 mm broad, 7- to 21-nerved 24
22. Spikes uniform; stipules free from base of leaf; fruits 1.6–5.0 mm long
.... 23
22. Spikes of two or more kinds, those in the axils of the submersed leaves
subgloboid, those in the axils of the floating leaves elongate; stipules
adnate to base of leaf; fruits 1.0–1.5 mm long ...... 13. *P. diversifolius*
23. Submersed leaves 0.1–0.5 mm broad, 1- to 3-nerved; fruits 1.6–2.2 mm
long; floating leaves elliptic to narrowly obovate .... 11. *P. vaseyi*
23. Submersed leaves 0.5–2.0 mm broad, 3- to 5-nerved; fruits 3–5 mm long;
floating leaves ovate .... 12. *P. natans*
24. Stems compressed; submersed leaves linear, 5–10 mm broad; fruit with
3 sharp keels, capped with a minute, tooth-like beak .. 14. *P. epihydrus*
24. Stems terete; submersed leaves lanceolate to ovate, 10–75 mm broad;
fruit not sharply 3-keeled, capped with a prominent beak 25
25. Fruiting spike 4 cm long or longer; stipules of submersed leaves subper-
sistent, the middle and upper submersed leaves broadly lanceolate to
ovate, never crisped; fruit tapering to base; some nerves of floating
leaves more prominent than other nerves .... 15. *P. amplifolius*
25. Fruiting spike 2.0–3.5 cm long; stipules of submersed leaves falling away
early; middle and upper submersed leaves lanceolate to linear-lanceolate, sometimes crisped; fruit rounded at base; all nerves of floating
leaves uniform .... 16. *P. pulcher*
26. Fruits (including beak) 3.5–4.3 mm long .... 17. *P. nodosus*
26. Fruits (including beak) 1.7–3.5 mm long 27
27. Stems simple or once-branched; stipules strongly keeled 28
27. Stems repeatedly branched; stipules faintly keeled 29
28. Floating leaves 2.0–6.5 cm wide; fruits normally developed ....
.... 18. *P. illinoensis*
28. Floating leaves 0.5–1.0 cm wide; fruits not maturing 19. *P.* X *hagstromii*
29. Floating leaves elliptic to ovate, the petioles usually as long as or longer
than the blades; stipules obtuse 30
29. Floating leaves oblong- to linear-lanceolate, the petioles shorter than the
blades; stipules acuminate .... 22. *P.* X *rectifolius*
30. Submersed leaves acute at apex; fruit strongly 3-keeled, well-developed
.... 20. *P. gramineus*
30. Submersed leaves cuspidate at apex; fruit obscurely 3-keeled, poorly developed ....21. *P.* X *spathulaeformis*

    1. P. pectinatus L. Fennel-leaved Pondweed. May–Sept. Calcareous
water; occasional throughout the state.
    2. P. robbinsii Oakes. Pondweed. June–Sept. Stagnant water, very
rare; Lake Co.
    3. P. crispus L. Curly Pondweed. May–Sept. Native of Europe; natu-
ralized in muddy to calcareous ponds and streams; occasional in the
state.
    4. P. praelongus Wulfen. Pondweed. June–Sept. Lakes and rivers,
rare; Cook, Lake, and McHenry cos.
    5. P. richardsonii (Benn.) Rydb. Pondweed. June–Sept. Lakes and
rivers, rare; Cook, Kankakee, Lake, and McHenry cos.

6. P. zosteriformis Fern. Pondweed. June–Sept. Ponds, lakes, and streams, not common; known only from the n. half of the state.

7. P. foliosus Raf. Pondweed. June–Sept. Ponds, lakes, rivers, and streams; occasional throughout the state. (Including var. *macellus* Fern.)

8. P. friesii Rupr. Pondweed. June–Sept. Ponds and streams, rare; Cook and Lake cos.

9. P. strictifolius Benn. Pondweed. June–Sept. Lakes, very rare; Cook Co.

10. P. pusillus L. Pondweed. June–Sept. Ponds, lakes, and streams; occasional throughout the state, but rarer in the s. cos. (Including var. *minor* [Biv.] Fern. & Schub.; *P. berchtoldii* Fieber; *P. berchtoldii* Fieber var. *tenuissimus* [Mert. & Koch] Fern.)

11. P. vaseyi Robbins. Pondweed. June–Sept. Lakes and ponds, rare; Grundy and McHenry cos.

12. P. natans L. Pondweed. June–Sept. Lakes and streams; occasional throughout the state, except for the extreme s. cos.

13. P. diversifolius Raf. Pondweed. May–Oct. Quiet waters; occasional throughout the state.

14. P. epihydrus Raf. Pondweed. June–Sept. Quiet ponds and lakes, rare; Fulton, Hancock, and Lake cos.

15. P. amplifolius Tuckerm. Pondweed. June–Oct. Lakes and roadside ditches, rare; Cook, DuPage, and Lake cos.

16. P. pulcher Tuckerm. Pondweed. June–Sept. Shallow water, rare; Jackson, Mason, and Menard cos.

17. P. nodosus Poir. Pondweed. May–Oct. Ponds and streams; occasional throughout the state. *P. americanus* Cham. & Schlecht.—J.

18. P. illinoensis Morong. Pondweed. June–Oct. Ponds, lakes, and streams; occasional in the n. half of the state, rare in the s. half.

19. P. X hagstromii Benn. Pondweed. July–Sept. Lakes, very rare; Cook Co. Reputed to be a hybrid between *P. richardsonii* and *P. gramineus.*

20. P. gramineus L. Pondweed. June–Sept. Lakes, ponds, and streams, rare; Cook and Wabash cos.

21. P. X spathulaeformis (Robbins) Morong. Pondweed. June–Sept. Ponds and streams, very rare; Wabash Co. Reputed to be a hybrid between *P. illinoensis* and *P. gramineus.*

22. P. X rectifolius Benn. Pondweed. June–Sept. Ditches, in 1–3 feet of water; Cook Co. Reputed to be a hybrid between *P. nodosus* and *P. richardsonii.*

21. JUNCAGINACEAE

1. Inflorescence a spike-like raceme, without bracts; leaves all basal . . . . . .
. . . . 1. *Triglochin*
1. Inflorescence loosely racemose, bracteate; leaves basal and alternate . .
. . . . 2. *Scheuchzeria*

1. *Triglochin* L.   Arrow-grass

1. Carpels usually 6, the axis between them slender . . . . 1. *T. maritima*
1. Carpels 3, the axis between them broadly 3-winged . . . . . 2. *T. palustris*

1. T. maritima L. Arrow-grass. May–Aug. Sandy shores, swamps, wet ditches; known only from the extreme ne. cos.; also Peoria Co.

2. T. palustris L. Arrow-grass. May–July. Low areas; known only from the ne. cos.; also Peoria and Tazewell cos.

2. Scheuchzeria L.   Arrow-grass

1. S. palustris L. var. americana Fern. Arrow-grass. May–July. Bogs; Fulton, Lake, McHenry, and Menard cos. S. palustris L—G; S americana (Fern.) G. N. Jones—J.

22. ALISMACEAE

1. Receptacle convex, bearing several rows of pistils; stamens 12–numerous (6–9 in E. tenellus var. parvulus); flowers unisexual or perfect
.... 2
1. Receptacle flat, bearing a single row of pistils; stamens 6–9; flowers perfect .... 3. Alisma
2. Achenes not winged; base of whorled inflorescence branches bearing 3 bracts and several bracteoles; flowers perfect; stamens never more than 21 .... 1. Echinodorus
2. Achenes winged; base of whorled inflorescence branches bearing 3 bracts and no bracteoles, flowers mostly unisexual; stamens usually more than 21 .... 2. Sagittaria

1. Echinodorus Rich.   Burhead

1. Plants erect, less than 10 cm tall; leaves linear to lanceolate; flowers at most only 6 mm broad; stamens 6–9; achenes 10–15, beakless or nearly so .... 1. E. tenellus var. parvulus
1. Plants erect, usually more than 10 cm tall, or plants creeping or arching; leaves broadly ovate, rarely lanceolate; flowers at least 8 mm broad; stamens 12–21; achenes more than 40, beaked 2
2. Scape erect; stamens 12; style longer than the ovary; beak of achene straight .... 2. E. berteroi var. lanceolatus
2. Scape creeping or arching; stamens 21; style shorter than the ovary; beak of achene incurved .... 3. E. cordifolius

1. E. tenellus (Mart.) Buchenau var. parvulus (Engelm) Fassett. Small Burhead. Aug.–Sept. Wet ditches, rare; Cass and St. Clair cos. E. tenellus (Mart.) Buchenau—F; E. parvulus Engelm.—G, J.

2. E. berteroi (Spreng.) Fassett var. lanceolatus (Wats. & Coult.) Fassett. Burhead. July–Sept. Wet ditches, edges of swamps; scattered in the s. three-fourths of the state, very rare elsewhere. E. rostratus (Nutt.) Engelm.—F, G, J.

3. E. cordifolius (L.) Griseb. Creeping Burhead. July–Sept. Wet ditches, edges of swamps; mostly restricted to the w. cos.; also Crawford Co.

2. Sagittaria L.   Arrowleaf

1. Filaments roughened with minute scales 2
1. Filaments glabrous 4
2. Peduncle and pedicel inflated; sepals tightly appressed to fruiting head

.... 1. *S. calycina*
2. Peduncle and pedicel slender; sepals usually reflexed 3
3. Pistillate flowers and fruiting heads sessile or subsessile ... 2. *S. rigida*
3. Pistillate flowers and fruiting heads long-pedunculate . 3. *S. graminea*
4. Mature achene with 1–3 facial wings; beak of achene inserted apically 5
4. Mature achene with marginal wings only; beak of achene inserted laterally .... 7. *S. latifolia*
5. Beak of achene less than 0.5 mm long, straight, with 1 small, narrow facial wing .... 4. *S. cuneata*
5. Beak of achene more than 0.5 mm long, usually reflexed at apex, usually with 2–3 facial wings with 1 well developed 6
6. Receptacle not echinate; flowers in whorls of 3 or 4 .. 5. *S. brevirostra*
6. Receptacle echinate; flowers in whorls of more than 4 6. *S. longirostra*

1. S. calycina Engelm. Arrowleaf. June–Sept. Marshes, pond margins, shorelines, shallow water; locally throughout cent. and s. Ill., apparently absent elsewhere. (Including var. *maxima* Engelm.) *S. montevidensis* Cham. & Schlecht.—G; *Lophotocarpus calycinus* (Engelm.) J. G. Sm.—F.

2. S. rigida Pursh. Arrowleaf. May–Oct. Shallow water, muddy sand, in swamps, margins of ponds, or along waterways; not common but scattered in Ill.

3. S. graminea Michx. Narrow-leaved Arrowleaf. May–Sept. Swamps, mud, sand, or shallow water; uncommon but scattered in the n. half of the state, rare in the s. half.

4. S. cuneata Sheld. Arrowleaf. June–Sept. Mud or water in sloughs, and along waterways; confined to the n. half of the state.

5. S. brevirostra Mack. & Bush. Arrowleaf. June–Sept. Sloughs, shorelines, and shallow water; occasional throughout the state. *S. engelmanniana* J. G. Sm. ssp. *brevirostra* (Mack. & Bush) Bogin—G.

6. S. longirostra (Micheli) J. G. Sm. Arrowleaf. July–Sept. Springy woods, rare; Pope Co. *S. engelmanniana* J. G. Sm. ssp. *longirostra* (J. G. Sm.) Bogin.—G; *S. australis* (J. G. Sm.) Small.

7. S. latifolia Willd. Common Arrowleaf. June–Oct. Swamps, sloughs, ponds, shorelines, shallow water, mud; occasional to common throughout the state. (Including f. *diversifolia* [Engelm.] B. L. Robins.)

3. Alisma L.   Water Plantain

1. Petals 3.5–6.0 mm long; flowers at least 7 mm broad; styles at anthesis as long as the ovaries; achenes at least 2.5 mm long ....
.... 1. *A. plantago-aquatica* var. *americanum*
1. Petals 1.0–2.5 mm long; flowers at most 3.5 mm broad; styles at anthesis less than one-half as long as the ovaries; achenes less than 2.5 mm long
.... 2. *A. subcordatum*

1. A. plantago-aquatica L. var. americanum Roem. & Schultes. Water Plantain. July–Aug. Shallow water, marshes, ditches not common; confined primarily to the n. half of the state; also Wabash Co. *A. triviale* Pursh—F, J.

2. A. subcordatum Raf. Small-flowered Water Plantain. July–Aug. Shallow water, marshes, ditches; common throughout the state.

23. BUTOMACEAE

1. Butomus L. Flowering Rush

1. B. umbellatus L. Flowering Rush. July–Sept. Native of Europe; rarely naturalized in Ill.; Cook Co.

24. HYDROCHARITACEAE

1. Leaves borne along the stem, the longest not more than 3 cm, the broadest not more than 5 mm .... 1. *Elodea*
1. Leaves all basal, 2–200 cm long, 5–25 mm broad 2
2. Leaves narrow, elongate; fertile stamens 2, the filaments free ....
.... 2. *Vallisneria*
2. Leaves ovate to orbicular; fertile stamens 6–12, the filaments united into a column .... 3. *Limnobium*

1. Elodea Michx. Waterweed

1. Leaves in whorls of 4 or 6, generally more than 2 cm long; petals 9–12 mm long .... 1. *E. densa*
1. Leaves opposite or in whorls of 3, generally less than 2 cm long; petals minute or up to 5 mm long 2
2. Staminate flowers pedicellate, not liberated at maturity, with sepals 3.5–5.0 mm long; leaves 1–5 mm broad .... 2. *E. canadensis*
2. Staminate flowers sessile, liberated at maturity, with sepals 1.5–2.0 mm long; leaves smaller, usually 0.5–1.5 mm broad .... 3. *E. nuttallii*

1. E. densa (Planch.) Caspary. Elodea; Waterweed. July–Sept. Native to S. Am.; adventive in mine ponds; Franklin and Williamson cos. *Anacharis densa* (Planch.) Vict.—G.

2. E. canadensis Michx. Elodea; Waterweed. July–Sept. Quiet water; occasional in the n. half of the state, rare in the s. half. *Anacharis canadensis* (Michx.) Rich.—G.

3. E. nuttallii (Planch.) St. John. Elodea; Waterweed. July–Sept. Quiet water; occasional in the n. cos., not common in the s. cos. *Anacharis nuttallii* Planch.—G; *E. occidentalis* (Pursh) St. John–J.

2. Vallisneria L. Eelgrass

1. V. americana Michx. Eelgrass. June–Sept. Quiet water; restricted to the n. one-half of the state; also Alexander Co.

3. Limnobium Rich. Frog's-bit

1. L. spongia (Bosc) Steud. Sponge Plant. July–Aug. Swamps, rare; Alexander, Johnson, Massac, and Union cos.

25. POACEAE

1. Culms woody .... 80. *Arundinaria*
1. Culms herbaceous 2
2. Spikelets enclosed by a spiny bur .... 49. *Cenchrus*
2. Spikelets not enclosed by a spiny bur 3
3. Spikelets with one or more perfect florets 4
3. Spikelets unisexual (i.e., either all staminate or all pistillate) .. *Group G*
4. Inflorescence solitary, racemose, paniculate, or spicate, but not digitate
.... 5
4. Inflorescence digitate (the spikes and racemes radiating from near the
same point) .... *Group F*
5. Inflorescence spicate or spike-like, with one spike per culm .. *Group A*
5. Inflorescence solitary, racemose, or paniculate, but not composed of
single spikes 6
6. Each spikelet with 2 or more perfect florets 7
6. Each spikelet with one perfect floret (sterile or staminate lemmas may be
present, in addition) 8
7. Some part of the spikelet awned .... *Group B*
7. Spikelet without any awns .... *Group C*
8. Some part of the spikelet awned .... *Group D*
8. Spikelet without awns .... *Group E*

Group A

Inflorescence spicate or spike-like, with one spike per culm; spikelets with
one or more perfect florets.

1. Spikelets cylindrical, borne at swollen rachis joints, the entire spikelet
falling at maturity; each glume with one awn and one tooth 32. *Triticum*
1. Spikelets not as above; rachis joints not swollen; glumes awned or awnless, but not with one awn and one tooth 2
2. Spikelets borne edgewise to the rachis; inner glume absent, except in
the terminal spikelet .... 4. *Lolium*
2. Spikelets borne flatwise to the rachis; glumes present on all spikelets   3
3. Each spikelet subtended and usually surpassed by one or more sterile
bristles (not be be confused with awns) .... 48. *Setaria*
3. Each spikelet not subtended by bristles 4
4. Each spikelet with two or more perfect florets (*Anthoxanthum* and *Phalaris* have three lemmas, but two of them are sterile) 5
4. Each spikelet with a single perfect floret (1–2 sterile lemmas present in
addition in *Anthoxanthum* and *Phalaris;* 1 staminate lemma present in
addition in *Holcus*) 19
5. At least some part of the spikelet awned 6
5. Spikelets awnless throughout 15
6. Upper spikelets paired, the lowermost solitary ..... 30. X *Agrohordeum*
6. Spikelets either all paired, all borne in threes, or all solitary 7
7. Spikelets either all paired or all borne in threes 8
7. Spikelets solitary 10
8. Spikelets in threes .... 29. *Hordeum*
8. Spikelets paired 9
9. Glumes to 4 cm long; axis of inflorescence rarely breaking apart at maturity .... 27. *Elymus*

9. Glumes 6–8 cm long; axis of inflorescence breaking apart at maturity .. .. 28. *Sitanion*
10. Glumes awned 11
10. Glumes awnless 13
11. Glumes 3-nerved; annuals .... 32. *Triticum*
11. Glumes often 1-nerved; perennials 12
12. Spikelets more than 3 cm long; annuals .... 33. *Secale*
12. Spikelets up to 3 cm long; perennials .... 31. *Agropyron*
13. Blades 10–20 mm broad; annuals .... 32. *Triticum*
13. Blades 1–10 mm broad; perennials 14
14. Awn of lemma more than 1 mm long .... 31. *Agropyron*
14. Awn of lemma up to 1 mm long .... 9. *Koeleria*
15. Annuals .... 32. *Triticum*
15. Perennials 16
16. Spikelets paired .... 27. *Elymus*
16. Spikelets solitary 17
17. Spikelets borne flatwise to the continuous rachis .... 31. *Agropyron*
17. Spikelets borne all around the articulated (jointed) rachis 18
18. Blades 3–10 mm broad; lemmas densely pubescent on the nerves...... .... 61. *Tridens*
18. Blades 1–3 mm broad; lemmas merely scabrous .... 9. *Koeleria*
19. Upper spikelets paired, the lowermost solitary ..... 30. X *Agrohordeum*
19. Spikelets either all paired, all borne in threes, or all solitary 20
20. Spikelets paired or in groups of three; glumes long-awned 21
20. Spikelets solitary 22
21. Spikelets paired .... 27. *Elymus*
21. Spikelets in threes .... 30. *Hordeum*
22. Some part of the spikelet awned 23
22. Spikelets not awned 28
23. Lemmas awned; glumes awned .... 24. *Phleum*
23. Lemmas awned; glumes awned or awnless 24
24. Lemma awned from the middle .... 16. *Calamagrostis*
24. Lemma awned from the tip 25
25. Lemma I per spikelet, perfect 26
25. Lemmas 2–3 per spikelet, but only one perfect 27
26. Glumes united near the base .... 23. *Alopecurus*
26. Glumes free at the base .... 65. *Muhlenbergia*
27. Spikelets 5–10 mm long, each with one perfect floret and two empty lemmas .... 20. *Anthoxanthum*
27. Spikelets 3.5–5.0 mm long, each with one perfect floret and one staminate floret .... 15. *Holcus*
28. Glumes 9–15 mm long; lemma 7–14 mm long .... 17. *Ammophila*
28. Glumes to 7 (–10 in *Phalaris*) mm long; lemmas to 7 mm long 29
29. Each spikelet with one perfect floret and (1–)2 empty lemmas......... .... 22. *Phalaris*
29. Each spikelet with one perfect floret only 30
30. Lemma 3-nerved .... 65. *Muhlenbergia*
30. Lemma 1-nerved 31
31. Spikes broad, one-fourth to nearly as broad as long .... 67. *Crypsis*
31. Spikes slender, one-fifth or less as broad as long ...... 66. *Sporobolus*

Group B

Inflorescence solitary, racemose, or paniculate, but not spicate of digitate; spikelets with 2 or more perfect flowers; some part of the spikelet awned.

1. Lemmas 2-toothed at the apex 2
1. Lemmas acute or obtuse at the apex, not 2-toothed 9
2. Awn of lemma arising from between the teeth 3
2. Awn of lemma arising near the middle or base of the lemma 7
3. Lemmas 5- to 9-nerved 4
3. Lemmas 3-nerved 6
4. Callus of lemmas bearded .... 36. *Schizachne*
4. Callus of lemmas not bearded 5
5. Glumes much shorter than the entire spikelet .... 1. *Bromus*
5. Glumes equalling or longer than the uppermost floret .. 86. *Danthonia*
6. Panicles 3–5 (–8) cm long; spikelets 2- to 5-flowered ...... 62. *Triplasis*
6. Panicles 10–20 cm long; spikelets 6- to 12-flowered .... 70. *Leptochloa*
7. Glumes 17–30 mm long; awn of lemmas up to 40 mm long .. 13. *Avena*
7. Glumes 2.5–5.0 mm long; awn of lemmas 2.5–6.0 mm long 8
8. Awn arising from near the middle of the lemma; lemmas 3-nerved ...
  .... 11. *Aira*
8. Awn arising from near base of lemma; lemmas 5-nerved ....
  .... 12. *Deschampsia*
9. Lemmas 3-nerved .... 70. *Leptochloa*
9. Lemmas 5-nerved (all the nerves sometimes obscure in *Festuca*) ... 10
10. Blades involute, about 1 mm in diameter, if flat, less than 2 mm broad
  11
10. Blades flat, 2–8 mm broad 12
11. Plants annual; stamen 1 .... 2. *Vulpia*
11. Plants perennial; stamens 3 .... 3. *Festuca*
12. Lemmas glabrous; spikelets not crowded in 1-sided panicles 3. *Festuca*
12. Lemmas ciliate along the keel; spikelets crowded in 1-sided panicles ...
  .... 8. *Dactylis*

Group C

Inflorescence solitary, racemose, or paniculate, but not spicate or digitate; spikelets with 2 or more perfect flowers; spikelets awnless.

1. Lemmas distinctly 2-toothed at the apex 2
1. Lemmas acute to obtuse at the apex, not 2-toothed 3
2. Perennial; blades to 3 mm broad; panicle branches erect or spreading; spikelets 5- to 12-flowered; glumes 1 cm long .... 1. *Bromus*
2. Annual; blades 5–15 mm broad; panicle branches lax; spikelets 2-flowered; glumes 1.5–2.5 cm long .... 13. *Avena*
3. Glumes at least 15 mm long, as long as the spikelets .... 13. *Avena*
3. Glumes less than 10 mm long, shorter than the spikelets 4
4. Rachilla with long silky hairs, the hairs longer than the spikelets; culms to 4 m tall .... 84. *Phragmites*
4. Rachilla without long silky hairs longer than the spikelets; culms to 1.5 m tall 5
5. Lemmas 3-nerved 6
5. Lemmas 5- to many-nerved, or apparently nerveless, with only the mid-nerve conspicuous 13
6. Lemmas 6–10 mm long; grain beaked .... 40. *Diarrhena*
6. Lemmas 1.5–5.0 mm long; grain not beaked 7
7. Lemmas glabrous .... 60. *Eragrostis*
7. Lemmas pubescent 8
8. Lemmas densely hairy at base, frequently with a tuft of hairs 9
8. Lemmas pubescent only on the nerves 11

9. Lemmas villous at the base, but without a tuft of cobwebby hairs ... 10
9. Lemmas with a tuft of cobwebby hairs at the base, pubescent on the
nerves .... 6. *Poa*
10. Lemmas retuse or obtuse, 3.5–4.0 mm long .... 61. *Tridens*
10. Lemmas acute and mucronate, 4.5 mm long .... 63. *Redfieldia*
11. Lemmas keeled .... 60. *Eragrostis*
11. Lemmas rounded on the back 12
12. Lemmas 1.0–2.5 mm long; spikelets 1–5 mm long ...... 70. *Leptochloa*
12. Lemmas 4.0 mm long; spikelets 5–8 mm long .... 61. *Tridens*
13. Lemmas apparently nerveless 14
13. Lemmas obviously nerved 16
14. Spikelets disarticulating above the glumes 15
14. Spikelets disarticulating below the glumes .... 10. *Sphenopholis*
15. Glumes 2.0–4.5 mm long; plants of moist or dry woods .... 3. *Festuca*
15. Glumes up to 2 mm long; plants of waste ground ...... 5. *Puccinellia*
16. Lemmas 4–10 mm long 17
16. Lemmas 1.5–5.0 mm long 22
17. Lemmas as broad as long; inflorescence with up to eight spikelets .....
.... 7. *Briza*
17. Lemmas longer than broad; inflorescence with more than eight spikelets
.... 18
18. Lemmas 4–10 mm long, with nine or more nerves 19
18. Lemmas to 7 (–8) mm long, 5- to 7-nerved 20
19. Spikelets compressed, 6- to 18-flowered .... 87. *Chasmanthium*
19. Spikelets not compressed, 2- to 3-flowered .... 34. *Melica*
20. Lemmas obscurely 7-nerved; spikelets 10–20 mm long .... 35. *Glyceria*
20. Lemmas 5-nerved; spikelets less than 10 mm long 21
21. Spikelets not crowded in 1-sided panicles, not compressed   3. *Festuca*
21. Spikelets crowded in 1-sided panicles, compressed .... 8. *Dactylis*
22. Lemmas distinctly keeled .... 6. *Poa*
22. Lemmas rounded on the back 23
23. Nerves of lemma parallel to the summit 24
23. Nerves of lemma converging toward the summit 25
24. Sheaths closed; lodicules united .... 35. *Glyceria*
24. Sheaths open; lodicules free from each other .... 5. *Puccinellia*
25. Lemmas glabrous 26
25. Lemmas pubescent, at least on the nerves or the keel or at the base ...
.... 6. *Poa*
26. Plants annual; stamen 1 .... 2. *Vulpia*
26. Plants perennial; stamens 3 .... 3. *Festuca*

Group D

Inflorescence solitary, racemose, or paniculate, but not composed of single
spikes; each spikelet with one perfect floret (sterile or staminate lemmas
may be present, in addition); some part of the spikelet awned.

1. Spikelets borne singly (i.e., not paired) 2
1. Spikelets borne in pairs 23
2. Lemmas 3-awned 3
2. Lemmas 1-awned or awnless 4
3. Spikelets borne on one side of a long, arching raceme; lemma rounded
on the back; spikelets with one perfect lemma and 1–2 sterile ones ....
.... 75. *Bouteloua*

3. Spikelets borne in a more or less erect inflorescence, not 1-sided; lemma inrolled around the palea; no sterile lemma present   79. *Aristida*
4. First glume reduced to a sheath, united with the lowest, swollen joint of the rachilla .... 44. *Eriochloa*
4. First glume not reduced to a sheath and not united with the rachillar joint 5
5. Lemma awnless; glumes awned 6
5. Lemma awned; glumes awnless or awned 7
6. Plants over 1 m tall; lemmas 7–10 mm long .... 77. *Spartina*
6. Plants less than 1 m tall; lemmas 2–4 mm long ...... 65. *Muhlenbergia*
7. Spikelets arranged in 4 or more crowded ranks, each spikelet composed of one fertile and one sterile floret .... 47. *Echinochloa*
7. Spikelets not arranged in 4 or more crowded ranks, each spikelet composed of one fertile floret (each spikelet composed of one fertile floret and one staminate floret in *Arrhenatherum* or one sterile floret sometimes in *Gymnopogon*) 8
8. Blades 1–3 mm broad 9
8. Blades 3 mm broad or broader 14
9. First glume less than 1 mm long .... 65. *Muhlenbergia*
9. First glume at least 1.5 mm long 10
10. At least one awn of lemma 2–4 cm long 11
10. Awn of lemma up to 2 cm long 12
11. Tufted annuals from a cluster of fibrous roots .... 79. *Aristida*
11. Cespitose or stout perennials .... 37. *Stipa*
12. Lemma 2.0–4.5 mm long 13
12. Lemma 0.5–1.6 mm long .... 18. *Agrostis*
13. Glumes 5-nerved; lemmas indurated .... 38. *Oryzopsis*
13. Glumes 1-nerved; lemmas not indurated .... 65. *Muhlenbergia*
14. Plants 2 m tall or taller; all leaves 2 cm broad or broader .. 85. *Arundo*
14. Plants usually less than 2 m tall; none or only a few of the leaves as much as 2 cm broad 15
15. Second glume 5- to 7-nerved 16
15. Second glume 1- to 3-nerved 17
16. Awns straight or curved, not twisted near base; second glume 7-nerved
.... 38. *Oryzopsis*
16. Awns twisted near base; second glume 5-nerved .... 37. *Stipa*
17. Lemma (excluding awns) 5–10 mm long 18
17. Lemma (excluding awns) 1.5–5.0 mm long 20
18. First glume less than 1 mm long .... 39. *Brachyelytrum*
18. First glume, 2.5–8.0 mm long 19
19. Awn 10–20 mm long; spikelet (excluding awns) 7–10 mm long; lemma 5- to 7-nerved .... 14. *Arrhenatherum*
19. Awn to 1.5 mm long; spikelet (excluding awns) 2.5–6.5 mm long; lemma 3-nerved .... 19. *Cinna*
20. Spikelets remote along one side of a slender rachis, forming very slender unilateral spikes .... 71. *Gymnopogon*
20. Spikelets in contracted or open panicles 21
21. Lemma with a tuft of hairs at the base (on the callus), awned from near the middle .... 16. *Calamagrostis*
21. Lemma glabrous or pubescent, but without a large tuft of hairs on the callus, awned from the tip 22
22. Plants usually at least 1 m tall; spikelets disarticulating below the glumes .... 19. *Cinna*
22. Plants up to 1 m tall, usually smaller; spikelets disarticulating above the glumes .... 65. *Muhlenbergia*

23. Both spikelets pedicellate, the pedicels unequal in length ....
.... 50. *Miscanthus*
23. One spikelet sessile, the other pedicellate (or represented merely by the pedicel) 24
24. Pedicellate spikelet represented only by the pedicel . 53. *Sorghastrum*
24. Pedicellate spikelet present 25
25. Both spikelets of the pair with perfect florets .... 51. *Erianthus*
25. Only the sessile spikelet of the pair perfect 26
26. Inflorescence racemose or nearly spicate 27
26. Inflorescence paniculate .... 52. *Sorghum*
27. Flowering culms much branched into many short leafy branchlets terminated by 1–6 racemes 28
27. Flowering culms unbranched .... 56. *Bothriochloa*
28. Racemes 2 or more from the sheaths .... 54. *Andropogon*
28. Raceme solitary at the tip of the peduncle .... 57. *Schizachyrium*

Group E

Inflorescence solitary, racemose, or paniculate, but not spicate; each spikelet with one perfect floret (sterile or staminate lemmas may be present, in addition); no part of the spikelet awned.

1. Spikelets borne in pairs 2
1. Spikelets solitary (i.e., not borne in pairs) 5
2. One spikelet of the pair sessile, the other pedicellate   55. *Microstegium*
2. Both spikelets either sessile or pedicellate 3
3. First glume as long as or longer than the lemmas; plants 2.5–4.0 m tall
.... 50. *Miscanthus*
3. First glume absent or up to 0.5 mm long, much shorter than the lemmas; plants up to 1.5 m tall 4
4. Spikelets with long, tawny hairs longer than the spikelets...............
.... 42. *Trichachne*
4. Spikelets without long, tawny hairs exceeding the spikelets............
.... 45. *Paspalum*
5. First glume reduced to a sheath and united with the lowest, swollen joint of the rachilla .... 44. *Eriochloa*
5. First glume absent, reduced, or normal, neither sheath-like nor united with a swollen rachillar joint 6
6. Both glumes absent .... 81. *Leersia*
6. Both glumes present, although the first often much reduced or, if absent, the plants not producing seeds 7
7. First glume absent; plants with creeping rhizomes, rarely producing seeds .... *Zoysia* *
7. First glume present, although occasionally strongly reduced; rhizome present or absent; plants producing seeds 8
8. First glume up to one-half (to ⅔ in a few species of *Panicum*) as long as second glume 9
8. First glume nearly as long as the second glume, not conspicuously different in size 14
9. Each floret subtended by one or more bristles .... 48. *Setaria*
9. Florets not subtended by bristles 10

---

* *Zoysia* is frequently planted in Illinois as a choice lawn grass, but no collections have ever been made of it as an adventive. Therefore, it is excluded from the text.

Group F

Inflorescence digitate (the spikes and racemes radiating from near the same point).

Group G

Spikelets unisexual (i.e., either all staminate or all pistillate).

1. Plants to 40 cm tall, dioecious; staminate spikelets 3- to 75-flowered 2
1. Plants 1–4 m tall, monoecious; staminate spikelets 1- to 2-flowered .. 5
2. Lemmas with a tuft of cobwebby hairs at base .... 6. *Poa*
2. Lemmas without a tuft of cobwebby hairs at base 3
3. Both staminate and pistillate spikelets 10- to 75-flowered 60. *Eragrostis*
3. Staminate spikelets (2-) 3- to 15-flowered, pistillate spikelets 1- to 9-flowered 4
4. Staminate spikelets (2-) 3-flowered; pistillate spikelets 1-flowered ......
.... 76. *Buchloë*
4. Staminate spikelets 8- to 15-flowered; pistillate spikelets 7- to 9-flowered
.... 78. *Distichlis*
5. Staminate spikelets 2-flowered; glumes membranous or coriaceous . 6
5. Staminate spikelets 1-flowered; glumes none 7
6. Annual; staminate and pistillate spikelets in different inflorescences; pistillate spikelets borne in pairs .... 59. *Zea*
6. Perennial; staminate and pistillate spikelets in the same inflorescence; pistillate spikelets solitary .... 58. *Tripsacum*
7. Pistillate spikelets confined to the uppermost erect branches of the inflorescence, the staminate spikelets confined to the lower spreading branches; margin of leaf more or less smooth .... 82. *Zizania*
7. Pistillate and staminate spikelets on the same branches of the inflorescence; margin of leaf harsh and cutting .... 83. *Zizaniopsis*

1. Bromus L. Brome Grass

1. Some or all of the awns over 12 mm long; teeth of lemmas 2–5 mm long
.... 2
1. All or most of the awns less than 12 mm long; teeth of lemmas usually less than 2 mm long 3
2. Lemmas 16–21 mm long, scabrous or puberulent on the back; first glume 8–12 mm long; second glume 13–18 mm long; awns 20–30 mm long; blades and sheaths glabrous or short-pubescent .... 1. *B. sterilis*
2. Lemmas 10–12 mm long, villous throughout on the back, becoming hispidulous at the summit, rarely entirely glabrous; first glume 4–7 mm long; second glume 8–10 mm long; awns (10–) 12–15 mm long; blades and sheaths soft-pubescent .... 2. *B. tectorum*
3. First glume 3- to 5-nerved (1-nerved in *B. nottowayanus*); second glume 5- to 7-nerved; annuals or perennials 4
3. First glume 1-nerved; second glume 3- to 5-nerved; perennials. (*Bromus nottowayanus* has the first glume 1-nerved, the second glume 5- to 7-nerved.) 15
4. Perennials from rhizomes; blades 6–13 mm broad (occasionally 5 mm broad in *B. kalmii*) 5
4. Annuals from fibrous roots; blades 2–6 mm broad (to 8 mm broad only in *B. secalinus*) 7
5. Lemmas keeled; inflorescence erect .... 3. *B. marginatus*
5. Lemmas rounded on the back; inflorescence drooping 6
6. Awns 5–8 mm long; cauline leaves 6–8 .... 4. *B. nottowayanus*
6. Awns 1–3 mm long; cauline leaves 3–5 (–6) .... 5. *B. kalmii*
7. Lemmas strongly keeled on the back, 12–15 mm long .. 6 *B. willdenovii*

7. Lemmas rounded on the back, 5–11 (–12) mm long 8
8. Awns 0–6 mm long 9
8. Most or all the awns over 6 mm long 10
9. Blades softly villous; lemmas 9–11 (–12) mm long; awns 0–1 mm long ..
.... 8. *B. brizaeformis*
9. Blades harshly pubescent above, or glabrous; lemmas 5–8 mm long; awns 1–6 mm long, rarely absent .... 7. *B. secalinus*
10. Inflorescence erect or ascending 11
10. Inflorescence spreading or drooping 13
11. Lemmas plicate, conspicuously nerved; inflorescence compact ....
.... 9. *B. mollis*
11. Lemmas not plicate, faintly nerved; inflorescence open 12
12. Lower lemmas 7–9 mm long; branches of inflorescence solitary or paired, usually shorter than the spikelets; anthers 2.0–2.5 mm long.....
.... 10. *B. racemosus*
12. Lower lemmas 9–11 mm long; branches of inflorescence 2–6, usually much longer than the spikelets; anthers 1.5–2.0 mm long ....
.... 11. *B. commutatus*
13. Awn straight or nearly so; rachilla not exposed at maturity 14
13. Awn flexuous; rachilla exposed at maturity .... 13. *B. japonicus*
14. Lemmas all nearly the same length; anthers 4 mm long   12. *B. arvensis*
14. Lowest lemmas longer than the upper; anthers 2.0–2.5 mm long ....
.... 10. *B. racemosus*
15. Awns absent or up to 2 mm long; blades and sheaths glabrous ....
.... 14. *B. inermis*
15. Awns 2–8 mm long; blades and sheaths (particularly the lower) pubescent, rarely glabrous 16
16. Inflorescence narrow, erect; blades 2–3 mm broad ...... 15. *B. erectus*
16. Inflorescence spreading or drooping; blades (3–) 4–17 mm broad ... 17
17. Leaves 10–20 per culm, the blades auriculate at base .. 16. *B. purgans*
17. Leaves 5–8 per culm, the blades not auriculate 18
18. Lemmas pubescent throughout on the back, or glabrous ....
.... 17. *B. pubescens*
18. Lemmas pubescent only on the margins in the lower one-half to three-fourths of the lemma .... 18. *B. ciliatus*

1. B. sterilis L. Brome Grass. May–June. Native of Europe; adventive in waste ground; Champaign, Cook, and McDonough cos.

2. B. tectorum L. Downy Chess. April–July. Native of Europe; naturalized in waste areas, common; in every co.

3. marginatus Nees. Brome Grass. May–Aug. Native of the U.S.; adventive in waste ground; Cook and Kane cos.

4. B. nottowayanus Fern. Brome Grass. June–Aug. Moist, wooded ravines; Cook, Peoria, Stark, and Woodford cos.

5. B. kalmii Gray. Brome Grass. June–July. Dry woodlands, sandy soil, not common; confined to the n. half of the state.

6. B. willdenovii Kunth. Rescue Grass. May–July. Native to tropical America; introduced in Champaign Co. *B. catharticus* Vahl—F, G.

7. B. secalinus L. Chess. May–Aug. Native of tropical America; waste ground, common; probably in every co.

8. B. brizaeformis Fisch. & Mey. Rattlesnake Chess. May–July. Native of Europe; waste ground, rare; Richland and Washington cos.

9. B. mollis L. Soft Chess. May–July. Native of Europe; waste ground, not common; DuPage, Jackson, and Lawrence cos.

10. B. racemosus L. Chess. May–Aug. Native of Europe; waste ground; occasional throughout the state.

11. B. commutatus Schrad. Hairy Chess. May–Aug. Native of Europe; waste ground; common throughout the state. *B. racemosus* L.—G.

12. B. arvensis L. Chess. May–July. Native of Europe; waste ground, not common; scattered in the w. cent. cos.

13. B. japonicus Thunb. Japanese Chess. May–July. Native of Europe and Asia; waste ground, not common; scattered in the state.

14. B. inermis Leyss. Awnless Brome Grass. May–July. Native of Europe; roadsides, fields, waste ground, common; probably in every co. (Including f. *aristatus* [Schur] Fern.)

15. B. erectus Huds. Erect Brome Grass. July. Native of Europe; waste ground, rare; St. Clair Co.

16. B. purgans L. Brome Grass. June–Sept. Moist, open woods; occasional in the n. half of the state, rare in the s. half. *B. latiglumis* (Shear) Hitchcock—F, G, J. (Including *B. latiglumis* f. *incanus* [Shear] Fern.)

17. B. pubescens Muhl. Canada Brome Grass. June–Aug. Moist, open woods; common throughout the state. *B. purgans* L.—F, G, J. (Including *B. purgans* f. *laevivaginatus* Wieg.)

18. B. ciliatus L. Canada Brome Grass. June–Sept. Open woodlands; common throughout the state. (Including var. *intonsus* Fern.)

2. *Vulpia* K. C. Gmel.

1. Awns of lemma, if present, up to 5.5 mm long .... 1. *V. octoflora*
1. Some or all the awns of the lemma 1 cm long or longer 2
2. Second glume at least twice as long as the first glume; first glume less than 3 mm long .... 2 *V myuros*
2. Second glume not twice as long as the first glume; first glume more than 3 mm long .... 3. *V. dertonensis*

1. V. octoflora (Walt.) Rydb. Six-weeks Fescue. May–July. Inflorescence appearing racemose; lower glume 3.5–4.5 mm long; awns 3.5–5.5 mm long. Dry soil; scattered throughout the state. *F. octoflora* Walt.—G, J.

var. tenella (Willd.) Fern. Six-weeks Fescue. May–July. Inflorescence loosely spictate; lower glume 2.5–4.0 mm long; awns 1–3 mm long. Dry soil; occasional throughout the state. *Festuca octoflora* Walt.—J.

var. glauca (Nutt.) Fern. Six-weeks Fescue. May–July. Inflorescence densely spicate; lower glume 1.5–3.0 mm long; awns absent or up to 2 mm long. Dry soil, not common; scattered in Ill.

2. V. myuros (L.) K. Gmel. Foxtail Fescue. June–July. Native of Europe; waste ground, rare; Johnson and Massac cos. *F. myuros* L.—G.

3. V. dertonensis All. June–July. Native of Europe; waste ground; Massac Co. *F. dertonensis* (All.) Aschers. & Graebn.—G.

3. *Festuca* L.   Fescue

1. Leaf blades involute or plicate, 0.4–1.2 mm broad 2
1. Leaf blades flat, 3–11 mm broad 4

2. First glume 1–2 mm long; second glume 1.8–3.0 mm long; lemmas 2.5–3.5 mm long; awns absent, or to 0.5 mm long ...... 1. *F. capillata*

2. First glume 2.5–4.5 mm long; second glume 3.5–5.5 mm long; lemmas 4–7 mm long; awns 1.0–3.5 mm long 3

3. Lowest sheaths whitish, not becoming fibrous; lemmas essentially nerveless .... 2. *F. ovina*

3. Lowest sheaths brown or reddish, becoming fibrous; lemmas 3- to 5-nerved .... 3. *F. rubra*

4. Lemmas 5.5–10.0 mm long 5

4. Lemmas 3.3–5.2 mm long 6

5. Inflorescence 6- to 11-flowered; lemmas 5.5–8.0 mm long 4. *F. pratensis*

5. Inflorescence 4- to 5-flowered; lemmas 7–10 mm long 5. *F. arundinacea*

6. Inflorescence spreading at maturity; spikelets to 4 mm broad; lemmas acute or subacute .... 6. *F. obtusa*

6. Inflorescence ascending at maturity; spikelets about 5 mm broad; lemmas obtuse .... 7. *F. paradoxa*

1. *F. capillata* Lam. Slender Fescue. May–July. Native of Europe; waste ground, not common; scattered in metropolitan areas. *F. ovina* L. var. *capillata* (Lam.) Alef.—G.

2. *F. ovina* L. var. *duriuscula* (L.) Koch. Sheep Fescue. May–June. Native of Europe; waste ground, not common; scattered throughout the state. *F. ovina* L.—J.

3. *F. rubra* L. Red Fescue. June–July. Waste ground; infrequent throughout the state.

4. *F. pratensis* Huds. Meadow Fescue. May–Aug. Native of Europe; waste ground, common; in every co. *F. elatior* L.—F, G, J. (Including *F. elatior* f. *aristata* Holmb.)

5. *F. arundinacea* Schreb. Large Fescue. May–Aug. Native of Europe; waste ground, rare; scattered in Ill. *F. elatior* L. var. *arundinacea* (Schreb.) Wimmer—G, J.

6. *F. obtusa* Bieler. Nodding Fescue. May–July. Moist woodlands, occasional; scattered throughout the state.

7. *F. paradoxa* Desv. Fescue. May–July. Dry or moist woodlands; occasional in the s. three-fourths of the state, absent elsewhere.

4. *Lolium* L.   Rye Grass

1. Glume as long as or longer than the spikelet, 15–20 mm long .... 1. *L. temulentum*

1. Glume shorter than the spikelet, 4–12 mm long 2

2. Lemmas (or at least the uppermost) awned; spikelets 10- to 20-flowered; annual .... 2. *L. multiflorum*

2. Lemmas awnless; spikelets 6- to 10-flowered; perennial .. 3. *L. perenne*

1. *L. temulentum* L. Darnel. June–July. Native of Europe; waste ground, fields, not common; scattered in Ill.

2. *L. multiflorum* Lam. Italian Rye Grass. May–Sept. Native of Europe; waste ground, fields; occasional throughout the state.

3. *L. perenne* L. English Rye Grass. June–Sept. Native of Europe; waste ground, fields, lawns, common; probably in every co.

5. *Puccinellia* Parl.

1. Lemmas obscurely nerved .... 1. *P. distans*
1. Lemmas conspicuously nerved .... 2. *P. pallida*

    1. P. distans (L.) Parl. Alkali Grass. June–Oct. Native of Europe; waste ground; Cook Co.
    2. P. pallida (Torr.) Clausen. May–Aug. Swamps, very rare; Union Co. *Glyceria pallida* (Torr.) Trin.—F, G, J.

6. *Poa* L. *Bluegrass*

1. Plants dioecious; pistillate spikelets woolly; staminate spikelets glabrous or nearly so .... 4. *P. arachnifera*
1. Plants monoecious; spikelets perfect, variously pubescent or glabrous 2
2. Some spikelets transformed into bulblets .... 16. *P. bulbosa*
2. Spikelets normal 3
3. Lemmas without a tuft of cobwebby hairs at the base 4
3. Lemmas with a tuft of cobwebby hairs at the base 5
4. Tufted annual to about 30 cm tall, sometimes rooting at the lower nodes; lemmas elliptic to ovate .... 1. *P. annua*
4. Tufted perennial to 75 cm tall, not rooting at the lower nodes; lemmas oblong... 3. *P. autumnalis*
5. Nerves and keel of the lemma glabrous (except for the cobwebby tuft) .... 8. *P. languida*
5. Keel and sometimes the nerves of the lemma pubescent 6
6. Keel of the lemma pubescent, the nerves glabrous 7
6. Keel and at least some of the nerves of the lemma pubescent 8
7. Culms beneath the panicle and the sheaths scabrous; lemmas sharply nerved; ligule of upper leaves 4–8 mm long .... 9. *P. trivialis*
7. Culms beneath the panicle and the sheaths usually glabrous; lemmas obscurely nerved; ligule of upper leaves about 1 mm long .... 10. *P. alsodes*
8. Marginal nerves of the lemma pubescent, the intermediate nerves glabrous 9
8. All nerves of the lemma pubescent 13
9. Plants with rhizomes; lemmas with 5 prominent nerves 10
9. Plants without rhizomes; lemmas with 3 prominent nerves and 2 obscure nerves 11
10. Basal leaves flat, at least as broad as the culm; culm compressed at base, 2–3 mm thick at base; glumes broadly lanceolate, straight .... 5. *P. pratensis*
10. Basal leaves involute or filiform, narrower than the culm; culm terete at base, 1–2 mm thick at base; glumes narrowly lanceolate, arching .... 6. *P. angustifolia*
11. Culms very weak, solitary or in small tufts; sheaths scabrous; lowest branches of the panicle mostly paired .... 11. *P. paludigena*
11. Culms more firm, usually densely tufted; sheaths usually glabrous; lowest branches of the panicle in clusters of 3–5 12
12. Ligule 0.5–1.0 mm long; anthers 1.2–1.6 mm long ..... 12. *P. nemoralis*
12. Ligule 2–5 mm long; anthers up to 1 mm long .... 13. *P. palustris*
13. First glume 2.5–3.5 mm long; lemma 3.5–4.5 mm long; blades 1–2 mm broad .... 14. *P. wolfii*

13. First glume 1.5–2.5 (–2.7) mm long; lemma 1.5–3.5 mm long; blades 2–5 mm broad 14
14. Tufted perennial; inflorescence reflexed or spreading, 10–20 cm long; lemmas with 5 distinct nerves .... 15. *P. sylvestris*
14. Tufted annual or rhizomatous perennial; inflorescence ascending; lemmas with 3 distinct nerves and 2 obscure nerves 15
15. Tufted annual to 30 cm tall; culms terete; anthers 0.1–0.2 mm long .....
.... 2. *P. chapmaniana*
15. Rhizomatous perennial to 70 cm tall; culms compressed; anthers about 1 mm long .... 7. *P. compressa*

1. P. annua L. Annual Bluegrass. March–July. Native of Europe and Asia; mostly in moist waste ground; probably in every co.

2. P. chapmaniana Scribn. Annual Bluegrass. April–Sept. Fields and waste ground; occasional in the s. half of the state, less common in the n. half.

3. P. autumnalis Muhl. Bluegrass. March–June. Moist woodlands, very rare; Pope Co.

4. P. arachnifera Torr. Texas Bluegrass. Native to the sw. U.S.; roadsides and pastures; Winnebago Co.

5. P. pratensis L. Kentucky Bluegrass. April–July. Native of Europe and Asia; woods, fields, waste ground, common; in every co.

6. P. angustifolia L. Bluegrass. June–Sept. Woods and clearings, very rare; Union Co.

7. P. compressa L. Canadian Bluegrass. May–Aug. Native of Europe and Asia; dry soil, common; in every co.

8. P. languida Hitchc. Woodland Bluegrass. June–Sept. Moist woodlands, rare; Cook and JoDaviess cos.

9. P. trivialis L. Meadow Grass. May–Aug. Native of Europe; waste ground, rare; Cook and Stark cos.

10. P. alsodes Gray. Woodland Bluegrass. May–June. Moist woodlands, rare; Jackson and St. Clair cos.

11. P. paludigena Fern. & Wieg. Marsh Bluegrass. June–July. Bogs, rare; Kane Co.

12. P. nemoralis L. Woodland Bluegrass. June–Aug. Native of Europe; open woodlands, waste ground, rare; Champaign and Peoria cos.

13. P. palustris L. Fowl Bluegrass. June–Sept. Wet soil; occasional in the n. half of the state, rare in the s. half.

14. P. wolfii Scribn. Meadow Bluegrass. April–June. Meadows and woodlands, rare; Fulton, Henderson, and Peoria cos.

15. P. sylvestris Gray. Woodland Bluegrass. May–July. Moist woodlands; occasional throughout the state.

16. P. bulbosa L. Bulbous Blue Grass. Native of Europe; adventive in Ill.; DuPage Co.

7. *Briza* L.   *Quaking Grass*

1. B. maxima L. Big Quaking Grass. June–July. Native of Europe; waste ground, rare; St. Clair Co.

8. *Dactylis* L.   *Orchard Grass*

1. D. glomerata L. Orchard Grass. May–July. Native of Europe; waste ground, common; in every co.

9. Koeleria Pers. June Grass

1. K. macrantha (Ledeb.) Spreng. June Grass. June–Sept. Prairies, sandy black-oak woods; occasional throughout the state. *K. cristata* Pers.—F, G, J.

10. Sphenopholis Scribn. Wedge Grass

1. First glume subulate, less than 0.5 mm broad; lemmas smooth to scabrous; anthers less than 1 mm long .... 1. *S. obtusata*
1. First glume narrowly oblong, at least 0.5 mm broad; second lemma scabrous; anthers 1.0–1.5 mm long .... 2. *S. nitida*

1. S. obtusata (Michx.) Scribn. Wedge Grass. May–July. Panicle dense and spike-like; second glume firm, rounded or truncate at the apex. Woods and prairies, occasional; throughout the state.
var. major (Torr.) Erdman. Wedge Grass. May–July. Panicle loose; second glume scarious, acute to apiculate at the apex. Moist woods, moist prairies, occasional; throughout the state. *S. intermedia* (Rydb.) Rydb.—F, G, J.
2. S. nitida (Biehler) Scribn. Shining Wedge Grass. May–July. Dry woods, prairies, not common; scattered in Ill.

11. Aira L. Hairgrass

1. A. caryophyllaea L. Slender Hairgrass. May–June. Native of Europe; waste ground, rare; Piatt Co.

12. Deschampsia Beauv. Hairgrass

1. D. cespitosa (L.) Beauv. var. glauca (Hartm.) Lindm. Tufted Hairgrass. June–July. Along creeks and in swamps, rare; confined to extreme ne. Ill. *D. cespitosa* (L.) Beauv.—J.

13. Avena L. Oats

1. Spikelet 3-flowered; lemmas pubescent, with a bent awn .... 1. *A. fatua*
1. Spikelet 2-flowered; lemmas glabrous, awnless or with a straight awn ..
.... 2. *A. sativa*

1. A. fatua L. Wild Oats. May–Sept. Native of Europe; occasionally escaped into waste ground.
2. A. sativa L. Oats. May–Aug. Native of Europe and Asia; occasionally escaped into waste ground.

14. Arrhenatherum Beauv. Oat Grass

1. A. elatius (L.) Presl. Tall Oat Grass. July–Sept. Native of Europe; occasionally escaped into waste ground; widely scattered in Ill.

15. Holcus L. Velvet Grass

1. H. lanatus L. Velvet Grass. June–July. Native of Europe; adventive in waste ground; not common but scattered in Ill.

16. Calamagrostis Adans.   Reed Grass

1. Callus hairs of lemma shorter than or equalling the lemma; spikelets up to 4.5 mm long; awn of lemma straight 2
1. Callus hairs of lemma exceeding the lemma; spikelets 5–8 mm long; awn of lemma slightly bent .... 3. *C. epigeios*
2. Blades flat (at least when fresh), 4–8 mm broad; panicle open, more or less nodding; glumes spreading in fruit; lemmas translucent at tip .... 1. *C. canadensis*
2. Blades involute, 2–4 mm broad when unrolled (rarely broader); panicle contracted, spike-like, erect; glumes connivent at tip in fruit; lemmas firm throughout .... 2. *C. inexpansa*

   1. C. canadensis (Michx.) Beauv. Bluejoint Grass. June–July. Panicle loosely flowered; spikelets 2.8–3.8 mm long; glumes distinctly exceeding the lemma, acute to acuminate; moist soil; occasional in the n. half of the state, rare in the s. half, absent in the extreme s.
   var. macouniana (Vasey) Stebbins. Bluejoint Grass. Aug. Panicle densely flowered; spikelets 2.2–2.8 mm long; glumes nearly equalled by the lemma, obtuse to acute; wet ditch, very rare; Henry Co.
   2. C. inexpansa Gray var. brevior (Vasey) Stebbins. Northern Reed Grass. June–July. Wet ground, rare; Cook, JoDaviess, Lake, and Winnebago cos. *C. inexpansa* Gray—J.
   3. C. epigeios (L.) Roth. Feathertop. June–July. Native of Europe; adventive in strip mine, rare; Randolph Co.

17. Ammophila Host.   Beach Grass

   1. A. breviligulata Fern. Beach Grass. July–Sept. Sand dunes; Cook and Lake cos.

18. Agrostis L.   Bent Grass

1. Annuals; lemma sharply nerved, with a flexuous awn to 10 mm long ... . 1. *A. elliottiana*
1. Perennials; lemma obscurely nerved, awnless (rarely a very short awn present in a variety of *A. scabra*) 2
2. Tufted perennials without rhizomes; palea absent, or minute and nerveless 3
2. Rhizomatous or stoloniferous perennials; palea at least one-half as long as the lemma, 2-nerved 5
3. Flat blades 1–2 mm broad; spikelets 1.2–2.0 mm long; lemma 0.5–1.0 mm long .... 2. *A. hyemalis*
3. Flat blades 2–6 mm broad; spikelets 2–3 mm long; lemma 1.3–2.0 mm long 4
4. Panicle branches harshly scabrous, bearing florets only near the tip ... . 3. *A. scabra*
4. Panicle branches glabrous or nearly so, bearing florets from near the middle to the tip .... 4. *A. perennans*
5. Some of the panicle branches bearing florets to base; ligule 2–6 mm long .... 5. *A. alba*
5. None of the panicle branches bearing florets to the base; ligule less than 2 mm long .... 6. *A. tenuis*

1. A. elliottiana Schult. Awned Bent Grass. May–July. Dry soil, particularly on bluffs; restricted to the s. one-half of Ill.

2. A. hyemalis (Walt.) BSP. Tickle Grass. March–June. Woods and fields, common; probably in every co.

3. A. scabra Willd. Tickle Grass. June–Sept. Lemmas awnless. Moist or dry, usually open, soil; occasional in the n. cos., rare elsewhere.

f. tuckermanii Fern. June–Sept. Lemmas awned. Moist ground, rare; Cook Co.

4. A. perennans (Walt.) Tuckerm. Upland Bent Grass. June–Sept. Dry woodlands, common in the s. half of the state, becoming increasingly less common northward. (Including var. *aestivalis* Vasey.)

5. A. alba L. Red Top. June–Sept. Blades 5–10 mm broad; rhizomes present; culms erect; panicle purple, the branches spreading. Fields, common; in every co. *A. stolonifera* L. var. *major* (Gaud.) Farw.—G.

var. palustris (Huds.) Pers. Creeping Bent Grass. June–Sept. Blades 1–5 mm broad; stolons present; culms decumbent; panicle straw-colored, the branches ascending. Wet ground; occasional throughout the state. *A. stolonifera* L. var. *major* (Gaud.) Farw.—G; *A. palustris* Huds.—J.

6. A. tenuis Sibth. Rhode Island Bent. June–Sept. Native of Europe; escaped from lawns into waste ground; scattered in some n. cos.

19. Cinna L.   Wood Reed

1. Panicle gray-green, the branches mostly ascending; spikelets 4.0–6.5 mm long; second glume 3-nerved; awn less than 0.5 mm long . . . .
. . . . 1. *C. arundinacea*
1. Panicle green, the branches mostly spreading; spikelets 2.5–4.0 mm long; second glume 1-nerved; awn 0.5–1.5 mm long . . . . . 2. *C. latifolia*

1. C. arundinacea L. Stout Wood Reed. July–Sept. Moist woodlands, damp soil; occasional throughout the state.

2. C. latifolia (Trev.) Griseb. Drooping Wood Reed. July–Sept. Moist woods, rare; DeKalb and Winnebago cos.

20. Anthoxanthum L.   Vernal Grass

1. Spikelets brownish-green, 8–10 mm long; glumes pubescent; awns of empty lemmas included or barely exserted; perennials to nearly 1 m tall
. . . . 1. *A. odoratum*
1. Spikelets whitish-green, 5–7 mm long; glumes glabrous; awns of empty lemmas long-exserted; annuals to 35 cm tall . . . . 2. *A. aristatum*

1. A. odoratum L. Sweet Vernal Grass. June. Native of Europe; fields, rare; Cook, DuPage, and Lake cos.

2. A. aristatum Boiss. Annual Sweet Grass. Aug. Native of Europe; waste ground; Rock Island Co. *A. puellii*—F.

21. Hierochloë R. Br.   Sweet Grass

1. H. odorata (L.) Beauv. Sweet Grass. May–June. Meadows; occasional in the n. one-fifth of the state.

22. Phalaris L. Canary Grass

1. Keel of glumes wingless, the glumes, 4.5–6.5 mm long; sterile florets 1–2 mm long; perennial from creeping rhizomes .... 1. *P. arundinacea*
1. Keel of glumes broadly winged, the glumes 7–10 mm long; sterile florets 2.5–4.5 mm long; annual .... 2. *P. canariensis*

    1. P. arundinacea L. Reed Canary Grass. May–July. Meadows and similar moist situations; occasional throughout the state. (Including f. *picta* [L.] Asch. & Graebn.)
    2. P. canariensis L. Canary Grass. June–July. Native of Europe; waste ground; occasional throughout the state.

23. Alopecurus L. Foxtail

1. Spike-like panicle 6–10 mm thick; spikelets 4.0–5.5 mm long; glumes acute, 4.0–5.5 mm long; lemma 4.0–5.5 mm long, the awn exserted 4–5 mm .... 1. *A. pratensis*
1. Spike-like panicle to 5.5 mm thick; spikelets 2.0–2.7 mm long; glumes obtuse, 2.0–2.7 mm long; lemma 2.0–2.7 mm long, the awn exserted only up to 4 mm 2
2. Perennial from slender rhizomes; awn exserted up to 1 mm .... 2. *A. aequalis*
2. Tufted annual; awn exserted from 1.5–4.0 mm .... 3. *A. carolinianus*

    1. A. pratensis L. Meadow Foxtail. June. Native of Eurasia; along railroads; known from a few cos. in the n. half of the state.
    2. A. aequalis Sobol. Foxtail. May–July. Wet ground, occasionally in shallow water, not common; scattered throughout the state.
    3. A. carolinianus Walt. Common Foxtail. April–July. Moist ground; occasional throughout the state.

24. Phleum L. Timothy

    1. P. pratense L. Timothy. June–Aug. Native of Europe; waste ground, fields, common; in every co.

25. Milium L. Millet Grass

    1. M. effusum L. Millet Grass. June–Aug. Moist woodlands, rare; Kane and Tazewell cos.

26. Beckmannia Host Slough Grass

    1. B. syzigachne (Steud.) Fern. American Slough Grass, July–Sept. Wet ground, very rare; Cook and Lake cos.

27. Elymus L. Wild Rye

1. Rhizomes present; glumes 3–4 mm broad; lemmas awnless .... 1. *E. arenarius*
1. Rhizomes absent; glumes up to 2.5 mm broad; lemmas awned (awnless in one variety of *E. virginicus*) 2

2. Glumes reduced to unequal filiform bristles, or absent; spikelets widely spreading .... 2. *E. hystrix*
2. Glumes subequal in length; spikelets ascending 3
3. Base of glumes indurated; paleas up to 8.5 mm long 4
3. Base of glumes thin or indurated for only 1 mm; paleas 8.5–12.0 (–14.0) mm long .... 6. *E. canadensis*
4. Glumes (0.8–) 1.0–2.0 mm wide, swollen on at least half or all of the adaxial surface .... 3. *E. virginicus*
4. Glumes 0.2–1.0 mm wide, indurated for 1–3 mm adaxially 5
5. Paleas 7.0–8.5 mm long, the apices bidentate .... 4. *E. riparius*
5. Paleas 5.5–6.5 mm long, the apices obtuse .... 5. *E. villosus*

1. E. arenarius L. Wild Rye. July–Oct. Sandy shores of Lake Michigan, rare; Cook and Lake cos. *E. arenarius* L. var. *villosus* Mey.—F; *E. mollis* Trin.—G.

2. E. hystrix L. Bottlebrush Grass. June–Aug. Lemmas glabrous. Woodlands, common; in every co. *Hystrix patula* Moench—F, G, J.

var. bigeloviana (Fern.) Mohlenbrock. Bottlebrush Grass. June–Aug. Lemmas pubescent. Woodlands; occasional throughout the state. *Hystrix patula* Moench var. *bigeloviana* (Fern.) Deam—F; *H. patula* Moench f. *bigeloviana* (Fern.) Gl.—G.

3. E. virginicus L. Virginia Wild Rye. June–Sept. Lemmas to 30 mm long, glumes to 25 mm long, both glabrous or scabrous. Fields and woodlands, low ground; common throughout the state. (Including var. *jejunus* [Ramaley] Bush.) Forma *hirsutiglumis* (Scribn.) Fern. has villous glumes and lemmas.

var. submuticus Hook. June–Sept. Glumes and lemmas acuminate or subulate-tipped. Fields and woodlands, not common.

var. glabriflorus (Vasey) Bush. June–Sept. Lemmas 35 mm long or longer, glumes 27 mm long or longer, both glabrous or scabrous. Fields and woodlands, occasional throughout the state. Forma *australis* (Scribn. & Ball) Fern. has hirsute glumes and lemmas.

4. E. riparius Wiegand. Wild Rye. July–Sept. Woodlands; not common but scattered.

5. E. villosus Muhl. Slender Wild Rye. June–Sept. Glumes and lemmas hispid to hirsute. Woodlands, common; in every co.

f. arkansanus (Scribn. & Ball) Fern. June–Sept. Glumes scabrous; lemmas glabrous or scabrous. Woodlands, rare; DuPage, Henry, Jackson, and Stark cos.

6. E. canadensis L. Nodding Wild Rye. June–Sept. Woods, roadsides, dry prairies, common; in every co. (Including var. *glaucifolius* [Muhl.] Torr.; var. *brachystachys* [Scribn. & Ball] Farw.; var. *crescendus* [Ramaley] Bush.)

28. Sitanion Raf.

1. S. hystrix (Nutt.) J. G. Sm. Squirrel-tail. July–Sept. Adventive in Ill. along a railroad; Mason Co.

29. Hordeum L. Barley

1. Lateral spikelets of each group of three pedicellate and sterile 2
1. Lateral spikelets of each group of three sessile and perfect 4
2. Fertile spikelet subtended by awns less than 2 cm long; spikes less than 2 cm broad 3

2. Fertile spikelet subtended by awns exceeding 2 cm long; spikes 2–8 cm
broad .... 3. *H. jubatum*
3. Four glumes of each group of 3 spikelets setiform, the other two glumes
dilated at the base; awn of lemma of central spikelet 8–15 mm long ....
.... 1. *H. pusillum*
3. All glumes setiform; awn of lemma of central spikelet 5–6 mm long ....
.... 2. *H. brachyantherum*
4. Annual; spikes erect; glumes flat, about 1 mm broad; awns of lemmas
60–150 mm long, or absent and replaced by a three-lobed appendage ..
.... 4. *H. vulgare*
4. Perennial; spikes nodding; glumes setiform; awns of lemmas 15–35 mm
long .... 5. *H.* X *montanense*

1. H. pusillum Nutt. Little Barley. May–July. Fields, waste ground;
common in the s. three-fourths of the state, rare in the n. one-fourth.

2. H. brachyantherum Nevski. Meadow Barley. May–July. Native to
N. Am.; adventive in a lawn; Jackson Co.

3. H. jubatum L. Squirrel-tail Grass. June–Sept. Fields; common in
the n. one-half of the state, occasional elsewhere.

4. H. vulgare L. Common Barley. June–Sept. Lemmas awned. Native
of Asia; occasional along country roads.

var. trifurcatum (Schlecht.) Alefeld. Pearl Barley. June–Sept.
Lemmas three-lobed at the apex, awnless. Native of Asia; occasional
along county roads.

5. H. X montanense Scribn. Barley. June–July. Prairies and roadsides, not common; Marshall, Peoria, and Stark cos. Reputed to be a
hybrid between *H. jubatum* and *Elymus virginicus*.

30. X Agrohordeum G. Camus

1. A. X macounii (Vasey) Lepage. Macoun's Wild Rye. June–July.
Native to w. N. Am.; adventive in Ill.; Cook Co. *Elymus macounii*
Vasey—F, G. Reputed to be a hybrid between *Agropyron trachycau-
lum* (Link) Malte and *Hordeum jubatum* L.

31. Agropyron Gaertn. Wheat Grass

1. Lemmas 5–7 mm long; glumes 2–5 mm long; spikelets pectinately
arranged; tufted plants 2
1. Lemmas 8–25 mm long; glumes 8–18 mm long; spikelets not pectinately
arranged; tufted or rhizomatous plants 3
2. Blades flat; spikelets 8–12 mm long; glumes abruptly tapering to the 2–3
mm long awn .... 1. *A. desertorum*
2. Blades involute (at least when dry); spikelets 5–7 mm long; glumes gradually tapering to the 2–5 mm long awn .... 2. *A. cristatum*
3. Tufted plants; spikelets at maturity (and on herbarium specimens) readily breaking up into individual florets when touched 4
3. Rhizomatous plants; spikelets at maturity falling in their entirety 5
4. Awn of lemmas 10–30 mm long; spikes dense; glumes 12–18 mm long
.... 3. *A. subsecundum*
4. Awn of lemmas absent or up to 2 mm long; spikes more slender; glumes
8–12 mm long .... 4. *A. trachycaulum*
5. Blades flat, 5–10 mm broad; sheaths pubescent (in Ill.) .... 5. *A. repens*
5. Blades involute when dry, 2–5 mm broad; sheaths glabrous (in Ill.) .....
.... 6. *A. smithii*

1. A. desertorum (Fisch.) Schult. Wheat Grass. June–Aug. Native of Russia; introduced in Ill.; Cook and JoDaviess cos.

2. A. cristatum (L.) Gaertn. Crested Wheat Grass. June–Aug. Native of Russia; adventive in Ill.; Fulton and JoDaviess cos.

3. A. subsecundum (Link) Hitchcock. Bearded Wheat Grass. July–Aug. Woodlands, fields, rare; confined to the extreme n. cos. *A. trachycaulum* (Link) Malte var. *unilaterale* (Cassidy) Malte—F.

4. A. trachycaulum (Link) Malte. Slender Wheat Grass. June–Aug. Fields and waste ground and along railroads, rare; known from a few n. cos.

5. A. repens (L.) Beauv. Quack Grass. June–July. Lemmas awnless. Native of Europe and Asia; adventive in fields and waste ground; common in the n. three-fourths of the state, rare in the s. one-fourth.

f. aristatum (Schum.) Holmb. Quack Grass. June–July. Lemmas with an awn up to 10 mm long. Native of Europe and Asia; adventive in waste ground; Cook Co.

6. A. smithii Rydb. Western Wheat Grass. June–Aug. Lemmas glabrous, scabrous, or pubescent near base. Native of the w. U.S.; adventive along railroads; occasional in the n. two-thirds of the state, rare in the s. one-third.

var. molle (Scribn. & Smith) Jones. Western Wheat Grass. June–Aug. Lemmas short-pilose throughout. Native of the w. U.S.; adventive along railroads; Cook, DuPage, and Will cos. *A. dasystachyum* (Hook.) Scribn.—F; *A. molle* (Scribn. & Smith) Rydb.—J.

32. Triticum L. Wheat

1. Spikelets compressed, disarticulating above the glumes; joints of rachis not swollen; blades 10–20 mm broad .... 1. *T. aestivum*
1. Spikelets cylindrical, falling in their entirety; joints of rachis swollen; blades 2–3 mm broad .... 2. *T. cylindricum*

1. T. aestivum L. Wheat. June–Aug. Sporadic in fields and along roadsides; several cultivated varieties may be found, including Bearded Wheat, with long awns on the lemmas.

2. T. cylindricum (Host) Ces. Jointed Goat Grass. June–Sept. Native of Europe; adventive in waste ground; scattered throughout the state. *Aegilops cylindrica* Host—F, G, J.

33. Secale L. Rye

1. S. cereale L. Rye. May–July. Occasionally adventive in waste ground; frequently planted by the highway department along new road right-of-ways.

34. Melica L. Melic Grass

1. Cauline leaves 3–4, 2–5 mm broad; sheaths scabrous; glumes nearly equal in length; first glume oblong, at least twice as long as broad; fertile lemmas usually 2 .... 1. *M. mutica*
1. Cauline leaves 5–8, 5–12 mm broad; sheaths glabrous; glumes unequal in length; first glume ovate, less than twice as long as broad; fertile lemmas usually 3 .... 2. *M. nitens*

1. M. mutica Walt. Two-flowered Melic Grass. May–June. Rocky woodlands, occasional; scattered throughout the state, but rare in the ne. cos.

2. M. nitens (Scribn.) Nutt. Three-flowered Melic Grass. May–July. Rocky woods, prairies; occasional throughout the state, but rare in the ne. cos.

35. *Glyceria R. Br. Manna Grass*

1. Spikelets at least 10 mm long; sheaths compressed 2
1. Spikelets 2–8 mm long; sheaths terete or subterete 4
2. Principal leaves 2–5 mm broad; lemmas shining, scabrous only on the nerves; pedicels very slender, all one-fourth to two-thirds the length of the spikelets .... 1. *G. borealis*
2. Principal leaves 6–18 mm broad; lemmas dull, scabrous between the nerves; pedicels thickened upward, less than one-fourth the length of the spikelets (except for the terminal ones) 3
3. Principal blades 6–12 mm broad; lemmas obscurely nerved, scabrous, 3.5–5.5 mm long; anthers over 1 mm long .... 2. *G. septentrionalis*
3. Principal blades 10–18 mm broad; lemmas sharply nerved, hirtellous, 2.5–3.0 mm long; anthers less than 1 mm long .... 3. *G. arkansana*
4. Lemmas 3–4 mm long, obscurely nerved; spikelets 3–4 mm broad ......
.... 4. *G. canadensis*
4. Lemmas 1.5–2.7 mm long, sharply nerved; spikelets 2.0–2.5 mm broad   5
5. Inflorescence 5–20 cm long; spikelets 2.0–4.5 mm long; first glume 0.5–1.0 mm long; second glume 0.8–1.3 mm long; lemmas 1.5–2.0 mm long .... 5. *G. striata*
5. Inflorescence 20–40 cm long; spikelets 5–6 mm long; first glume 1.2–2.0 mm long; second glume 1.5–2.5 mm long; lemmas 2.0–2.7 mm long ....
.... 6. *G. grandis*

1. G. borealis (Nash) Batchelder. Northern Manna Grass. June–July. Shallow water, rare; JoDaviess, Lake, and Stephenson cos.

2. G. septentrionalis Hitchc. Manna Grass. May–Aug. Shallow water, marshy soil, swampy meadows, occasional; scattered throughout the state.

3. G. arkansana Fern. Manna Grass. May–June. Shallow water of swamps, very rare; Union Co. *G. septentrionalis* Hitchc.—G, J.

4. G. canadensis (Michx.) Trin. Rattlesnake Manna Grass. June–Sept. Wet ground, rare; Cook and Peoria cos.

5. G. striata (Lam.) Hitchcock. Fowl Manna Grass. May–Aug. Spikelets green; uppermost branches of the panicle more or less nodding; lemmas with a minutely scarious apex. Moist soil, common; in every co.

var. stricta (Scribn.) Fern. Fowl Manna Grass. May–Aug. Spikelets purple; uppermost branches of the panicle ascending; lemmas with a broadly scarious apex. Wet ground; occasional in the n. one-third of the state, absent elsewhere.

6. G. grandis S. Wats. American Manna Grass. June–Aug. Wet ground, not common; restricted to the extreme n. cos.

36. *Schizachne Hack. False Melic Grass*

1. S. purpurascens (Torr.) Swallen. False Melic Grass. June. Moist wooded slope, very rare; JoDaviess Co.

37. Stipa L. Needle Grass

1. Sheaths villous on the margins and at the summit; ligule less than 1 mm long; glumes 5–11 mm long; lemma 4.5–6.0 mm long, pubescent throughout, the awn 2–4 cm long ....— 1. *S. viridula*
1. Sheaths more or less glabrous; ligule (at least of the upper leaves) 3–6 mm long; glumes 15–40 mm long; lemma 9–25 mm long, pubescent at base, becoming glabrate above, the awn 10–20 cm long 2
2. Glumes 15–28 mm long; lemma 9–13 mm long, the flexuous but obscurely geniculate awn 10–15 cm long; ligule of upper leaves 3–4 mm long .... 2. *S. comata*
2. Glumes 28–42 mm long; lemma 16–25 mm long, the twice geniculate awn 12–20 cm long; ligule of upper leaves 4–6 mm long . 3. *S. spartea*

1. S. virdula Trin. Feather Grass. June–July. Edge of woods and along railroad track near pond, rare; Kane and McHenry cos.
2. S. comata Trin, & Rupr. Needle Grass. July–Aug. Dry or loamy soil, usually in prairies, rare; Cook, Kane, and Winnebago cos.
3. S. spartea Trin. Porcupine Grass. May–June. Sandy soil, particularly in prairies; occasional in the n. two-thirds of the state, nearly absent in the s. third.

38. Oryzopsis Michx. Rice Grass

1. Blades flat, 5–15 mm broad; spikelets (excluding the awn) 6–9 mm long; glumes acute to acuminate, 7–9 mm long, conspicuously 7-nerved; lemma 5.5–8.5 mm long, with the awn 5–25 mm long 2
1. Blades involute, 1–2 mm broad; spikelets (excluding the awn) 3–4 mm long; glumes obtuse, 3.5–4.0 mm long, obscurely 5-nerved; lemma 3.5–4.0 mm long, with the awn 1–2 mm long .... 3. *O. pungens*
2. Upper leaves longer than lower leaves; lemma dark brown to blackish, the awn 12–25 mm long .... 1. *O. racemosa*
2. Upper leaves shorter than lower leaves; lemma pale green to yellowish, the awn 5–10 mm long .... 2. *O. asperifolia*

1. O. racemosa (J. E. Smith) Ricker. Rice Grass. July–Sept. Rich, rocky woodlands, rare; Grundy, LaSalle, Mason, Peoria, Vermilion, and Winnebago cos.
2. O. asperifolia Michx. Rice Grass. April–July. Dry woodlands, very rare; Cook Co.
3. O. pungens (Torr.) Hitchcock. Rice Grass. April–June. Dry soil, very rare; Menard Co.

39. Brachyelytrum Beauv.

1. B. erectum (Schreb.) Beauv. May–Aug. Moist or occasionally dry woodlands; occasional throughout the state.

40. Diarrhena Beauv.

1. D. americana Beauv. var. obovata Gleason. June–Sept. Low, shaded woods, moist ledges, base of limestone cliffs; throughout the state, except for the ne. cos. *D. americana* Beauv.—F, J.

41. *Digitaria* Heist. *Finger Grass*

1. Culms rooting at the lower nodes, decumbent at the base; rachis broadly winged, about 1 mm broad 2
1. Culms erect, not rooting at the lower nodes; rachis narrowly winged, less than 1 mm broad 3
2. Sheaths (at least the lower) papillose-pilose; blades pilose to scabrous; spikelets 2.5–3.5 mm long; second glume about half as long as spikelet, usually 1.2–1.6 mm long; fertile lemma greenish-brown 1. *D. sanguinalis*
2. Sheaths glabrous; blades glabrous; spikelets 1.7–2.2 mm long; second glume about as long as spikelet, 1.7–2.2 mm long; fertile lemma dark brown to blackish .... 2. *D. ischaemum*
3. Racemes less than 10 cm long; spikelets 1.5–1.7 (–2.0) mm long; second glume and sterile lemma more or less glabrous to short-pubescent, 1.5–1.7 (–2.0) mm long .... 3. *D. filiformis*
3. Racemes over 10 cm long; spikelets 2.0–2.5 mm long; second glume and sterile lemma long-pubescent, 2.0–2.5 mm long .... 4. *D. villosa*

1. D. sanguinalis (L.) Scop. Crab Grass. June–Oct. Spikelets 2.5–3.0 mm long; sterile lemma appressed-pubescent. Native of Europe and Asia; adventive in waste ground and lawns; very common in every co.
var. ciliaris (Retz.) Parl. Crab Grass. June–Oct. Native of Europe and Asia; adventive in waste ground; Perry Co.
2. D. ischaemum (Schreb.) Muhl. Smooth Crab Grass. July–Oct. Native of Europe; adventive in waste ground; throughout the state.
3. D. filiformis (L.) Koel. Slender Crab Grass. Aug.–Sept. Sandy soil; occasional in cent. Ill., absent in the extreme n., rare in the extreme s.
4. D. villosa (Walt.) Pers. Hairy Finger Grass. July–Sept. Sandy soil, rare; Jackson Co. *D. filiformis* (L.) Koel. var. *villosa* (Walt.) Fern.—F, G; *D. filiformis* (L.) Koel.—J.

42. *Trichachne* Nees *Sour Grass*

1. T. insularis (L.) Nees. Sour Grass. July–Sept. Native of the se. U.S., the W.I., and S.Am.; adventive in a roadside ditch; Williamson Co.

43. *Leptoloma* Chase *Fall Witch Grass*

1. L. cognatum (Schult.) Chase. Fall Witch Grass. July–Sept. Sandy soil; occasional throughout the state.

44. *Eriochloa* HBK. *Cup Grass*

1. Pedicels and rachis villous; spikelets about 5 mm long ... 1. *E. villosa*
1. Pedicels and rachis short-pilose; spikelets 3.5–5.0 mm long 2
2. Grain 2.0–2.5 mm long, with an awn to 1 mm long; blades pubescent, 3–7 mm broad .... 2. *E. contracta*
2. Grain 3 mm long, apiculate; blades glabrous, 5–10 mm broad ....
.... 3. *E. gracilis*

1. E. villosa (Thunb.) Kunth. Cup Grass. Aug. Native of Asia; adventive in waste ground; Cook, DuPage, and Livingston cos.

2. E. contracta Hitchcock. Prairie Cup Grass. June–Sept. Moist soil; native of the w. U.S.; Jackson and Union cos.

3. E. gracilis (Fourn.) Hitchcock. Cup Grass. July–Oct. Field border; native of the w. U.S.; Union Co.

45. *Paspalum* L.   Bead Grass

1. Rachis foliaceous, the margins folded over and clasping the spikelets or their bases 2
1. Rachis firm, narrow or broad, but the margins not folded over the rows of spikelets 3
2. Racemes of each inflorescence 1–5; rachis shorter than the rows of spikelets .... 1. *P. dissectum*
2. Racemes of each inflorescence 5–50, usually more than 10; rachis longer than the rows of spikelets .... 2. *P. fluitans*
3. Rachis broad, over 1.5 mm wide; spikelets arranged in 4 rows .... 3. *P. pubiflorum* var. *glabrum*
3. Rachis narrower, less than 1.5 mm wide (about 1.5 mm wide in *P. lentiferum*); spikelets in 2 rows (4 in some racemes of *P. floridanum*) ... 4
4. Spikelets 3.6 mm long or longer; culms robust, 1–2 m tall .... 4. *P. floridanum*
4. Spikelets less than 3.2 mm long; culms slender, usually less than 1 m tall (occasionally to 1.5 m in *P. lentiferum*) 5
5. Spikelets 2.5–3.2 (–3.4) mm long; sterile lemma 5-nerved, with lateral nerves approximate at the margins 6
5. Spikelets 1.8–2.4 mm long; sterile lemma 3-nerved, the marginal nerves obscure at maturity 7
6. Spikelets solitary; leaves glabrous to sparsely pilose .... 5. *P. laeve*
6. Spikelets paired, or paired and solitary on the same raceme; leaves pilose, becoming villous at base .... 6. *P. lentiferum*
7. Spikelets glabrous; nodes of culms glabrous; leaves glabrous or variously pubescent, but not velvety on both surfaces 7. *P. ciliatifolium*
7. Spikelets pubescent, often densely so; nodes of culms pubescent; leaves velvety on both surfaces .... 8. *P. bushii*

1. P. dissectum (L.) L. Bead Grass. June–Sept. Moist soil, edges of shallow swamps, very rare; Perry, Pulaski, and St. Clair cos.

2. P. fluitans (Ell.) Kunth. Swamp Bead Grass. June–Sept. Floating in shallow standing water; occasional in the s. two-thirds of the state.

3. P. pubiflorum Rupr. var. glabrum (Vasey) Vasey. Bead Grass. June–Sept. In moist soil in ditches, along roadsides, and along streams; occasional in the s. one-half of the state, absent elsewhere. *P. geminum* Nash—J.

4. P. floridanum Michx. Giant Bead Grass. June–Sept. Low, moist sandy soil; not common in the s. one-third of the state, absent elsewhere. *P. glabratum* (Engelm.) Mohr.—J.

5. P. laeve Michx. Bead Grass. June–Sept. Moist soil of roadside ditches, meadows, and stream borders; common in the s. one-half of the state, rare elsewhere. (Including var. *circulare* [Nash] Stone.) *P. circulare* Nash—J.

6. P. lentiferum Lam. Bead Grass. June–Sept. Wet, roadside ditches, low post-oak flats, rare; Massac and Pulaski cos.

7. P. ciliatifolium Michx. Bead Grass. June–Sept. Dry or moist sandy soils, open woods; common in the s. two-thirds of the state,

rare elsewhere. (Including var. *muhlenbergii* [Nash] Fern. var. *stra-mineum* [Nash] Fern., *P. pubescens* Muhl., and *P. stramineum* Nash.)

    8. P. bushii Nash. Hairy Bead Grass. June–Sept. Fields, along edge of woods; restricted to a few sw. cos. *P. ciliatifolium* Michx.—G.

46. Panicum L. Panic Grass

1. Spikelets 1.0–1.9 mm long 2
1. Spikelets 2.0 mm long or longer 3
2. Spikelets glabrous .... Group I
2. Spikelets pubescent .... Group II
3. Spikelets 2.0–2.9 mm long 4
3. Spikelets 3.0 mm long or longer 5
4. Spikelets glabrous .... Group III
4. Spikelets pubescent .... Group IV
5. Spikelets glabrous .... Group V
5. Spikelets pubescent .... Group VI

Group I

Spikelets 1.0–1.9 mm long, glabrous.

1. Sheaths papillose-hispid 2
1. Sheaths glabrous or variously pubescent, but not papillose 3
2. Fruits stramineous; spikelets 0.9–1.0 mm broad; grain 0.8 mm broad; blades 6–10 mm broad .... 3. *P. gattingeri*
2. Fruits nigrescent; spikelets 0.7 mm broad; grain 0.6 mm broad; blades 2–8 mm broad .... 4. *P. philadelphicum*
3. Grain 1.3–1.4 mm long 4
3. Grain 1.8–1.9 mm long 5
4. First glume up to ¼ as long as the spikelet; nodes of culms with reflexed hairs; spikelets 1.5–1.8 mm long .... 17. *P. microcarpon*
4. First glume nearly ½ as long as the spikelet; nodes of culms glabrous; spikelets at least 1.8 mm long .... 8. *P. rigidulum*
5. Culms very slender and wiry; first glume ½ as long as the spikelet; grain 0.7 mm broad; sterile palea enlarging at maturity, and forcing open the spikelet .... 12. *P. hians*
5. Culms not wiry; first glume ¼–⅓ as long as the spikelet; grain 0.8–1.0 mm broad; sterile palea absent, or at least not enlarging at maturity . 6
6. Annual; basal and culm leaves similar; basal winter rosette absent; spikelets short-beaked .... 1. *P. dichotomiflorum*
6. Tufted perennial; basal leaves crowded and shorter than the culm leaves; spikelets subacute to acute .... 20. *P. dichotomum*

Group II

Spikelets 1.0–1.9 mm long, pubescent.

1. Some or all the sheaths papillose-pilose; spikelets obtuse; blades pubescent throughout on the upper surface (except in some specimens of *P. lanuginosum*) 2
1. None of the sheaths papillose-pilose; spikelets obtuse or acute; upper surface of blades glabrous except near the base (rarely puberulent throughout in *P. columbianum* and *P. laxiflorum*) 7

2. Culms pilose with horizontally spreading hairs ...... 25. *P. praecocius*
2. Culms variously pubescent, but the hairs not horizontally spreading . 3
3. Upper surface of blades glabrous except for long hairs at the base; autumnal form matted .... 24. *P. lanuginosum*
3. Upper surface of blades pilose or appressed-pubescent; autumnal form erect or spreading (becoming matted in a variety of *P. meridionale*) . 4
4. Spikelets 1.3–1.5 mm long; grain 1.2–1.3 mm long 5
4. Spikelets 1.6–1.9 mm long; grain 1.5–1.6 mm long 6
5. Panicle branches (at least the lower) drooping, the axes long-pilose ....
.... 24. *P. lanuginosum*
5. Panicle branches ascending, the axes glabrous or puberulent ....
.... 23. *P. meridionale*
6. Upper surface of leaves pilose, the hairs 3–5 mm long; first glume acute or acuminate, about ½ as long as the spikelet ...... 26. *P. subvillosum*
6. Uppersurface of leaves short-pubescent, the hairs less than 3 mm long; first glume obtuse and truncate, ¼–⅓ as long as the spikelet ....
.... 24. *P. lanuginosum*
7. Ligule 3–5 mm long .... 24. *P. lanuginosum*
7. Ligule less than 2 mm long 8
8. Grain 1.7–1.8 mm long 9
8. Grain 1.3–1.5 mm long 10
9. Panicle branches ascending, viscid; sheaths viscid; autumnal culms leafy to the base .... 18. *P. nitidum*
9. Panicle branches spreading, not viscid; sheaths not viscid; autumnal culms essentially leafless below the middle .... 20. *P. dichotomum*
10. Sheaths retrorsely pilose .... 16. *P. laxiflorum*
10. Sheaths glabrous or ciliate or ascending-pilose 11
11. Sheaths ascending-pilose, with long, soft hairs intermingled with short, crisp ones; blades up to 7 mm broad .... 29. *P. columbianum*
11. Sheaths glabrous or ciliate, without two types of hairs; blades 7–25 mm broad 12
12. Spikelets ellipsoid, about 0.7 mm broad; nodes with reflexed hairs or glabrous 13
12. Spikelets obovoid-spherical, 1.0–1.3 mm broad; nodes appressed-puberulent or glabrous 14
13. Spikelets 1.5–1.6 mm long; nodes with reflexed hairs 17. *P. microcarpon*
13. Spikelets 1.8 mm long or longer; nodes glabrous ...... 8. *P. rigidulum*
14. Panicle nearly as broad as long; culms spreading; nodes appressed-puberulent .... 30. *P. sphaerocarpon*
14. Panicle ¼–½ as broad as long; culms erect; nodes more or less glabrous
.... 31. *P. polyanthes*

Group III

Spikelets 2.0–2.9 mm long, glabrous.

1. Spikelets 1.2–1.7 mm broad 2
1. Spikelets 0.7–1.0 mm broad 3
2. Blades 3–6 mm broad; spikelets 2.1–2.4 mm long, 1.2–1.3 mm broad
.... 15. *P. linearifolium*
2. Blades 6–12 mm broad; spikelets 2.9–3.0 mm long, 1.7 mm broad
.... 34. *P. oligosanthes*
3. At least some of the sheaths papillose-hispid 4
3. None of the sheaths papillose-hispid 7

4. Spikelets 2.5–4.0 mm long, acuminate .... 5. *P. capillare*
4. Spikelets up to 2.5 mm long, acute 5
5. Panicle at least half the entire length of the plant .... 5. *P. capillare*
5. Panicle up to ⅓ the entire length of the plant 6
6. Fruits stramineous; blades 6–10 mm broad; spikelets 0.9–1.0 mm broad; culms stout, to 100 cm tall .... 3. *P. gattingeri*
6. Fruits nigrescent; blades 2–8 mm broad; spikelets 0.7 mm broad; culms slender, to 50 cm tall .... 4. *P. philadelphicum*
7. At least the lower nodes bearded; sheaths softly pubescent ....
.... 20. *P. dichotomum*
7. Nodes glabrous or sparsely pilose; sheaths glabrous, ciliate, pilose, or appressed-pubescent, but not softly pubescent 8
8. First glume ⅕–¼ as long as the spikelet .... 1. *P. dichotomiflorum*
8. First glume ⅓–½ as long as the spikelet 9
9. Ligule ciliate, 2–3 mm long .... 10. *P. longifolium*
9. Ligule 1 mm long or less 10
10. Spikelets 2.5–2.8 mm long; grain short-stipitate .... 9. *P. stipitatum*
10. Spikelets up to 2.5 mm long (rarely 2.5 mm long in *P. yadkinense*); grain not stipitate 11
11. Blades pilose on the upper surface near base 12
11. Blades glabrous, except sometimes for marginal cilia 13
12. Blades 1–5 mm broad; palea indurate at maturity, enlarging and dilating the spikelet .... 12. *P. hians*
12. Blades 5–10 mm broad; palea not indurate or enlarged at maturity .....
.... 8. *P. rigidulum*
13. Spikelets 2.2–2.5 mm long 14
13. Spikelets 2.0–2.2 mm long 15
14. Sheaths with pale glandular spots; spikelets acute; first glume about ⅓ as long as spikelet .... 22. *P. yadkinense*
14. Sheaths without pale grandular spots; spikelets acuminate; first glume about ½ as long as spikelet .... 8. *P. rigidulum*
15. First glume about ½ as long as the spikelet; second glume and sterile lemma longer than the grain; grain 1.3 mm long, 0.6 mm broad ....
.... 8. *P. rigidulum*
15. First glume about ⅓ as long as the spikelet; second glume and sterile lemma shorter than the grain; grain 1.8 mm long, 0.9 mm broad
.... 20. *P. dichotomum*

Group IV

Spikelets 2.0–2.9 mm long, pubescent.

1. Nodes bearded, but with a sticky ring immediately beneath them
.... 37. *P. scoparium*
1. Nodes beardless or, if with a beard, then without a sticky ring immediately beneath the nodes 2
2. Spikelets up to 2.5 mm long 3
2. Spikelets 2.5–2.9 mm long 13
3. Ligule 2–5 mm long; sheaths papillose-pubescent 4
3. Ligule 1 mm long or less; sheaths not papillose-pubescent (except *P. linearifolium*) 6
4. Pubescence of culms horizontally spreading; autumnal form freely branched; ligule 4–5 mm long .... 27. *P. villosissimum*

4. Pubescence of culms appressed or ascending; autumnal form sparsely branched; ligule 2–3 mm long 5

5. Pubescence of blades and sheaths silky; upper surface of blades pubescent along both margins .... 27. *P. villosissimum*

5. Pubescence of blades and sheaths short and stiff; upper surface of blades more or less glabrous .... 28. *P. scoparioides*

6. Spikelets 1.2–1.5 mm broad (occasionally narrower in *P. mattamuskeetense*); grain 2.0–2.1 mm long, 1.1–1.2 mm broad 7

6. Spikelets 0.9–1.1 mm broad; grain 1.5–1.9 mm long, 0.9–1.0 mm broad 10

7. Sheaths papillose-pilose; spikelets 1.3–1.5 mm broad .... 15. *P. linearifolium*

7. Sheaths puberulent, velvety, or glabrous; spikelets 1.2–1.3 mm broad   8

8. Lowermost nodes bearded; lower sheaths velvety-pubescent .... 21. *P. mattamuskeetense*

8. Nodes without a beard; lower sheaths glabrous or puberulent, but rarely velvety 9

9. Sheaths puberulent; culms crisp-puberulent; spikelets at least 2.4 mm long .... 38. *P. commutatum*

9. Sheaths glabrous; culms sparsely pilose on the nodes; spikelets 2.1–2.4 mm long .... 15. *P. linearifolium*

10. Sheaths retrorsely pilose; spikelets papillose-pilose; grain 1.5 mm long .... 16. *P. laxiflorum*

10. Sheaths puberulent or merely ciliate; spikelets not papillose; grain 1.7–1.9 mm long 11

11. Upper sheaths viscid-spotted; nodes with reflexed hairs   18. *P. nitidum*

11. Upper sheaths not viscid; nodes glabrous or sparsely pilose 12

12. Second glume and sterile lemma as long as the grain; plants green .... 19. *P. boreale*

12. Second glume and sterile lemma shorter than the grain; plants often purplish .... 20. *P. dichotomum*

13. Sheaths with papillose hairs 14

13. Sheaths without papillose hairs 19

14. Blades 12–30 mm broad .... 40. *P. clandestinum*

14. Blades 2–12 mm broad 15

15. Blades glabrous or scabrous on the upper surface 16

15. Blades hirsute or velvety on the upper surface 18

16. Blades 2–5 mm broad 17

16. Blades 6–12 mm broad .... 34. *P. oligosanthes*

17. Spikelets 1.6–1.7 mm broad; grain 2.4 mm long, 1.5–1.6 mm broad .... 14. *P. perlongum*

17. Spikelets 1.3–1.5 mm broad; grain 2.0–2.1 mm long, 1.2 mm broad .... 15. *P. linearifolium*

18. Blades long-hirsute on both surfaces; grain 2.4–2.5 mm long .... 32. *P. wilcoxianum*

18. Blades velvety on both surfaces; grain 2.2 mm long .... 33. *P. malacophyllum*

19. Lowermost nodes bearded; lower sheaths velvety-pubescent .... 21. *P. mattamuskeetense*

19. None of the nodes bearded; sheaths glabrous or puberulent, but not velvety 20

20. Spikelets 2.6–2.8 mm long, obtuse to subacute; blades firm, more or less cordate at base .... 38. *P. commutatum*

20. Spikelets more than 2.8 mm long, abruptly short-pointed; blades thin, narrowed or slightly rounded at base .... 39. *P. joori*

Group V

Spikelets 3.0 mm long or longer, glabrous.

1. Spikelets obtuse .... 34. *P. oligosanthes*
1. Spikelets acute to acuminate 2
2. First glume over ½ as long as the spikelet; grain 2.4–3.0 mm long 3
2. First glume up to ½ as long as the spikelet; grain 1.7–2.3 mm long .....
.... 4
3. Sheaths papillose-hispid; panicle more or less nodding 6. *P. miliaceum*
3. Sheaths ciliate or villous at the throat; panicle ascending or spreading
.... 7. *P. virgatum*
4. Grain 1.4–1.5 mm broad; spikelets 1.5–1.7 mm broad ....
.... 13. *P. depauperatum*
4. Grain 0.8–1.0 mm broad; spikelets 0.9–1.2 mm broad 5
5. Sheaths glabrous, except for the ciliate margins; first glume ⅕–¼ as
long as the spikelet .... 1. *P. dichotomiflorum*
5. Sheaths papillose-pubescent (occasionally glabrous in *P. anceps*); first
glume ⅓–½ as long as the spikelet 6
6. Ligule less than 1 mm long; second glume and sterile lemma beaked-
acuminate; culms glabrous .... 11. *P. anceps*
6. Ligule 1–3 mm long; second glume and sterile lemma acuminate, but
not conspicuously beaked; culms usually pubescent, at least on the
nodes 7
7. Axis of panicle glabrous; grain 2.0 mm long .... 2. *P. flexile*
7. Axis of panicle short-pilose; grain 1.7–1.8 mm long ..... 5. *P. capillare*

Group VI

Spikelets 3.0 mm long or longer, pubescent.

1. Grain 2.1–2.5 mm long 2
1. Grain 2.8–3.5 mm long 8
2. Some or all the blades at least 12 mm broad 3
2. Blades 2–12 mm broad 4
3. Culms and usually the sheaths papillose-hispid; spikelet subacute to
acute .... 40. *P. clandestinum*
3. Culms and sheaths glabrous; spikelet abruptly short-pointed 39. *P. joori*
4. Blades glabrous above, glabrous or sparsely pilose beneath 5
4. Blades long-hirsute or velvety above and beneath 7
5. Blades 6–12 mm broad .... 34. *P. oligosanthes*
5. Blades 2–5 mm broad 6
6. Spikelets acute, the second glume and sterile lemma beaked ....
.... 13. *P. depauperatum*
6. Spikelets obtuse, the second glume and sterile lemma not beaked .....
.... 14. *P. perlongum*
7. Blades long-hirsute on both surfaces; grain 2.4–2.5 mm long ....
.... 32. *P. wilcoxianum*
7. Blades velvety on both surfaces; grain 2.2 mm long ....
.... 33. *P. malacophyllum*
8. Blades papillose-pubescent on both surfaces; first glume a little more
than ½ as long as the spikelet .... 36. *P. leibergii*
8. Blades not papillose-pubescent; first glume up to ½ as long as the
spikelet 9
9. Pubescence of sheath not papillose 10

   9. Pubescence of sheath papillose 11
  10. Nodes glabrous or sparsely pilose; spikelets 3.0–3.7 mm long ....
       .... 41. *P. latifolium*
  10. Nodes retrorsely bearded; spikelets 3.8–5.0 mm long ..... 42. *P. boscii*
  11. Nodes not retrorsely bearded .... 34. *P. oligosanthes*
  11. Nodes retrorsely bearded 12
  12. Ligule 3–4 mm long .... 35. *P. ravenelii*
  12. Ligule 1 mm long .... 42. *P. boscii*

   1. P. dichotomiflorum Michx. Fall Panicum. June–Oct. Spikelets 2.4–3.5 mm long; at least some of the leaves over 5 mm broad; culms mostly upright, not geniculate, the nodes not swollen; sheaths not inflated. Fields, waste ground; common throughout the state.
   var. geniculatum (Muhl.) Fern. Fall Panicum. June–Oct. Spikelets 2.4–3.5 mm long; at least some of the blades over 5 mm broad; culms mostly spreading, geniculate, some of the nodes swollen; sheaths inflated. Low waste areas, fields; scattered throughout the state.
   var. puritanorum Svenson. Panic Grass. June–Sept. Spikelets 1.7–2.3 mm long; blades up to 5 (–8) mm broad. Along streams, very rare; Cook Co.
   2. P. flexile (Gattinger) Scribn. Slender Panic Grass. July–Oct. Moist, sandy soil; occasional in the s. one-third of the state, rare elsewhere.
   3. P. gattingeri Nash. Panic Grass. July–Oct. Dry or moist open ground, waste places; scattered throughout the state, except for the n. three tiers of cos. *P. capillare* L. var. *campestre* Gatt.—G.
   4. P. philadelphicum Bernh. Panic Grass. July–Oct. Dry, usually sandy soil; locally scattered in the s. three-fifths of the state; also DeKalb Co.
   5. P. capillare L. Witch Grass. July–Oct. Lowest branches of panicle included at the base; spikelets mostly 2.0–2.5 mm long; grain about 1.5 mm long. Fields, waste ground; common throughout the state. *P. capillare* L. var. *agreste* Gatt.—G.
   var. occidentale Rydb. Witch Grass. July–Oct. Open ground, very rare; Henderson Co.
   6. P. miliaceum L. Broomcorn Millet. June–Oct. Native of Europe and Asia; escaped from cultivation into waste ground; scattered in Ill.
   7. P. virgatum L. Switch Grass. June–Oct. Fields, waste ground, prairies, rocky stream beds, woods; rather common throughout the state.
   8. P. rigidulum Bosc. Munro Grass. June–Oct. Panicle branches spreading to ascending; spikelets 1.8–2.2 mm long. Moist soil in prairies, along ponds or creeks, low woodlands; occasional in the s. one-third of the state, less common elsewhere. *P. agrostoides* Spreng.—F, G, J.
   var. condensum (Nash) Mohlenbrock. Panic Grass. June–Oct. Panicle branches erect; spikelets 2.2–2.5 mm long. Moist soil, rare; Johnson and Massac cos. *P. agrostoides* Spreng, var. *condensum* (Nash) Fern—F; *P. agrostoides* Spreng.—G.
   9. P. stipitatum Nash. Panic Grass. June–Oct. Moist soil, rare; Johnson Co. *P. agrostoides* Spreng. var. *elongatum* Scribn.—G.
   10. P. longifolium Torr. Panic Grass. June–Oct. Rocky ledges in wooded ravines, very rare; Monroe Co.
   11. P. anceps Michx. Panic Grass. June–Oct. Moist soil, woodlands,

prairies, roadside ditches, stream banks; common in the s. one-third of the state, rare to absent elsewhere.

12. P. hians Ell. Panic Grass. June–Oct. Low roadside ditch, rare; Alexander Co.

13. P. depauperatum Muhl. Panic Grass. May–Sept. Dry, open woodlands, prairies, waste ground; occasional throughout the state.

14. P. perlongum Nash. Panic Grass. May–Sept. Dry soil, particularly in prairies or upland woods; occasional in the n. half of the state, rare in the s. half.

15. P. linearifolium Scribn. Panic Grass. May–Sept. Sheaths papillose-pilose. Dry woods; occasional in the s. one-fourth of the state, rare elsewhere.

var. werneri (Scribn.) Fern. Panic Grass. May–Sept. Sheaths glabrous. Wooded slopes, very rare; LaSalle Co.

16. P. laxiflorum Lam. Panic Grass. May–Sept. Woodlands; common in the s. tip of the state, rare or absent elsewhere. *P. xalapense* HBK.—J.

17. P. microcarpon Muhl. Panic Grass. May–Oct. Wet ground, often in woods; occasional in the s. one-third of the state, absent elsewhere except Peoria Co. *P. nitidum* Lam. var. *ramulosum* Torr.—G.

18. P. nitidum Lam. Panic Grass. June–Sept. Xeric limestone blufftop, very rare; Jackson Co.

19. P. boreale Nash. Northern Panic Grass. June–Oct. Moist sand, very rare; Lake Co.

20. P. dichotomum L. Panic Grass. May–Oct. Nodes glabrous or sparsely pubescent, not bearded; grain slightly exserted. Dry soil, usually in woodlands; rather common in the s. one-fourth of the state, occasional to rare elsewhere.

var. barbulatum (Michx.) Wood. Panic Grass. May–Oct. At least the lowermost nodes bearded; grain included. Dry, usually rocky woods; occasional in the s. one-third of the state. *P. dichotomum* L.—G, J.

21. P. mattamuskeetense Ashe. Panic Grass. May–Sept. Along a levee, rare; Massac Co.

22. P. yadkinense Ashe. Panic Grass. May–Sept. Moist ground, rare; restricted to a few s. cos.

23. P. meridionale Ashe. Panic Grass. May–Sept. Vernal blades 1.5–4.0 cm long; plants greenish-yellow; autumnal form nearly erect. Sandy woodlands, rare; known from the extreme n. cos. and two s. cos.

var. albemarlense (Ashe) Fern. Panic Grass. May–Sept. Vernal blades 4.5–7.0 cm long; plants grayish; autumnal form spreading to ascending, but eventually forming mats. Margin of wet, peaty meadow, very rare; Kankakee Co. *P. meridionale* Ashe—G.

24. P. lanuginosum Ell. Panic Grass. May–Sept. Axis of panicle branches with some long-spreading hairs; upper surface of blades glabrous except for long hairs at the base; spikelets 1.6–1.9 mm long. Usually low, moist, open situations; common throughout the state. *P. tennesseense* Ashe—J.

var. implicatum (Scribn.) Fern. Panic Grass. May–Sept. Axis of panicle branches with some long-spreading hairs; upper surface of blades pubescent; spikelets 1.3–1.4 mm long. Swampy soil or in moist depressions of sandstone cliffs; occasional throughout the state. *P. implicatum* Scribn.—J.

var. lindheimeri (Nash) Fern. Panic Grass. May–Sept. Axis of panicle branches either glabrous or with appressed hairs; spikelets 1.4–1.6 mm long. Sandy soil, usually in woodlands; occasional throughout the state. *P. lindheimeri* Nash—J.

var. septentrionale (Fern.) Fern. Panic Grass. May–Sept. Axis of panicle branches either glabrous or with appressed hairs; spikelets 1.6–1.9 mm long. Moist soil; occasional in the s. cos.

25. P. praecocius Hitchc. & Chase. Panic Grass. May–Sept. Dry soil, often in prairies; occasional throughout the state, becoming rare in the extreme s. cos.

26. P. subvillosum Ashe. Panic Grass. June–Oct. Sandy soil in an open field, very rare; Lake Co.

27. P. villosissimum Nash. Hairy Panic Grass. May–Sept. Culms and sheaths spreading papillose-pilose; ligule 4–5 mm long; upper surface of blades appressed long-pilose. Sandy soil, often in woodlands; occasional throughout the state.

var. pseudopubescens (Nash) Fern. Panic Grass. May–Sept. Culms and sheaths appressed papillose-pilose; ligule 2–3 mm long; upper surface of blades glabrous or short-pilose. Sandy soil; occasional throughout the state, becoming rare in the s. cos. *P. pseudopubescens* Nash—J.

28. P. scoparioides Ashe. Panic Grass. June–Oct. Dry field, very rare; Lake Co.*P. villosissimum* Nash var. *scoparioides* (Ashe) Fern.—F., G.

29. P. columbianum Scribn. Panic Grass. June–Oct. Sandy woodlands; rare in the n. one-fourth of the state.

30. P. sphaerocarpon Ell. Panic Grass. June–Oct. Sandy soil; occasional in the s. two-fifths of the state, absent elsewhere, except for Christian Co.

31. P. polyanthes Schult. Panic Grass. June–Oct. Low ground, chiefly in woodlands; occasional in the s. one-third of the state; also Peoria Co.

32. P. wilcoxianum Vasey. Panic Grass. June–Sept. Dry soil, mostly in prairies; confined to the extreme nw. cos. and Pope Co.

33. P. malacophyllum Nash. Panic Grass. June–Oct. Limestone bluffs, dry woods, rare; Jackson and Pope cos.

34. P. oligosanthes Schult. Panic Grass. June–Oct. Spikelets 3.2–4.0 mm long; culms appressed-pubescent; first glume sparsely hirsute; grains 2.8–3.0 mm long. Sandy soil, mostly in woodlands; occasional throughout the state.

var. scribnerianum (Nash) Fern. Panic Grass. June–Oct. Spikelets 3.2–3.3 mm long; culms glabrous or spreading-pubescent; first glume glabrous; grains 2.8–2.9 mm long. Dry, sandy soil, often in prairies; occasional throughout the state. *P. scribnerianum* Nash—G, J.

var. helleri (Nash) Fern. Panic Grass. May–Sept. Spikelets 2.9–3.0 mm long; culms appressed-pubescent below; first glume glabrous; grains 2.4–2.5 mm long. Limestone ledge, rare; Randolph Co. *P. helleri* Nash—G.

35. P. ravenelii Scribn. & Merr. Ravenel's Panic Grass. June–Oct. Cherty ravines, sandstone ledges, rare; Hardin and Union cos.

36. P. leibergii (Vasey) Scribn. Leiberg's Panic Grass. June–Oct. Dry soil, mostly in sandstone woodlands; occasional in the n. half of the state, rare in the s. half.

37. P. scoparium Lam. Panic Grass. June–Oct. Moist fields, roadsides, rare; Pope Co.

38. P. commutatum Schult. Panic Grass. June–Oct. Leaves to 25 mm

broad, heart-shaped at base; spikelets 2.6–2.8 mm long. Mostly dry woodlands; occasional in the s. one-third of the state, absent elsewhere.

var. ashei Fern. Panic Grass. June–Oct. Leaves to 10 mm broad, narrowed at base; spikelets 2.4–2.7 mm long. Dry woodlands, rare; Jackson Co.

39. P. joori Vasey. Panic Grass. June–Oct. Low swampy woods, rare; Johnson Co. *P. commutatum* Schult. var. *joorii* (Vasey) Fern.—F; *P. commutatum* Schult.—G.

40. P. clandestinum L. Broad-leaved Panic Grass. June–Oct. Moist, sandy soil; common in the s. half of the state, becoming less frequent northward.

41. P. latifolium L. Broad-leaved Panic Grass. June–Oct. Dry, rocky woodlands; occasional throughout the state.

42. P. boscii Poir. Large-fruited Panic Grass. June–Oct. Leaves, sheaths, and culms glabrous or sparsely pubescent. Woodlands; common in the s. half of the state, rare or absent elsewhere.

var. molle (Vasey) Hitchc. & Chase. Large-fruited Panic Grass. June–Oct. Leaves, sheaths, and culms softly pubescent. Woodlands; occasional in the s. one-fourth of the state.

47. Echinochloa Beauv.   Barnyard Grass

1. Second glume with an awn 2–10 mm long; sheaths papillose-hirsute (glabrous in f. *laevigata*); grain about three times longer than broad . . . .
. . . . 1. *E. walteri*
1. Second glume awnless or with an awn less than 2 mm long; sheaths glabrous or scabrous; grain at most about twice as long as broad, usually shorter 2
2. Racemes of panicle slender, distant; grain 2.0–2.5 mm long; blades 2.–6 mm broad . . . . 2. *E. colonum*
2. Racemes of panicle broader, more crowded; grain 2.5–3.5 mm long; blades 5–25 mm broad 3
3. Fertile lemma with a weak, easily broken tip, with a ring of minute setae just below the tip; rachis and axis of panicle setose 4
3. Fertile lemma without a ring of setae just below the tip; rachis and axil of panicle not setose . . . . 5. *E. pungens*
4. Panicle green or purple; sterile lemma short-awned; pubescence of second glume and sterile lemma hispidulous or setose . . . . . 3. *E. crus-galli*
4. Panicle dark brownish-purple; sterile lemma awnless; pubescence of second glume and sterile lemma appressed . . . . 4. *E. frumentacea*

1. E. walteri (Pursh) Heller. Aug.–Oct. Sheaths (at least the lower) papillose-hirsute. Low ground, swamps, rarely in standing water; not common, but scattered throughout the state.

f. laevigata Wieg. Aug.–Oct. Sheaths glabrous. Swamps, rare; Union Co.

2. E. colonum (L.) Link. Jungle Rice. Aug.–Oct. Native of Europe and Asia; adventive in waste ground; Cook Co.

3. E. crus-galli (L.) Beauv. Barnyard Grass. Aug.–Oct. Native of Europe and Asia; adventive in waste ground; occasional throughout the state.

4. E. frumentacea (Roxb.) Link. Billion Dollar Grass. Aug.–Oct. Native of Asia; occasionally escaped from cultivation throughout the

state. *E. crusgalli* (L.) Beauv. var. *frumentacea* (Roxb.) W. Wight—G, J.

5. E. pungens (Poir.) Rydb. Barnyard Grass. Aug.–Oct. Spikelets and grain 3.5–4.5 mm long; awn of sterile lemma more than 3 mm long. Low ground; occasional throughout the state. *E. crusgalli* (L.) Beauv.—J.

var. microstachya (Wieg.) Mohl. Barnyard Grass. Aug.–Oct. Spikelets and grain less than 3.5 mm long; awn of sterile lemma less than 3 mm long or absent; panicle purple; second glume and sterile lemma papillose-hispid. Low ground; Lake Co. *E. microstachya* (Wieg.) Rydb.—G.

var. wiegandii Fassett. Barnyard Grass. Aug.–Oct. Spikelets and grain less than 3 mm long; awn of sterile lemma less than 3 mm long or absent; panicle green; second glume and sterile lemma minutely pubescent. Low ground; occasional throughout the state.

48. Setaria Beauv. Foxtail Grass

1. Perennial from short, knotty rhizomes; second glume 1.0–1.5 mm long .... 1. *S. geniculata*
1. Annuals from fibrous roots; second glume 1.8–2.7 mm long 2
2. Spike with retrorsely scabrous bristles .... 2. *S. verticillata*
2. Spike with antrorsely scabrous bristles. 3
3. Each spikelet subtended by 5–20 bristles; blades more or less twisted .. .. 3. *S. lutescens*
3. Each spikelet subtended by 1–3 bristles; blades usually not twisted .. 4
4. Spike strongly arching; blades strigose above, puberulent below .... 4. *S. faberi*
4. Spike erect or scarcely arching; blades glabrous (sometimes ciliate) . 5
5. Spike usually not green, often lobed or interrupted; first glume short-acuminate, 1.2–1.5 mm long; fertile lemma smooth .... 5. *S. italica*
5. Spike usually green, unlobed (except var. *major*); first glume acute, 0.7–1.0 mm long; fertile lemma rugose or rugulose .... 6. *S. viridis*

1. S. geniculata (Lam.) Beauv. Perennial Foxtail. July–Sept. Fields and roadsides, not common; apparently absent from the n. one-third of the state.

2. S. verticillata (L.) Beauv. Bristly Foxtail. July–Sept. Native of Europe and Asia; adventive in waste ground; occasional in the n. half of the state, rare in the s. half.

3. S. lutescens (Weigel) Hubb. Yellow Foxtail. June–Sept. Waste ground, common; in every co. *S. glauca* (L.) Beauv.—F, G.

4. S. faberi Herrm. Giant Foxtail. July–Oct. Native of Asia; adventive in waste ground, very common; in every co.

5. S. italica (L.) Beauv. Italian Millet. July–Sept. Native of Europe and Asia; adventive in waste ground, not common.

6. S virdis (L.) Beauv. Green Foxtail. June–Sept. Culms to 75 cm tall; blades to 12 mm broad; panicle unlobed. Native of Europe and Asia; adventive in waste ground, common; in every co.

var. major (Gaudin) Pospichal. Giant Green Foxtail. June–Sept. Culms usually at least 1.5 m tall; blades to 25 mm broad; panicle appearing lobed near the base. Native of Europe; adventive in cultivated fields; Champaign and LaSalle cos.

49. Cenchrus L.   Sand Bur

1. C. longispinus (Hack.) Fern. Sand Bur. July–Sept. Sandy soil; occasional throughout the state. *C. pauciflorus* Benth.—G, J.

50. Miscanthus Anderss.   Eulalia

1. Fertile lemma awned; blades up to 10 mm broad .... 1. *M. sinensis*
1. Fertile lemma awnless; blades 10–18 mm broad .. 2. *M. sacchariflorus*

    1. M. sinensis Anderss. Eulalia. Sept.–Oct. Native of Asia; escaped along roads; Cass and Jackson cos.
    2. M. sacchariflorus (Maxim.) Hack. Plume Grass. Aug.–Oct. Native of Asia; adventive in waste ground; known from four n. cos.

51. Erianthus Michx.   Plume Grass

1. Awn of fertile lemma up to 6 mm long; blades usually not more than 10 mm broad .... 1. *E. ravennae*
1. Awn of fertile lemma 10–16 mm long; blades usually 10–25 mm broad   2
2. Panicle silvery, the axis villous; culms sericeous below the panicle; hairs subtending the spikelets much exceeding the spikelets; awn of fertile lemma twisted below .... 2. *E. alopecuroides*
2. Panicle tawny brown, the axis more or less glabrous; culms glabrous; hairs subtending the spikelets not exceeding the spikelets; awn of fertile lemma straight .... 3. *E. brevibarbis*

    1. E. ravennae (L.) Beauv. Ravenna Grass. Aug.–Oct. Native of Europe; escaped along roads; Jackson, Monroe, and Randolph cos.
    2. E. alopecuroides (L.) Ell. Plume Grass. Sept.–Oct. Open woods; occasional in the s. tip of the state.
    3. E. brevibarbis Michx. Brown Plume Grass. Aug.–Oct. Dry hills, very rare; sw. Ill.

52. Sorghum Moench   Sorghum

1. Perennials from rhizomes; awn of fertile lemma twisted near base ... 2
1. Annuals; awn of fertile lemma usually not twisted near base 3
2. Culms up to 2 m tall; blades 10–20 mm broad; awn of fertile lemma 10–15 mm long .... 1. *S. halepense*
2. Culms more than 2 m tall; some or all of the blades 25–50 cm broad; awn of fertile lemma usually less than 10 mm long .... 2. *S. almum*
3. Sessile spikelet 4.5–5.5 mm long, shorter than the pedicellate spikelet; awn of fertile lemma falling away early .... 3. *S. bicolor*
3. Sessile spikelet 6–7 mm long, as long as the pedicellate spikelet; awn of fertile lemma persistent .... 4. *S. sudanense*

    1. S. halepense (L.) Pers. Johnson Grass. June–Oct. Native of Europe and Asia; adventive in waste ground; occasional to common in the s. half of the state, becoming less common northward.
    2. S. almum Parodi. Sorghum Grass. July–Oct. Cultivated species of Argentina; adventive along roads and edges of fields; Jackson Co.
    3. S. bicolor (L.) Moench. Sorghum. Aug.–Oct. Native of Asia; occasionally escaped from cultivation; occasional throughout the state.

S. vulgare Pers.—F, G, J. Several cultivated varieties occasionally are adventive—var. *drummondii* (Nees) Mohl., var. *caffrorum* (Retz.) Mohl., and var. *saccharatum* (L.) Mohl.

4. S. sudanense (Piper) Stapf. Sudan Grass. Sept.–Oct. Native of Africa; adventive along a highway; Jackson Co.

53. Sorghastrum Nash   Indian Grass

1. S. nutans (L.) Nash. Indian Grass. Aug.–Oct. Prairies, fields, open woodlands; occasional to common throughout the state.

54. Andropogon L.   Beardgrass

1. Pedicellate spikelet staminate, 3–10 mm long; upper sheaths not inflated; sessile spikelet 4.5–10.0 mm long .... 1. *A. gerardii*
1. Pedicellate spikelet undeveloped, with only the long-villous pedicels present; upper sheaths inflated; sessile spikelet 2.5–5.0 mm long 2
2. Upper sheaths somewhat inflated, 2–6 cm long; culms more or less glabrous; racemes enclosed only at their base, the peduncles 2–10 mm long; awn straight or nearly so .... 2. *A. virginicus*
2. Upper sheaths greatly inflated, 6–12 mm long; culms villous at the upper nodes; racemes nearly entirely enclosed by the sheath, the peduncles more than 10 mm long; awn twisted or curved near base   . 3. *A. elliottii*

1. A. gererdii Vitman. Big Bluestem. July–Sept. Prairies; occasional to common throughout the state. *A. furcatus* Muhl.—J.

2. A. virginicus L. Broom Sedge. Aug.–Oct. Fields and open woods; occasional or common in the s. half of the state, absent in the n. half, except for Cook, DuPage, Grundy, Kankakee, and Will cos.

3. A. elliottii Chapm. Elliott's Broom Sedge. Sept.–Oct. Fields, open woodlands; restricted to the s. tip of the state.

55. Microstegium Nees

1. M. vimineum (Trin.) A. Camus. Eulalia. Sept.–Oct. Native of Asia; adventive in waste ground; Massac and Pope cos. *Eulalia viminea* (Trin.) Kuntze—F, G.

56. Bothriochloa Kuntze

1. B. saccharoides (Swartz) Rydb. Silver Beardgrass. Native to the s. and w. of Ill.; adventive in waste ground; Alexander, Johnson, and Sangamon cos. *Andropogon saccharoides* Sw.—F, G.

57. Schizachyrium Nees

1. S. scoparium (Michx.) Nash. Little Bluestem. Aug.–Oct. Prairies, fields, open woodlands; occasional to common throughout the state. *Andropogon scoparius* Michx.—F, G, J.

58. Tripsacum L.   Gama Grass

1. T. dactyloides (L.) L. Gama Grass. May–Sept. Low ground; occasional to rare in the s. two-thirds of the state.

59. *Zea* L. Corn

1. Z. mays L. Corn. June–Sept. Escaped from cultivation along roads.

60. *Eragrostis* Beauv. Love Grass

1. Plants forming mats, the culms rooting at the nodes 2
1. Plants erect or ascending, not forming mats 3
2. Sheaths more or less glabrous; inflorescence 2–8 cm long, the peduncle glabrous; lemmas 1.5–2.0 mm long, acute, glabrous; anthers 0.2–0.5 mm long; plants monoecius .... 1. *E. hypnoides*
2. Sheaths pubescent; inflorescence 10–25 cm long, the peduncle villous; lemmas 2–4 mm long, acuminate, sparsely villous along the nerves; anthers 1.5–2.0 mm long; plants dioecious .... 2. *E. reptans*
3. All lemmas 2.5–3.5 mm long; perennials .... 3. *E. trichodes*
3. Lemmas (or most of them) 1.0–2.5 mm long; annuals (except nos. 6, 7, & 9) 4
4. Leaf-blades with small, rounded, wart-like projections (glands) along the margin; lemmas often glandular along the keel 5
4. Leaf-blades and lemmas without wart-like projections (glands) 6
5. Spikelets 2.2–3.0 mm broad; lemmas 2.0–2.5 mm long, glandular along the keel; anthers 0.5 mm long .... 4. *E. cilianensis*
5. Spikelets 1.5–2.0 mm broad; lemmas 1.5–2.0 mm long, glandular or eglandular along the keel; anthers 0.2 mm long .... 5. *E. poaeoides*
6. Plants perennial from short rhizomes 7
6. Plants annual (perennial in no. 9), tufted, without rhizomes 8
7. Spikelets purplish, 3–8 mm long; branches of the panicle spreading or reflexed; blades flat or folded .... 6. *E. spectabilis*
7. Spikelets grayish-green, 8–10 mm long; branches of the panicle ascending; blades involute .... 7. *E. curvula*
8. Lemmas 1.8–2.5 mm long; second glume 1.6–2.0 mm long 9
8. Lemmas 1.0–1.8 mm long; second glume 0.8–1.6 mm long 10
9. Spikelets 8- to 12-flowered; pubescence of lower sheaths not papillose-based .... 8. *E. neomexicana*
9. Spikelets 2- to 3-flowered; pubescence of lower sheaths papillose-based
.... 9. *E. hirsuta*
10. Lemmas conspicuously 3-nerved; spikelets 4–8 mm long 11
10. Lemmas obscurely 3-nerved; spikelets 1.5–4.0 mm long 12
11. Panicle 15–25 (–30) cm long .... 10. *E. pectinacea*
11. Panicle 30–50 cm long .... 11. *E. diffusa*
12. Panicle at least two-thirds as long as the entire plant; first glume 1.0–1.5 mm long .... 12. *E. capillaris*
12. Panicle not more than one-half as long as the entire plant; first glume 0.5–1.2 mm long 13
13. Panicle broadest near base; second glume one-half to three-fourths as long as the lowest lemma .... 13. *E. pilosa*
13. Panicle broadest near middle; second glume as long as the lowest lemma .... 14. *E. frankii*

1. E. hypnoides (Lam.) BSP. Pony Grass. July–Oct. Wet ground, usually in sandy or muddy areas; throughout the state.
2. E. reptans (Michx.) Nees. Pony Grass. July–Oct. Sandy soil; scattered throughout the state.

3. E. trichodes (Nutt.) Wood. Thread Love Grass. July–Oct. Inflorescence purplish; spikelets 3- to 6-flowered, 4–7 mm long; glumes and lemmas to 3 mm long. Open, sandy areas; restricted to the w. half of the state.

var. pilifera (Scheele) Fern. Thread Love Grass. July–Oct. Inflorescence yellowish; spikelets 8- to 15-flowered, 8–12 mm long; glumes and lemmas 3.0–3.5 mm long. Sand prairies; not common and confined to the Illinois River banks. *E. trichodes* (Nutt.) Wood—G, J.

4. E. cilianensis (All.) Mosher. Stinking Love Grass. May–Oct. Native of Europe and Asia; introduced in fields and waste ground; in every co. *E. megastachya* (Koel.) Link—F.

5. E. poaeoides Beauv. Love Grass. June–Oct. Native of Europe; adventive in fields, waste ground; probably in every co.

6. E. spectabilis (Pursh) Steud. Tumble-grass. June–Oct. Sandy soil; probably in every co. (Including var. *sparsihirsuta* Farw.)

7. E. curvula (Schrad.) Nees. Weeping Love Grass. Aug.–Oct. Native of Africa; introduced into waste ground; Morgan Co.

8. E. neomexicana Vasey. Love Grass. July–Oct. Native to the sw. U.S.; adventive along a railroad; St. Clair Co. *E. mexicana* (Hornem.) Link—G.

9. E. hirsuta (Michx.) Nees. Aug.–Sept. Dry soil, very rare; Massac Co.

10. E. pectinacea (Michx.) Nees. Love Grass. July–Oct. Waste ground, fields; in every co.

11. E. diffusa Buckl. Western Love Grass. July–Oct. Native of the w. U.S.; adventive along a road; Menard Co. *E. pectinacea* (Michx.) Nees—G.

12. E. capillaris (L.) Nees. Lace Grass. July–Sept. Dry, rocky soil, particularly in woodlands; occasional throughout the state.

13. E. pilosa (L.) Beauv. Love Grass. July–Sept. Native of Europe; adventive in waste ground; restricted to a few s. cos. *E. pectinacea* (Michx.) Nees—J.

14. E. frankii C. A. Meyer. Love Grass. July–Oct. Spikelets 2- to 5-flowered, 2–3 mm long. Sandy soil, occasional; scattered throughout the state.

var. brevipes Fassett. Love Grass. July–Oct. Spikelets 6- to 7-flowered, 3–4 mm long. Sandy soil, rare; Henderson Co.

61. Tridens Roem. & Schultes

1. Panicle loose, open; glumes oblong to ovate; lemmas 3.5–4.0 mm long
. . . . 1. *T. flavus*
1. Panicle contracted, spike-like; glumes linear-lanceolate; lemmas 2.5–3.0 mm long . . . . 2. *T. strictus*

1. T. flavus (L.) Hitchcock. Purple-top. June–Sept. Fields and edge of woodlands; common, except in the n. one-fifth of the state where it is apparently very rare. (Including f. *cuprea* Fosb.) *Triodia flava* (L.) Smyth—F, G.

2. T. strictus (Nutt.) Nash. July–Oct. Waste ground and fields; not common and scattered. *Triodia stricta* (Nutt.) Benth.—F, G.

62. *Triplasis Beauv.* Sand Grass

1. T. purpurea (Walt.) Chapm. Sand Grass. Aug.–Oct. Sandy soil, occasional in the n. three-fourths of the state, absent in the s. one-fourth.

63. *Redfieldia Vasey* Blowout Grass

1. R. flexuosa (Thurb.) Vasey. Blowout Grass. July–Oct. Native of the w. U.S.; adventive in Ill.; Hancock Co.

64. *Calamovilfa (Gray) Hack.* Sand Reed

1. C. longifolia (Hook.) Scribn. Sand Reed. July–Sept. Sandy areas; scattered in the n. half of the state; also St. Clair Co.

65. *Muhlenbergia Schreb.* Muhly

1. Panicle diffuse, open, at least 5 cm across; spikelets on pedicels longer than the lemmas 2
1. Panicle narrower, contracted to less than 2 cm thick 3
2. Rhizomes present; panicle 5–20 cm long; spikelets about 1.5 mm long; lemma glabrous, 1.3–1.7 mm long, awnless .... 1. *M. asperifolia*
2. Rhizome lacking; panicle 20–45 cm long; spikelets (excluding awns) 3.0–4.5 mm long; lemma scabrous, 3.0–4.5 mm long, with an awn 5–20 mm long .... 2. *M. capillaris*
3. Plants densely tufted from firm bases or declined and rooting at the lower nodes, without rhizomes 4
3. Plants rhizomatous, not rooting at the lower nodes 6
4. Lower portion of culms decumbent, rooting at the nodes; blades (1–) 2–4 mm broad, flat; glumes 0.1–1.5 (–1.7) mm long; lemma awned 5
4. Culms stiffly erect, tufted from firm bases; blades 1–2 mm broad, flat or involute; glumes 1.7–2.8 mm long; lemma acuminate, awnless .... 3. *M. cuspidata*
5. Glumes obtuse, the first 0.1–0.2 mm long, the second 0.1–0.3 mm long; awn of lemma 1.5–4.0 mm long .... 4. *M. schreberi*
5. Glumes acute to aristate, the first 0.5 mm long, the second 1.0–1.5 (–1.7) mm long; awn of lemma 0.5–1.5 mm long .... 5. *M. X curtisetosa*
6. Internodes glabrous or minutely scabrous, but not puberulent or pilose (*M. racemosa* rarely puberulent at summit of internodes) 7
6. Internodes puberulent or pilose 10
7. Panicle to 4 mm broad; glumes ovate-lanceolate, 1.3–2.0 (–2.5) mm long 8
7. Panicle 5–15 mm broad; glumes lance-subulate, 1.6–8.0 mm long ... 9
8. Body of lemma 1.7–2.3 mm long, awnless or with an awn 1–2 (–4) mm long .... 6. *M. sobolifera*
8. Body of lemma 2.5–3.3 mm long, with an awn 2–7 mm long 7. *M. bushii*
9. Spikelets 2–4 mm long; glumes 2–3 mm long .... 8. *M. frondosa*
9. Spikelets 4–8 mm long; glumes 4–8 mm long. .... 9. *M. racemosa*
10. Lemma glabrous at base .... 10. *M. glabrifloris*
10. Lemma pilose at base 11
11. Glumes (including awns) (3.2–) 4.5–8.0 mm long ..... 11. *M. glomerata*
11. Glumes 1.5–3.5 mm long 12

12. Glumes ovate-laneolate; anthers 1.0–1.5 mm long; mature grain 2.0–2.3 mm long .... 12. *M. tenuiflora*
12. Glumes linear-lanceolate; anthers 0.3–0.8 mm long; mature grain 1.3–1.8 mm long 13
13. Glumes silvery or whitish; ligules 1.0–2.5 mm long; anthers 0.5–0.8 mm long .... 13. *M. sylvatica*
13. Glumes green or purplish; ligules 0.5–1.0 mm long; anthers 0.3–0.5 mm long .... 14. *M. mexicana*

1. M. asperifolia (Nees & Meyen) Parodi. Scratch Grass. June–Sept. Sandy soil; scattered in the n. two thirds of the state, apparently absent elsewhere.

2. M. capillaris (Lam.) Trin. Hair Grass. Sept.–Oct. Sandy woods; confined to the s. cos.

3. M. cuspidata (Torr.) Rydb. Muhly. Aug.–Oct. Gravelly soil, rare; Kankakee and Will cos.

4. M. schreberi J. F. Gmel. Nimble Will. July–Oct. Waste ground, woodlands; in every co.

5. M. X curtisetosa (Scribn.) Pohl. Muhly. July–Oct. Woods, rare; Champaign and Fulton cos. The Fulton Co. specimen is the type.

6. M. sobolifera (Muhl.) Trin. Muhly. July–Oct. Lemma awnless. Dry, rocky woodlands; occasional throughout the state.

f. setigera (Scribn.) Deam. Muhly. July–Oct. Lemma with an awn to 3 mm long. Dry woodlands; rare; Wabash Co.

7. M. bushii Pohl. Muhly. Aug.–Oct. Low woods, not common. *M. brachyphylla* Bush—F, G, J.

8. M. frondosa (Poir.) Fern. Muhly. Aug.–Sept. Lemma awnless. Moist woodlands, roadsides, fields; common throughout the state.

f. commutata (Scribn.) Fern. Muhly. Aug.–Sept. Lemma with an awn 4–11 mm long. Moist woodlands; occasional throughout the state.

9. M. racemosa (Michx.) BSP. Marsh Muhly. Aug.–Sept. Moist soil, occasional; not common in the e. cos.

10. M. glabrifloris Scribn. Muhly. July–Oct. Moist woodlands; occasional in the cent. and s. cos. absent from the n. cos. *M. glabriflora* Scribn.—F, G, J.

11. M. glomerata (Willd.) Trin. Muhly. Aug.–Sept. Dry or wet ground, rare; Cook, DuPage, Kane, Kendall, Lake, McHenry, and St. Clair cos.

12. M. tenuiflora (Willd.) BSP. Slender Muhly. July–Oct. Rocky woods, moist bluffs; occasional throughout the state.

13. M. sylvatica (Torr.) Torr. Muhly. Aug.–Oct. Rich, often rocky woods; occasional throughout the state.

14. M. mexicana (L.) Trin. Muhly. Aug.–Oct. Lemma awnless. Moist soil, usually in woods; occasional throughout the state.

f. ambigua (Torr.) Fern. Muhly. Aug.–Oct. Lemma with an awn to 9 mm long. Moist soil, usually in woods, rare; Lake Co.

66. *Sporobolus R. Br.   Dropseed*

1. Panicle more or less open, spreading, over 10 cm long; perennials .. 2
1. Panicle contracted, spike-like, less than 10 cm long (except in *S. asper*); perennials or annuals 3
2. Sheaths densely villous, at least near the top; spikelets 1.8–2.5 mm long; glumes acute, the first 1.0–1.5 mm long, the second 2.0–2.5 mm long; lemma 2.0–2.5 mm long .... 1. *S. cryptandrus*

2. Sheaths glabrous or nearly so; spikelets 4–6 mm long; glumes subulate or acuminate, the first 2–4 mm long, the second 4–6 mm long; lemma 3.5–5.5 mm long .... 2. *S. heterolepis*
3. Perennials; blades (before drying) 2.5 mm broad; panicle 5–25 cm long; only the upper sheaths swollen 4
3. Annuals; blades 1–2 mm broad; panicle 1–5 cm long; all or nearly all sheaths swollen 5
4. Lemma glabrous; blade pilose on upper surface near base . 3. *S. asper*
4. Lemma sparsely villous; blade scabrous on margin and tip, but not pilose .... 4. *S. clandestinus*
5. Spikelets 3.5–6.5 mm long; first glume 2.8–4.0 mm long; second glume 3.0–4.5 mm long; lemma 3–5 mm long, minutely villous ....
.... 5. *S. vaginiflorus*
5. Spikelets 2–3 mm long; first glume 1.5–2.5 mm long; second glume 1.7–2.7 mm long; lemma 2–3 mm long, mostly glabrous   6. *S. neglectus*

1. S. cryptandrus (Torr.) Gray. Sand Dropseed. Aug.–Oct. Sandy soil; occasional in the n. third of the state and in the w. cos.

2. S. heterolepis (Gray) Gray. Prairie Dropseed. Aug.–Sept. Dry soil, often prairies; occasional in the n. half of the state, rare in the s. half.

3. S. asper (Michx.) Kunth. Dropseed. Sept.–Oct. Dry, often sandy soil; occasional throughout the state.

4. S. clandestinus (Biehler) Hitchc. Dropseed. Aug.–Oct. Sandy soil, not common; scattered in Ill.

5. S. vaginiflorus (Torr.) Wood. Poverty Grass. Sept.–Oct. Dry soil; occasional throughout the state. (Including var. *inaequalis* Fern.)

6. S. neglectus Nash. Sheathed Dropseed. Sept.–Oct. Dry soil; occasional throughout the state.

67. Crypsis Ait.

1. C. schoenoides (L.) Lam. Aug.–Sept. Native of Europe; introduced into waste ground; Cook, Grundy, and St. Clair cos. *Heleochloa schoenoides* (L.) Host—F, G, J.

68. Eleusine Gaertn.   Goose Grass

1. E. indica (L.) Gaertn. Goose Grass. June–Oct. Native of Eurasia; adventive in lawns and waste ground; probably in every co.

69. Dactyloctenium Willd.   Crowfoot Grass

1. D. aegyptium (L.) Beauv. Crowfoot Grass. Aug. Native of the Old World; adventive in waste ground; St. Clair Co.

70. Leptochloa Beauv.   Sprangletop

1. Spikelets 2- to 4-flowered; lemmas 0.7–1.5 mm long; sheaths papillose-pilose; ligules 1–2 mm long 2
1. Spikelets 5- to 10-flowered; lemmas 2–8 mm long; sheaths glabrous or nearly so; ligules 3–7 mm long 3
2. Lemmas 1.0–1.5 mm long; 0.7–0.9 mm long; glumes acute 1. *L. filiformis*
2. Lemmas 0.7–1.0 mm long; grain 0.4–0.5 mm long; glumes aristate .....
.... 2. *L. attenuata*

3. Blades becoming involute, 1–5 mm broad; lemma with an awn 0.5–1.0 mm long 4
3. Blades flat, 5–10 mm broad; lemma apiculate, awnless   5. *L. panicoides*
4. Lemmas up to 4 mm long, pale green .... 3. *L. fascicularis*
4. Lemmas 5–8 mm long, purplish to lead-colored .... 4. *L. acuminata*

1. L. filiformis (Lam.) Beauv. Red Sprangletop. July–Sept. Low, sandy soil, particularly along rivers; occasional in the s. two-fifths of the state.

2. L. attenuata (Nutt.) Steud. Sprangletop. July–Sept. Sandy shores; occasional in the s. tip of the state. *L. filiformis* (Lam.) Beauv.—J.

3. L. fascicularis (Lam.) Gray. Salt Meadow Grass. July–Sept. Wet soil; scattered in the s. three-fifths of the state; also Cook Co. *Diplachne fascicularis* (Lam.) Beauv.—F.

4. L. acuminata (Nash) Mohlenbr. Salt Meadow Grass. July–Sept. Native of the w. U.S.; adventive along a railroad and in saline waste ground; Cook, DuPage, Grundy, Kane, Kankakee, Kendall, Lake, and Will cos. *L. fascicularis* (Lam.) Gray var. *acuminata* (Nash) Gl.—G; *Diplachne acuminata* Nash—F.

5. L. panicoides (Presl) Hitchc. Salt Meadow Grass. July–Sept. Low areas, rare; Calhoun and Pike cos. *Diplachne halei* Nash—F.

71. *Gymnopogon* Beauv.   Beardgrass

1. G. ambiguus (Michx.) BSP. Beardgrass. July–Oct. Sandy or gravelly soil in open areas, very rare; Pope Co.

72. *Schedonnardus* Steud.   Tumble Grass

1. S. paniculatus (Nutt.) Trel. Tumble Grass. July–Sept. Salt licks, very rare; Hancock Co.

73. *Cynodon* Rich.   Bermuda Grass

1. C. dactylon (L.) Pers. Bermuda Grass. July–Oct. Native of Europe; adventive in lawns and waste ground; occasional in the s. half of the state, much less common in the n. half.

74. *Chloris* Swartz   Finger Grass

1. Blades 1–3 mm broad, obtuse; awn of fertile lemma 5–9 mm long; empty lemma 1, with an awn 3.5–5.0 mm long; plants to 40 cm tall ....
.... 1. *C. verticillata*
1. Blades 3–5 mm broad, tapering to a long, fine point; awn of fertile lemma 1–5 mm long; empty lemmas 2, the upper awnless; plants usually over 1 m tall .... 2. *C. gayana*

1. C. verticillata Nutt. Windmill Grass. June–Sept. Native of the w. U.S.; adventive in lawns; scattered in Ill.

2. C. gayana Kunth. Finger Grass. June–Sept. Native of Africa; adventive in a strip mine; Perry Co.

75. Bouteloua Lag.   Grama Grass

1. Spikes 10–50, spreading or nodding, to 2 cm long, falling entire at maturity; spikelets 7–10 mm long .... 1. *B. curtipendula*
1. Spikes 1–3 (–6), straight or curved backward, 2–5 cm long, the florets falling from the glumes; spikelets 5–6 mm long 2
2. Rachis of inflorescence projecting 2–5 mm beyond last spike; second glume papillose-hirsute on keel; empty lemma glabrous at base; blades papillose-pubescent .... 2. *B. hirsuta*
2. Rachis of inflorescence not projecting; second glume sparsely pilose-papillose on keel; empty lemma long-villous at base; blades glabrous or scabrous .... 3. *B. gracilis*

1. B. curtipendula (Michx.) Torr. Sideoats Grama. July–Sept. Prairies, dry hills; occasional in the n. half of the state, rather common on the bluffs bordering the Mississippi River from JoDaviess to Alexander cos.

2. B. hirsuta Lag. Grama Grass. July–Sept. Prairies; confined to the nw. one-fourth of the state.

3. B. gracilis (HBK.) Lag. Blue Grama. July–Sept. Sand flats; JoDaviess Co.; also adventive at a strip mine in Williamson Co.

76. Buchloë Engelm.   Buffalo Grass

1. B. dactyloides (Nutt.) Engelm. Buffalo Grass. Aug.–Sept. Prairies, rare; Peoria and Winnebago cos.

77. Spartina Schreb.   Cord Grass

1. S. pectinata Lind. Cord Grass. June–Sept. Wet prairies, marshes; occasional throughout the state, although more common in the n. cos.

78. Distichlis Raf.   Salt Grass

1. D. stricta (Torr.) Rydb. Salt Grass. July–Oct. Native of w. N. Am.; adventive along railroads and in saline waste areas; Champaign, Cook, DeKalb, Kane, and Kendall cos.

79. Aristida R. Br.   Three Awn

1. Awns of lemma twisted and united into a basal column 8–15 mm long; glumes 20–30 mm long .... 1. *A. tuberculosa*
1. Awns of lemma free to base or united into a basal column up to 3 mm long; glumes up to 20 mm long (except *A. oligantha* and *A. ramosissima*) 2
2. Central awn of most lemmas at least 20 mm long 3
2. Central awn of lemma up to 20 mm long, usually considerably shorter   8
3. Lateral awns of lemma at least 12 mm long 4
3. Lateral awns of lemma up to 6 mm long, or even absent .... 7. *A. ramosissima*
4. Awns of lemma united at base into a column 1–3 mm long, the awns articulated with the summit of the lemma; sheaths villous on the margins .... 2. *A. desmantha*

4. Awns of lemma free to base, the awns not articulated with the summit of the lemma; sheaths glabrous or nearly so 5
5. All awns of lemma 35–70 mm long; glumes 12–32 mm long; lemma 12–20 mm long .... 3. *A. oligantha*
5. Central awn of lemma 15–33 mm long, the lateral awns 12–24 mm long; glumes 6–14 mm long; lemma 5.5–8.5 mm long 6
6. Perennial; first glume longer than second .... 4. *A. purpurascens*
6. Annual; first glume equal to second or shorter 7
7. Glumes essentially equal in length; central awn of lemma longer than lateral awns .... 5. *A. intermedia*
7. Second glume longer than first glume; awns of lemma equal in length ....6. *A. necopina*
8. Lateral awns 12–20 mm long; central awn curved or divergent at base, not coiled 9
8. Lateral awns up to 12 mm long; central awn coiled at base (or bent conspicuously near base in *A. longespica*) 11
9. Central awn of lemma conspicuously bent near base; glumes obscurely 3-nerved .... 8. *A. longespica*
9. Awns of lemma not conspicuously bent near base; glumes sharply l-nerved 10
10. Glumes nearly equal in length; central awn of lemma longer than the lateral ones .... 5. *A. intermedia*
10. Second glume longer than first glume; awns of lemma equal ....6. *A. necopina*
11. Glumes essentially equal, the second 4–10 mm long; lemma 4.0–8.5 mm long; inflorescence usually reduced to a raceme 12
11. Second glume longer than first, usually by at least 2 mm, the second 7–15 mm long; lemma 7.5–10.5 mm long; inflorescence a slender panicle 13
12. Central awn conspicuously bent at base, not coiled, 6.5–20.0 mm long; lateral awns 3–15 mm long; glumes obscurely 3-nerved 8. *A. longespica*
12. Central awn loosely coiled at base, 3–10 mm long; lateral awns 0.7–3.3 mm long; glumes 1-nerved .... 9. *A. dichotoma*
13. Lateral awns of lemma 5–12 mm long; central awn 10–19 mm long .... 10. *A. basiramea*
13. Lateral awns of lemma 2–4 mm long; central awn 7.0–12.5 mm long .... 11. *A. curtissii*

1. A. tuberculosa Nutt. Needle Grass. Aug.–Oct. Sandy soil; occasional in the n. half of the state; also Union Co.

2. A. desmantha Trin. & Rupr. Three Awn. Aug.–Sept. Sandy soil, very rare; Cass, Mason, and Morgan cos.

3. A. oligantha Michx. Three Awn. Aug.–Oct. Dry soil of fields and woods; along railroads; common throughout the state.

4. A. purpurascens Poir. Arrowfeather. Aug.–Sept. Sandy soil; occasional throughout the state.

5. A. intermedia Scribn. & Ball. Three Awn. Sept.–Oct. Sandy soil, rare; Grundy, Henry, Lake, Lee, McHenry cos.

6. A. necopina Shinners. Three Awn. Sept.–Oct. Sandy soil, rare; Lee Co.

7. A. ramosissima Engelm. Slender Three Awn. Aug.–Sept. Lateral awns 0.5–0.6 mm long. Dry soil of fields; occasional in the s. half of the state, absent in the n. half.

f. uniaristata (Gray) Mohlenbr. Slender Three Awn. Aug.–Sept. Lateral awns absent, or less than 0.5 mm long. Dry soil of fields, rare; Marion and St. Clair cos.

8. A. longespica Poir. Three Awn. Aug.–Oct. Lateral awns 3–4 mm long; central awn 6.5–13.0 mm long; glumes 4–6 mm long. Sandy soil, particularly in fields and along highways; occasional throughout the state, although rare or absent in the n. cent. cos.

var. geniculata (Raf.) Fern. Three Awn. Aug.–Sept. Lateral awns 4–15 mm long; central awn 10–20 mm long; glumes 5–9 mm long. Sandy soil, rare; Henry and Jackson cos.

9. A. dichotoma Michx. Three Awn. Aug.–Oct. Dry fields and along highways; occasional in the s. half of the state, less common in the n. half.

10. A. basiramea Engelm. Three Awn. Aug.–Oct. Dry, sandy soil, not common; confined to the n. four-fifths of the state; also Gallatin and Jackson cos.

11. A. curtissii (Gray) Nash. Three Awn. Aug.–Oct. Dry soil, rare; Massac and Ogle cos. A. dichotoma Michx. var. curtissii Gray—F.

80. Arundinaria Michx. Cane

1. A. gigantea (Walt.) Chapm. Giant Cane. April–May. Low ground; occasional in the s. one-third of the state, absent elsewhere.

81. Leersia Sw. Cut Grass

1. Spikelets broadly rounded, 3–4 mm broad, over half as wide as long ... . 1. *L. lenticularis*
1. Spikelets oblongoid, 1–2 mm broad, less than half as wide as long .. 2
2. Sheaths conspicuously retrorse-scabrous; blades spinulose on the margins; lowest panicle branches whorled; stamens 3 . 2. *L. oryzoides*
2. Sheaths glabrous or scaberulous; blades scaberulous; lowest panicle branches solitary; stamens 1–2 .... 3. *L. virginica*

1. L. lenticularis Michx. Catchfly Grass. Aug.–Oct. Low woods, swamps, marshes; occasional throughout the state, except for the extreme n. cos.

2. L. oryzoides (L.) Swartz. Rice Cutgrass. Aug.–Oct. Low, moist soil; occasional throughout the state.

3. L. virginica Willd. White Grass. July–Sept. Moist woodlands; common throughout the state. (Including var. ovata [Poir.] Fern.)

82. Zizania L. Wild Rice

1. Z. aquatica L. Wild Rice. June–Sept. Pistillate lemma scabrous, slenderly nerved; aborted spikelets up to 1 mm broad, subulate, tapering into the awn. Shallow water, not common; scattered in Ill., except for the s. cos.

var. interior Fassett. Wild Rice. June–Sept. Pistillate lemma glabrous, broadly nerved; aborted spikelets 1.5–2.0 mm broad, linear, abruptly tapering into the awn. Shllow water, rare; Cook, Lake, and Union cos. Z. aquatica L.— J.

83. Zizaniopsis Doell & Aschers

1. Z. miliacea (Michx.) Doell & Aschers. Southern Wild Rice. July–Oct. Edge of lake, rare; Montgomery Co.

84. Phragmites Trin.   Reed

1. P. australis Trin. Reed. July–Sept. Moist soil; occasional in the n. half of the state, uncommon in the s. half. *P. communis* Trin.—F, G, J.

85. Arundo L.

1. A. donax L. Giant Reed. Sept.–Oct. Native along the Mediterranean Sea; adventive along a gravel road in Pope Co.

86. Danthonia Lam. & DC.   Wild Oat Grass

1. D. spicata (L) Beauv. Poverty Oat Grass; Curly Grass. May–Aug. Dry woods and bluffs; occasional to common throughout the state.

87. Chasmanthium Link   Sea Oats

1. C. latifolium (Michx.) Yates. Sea Oats. July–Oct. Moist soil; common in the s. half of the state, rare in the n. half. *Uniola latifolia* Michx.—F, G, J. The type is from Ill.

26. CYPERACEAE

1. Flowers bisexual; achenes not enclosed in a sac-like structure (perigynium) nor white in color (except for *Cyperus erythrorhizos*) 2
1. Flowers unisexual; achenes either enclosed in a sac-like structure (perigynium) or white or buff in color 14
2. Spikelets 3- or more-flowered (1-flowered in *Cyperus densicaespitosus* which has a strong sweet odor) 3
2. Spikelets 1- to 2-flowered 13
3. Spikelets flattened; scales 2-ranked 4
3. Spikelets not flattened; scales spirally arranged 5
4. Inflorescence terminal; bristles absent; achene without a tubercle......
.... 1. *Cyperus*
4. Inflorescence axillary; bristles present; achene with a tubercle ....
.... 2. *Dulichium*
5. Achene crowned by a tubercle 6
5. Achene without a tubercle 7
6. Spikelet one; leaves absent .... 3. *Eleocharis*
6. Spikelets 2-many; leaves present .... 4. *Bulbostylis*
7. Flowers not subtended by bristles or dilated scales (rudimentary scales only in *Lipocarpha*) 8
7. Flowers subtended by bristles or dilated scales although, in some species of *Scirpus*, the scales sometimes falling away early 10
8. Bract subtending the inflorescence 1 .... 8. *Scirpus*
8. Bracts subtending the inflorescence 2 or more 9

9. Styles the same thickness throughout .... 8. *Scirpus*
9. Styles dilated toward base .... 5. *Fimbristylis*
10. Flowers subtended by minute, dilated rudimentary perianth scales .....
.... 6. *Lipocarpha*
10. Flowers subtended by ovate-oblong scales or by bristles 11
11. Flowers subtended by ovate-oblong scales .... 7. *Fuirena*
11. Flowers subtended by bristles 12
12. Bristles pale or dark brown, reddish, or blackish at maturity ....
.... 8. *Scirpus*
12. Bristles white at maturity, forming "cottony" heads .... 9. *Eriophorum*
13. Achene crowned with a tubercle; bristles subtending the flower ....
.... 10. *Rhynchospora*
13. Achene without a tubercle; bristles absent .... 11. *Cladium*
14. Achene white or buff, not enclosed in a perigynium .... 12. *Scleria*
14. Achene not white, enclosed in a perigynium .... 13. *Carex*

1. Cyperus L.   Galingale

1. Stigmas 2; achenes lenticular 2
1. Stigmas 3; achenes trigonous 6
2. Spikelets 1-flowered; inflorescence of 1–3 sessile heads; plants with a
strong sweet odor when fresh .... 1. *C. densicaespitosus*
2. Spikelets 5- to 35-flowered; inflorescence of 1-several radiating sessile
spikelets and usually 1-several rays; plants without a distinctive sweet
odor when fresh 3
3. Achenes black, nearly as broad as long, with horizontal wrinkles; scales
straw-colored .... 2. *C. flavescens*
3. Achenes drab or gray, longer than broad, without horizontal wrinkles;
scales usually suffused with purple 4
4. Styles divided nearly to base, persistent and conspicuously exserted to 4
mm from scales .... 3. *C. diandrus*
4. Styles divided to about the middle, falling away early, hidden by the
scales or exserted to 2 mm from scales 5
5. Scales closely appressed, strongly suffused with purple (straw-colored
in *C. rivularis* f. *elutus*) .... 4. *C. rivularis*
5. Scales with tips somewhat spreading, the spikelets appearing serrate,
straw-colored or purple on the margins .... 5. *C. filicinus*
6. Scales with strongly recurved tips .... 6. *C. aristatus*
6. Scales with tips either appressed or slightly spreading 7
7. Cluster of spikelets spherical or globose, with spikelets radiating in all
directions 8
7. Cluster of spikelets hemispherical, cylindrical, ellipsoid or lanceoloid,
but not spherical or globose 11
8. Scales appressed, but with the tips shortly recurved; spikes to 8 mm
across 9
8. Scales appressed to spreading, their tips straight; some of the heads
over 1 cm across 10
9. Achenes linear, 1.0–1.4 mm long; perennial plants at least 35 cm tall at
maturity .... 7. *C. pseudovegetus*
9. Achenes ellipsoid to oblongoid, 0.5–1.0 mm long; annual plants never
more than 35 cm tall .... 8. *C. acuminatus*
10. Scales appressed; spikelets 2- to 3-flowered .... 9. *C. ovularis*
10. Scales spreading; spikelets 5- to several-flowered ... 10. *C. grayioides*
11. Spikelets arising from a common point on the axis 12

1. C. densicaespitosus Mattf. & Kükenth. July–Sept. Moist, open soil; restricted to the one-half of the state. *C. tenuifolius* (Steud.) Dandy.—F.

2. C. flavescens L. July–Oct. Wet open soil, rare in the s. one-sixth of the state; also Coles, McDonough, and Peoria cos. *C. flavescens* L. var. *poaeformis* (Pursh) Fern.—F.

3. C. diandrus Torr. June–Oct. Wet ground along banks and shores, not common; scattered in the w. half of Ill.

4. C. rivularis Kunth. July–Oct. Scales strongly suffused with red-brown or purple. Wet ground along banks and shores; throughout the state, but rare in the s. one-third. (Including f. *elongatus* Boeckl.)

f. elutus (C. B. Clarke) Kükenth. July–Oct. Scales stramineous throughout, except for the greenish midrib.

5. C. filicinus Vahl. July–Sept. Native to the s. U.S.; adventive in wet, roadside ditch; Jackson Co.

6. C. aristatus Rottb. May–Oct. Moist soil; scattered throughout the state. *C. inflexus* Muhl.—F, J.

7. C. pseudovegetus Steud. June–Oct. Moist soil; restricted to the s. one-fourth of the state. *C. virens* Gray—F.

8. C. acuminatus Torr. & Hook. June–Oct. Wet ground; scattered throughout the state except for very northernmost cos.

9. C. ovularis (Michx.) Torr. Round-headed Sedge. June–Sept. Dry sandy woods, old fields; common in the s. one-third of the state; also Peoria Co. (Including var. *robustus* Britt., var. *sphaericus* Boeckl., and var. *wolfii* [Wood] Kükenth.)

10. C. grayioides Mohlenbr. Aug. Sand prairies and blowouts, rare; Mason and Whiteside cos.

11. C. compressus L. Sept. Native of the s. U.S. and trop. Am.; adventive in roadside ditch; Lawrence Co.

12. C. schweinitzii Torr. June–Sept. Sandy soil; occasional in the n. half of the state; adventive in Jackson Co.

13. C. houghtonii Torr. June–Sept. Sandy soil, rare; Boone, Cook, Grundy, Kankakee Lake, and Will cos.

14. C. filiculmis Vahl. May–Oct. Plants with a central sessile glomerule and 3–6 (–8) well-developed rays; achenes (1.7–) 1.8–2.2 mm long. Rocky or sandy soil; common in the n. half of the state, s. to Madison Co. (Including *C.* X *mesochorus* Geise.)

var. macilentus Fern. May–Oct. Plants with only a central sessile glomerule or occasionally with 1–2 rays; achenes 1.2–1.8 (–2.0) mm long. Sandy soil; common in the n. half of the state, rare southward.

15. C. erythrorhizos Muhl. Aug.–Oct. Moist, often sandy soil; occasional throughout the state.

16. C. iria L. July–Oct. Native of Europe and Asia; adventive in wet meadows; Alexander Co.

17. C. engelmannii Steud. July–Oct. Wet ground; rare but scattered in Ill.

18. C. lancastriensis Porter. Aug.–Oct. Moist, sandy woods, rare; Massac and Pulaski cos.

19. C. esculentus L. Nut-grass. June–Oct. Spikelets to 15 mm long, 1.5–3.0 mm broad. Moist, frequently cultivated soil, commom; in every co.

var. leptostachyus Boeckl. June–Oct. Moist, often cultivated soil; scattered in Ill.

20. C. ferruginescens Boeckl. Aug.–Oct. Rich moist soil; scattered throughout the state.

21. C. strigosus L. June–Oct. Wet gorund, common; in every co. (Including f. *robustior* Kunth and var. *stenolepis* [Torr.] Kükenth.)

2. Dulichium Rich. ex Pers.   Three-way Sedge

1. D. arundinaceum (L.) Britt. Three-way Sedge. July–Oct. Swamps and low ground in woods; throughout the state, but not common.

3. Eleocharis R. Br.   Spike Rush

1. Spikelet about the same thickness as the culm; plants robust, often 1 m tall or taller 2
1. Spikelet usually conspicuously thicker than the culm; plants usually less than 1 m tall (except *E. rostellata*) 3
2. Culms round, septate by cross partitions; bristles shorter than the achene, or lacking . . . . 1. *E. equisetoides*

2. Culms 4-sided, sharply angled, not septate; bristles slightly longer than the achene .... 2. *E. quadrangulata*
3. Style 2-cleft; achenes obovoid-lenticular (rarely sometimes tufted in *E. olivacea*) 4
3. Style 3-cleft; achenes usually trigonous 9
4. Plants perennial, with rhizomes 5
4. Plants annual, with fibrous roots 8
5. Achenes at most 1.0 mm long, the small, conical tubercle $^1\!/_6$–$^1\!/_8$ as long .... 3. *E. olivacea*
5. Achenes averaging somewhat longer than 1.0 mm, the conical tubercle $^1\!/_2$–$^1\!/_4$ as long 6
6. Basal scales 1–3, never completely encircling the culm; culms averaging 1.3 mm wide 7
6. Basal scale 1, suborbicular, completely encircling the culm; culms averaging 0.6 mm wide .... 6. *E. erythropoda*
7. Basal sheaths of mature culms with prominent V-shaped sinuses; culms often soft and inflated .... 4. *E. smallii*
7. Basal sheaths truncate to slightly oblique at apex; culms usually rigid .. .. 5. *E. macrostachya*
8. Achenes lustrous black or purple; plants usually less than 15 cm tall ... . 7. *E. caribaea*
8. Achenes yellow to deep brown; plants usually 15–50 cm tall   8. *E. obtusa*
9. Tubercle conspicuously differentiated from achene; achenes up to 2 mm long 10
9. Tubercle confluent with top of achene, long-conical; achenes 2–3 mm long 14
10. Achenes with 8–19 longitudinal ribs 11
10. Achenes pitted or with papillae 12
11. Scales 1.5–2.2 mm long; culms 0.1–0.5 mm wide, capillary .... 9. *E. acicularis*
11. Scales 2.5–3.0 mm long; culms about 1 mm wide, often inrolled .... 10. *E. wolfii*
12. Tubercle not subulate; achene 1.0–1.2 mm long (including the tubercle); perennials 13
12. Tubercle subulate; achene about 1.5 mm long (including the tubercle); annual .... 13. *E. intermedia*
13. Culm with 5 vascular bundles and appearing (4-) 5-angled; mature achene usually olivaceous and warty .... 11. *E. tenuis*
13. Culm with 6–14 vascular bundles and appearing low-angular or compressed; mature achene usually yellow, brown, or orange-brown, reticulate or slightly warty .... 12. *E. elliptica*
14. Stems flat, to 2 mm wide; scales elliptic, obtuse; plants tufted .... 14. *E. rostellata*
14. Stems usually 3-angled, less than 1 mm wide; scales lanceolate, acute; plants with stoloniferous rhizomes .... 15. *E. pauciflora*

    1. E. equisetoides (Ell.) Torr. Horsetail Spike Rush. July–Oct. Wet ground or standing water, very rare; Cook Co.
    2. E. quadrangulata (Michx.) Roem. & Schultes. Square-stemmed Spike Rush. June–Oct. Shallow water in ponds and lakes, rare; primarily in the s. one-fourth of the state. (Including var. *crassior* Fern.)
    3. E. olivacea Torr. July–Sept. Wet sands, very rare; Cook Co.
    4. E. smallii Britt. June–Sept. Edge of swamps, sloughs, ponds, streams; occasional throughout the state. *E. palustris* (L.) Roem. & Schultes—G.

5. E. macrostachya Britt. June–Aug. Edge of swamps, sloughs, ponds, streams; occasional throughout the state. *E. palustris* (L.) Roem. & Schultes—G.

6. E. erythropoda Steud. June–Sept. Wet soil; common in the n. half of the state, rare elsewhere. *E. calva* (Gray) Torr.—F, G, J.

7. E. caribaea (Rottb.) Blake. June–Sept. Wet sand, very rare; Cook Co. (Including var. *dispar* [Hill] Blake.) *E. geniculata* (L.) Roem. & Schultes—F, G, J.

8. E. obtusa (Willd.) Schult. May–Oct. Tubercle more than ⅔ the width of the achene and ¼–½ the height of the achene. Wet ground; common throughout the state.

var. detonsa (Gray) Drap. & Mohlenbr. May–Oct. Tubercle more than ⅔ the width of the achene and up to ¼ the height of the achene. Wet ground; occasional throughout the state. *E. engelmannii* Steud.—F, G, J.

var. ovata (Roth) Drap. & Mohlenbr. May–Oct. Tubercle ½–⅔ the width of the achene. Wet ground; scattered but rare throughout the state. *E. ovata* (Roth) Roem. & Shultes—F, G, J.

9. E. acicularis (L.) Roem. & Schultes. July–Oct. Low wet ground; occasional throughout the state. (Including var. *gracilescens* Svenson and f. *inundata* Svenson.)

10. E. wolfii (Gray) Patterson. May–July. Wet ground; rare and scattered in Ill.

11. E. tenuis (Willd.) Schult. var. verrucosa (Svenson) Svenson. May–Sept. Low wet ground; moist crevices on dry bluffs; occasional throughout the state. *E. tenuis* (Willd.) Schult.—J.

12. E. elliptica Kunth. May–July. Culm with 6–8 vascular bundles and appearing with 6–8 low angles. Low areas; occasional throughout the state.

var. compressa (Sull.) Drap. & Mohlenbr. May–July. Culm with 9–14 vascular bundles and appearing compressed. Low areas; scattered throughout the state. *E. compressa* Sull.—F, G, J.

13. E. intermedia (Muhl.) Schult. July–Oct. River banks and swampy areas; fairly common along the Illinois River, rare or absent elsewhere.

14. E. rostellata (Torr.) Torr. July–Sept. Marshy, calcareous soils, rare; Lake and McHenry cos.

15. E. pauciflora (Lightf.) Link. July–Oct. Wet areas; restricted to the extreme ne. cos.

4. Bulbostylis (Kunth) C. B. Clarke

1. B. capillaris (L.) C. B. Clarke. July–Oct. Moist, sandy soil; occasional throughout the state. (Including var. *crebra* Fern. and var. *isopoda* Fern.)

5. Fimbristylis Vahl

1. Style branches 2; achene lenticular 2
1. Style branches 3; achene trigonous .... 4. *F. autumnalis*
2. Spikelets in a close head; achene 0.4–0.5 mm long; cespitose annuals
.... 1. *F. vahlii*
2. Spikelets on elongated rays, cymose; achene 1.0–1.8 mm long; tufted perennials (or annuals) with hardened bases 3

3. Outer scales puberulent; achene 1.3–1.7 mm long, reticulate ....
.... 2. *F. puberula* var. *drummondii*
3. All scales glabrous; achene 1.0–1.3 mm long, with horizontal and transverse ribs, or sometimes irregularly tuberculate .... 3. *F. baldwiniana*

1. F. vahlii (Lam.) Link. July–Oct. Moist soil, rare; Cass Co.
2. F. puberula (Michx.) Vahl var. drummondii (Boeckl.) Ward. May–
July. Moist or dry soil, rare; restricted to the n. half of the state; also
St. Clair Co. *F. drummondii* Boeckl.—F, J; *F. caroliniana* (Lam.)
Fern.—G.
3. F. baldwiniana (Schult.) Torr. June–Oct. Moist soil, rare; Alexander, Johnson, and Massac cos. *F. annua* (All.) Roem. & Schultes—G.
4. F. autumnalis (L.) Roem. & Schultes. June–Oct. Moist or rather
dry, sandy soil; occasional throughout the state. (Including var. *mu-
cronulata* [Michx.] Fern.)

6. *Lipocarpha R. Br.*

1. L. maculata (Michx.) Torr. June–Sept. Around a pond, very rare;
Cass Co.

7. *Fuirena Rottb. Umbrella Grass*

1. F. scirpoides Michx. July. Along edge of lake, very rare; Hamilton
Co.

8. *Scirpus L. Bulrush*

1. Inflorescence appearing lateral; lowest bract culm-like 2
1. Inflorescence appearing terminal; bracts leaf-like or scale-like 11
2. Achenes 1.5–4.0 mm long 3
2. Achenes 1.0–1.5 mm long .... 10. *S. koilolepis*
3. Spikelets sessile or subsessile; culm 3-angled 4
3. Spikelets (or most of them) on elongated pedicels; culm terete 9
4. Achene black (immature achenes whitish in *S. hallii*); plants with fibrous
roots 5
4. Achene white, stramineous, or pale brown (rarely nearly black in *S.
americanus*); plants with rhizomes 7
5. Achene smooth or pitted; scales obtuse or mucronulate 6
5. Achene transversely corrugated; scales awn-tipped .... 3. *S. hallii*
6. Achene smooth; culm scarcely angled .... 1. *S. smithii*
6. Achene pitted; culm 3-angled .... 2. *S. purshianus*
7. Scales entire; spikelets greenish or stramineous; culms slender, at most
2 mm thick; rhizome slender and weak 8
7. Scales ciliate, 2-cleft; spikelets reddish-brown; culms rather stout,
mostly over 2 mm thick; rhizome stout and strong ... 6. *S. americanus*
8. Culms 3-angled; leaves 2–3 .... 4. *S. torreyi*
8. Culms terete; leaves 4 or more .... 5. *S. subterminalis*
9. Style 2-cleft; spikelets ovoid, reddish-brown; achene plano-convex . 10
9. Style 3-cleft; spikelets usually ellipsoid, pale brown; achene trigonous
.... 9. *S. heterochaetus*
10. Scales viscid, red-dotted, mucronulate .... 7. *S. acutus*
10. Scales glabrous, not red-dotted, awned .... 8. *S. validus*

11. Involucral leaves 2 or more 12
11. A true involucre absent, but represented by an elongated outer scale of the spikelet 23
12. Bristles firm or none, straight, shorter than to twice as long as the achene 13
12. Bristles slender, curled, much longer than the achene 15
13. Spikelets 1–2 mm broad .... 11. *S. micranthus*
13. Spikelets 5–10 mm broad 14
14. Achene sharply trigonous, 4–5 mm long; bristles stout, persistent ......
.... 12. *S. fluviatilis*
14. Achene obtusely, if at all, trigonous, 2.5–4.0 mm long; bristles weak, deciduous .... 13. *S. paludosus*
15. Stolons present (sometimes apparently absent in *S. polyphyllus*); bristles concealed by the scales at maturity, retrorsely barbed 16
15. Stolens absent; bristles usually exceeding the scales at maturity, smooth or sparingly hairy 20
16. Culms with 10–20 leaves; scales reddish-brown ...... 14. *S. polyphyllus*
16. Culms with 2–10 leaves; scales brown or black 17
17. All sheaths greenish; bristles (if present) glabrous near base 18
17. Lowest sheaths reddish; bristles hairy throughout . 18. *S. microcarpus*
18. Bristles 0–3, shorter than the achenes .... 15. *S. georgianus*
18. Bristles usually 5 or 6, shorter than to slightly longer than the achenes
.... 19
19. Lower leaf blades and sheaths usually nodulose-septate; scales mostly brownish; longer bristles frequently exceeding the achenes ....
.... 16. *S. atrovirens*
19. Lower leaf blades and sheaths nearly smooth; scales mostly blackish; longer bristles usually shorter than or about equalling the achenes ....
.... 17. *S. hattorianus*
20. Bristles at maturity equalling or shorter than the scales; scales usually with prominent green midribs .... 19. *S. pendulus*
20. Bristles at maturity much longer than the scales; scales usually with inconspicuous midribs 21
21. Spikelets and involucels reddish-brown .... 20. *S. cyperinus*
21. Spikelets and involucels whitish-brown or blackish 22
22. Spikelets and involucels whitish-brown .... 21. *S. pedicellatus*
22. Spikelets and involucels blackish .... 22. *S. atrocinctus*
23. Culms smooth, round; bristles longer than the achenes ....
.... 23. *S. cespitosus*
23. Culms rough, triangular; bristles about as long as the achenes ....
.... 24. *S. verecundus*

1. S. smithii Gray. July–Oct. Wet shores, very rare; Cass Co.

2. S. purshianus Fern. Aug.–Oct. Shores of lakes; rare in the s. two-thirds of the state. (Including f. *williamsii* [Fern.] Fern.) *S. smithii* Gray var. *williamsii* (Fern.) Beetle—G.

3. S. hallii Gray. Aug.–Oct. Shores of ponds, rare; Cass, Mason, and Menard cos.

4. S. torreyi Olney. July–Sept. Shores of ponds, very rare; St. Clair Co.

5. S. subterminalis Torr. July–Oct. Shallow water, very rare; Cook and Lake cos.

6. S. americanus Pers. June–Sept. Along shores, in marshes; scattered throughout the state.

7. S. acutus Muhl. Aug.–Sept. Spikelets on divergent rays to 5 cm long. Shallow water; occasional throughout the state.

f. congestus (Farw.) Fern. Aug.–Sept. Spikelets all sessile, forming a dense glomerule. Shallow water, rare; Lake Co.

8. S. validus Vahl. June–Sept. Swamps, marshes, shallow water; common in the n. half of the state, less common elsewhere. *S. validus* Vahl var. *creber* Fern.—F. (Including var. *condensatus* [Peck] Beetle.)

9. S. heterochaetus Chase. July–Sept. Shores and swamps; scattered throughout the state, but not common.

10. S. koilolepis (Steud.) Gl. May–June. Pastures, fields, rare; confined to the s. tip of the state.

11. S. micranthus Vahl. Aug.–Oct. Scales narrowly obovate, spreading. Sandy banks; scattered throughout the state. *Hemicarpha micrantha* (Vahl) Pax—F, G, J. (Including var. *minor* [Schrad.] Friedland.)

var. drummondii (Nees) Mohlenbr. Aug.–Oct. Scales broadly obovate to suborbicular, appressed. Wet, sandy soil; Cook and Lake cos. *Hemicarpha drummondii* Nees—F, G, J.

12. S. fluviatilis (Torr.) Gray. July–Sept. Margins of streams and lakes; occasional throughout the state.

13. S. paludosus A. Nels. Bayonet-grass. July–Sept. Shores and marshes, rare; Cook Co. *S. maritimus* var. *paludosus* (A. Nels.) Gl.—G.

14. S. polyphyllus Vahl. July–Sept. Low woodlands; rare and scattered in Ill.

15. S. georgianus Harper. June–Sept. Moist soil; occasional throughout the state. *S. atrovirens* Willd. var. *georgianus* (Harper) Fern.—F, G, J.

16. S. atrovirens Willd. June–Aug. Wet soil, common; throughout the state.

17. S. hattorianus Mak. June–Aug. Ditches, very rare; Kankakee Co.

18. S. microcarpus Presl. Aug.–Sept. Marshes and swamps, rare; Lake co. *S. rubrotinctus* Fern.—F, G, J.

19. S. pendulus Muhl. June–Aug. Low woods and along streams, common; probably in every co. *S. lineatus* Michx.—F, G, J.

20. S. cyperinus (L.) Kunth. Aug.–Oct. Swamps and marshes; common throughout the state. (Including var. *pelius* Fern., var. *rubricosus* [Fern.] Gilly, *S. eriophorum* Michx.)

21. S. pedicellatus Fern. July–Aug. Wet soil, rare; Cook and Pope cos. *S. cyperinus* (L.) Kunth—J.

22. S. atrocinctus Fern. June–July. Swamps and boggy areas, very rare; Lake Co.

23. S. cespitosus L. var. callosus Bigel. June–Aug. Boggy situations, rare; Lake and McHenry cos. *S. cespitosus* L.—J.

24. S. verecundus Fern. April. Cherty slopes, very rare; Union Co.

9. *Eriophorum* L. Cotton Sedge

1. Leaves up to 1.5 mm wide, channeled throughout; involucral bract 1 ... 2
1. Leaves 1.5–8.0 mm wide, channeled only above the middle; involucral bracts 2–5 3
2. Culms glabrous throughout; sheath of leaf longer than blade; achene 1.5–2.0 mm long .... 1. *E. gracile*
2. Culms scabrous above; sheath of leaf shorter than blade; achene 2.5–3.1 mm long .... 2. *E. tenellum*

3. Stamens 3; scales one-nerved 4
3. Stamen 1; scales several-nerved .... 5. *E. virginicum*
4. Peduncles glabrous; anthers 2.5–5.0 mm long .... 3. *E. angustifolium*
4. Peduncles puberulent; anthers 1.0–1.3 mm long ... 4. *E. viridi-carinatum*

  1. E. gracile Koch. May–July. Bogs, rare; Peoria Co. *E. angustifolium* Honck.—J.
  2. E. tenellum Nutt. June–Sept. Bogs, rare; Cass Co. *E. angustifolium* Honck.—J.
  3. E. angustifolium Honck. June–Aug. Calcareous fens; scattered in the n. one-fourth of Ill. (Including var. *majus* Schultz.)
  4. E. viridi-carinatum (Engelm.) Fern. May–Aug. Swamps and bogs, rare; Lake and Rock Island cos.
  5. E. virginicum L. Aug.–Oct. Bogs, rare; Lake Co. (Including f. *album* [Gray] Wieg.)

10. Rhynchospora Vahl Beaked Rush

1. Leaves 6–20 mm broad; achene 4.7–5.5 mm long, 2.5–3.0 mm broad; beak 12–20 mm long 2
1. Leaves 0.3–6.0 mm broad (7.0 rarely in *R. glomerata*); achene 1.2–2.6 mm long, 0.8–1.6 mm broad; beak 0.3–1.6 mm broad; beak 0.3–1.8 mm long 3
2. At least one of the usually 6 bristles longer than the achene ....
.... 1. *R. macrostachya*
2. None of the usually 3–5 bristles longer than the achene 2. *R. corniculata*
3. Bristles antrorsely hairy, rarely smooth or entirely lacking; tubercle 0.3–0.6 mm high, broadly deltoid .... 3. *R. globularis*
3. Bristles retrorsely hairy, rarely smooth; tubercle 0.6–1.8 mm long, conical 4
4. Spikelets lanceoloid; bristles 3.6–4.3 mm long; leaves capillary, at most 0.4 mm broad .... 4. *R. capillacea*
4. Spikelets ovoid to lance-ovoid; bristles 2.0–3.5 mm long (occasionally to 4.0 mm long in *R. glomerata*); leaves 0.5–7.0 mm broad 5
5. Achene with a conspicuous margin, smooth 6
5. Achene emarginate, or nearly so, slightly rugulose .... 7. *R. alba*
6. Achene 1.2–1.4 mm wide, with broad shoulders; bristles 3–4 mm long ..
.... 5. *R. glomerata*
6. Achene 0.8–1.2 mm wide, pyriform; bristles 2.0–2.8 mm long ....
.... 6. *R. capitellata*

  1. R. macrostachya Torr. July–Oct. Swampy woods, very rare; Pulaski Co.
  2. R. corniculata (Lam.) Gray. June–Sept. Swamps and low roadside ditches; restricted to the s. tip of Ill. (Including var. *interior* Fern.)
  3. R. globularis (Chapm.) Small. June–July. Moist, sandy soil, very rare; Cook and Kankakee cos. (Including var. *recognita* Gale.)
  4. R. capillacea Torr. July–Sept. Bogs and marshes; rare in the n. tip of the state; also St. Clair Co. (Including f. *leviseta* [E. J. Hill] Fern.)
  5. R. glomerata (L.) Vahl. July–Oct. Moist, often sandy soil, rare; Kankakee and Pope cos.
  6. R. capitellata (Michx.) Vahl. July–Oct. Moist soil; not common in the n. one-third of the state; absent elsewhere except for Pope Co.

7. R. alba (L.) Vahl. July–Sept. Sphagnum bogs, rare; Cook, Lake, and Peoria cos.

11. Cladium P. Browne   Twig Rush

1. C. mariscoides (Muhl.) Torr. Aug.–Oct. Swamps and bogs; not common and restricted to the extreme ne. corner of Ill.

12. Scleria Berg.   Nut Rush

1. Achene smooth; tubercles of hypogynium none; coarse perennial ......
.... 1. *S. triglomerata*
1. Achene papillose, pitted, or wrinkled; tubercles of hypogynium 0, 6, or 9; slender perennials or annuals 2
2. Achene papillose; tubercles of hypogynium 6 or 9; perennial ....
.... 2. *S. pauciflora*
2. Achene pitted or wrinkled; tubercles none; annual 3
3. Achene pitted; hypogynium 3-lobed; inflorescence closely paniculate; achene 1.8–2.2 mm long .... 3. *S. reticularis*
3. Achene wrinkled; hypogynium absent; inflorescence verticillate; achene 1.5 mm long .... 4. *S. verticillata*

1. S. triglomerata Michx. June–Sept. Moist or dry woods, fields; occasional in the n. half of the state, rare elsewhere.
2. S. pauciflora Muhl. June–Sept. Culms and leaves glabrous or sparsely hirtellous. Dry soil, in woods or on bluffs; occasional in the s. tip of the state, rare elsewhere.
var. caroliniana (Willd.) Wood. June–Sept. Culms and leaves densely pilose. Dry soil, rare; Lee Co.
3. S. reticularis Michx. Aug.–Oct. Sandy shore of pond, very rare; Cass Co.
4. S. verticillata Muhl. July–Sept. Bogs and marshes, rare; scattered in the n. half of the state.

13. Carex L.   Sedge

1. Spikes not all alike; stigmas 2 or 3; achenes lenticular, suborbicular, or trigonous 2
1. Spikes essentially alike; stigmas 2; achenes lenticular or plano-convex
.... 7
2. Stigmas 2; achenes lenticular or suborbicular .... Group 1
2. Stigmas 3; achenes trigonous 3
3. Perigynia pubescent (sometimes scarcely so) .... Group 2
3. Perigynia glabrous 4
4. Perigynia few; scales of staminate spike united at the base .. Group 3
4. Perigynia many; scales of staminate spike free at the base 5
5. Foliage pubescent (leaves only rough or scabrous keyed elsewhere) ...
.... Group 4
5. Foliage glabrous 6
6. Beak of perigynia not bidentate, at most emarginate, or absent ....
.... Group 5
6. Beak of perigynia bidentate .... Group 6
7. Staminate flowers (or remnants) at the tips of some spikes (androgynous) .... Group 7

7. Staminate flowers (or remnants) at the base of some spikes (gynecandrous) ....Group 8

Group 1

Spikes not all alike; styles 2; achenes lenticular or subglobose.

1. Plants slender; culms rarely over 3.5 dm tall, but not more than 1 mm thick; lowest leaf-like bract long-sheathing; perigynia beakless, white-powdery or orange (drying brown), subglobose to plump-obovoid. .. 2
1. Plants stout; culms more than 3.0 dm tall, more than 1 mm thick; lowest leaf-like bract sheathless or short-sheathing; perigynia short-beaked, brown, tawny, or green, compressed-lenticular 3
2. Terminal spike gynecandrous; pistillate scales appressed; perigynia white-powdery .... 69. *C. garberi*
2. Terminal spike staminate; pistillate scales spreading; perigynia orange (drying brown) .... 70. *C. aurea*
3. Staminate spikes long-pedunculate (especially the lower ones), arching-pendulous; pistillate scales with long, rough awns, the scales 2–4 times longer than the perigynia with the margins hyaline; achene wrinkled or constricted on one side .... 71. *C. crinita*
3. Pistillate spikes sessile to short-pedunculate, erect to spreading; pistillate scales obtuse to long-acuminate, shorter than to slightly longer than the perigynia, with the margins brown or reddish; achene with a regular shape 4
4. Lower pistillate spikes spreading or nodding; perigynia tapering to a prominent beak, the beak bending or twisting with drying .. 76. *C. torta*
4. Pistillate spikes erect; perigynia beakless or straight, short-beaked .. 5
5. Pistillate scales long-acuminate, much exceeding the perigynia; perigynia compressed-obovate, widest above the middle 75. *C. haydenii*
5. Pistillate scales obtuse to acuminate, shorter than to slightly exceeding the perigynia; perigynia compressed-elliptic, widest at or below the middle 6
6. Lower leaf sheaths easily tearing ventrally and with conspicuous cross-connecting filaments, the upper sheath subscabrous ventrally .... 73. *C. stricta*
6. Lower leaf sheaths usually intact and without filaments, the upper ones smooth ventrally 7
7. Culms phyllopodic, with old leaves persisting; pistillate scales acuminate; perigynia 2.20–2.75 mm long, 1.50–1.75 mm wide .... 74. *C. emoryi*
7. Culms aphyllopodic, with the old leaves absent; pistillate scales obtuse to acute; perigynia 2.7–3.2 mm long, 1.50–2.75 mm wide .... 72. *C. substricta*

Group 2

Spikes not all alike; stigmas 3; achenes trigonous; perigynia pubescent (sometimes scarcely so).

1. Leaf blades or sheaths pubescent (only scabrous-puberulent in *C. debilis*) 2
1. Leaf blades and sheaths glabrous 5
2. Spikes of two kinds, the topmost usually composed of staminate flowers

but occasionally with 1–3 perigynia; perigynia with a conspicuous beak
.... 3

2. Spikes of one kind, the staminate scales conspicuous at the base and inconspicuous or absent at the apex; perigynia without a conspicuous
beak 4

3. Leaf blades and sheaths scabrous-puberulent; staminate spike elongate,
long-pedunculate, occasionally with 1–3 perigynia; perigynia stalked, the
beak long-tapering, scarcely pubesent or glabrous ...... 92. *C. debilis*

3. Leaf blades and sheaths conspicuously pubescent; staminate spike
short, sessile, without any perigynia; perigynia sessile, the beak abruptly
contracted, conspicuously pubescent .... 68. *C. hirtifolia*

4. Leaves usually shorter than the culms; pistillate spike linear, gradually
tapering to the base, 15–40 mm long .... 86. *C. virescens*

4. Leaves usually exceeding the culms; pistillate spikes cylindric to oblong,
abrupt or rounded at the base, 5–20 mm long .... 87. *C. swanii*

5. Perigynia beak less than 0.4 mm long or absent 6

5. Perigynia beak greater than 0.4 mm long 8

6. Pistillate scales obtuse, awnless, concealing the perigynia; bracts of the
inflorescence scale-like .... 66. *C. richardsonii*

6. Pistillate scales acute to acuminate or awned, the perigynia easily seen;
bracts of the inflorescence leaf-like 7

7. Perigynia 2.5–3.5 mm long, ellipsoid, short-stipitate; pistillate scales ½–¾
the length of the perigynia, acute to acuminate .... 109. *C. digitalis*

7. Perigynia 3.5–4.5 mm long, obovoid, long-stipitate; pistillate scales
equalling or exceeding the perigynia, awned ...... 65. *C. pedunculata*

8. Perigynia in a head, radiating in all directions, 12–18 mm long; leaves
5–14 mm wide .... 132. *C. grayi*

8. Perigynia in a spike, spreading to ascending; perigynia less than 10 mm
long; leaves seldom more than 6 mm wide 9

9. Most or all culms shorter than or hidden by the leaves or leaf bases;
remains of old fruiting culms often persistent among the leaves 10

9. Most culms as long as or longer than the leaves, not hidden among the
leaf bases; older fruiting culms not persistent 13

10. Peduncle one per culm, all similar, elongated to 20 cm long, producing
both staminate and pistillate flowers; body of perigynia ellipsoid-
oblongoid, ½–⅔ longer than wide .... 61. *C. nigromarginata*

10. Peduncles 2–4 per culm, not all alike, one of them elongated and with
staminate and pistillate flowers, the other 1–3 mostly short (up to 5 cm
long) and with only pistillate flowers hidden among the tufted leaf bases;
body of perigynia broadly ellipsoid or ellipsoid-ovoid to subglobose, ⁴/₅
to as long as wide 11

11. Leaf blades thin, soft, erect to ascending, 1.5–3.0 mm wide; perigynia
pubescent in the upper half, 2.25–4.00 mm long, membranaceous, with
the beak less than 2 mm long 12

11. Leaf blades thick, rigid, spreading at maturity, 2.0–4.5 mm wide;
perigynia glabrous to sparsely pubescent in the upper half, 3.50–4.75
mm long, subcoriaceous, with the beak at least 2 mm long 64. *C. tonsa*

12. Beak of perigynium 0.9–1.7 mm long, about ¾ as long as the perigynium
body; pistillate scales long-acuminate .... 62. *C. umbellata*

12. Beak of perigynium 0.5–1.0 mm long, about ½ as long as the perigynium
body; pistillate scales acute .... 63. *C. abdita*

13. Culms usually 4–12 dm tall, not densely tufted; pistillate spikes 10–90
mm long, 3–6 times longer than thick; staminate spike usually long-
pedunculate, 2–5 cm long; plants of wet places 14

13. Culms usually less than 3 dm tall (sometimes taller in *C. pensylvanica*), densely tufted; pistillate spikes 5–10 mm long, less than 2 times longer than wide; staminate spike sessile to short-pedunculate, usually less than 2 cm long; plants of dry places 17
14. Perigynia 6–10 mm long, lanceoloid, evenly tapering to the beak, prominently nerved, the teeth 1.5–2.0 mm long .... 127. *C. trichocarpa*
14. Perigynia 2.5–5.5 mm long, ovoid to lance-ovoid, more abruptly tapering into the beak, obscurely nerved (especially in the upper ⅓), the beak about ¼ the total length of the perigynia, the teeth not over 1 mm long 15
15. Perigynia ovoid-lanceoloid, sparsely pubescent, 5.00–5.25 mm long, the beak 1.50–1.75 mm long .... 128. *C.* X *subimpressa*
15. Perigynia ovoid to suborbicular, densely pubescent, 2.5–4.0 mm long, the beak 1.00–1.25 mm long 16
16. Leaf blades flat, some with revolute margins, 2–5 mm wide; achene obovoid to ovoid-ellipsoid; culms sharply triangular ..... 81. *C. lanuginosa*
16. Leaf blades involute-filiform, up to 2 mm wide; achene ellipsoid; culms obtusely triangular .... 80. *C. lasiocarpa*
17. Perigynium body suborbicular to ovoid, about as wide as long, slightly 3-angled 18
17. Perigynium body ellipsoid, longer than wide, noticeably 3-angled ... 19
18. Plants with long scaly rhizomes, the base coarsely fibrillose; ligule longer than wide; leaf blades 1–3 mm wide .... 56. *C. pensylvanica*
18. Plants without rhizomes, slightly or not at all fibrillose; ligule wider than long; leaf blades 3–7 mm wide .... 57. *C. communis*
19. Plants with elongated scaly rhizomes, slightly cespitose; perigynia whitish-green; staminate spike short-stalked ..... 60. *C. physorhyncha*
19. Plants with short stout rhizomes, densely cespitose; perigynia olive-green; staminate spike sessile 20
20. Culms weak, loosely spreading to reclining; all but the lowest spikes aggregated into an ovoid-ellipsoid head 1–2 cm long . 58. *C. emmonsii*
20. Culms stiffer, ascending; spikes interrupted or in a linear-cylindric head 1–5 cm long .... 59. *C. artitecta*

Group 3

Spikes not all alike; stigmas 3; achenes trigonous; perigynia glabrous; scales of staminate flowers united at the base; spikes androgynous.

1. Culms equalling or longer than the leaves; perigynia beakless, the body ellipsoid; lowest pistillate scale of normal size .... 54. *C. leptalea*
1. Culms shorter than the leaves; perigynia long-beaked, the body subglobose; lowest pistillate scale prolonged and leaf-like ..... 55. *C. jamesii*

Group 4

Spikes 2 or more; staminate scales free at the base; pubescence present on some of the foliage (scabrous leaves not included).

1. Staminate flowers (or remnants) and perigynia in different spikes, with the staminate spike overtopping the pistillate spike 2
1. Staminate flowers (or remnants) and perigynia in the same spike, with the staminate flowers conspicuous at the base 4
2. Perigynia 2.5–3.0 mm long, ellipsoid, beakless, entire   82. *C. pallescens*

2. Perigynia 4.5–12.0 mm long, ovoid to lanceoloid, beaked, deeply bidentate or nearly emarginate 3
3. Pistillate spikes (as well as some androgynous ones) 5–12 cm long; perigynia 8–12 mm long, the beak deeply bidentate .. 124. *C. atherodes*
3. Pistillate spikes 1.0–2.5 cm long; perigynia 4.5–5.0 mm long, the beak merely emarginate .... 102. *C. hitchcockiana*
4. Pistillate spikes over 2 cm long, narrowly cylindric; perigynia 3.5–6.0 mm long 5
4. Pistillate spikes less than 2 cm long, ovoid, short-cylindric; perigynia 2.00–3.75 mm long 6
5. Pistillate scales with an awn ¾ to as long as the main body of the scale, spreading; perigynia 4–6 mm long .... 91. *C. davisii*
5. Pistillate scales long-acuminate or awn-tipped, appressed; perigynia 3.5–4.0 mm long .... 90. *C. oxylepis*
6. Leaf blades and sheaths soft-pubescent 7
6. Leaf blades and sheaths sparsely hairy to nearly glabrous .... 84. *C. caroliniana*
7. Perigynia 2.50–3.75 mm long, shorter than the prominently awned scales .... 85. *C. bushii*
7. Perigynia 2.0–2.5 mm long, longer than the obtuse to short-pointed scales .... 83. *C. hirsutella*

Group 5

Spikes 2 or more; plants glabrous; beak of perigynia not bidentate, at most emarginate, or absent.

1. Culms without well-developed leaf blades .... 106. *C. plantaginea*
1. Culms with well-developed leaf blades 2
2. Basal leaves, especially those of the sterile shoots, 10–30 mm wide .. 3
2. Basal leaves less then 10 mm wide 9
3. Blade of uppermost leaf never more than 3 times as long as the sheath 4
3. Blade of uppermost leaf many times longer than the sheath 5
4. Perigynia 5–7 mm long; staminate scales red-tinged . 107. *C. careyana*
4. Perigynia 2.0–4.5 mm long; staminate scales pale or brown .... 108. *C. platyphylla*
5. Perigynia sharply triangular, with flattish sides, symmetrical .... 110. *C. laxiculmis*
5. Perigynia rounded-triangular (at least below), with swollen sides, asymmetrical 6
6. Bracteal sheaths with the edges smooth or slightly rough; perigynia tapering to a straight or only slightly curved beak .... 112. *C. laxiflora*
6. Bracteal sheaths with the edges strongly serrulate; perigynia more abruptly contracted into a strongly bent beak 7
7. Staminate spike sessile or subsessile 8
7. Staminate spike long-pedunculate, overtopping the pistillate spikes .... 113. *C. striatula*
8. Some or all leaves over 14 mm wide; pistillate scales obtuse; sterile shoots culmless .... 111. *C. albursina*
8. None of the leaves over 12 mm wide; pistillate scales acute to aristate; sterile shoots developing culms .... 114. *C. blanda*

9. Beak of the perigynium bent or with the straight beak 0.6–1.0 mm long
.... 10
9. Perigynia beakless, or with the nearly straight beak less than 0.5 mm
long 16
10. Beak of the perigynium straight or only slightly bent 11
10. Beak of the perigynium abruptly bent 13
11. Perigynia with several elevated nerves; pistillate scales acute to smooth-
awned 12
11. Perigynia with many impressed nerves; pistillate scales rough-awned
.... 101. *C. oligocarpa*
12. Perigynia abruptly contracted into a 0.5 mm long beak, obtusely triangular above and below .... 112. *C. laxiflora*
12. Perigynia evenly tapered into a 0.6–1.0 mm long beak, sharply triangular
(at least above) .... 116. *C. styloflexa*
13. Staminate spike sessile or subsessile 14
13. Staminate spike pedunculate, overtopping the pistillate spikes 15
14. Pistillate scales acute to aristate; sterile shoots developing culms
.... 114. *C. blanda*
14. Pistillate scales obtuse; sterile shoots culmless ...... 111. *C. albursima*
15. Staminate spike brown or purplish; culms reddish at the base ....
.... 115. *C. gracilescens*
15. Staminate spike green or whitish; culms green or brown at the base ...
.... 113. *C. striatula*
16. Terminal spike pistillate toward the apex, staminate at the base 17
16. Terminal (subterminal in *C. eburnea*) spike entirely staminate or at most
with a few perigynia 20
17. Pistillate spikes drooping on long, slender peduncles . 89. *C. gracillima*
17. Pistillate spikes erect to ascending, sessile to short-pedunculate ... 18
18. Leaves 2–4 at the base of the culm, pale green, glaucous; perigynia
whitish .... 77. *C. buxbaumii*
18. Leaves numerous at the base of the culm, dark green, not glaucous;
perigynia dark green to brown 19
19. Perigynia nerveless, the tips spreading outward; spikes 4–8 times longer
than thick .... 79. *C. shortiana*
19. Perigynia nerved, the tips ascending; spikes 1½–3 times longer than
thick .... 84. *C. caroliniana*
20. Pistillate spikes drooping on long, weak peduncles 21
20. Pistillate spikes erect, ascending, or somewhat spreading, sessile to pedunculate 23
21. Pistillate spikes 2–6, 1.5–6.0 cm long; pistillate scales up to ½ as long as
the perigynia and narrower than the perigynia, hyaline with a green center, lanceolate, acute to long-awned 22
21. Pistillate spikes 1–3, 1–2 cm long; pistillate scales as long as and as wide
as the perigynia, brown or reddish, ovate, obtuse to cuspidate ....
.... 78. *C. limosa*
22. Pistillate spikes 3–6 mm wide; perigynia lanceoloid, beaked, the tips
spreading to recurving .... 88. *C. prasina*
22. Pistillate spikes 2–3 mm wide; perigynia ellipsoid, beakless, the tips ascending .... 89. *C. gracillima*
23. Leaves capillary, 0.5 mm wide; perigynia 2 mm long ... 67. *C. eburnea*
23. Leaves flattened, more than 1 mm wide; perigynia more than 2 mm long
.... 24
24. Staminate spike conspicuous, mostly long-pedunculate and overtopping
the pistillate spikes 25

24. Staminate spike inconspicuous, mostly sessile or short-pedunculate, shorter than or equalling the terminal pistillate spike (sometimes overtopping in *C. glaucodea*) 30
25. Perigynia pale to dark green or glaucous, with many coarse or fine elevated nerves, tapering to the base, triangular in cross-section, closely enveloping the achene 26
25. Perigynia dark green or brown, impressed-nerved to nearly nerveless, rounded at the base, terete in cross-section, loosely enveloping the achene 29
26. Pistillate spikes loosely flowered; perigynia in 2 or 3 rows, the lower ones separated from one another 27
26. Pistillate spikes closely flowered; perigynia in 3–6 rows, mostly overlapping 28
27. Pistillate scales obtuse; most basal sheaths with poorly developed leaf blades; perigynia with a slight beak; achenes triangular-obovoid, stipitate .... 105. *C. woodii*
27. Pistillate scales acute to acuminate; most basal sheaths with well-developed leaf blades; perigynia barely short-beaked; achenes triangular-ellipsoid, not stipitate .... 109. *C. digitalis*
28. Pistillate spikes linear-cylindric, 3.0–4.5 mm thick; perigynia appressed or ascending, 2.5–3.5 mm long .... 104. *C. tetanica*
28. Pistillate spikes oblongoid, 5–8 mm thick; perigynia spreading, 3–5 mm long .... 103. *C. meadii*
29. Culms cespitose from fibrous or short root-stocks; leaves deep green; perigynia with many impressed nerves .... 96. *C. conoidea*
29. Culm solitary from a creeping root-stock; leaves blue-green, glaucous; perigynia with a few elevated nerves .... 95. *C. crawei*
30. Pistillate scales not more than ½ the perigynia length, acuminate with a short, smooth awn; leaves blue-green, often glaucous 31
30. Pistillate scales equalling or exceeding the perigynia, with a long, scabrous awn; leaves dark green, rarely glaucous 33
31. Perigynia beaked, spreading to recurving, becoming squarrose, 2–4 mm long, the body subglobose to ovoid .... 94. *C. granularis*
31. Perigynia beakless or slightly emarginate, ascending, 3–6 mm long, the body ellipsoid 32
32. Leaves somewhat glaucous; perigynia 4–6 mm long; staminate spike always shorter than the terminal pistillate spike; plants of wet places ... . 99. *C. flaccosperma*
32. Leaves conspicuously glaucous; perigynia 3–5 mm long; staminate spike often overtopping the terminal pistillate spike; plants of dry places .... 100. *C. glaucodea*
33. Base of plant brown or green; leaves 4–10 mm wide; mature perigynia inflated; pistillate spikes close together, none near the base of the culm .... 98. *C. grisea*
33. Base of plant reddish or purplish; leaves 2–4 mm wide; mature perigynia tightly enveloping the achene; pistillate spikes widely separated, some near the base of the culm ..:......................... 97. *C. amphibola*

Group 6

Spikes 2 or more; plants glabrous; beak of perigynium bidentate.

1. Mature perigynia (9–) 12–18 mm long; staminate spike solitary 2
1. Mature perigynia 2–10 mm long; staminate spike none to 2 or more (if solitary, then the perigynia less than 10 mm long) 7

2. Developed pistillate spikes globose to hemispherical, about as wide as long; style straightish to coiled near the middle; perigynia beak much shorter than the body 3

2. Developed pistillate spikes oblong to cylindric, longer than wide; style bent or contorted near the base; perigynia beak equalling or longer than the body 4

3. Mature perigynia radiating in all directions, cuneate at the base, dull ... . 132. *C. grayi*

3. Mature perigynia wide-spreading to ascending, rounded at the base, often lustrous .... 133. *C. intumescens*

4. Mature perigynia almost horizontally spreading; body of achene mostly concave-truncate at the summit, as wide as long to wider than long, the sides concave more than ¾ their width, the projecting angles flattened and winglike .... 137. *C. gigantea*

4. Mature perigynia ascending to wide-spreading (but not horizontally); body of the achene subtruncate to acute, as wide as long to much longer than wide, the sides concave for less than ¾ their width, the projecting angles plumpish 5

5. Pistillate spikes more than 2 times longer than thick, long-cylindric; mature perigynia wide-spreading .... 136. *C. lupuliformis*

5. Pistillate spikes not more than 2 times longer than thick, short-cylindric to oblong; mature perigynia ascending to spreading 6

6. Leaf blades 2.5–7.0 mm wide; pistillate spikes loosely flowered; perigynia spreading, 9–14 mm long; achenes widest at the middle; plants usually solitary .... 134. *C. louisianica*

6. Leaf blades 5–11 mm wide; pistillate spikes compactly flowered; perigynia appressed-ascending, 10–17 mm long; achenes widest at or below the middle; plants most often cespitose .... 135. *C. lupulina*

7. Flowering culms usually 1–10 cm tall, mostly shorter than the leaves; pistillate spikes hidden among the leaf bases .... 64. *C. tonsa*

7. Flowering culms usually greater than 20 cm tall, mostly as long as or longer than the leaves; pistillate spikes conspicuous near the summit, or overtopping the leaves 8

8. Beak of the perigynium 0.2–0.4 mm long; foliage strongly glaucous; achenes poorly developed; rhizomes with conspicuous long, narrow, overlapping, fibrillose scales; culm base strongly fibrillose .... 125. *C.* X *fulleri*

8. Beak of the perigynium mostly greater than 2 mm long; foliage not glaucous (somewhat blue-green in *C. hyalinolepis*); achenes well-developed; rhizomes without conspicuous fibrillose scales; culm base non-fibrillose 9

9. Terminal spikes with perigynia above and staminate flowers (or remnants) below 10

9. Terminal spike staminate (occasionally with 1–2 perigynia, or absent or aborted in *C. frankii*) 11

10. Perigynia horizontally spreading; pistillate spikes globose to short-cylindric; pistillate scales acute to short-awned; styles bent or contorted .... 120. *C. squarrosa*

10. Perigynia appressed-ascending; pistillate spikes long-cylindric; pistillate scales mostly obtuse; styles straight or slightly curved . 121. *C. typhina*

11. Staminate spike 1 (occasionally absent or aborted in *C. frankii*) 12

11. Staminate spikes 2 or more 20

12. Leaves inrolled, becoming filiform and subterete   141. *C. oligosperma*

12. Leaves flattened 13

13. Pistillate spikes linear-cylindric, 10–15 times longer than thick, often

drooping on long weak peduncles; perigynia more than 4 times longer
than thick .... 92. *C. debilis*

13. Pistillate spikes subglobose, thick-cylindric or oblong, 2–5 times longer
than thick, ascending to spreading from stout peduncles; perigynia less
than 4 times longer than thick 14

14. Body of perigynium obovoid; pistillate scales narrowly linear, much exceeding the perigynia; staminate spike occasionally absent or aborted
.... 119. *C. frankii*

14. Body of perigynium lanceoloid, ellipsoid, or ovoid; pistillate scales ovate
to lanceolate, much shorter than to barely exceeding the perigynia;
staminate spike conspicuous 15

15. Pistillate scales with a serrulate awn 16

15. Pistillate scales entire (ciliate in some), with or without an awn, if awned,
the awn not serrulate 18

16. Perigynia (at least the lowest) reflexed, the teeth 1.0–2.2 mm long and
widely divergent; plants greenish to brown at the base . 129. *C. comosa*

16. Perigynia widely spreading to ascending, the teeth less than 0.75 mm
long and appressed to slightly spreading; plants reddish at the base   17

17. Perigynia 5–7 mm long, 1.5–2.0 mm thick, 12- to 20-nerved, slightly inflated .... 130. *C. hystricina*

17. Perigynia 6–9 mm long, 2–4 mm thick, 7- to 10-nerved, strongly inflated
.... 131. *C. lurida*

18. Perigynia 2–3 mm long, the beak less than 1 mm long   . 118. *C. viridula*

18. Perigynia 3.5–6.0 mm long, the beak 2.0–2.5 mm long 19

19. Pistillate spikes subglobose, erect, sessile; perigynia 3.5–4.5 mm long,
the lower ones reflexed, about 10-ribbed .... 117. *C. cryptolepis*

19. Pistillate spikes elongate, drooping from long, weak peduncles;
perigynia 5–6 mm long, ascending to spreading, 2-ribbed ....
.... 93. *C. sprengelii*

20. Perigynia firm, light to dark green, dull 21

20. Perigynia papery, stramineous to brown, shiny 23

21. Teeth of the perigynia 1–3 mm long .... 126. *C. laeviconica*

21. Teeth of the perigynia about 0.5 mm long 22

22. Perigynia with many elevated nerves; lower sheaths filamentous ventrally; stems aphyllopodic, purplish at the base .... 122. *C. lacustris*

22. Perigynia with finely impressed nerves, or with a few elevated nerves;
lower sheaths not filamentous ventrally; stems phyllopodic, green or
brown at the base .... 123. *C. hyalinolepis*

23. Perigynia broadly ovoid, 5–6 mm wide; achenes notched on one angle
.... 140. *C. tuckermanii*

23. Perigynia narrowly ovoid, 2.5–3.5 mm wide; achenes with a regular
shape 24

24. Perigynia horizontally spreading to reflexed, the beak 2–3 mm long ....
.... 138. *C. retrorsa*

24. Perigynia ascending to slightly spreading, the beak 1–2 mm long ... 25

25. Teeth of the perigynia 1–3 mm long; styles straight   126. *C. laeviconica*

25. Teeth of the perigynia 0.5–1.0 mm long; styles curved to abruptly bent
.... 26

26. Perigynia widely spreading, with the lower ones sometimes reflexed;
culms obtusely angled and smooth below the spikes; ligule slightly if at
all longer than wide; lower sheaths not filamentous ventrally ....
.... 139. *C. rostrata*

26. Perigynia ascending; culms sharply angled and rough below the spikes;
ligule much longer than wide; lower sheaths somewhat filamentous ventrally... 142. *C. vesicaria*

Group 7

Spikes all essentially alike; stigmas 2; achenes lenticular; staminate flowers (or remnants) found at the tips of some of the spikes (androgynous).

1. Culms mostly solitary, from slender, elongated rhizomes or decumbent older culms 2
1. Culms several, cespitose, from fibrous roots or short, stout rhizomes 5
2. Inflorescence head-like, 8–17 mm long 3
2. Inflorescence elongate, 20–65 mm long 4
3. Culms erect, simple; apex of leaf sharply triangular; perigynia obscurely striate dorsally and ventrally, thin, with the margins serrate above ......
.... 1. *C. stenophylla*
3. Culms decumbent, often branching from the nodes; leaf flattened throughout; perigynia sharply nerved dorsally and ventrally, thick-spongy, with the margins entire throughout .... 3. *C. chordorrhiza*
4. Inflorescence 3–5 mm wide, green; staminate flowers inconspicuous; perigynia obscurely nerved dorsally, nerveless ventrally ....
.... 2. *C. praegracilis*
4. Inflorescence 5–12 mm wide, reddish-brown; staminate flowers conspicuous; perigynia distinctly nerved dorsally and ventrally . 4. *C. sartwellii*
5. Inflorescence branched near the base (some obscurely so), with several small spikelets on each branch 6
5. Inflorescence unbranched; spikes sessile on the main axis 15
6. Perigynia 4–7 mm long, attenuate, with the beak longer than the body, evenly tapering and coarsely serrate 7
6. Perigynia 2–4 (–5) mm long, ovoid to lanceoloid, with the beak shorter than the body, more or less abruptly contracted and finely serrate ... 9
7. Perigynia 6–7 mm long, obscurely nerved ventrally except at the base, the base abruptly enlarged and disklike, the beak 2–3 times the length of the body .... 29. *C. crus-corvi*
7. Perigynia 4–5 mm long, strongly nerved ventrally, rounded or slightly contracted at the base, the beak 1–2 times longer than the body 8
8. Sheaths with a thickened band at the orifice, smooth ventrally ....
.... 28. *C. laevivaginata*
8. Sheaths smooth at the orifice, cross-puckered ventrally .. 27. *C. stipata*
9. Flowering culms spongy, flattened when pressed; perigynia 3–4 (–5) mm long 10
9. Flowering culms hard, remaining angular when pressed; perigynia 2.0–3.5 mm long 11
10. Sheaths cross-puckered ventrally; spikes green; perigynia distinctly nerved dorsally, with the beak half as long as the body . 26. *C. conjuncta*
10. Sheaths smooth ventrally; spikes tawny; perigynia obscurely nerved dorsally, with the beak the length of the body .... 25. *C. alopecoidea*
11. Leaf sheaths cross-puckered ventrally; bracts filiform, conspicuous; pistillate scales awned 12
11. Leaf sheaths smooth ventrally; bracts inconspicuous or absent; pistillate scales acute to obtuse 13
12. Beak of the perigynium equalling the main body; leaves often longer than the culms .... 20. *C. vulpinoidea*
12. Beak of the perigynium shorter than the main body; leaves shorter than the culms .... 21. *C. annectens*
13. Leaves 3–8 mm wide; perigynia abruptly contracted into a 0.5 mm long beak, purple-black when mature, spongy at the base 22. *C. decomposita*

13. Leaves 1–3 mm wide; perigynia only slightly contracted or evenly tapering into a 1 mm long beak, tawny-brown when mature, solid at the base 14

14. Leaf sheaths striate-hyaline at the mouth; perigynia 2.0–2.5 mm long, convex ventrally, shiny brown .... 23. *C. diandra*

14. Leaf sheaths copper-tinged at the mouth; perigynia 2.5–3.5 mm long, flat to concave ventrally, dull tawny-brown .... 25. *C. prairea*

15. Spikelets with 1–3 perigynia and with 1–2 staminate flowers; perigynia ellipsoid, abruptly contracted into a beak 0.25 mm long   30. *C. disperma*

15. Spikelets usually with 5 or more perigynia and with at least four staminate flowers; perigynia linear-lanceoloid to ovoid-lanceoloid, abruptly contracted to a distinct beak at least 0.5 mm long 16

16. Perigynia corky-thickened or shrunken in the lower half, wide-spreading at maturity 17

16. Perigynia solid at the base, ascending at maturity 21

17. Beak of the perigynium smooth; pistillate scales deciduous, acuminate 18

17. Beak of the perigynium serrulate; pistillate scales persistent, acute . 19

18. Perigynia reflexed at maturity, broadly ovoid, gibbous and distinctly nerved in the lower ventral half, 1.50–1.75 mm wide   .... 6. *C retroflexa*

18. Perigynia widely spreading at maturity, ovoid-lanceoloid, spongy and wrinkled in the lower ventral half, 1.00–1.25 mm wide ... 7. *C. texensis*

19. Perigynia gradually tapering into the beak; stigmas long, slender, flexuous, not twisted, light reddish; leaves 1–2 mm wide 20

19. Perigyia abruptly contracted into the beak; stigmas stout, tightly twisted, dark red; leaves 2–3 mm wide .... 8. *C. convoluta*

20. Culms arising from a long-creeping, fibrillose rhizome; perigynia linear-lanceoloid, 3.6–4.0 mm long, 0.7–0.8 mm wide; pistillate scales acuminate to mucronate, ½ the length of the perigynia .... 10. *C. socialis*

20. Culms cespitose; perigynia lanceoloid, 3.0–3.5 mm long, 1.5 mm wide; pistillate scales obtuse to acute, ¾ the length of the perigynia .... 9. *C. rosea*

21. Leaf sheaths loose, easily friable ventrally, with cross-veins conspicuous dorsally 22

21. Leaf sheaths tight, mostly intact ventrally, with the veins all parallel dorsally 25

22. Inflorescence 1–3 cm long, ovoid to short-cylindric, with only the lowermost spikes slightly separated; leaf blades 3–7 mm wide; perigynia flat ventrally 23

22. Inflorescence 3–10 cm long, elongate, with most of the spikes remote and the lowermost widely separated; leaf blades 5–10 mm wide; perigynia concave ventrally .... 19. *C. sparganioides*

23. Perigynia light green to buff at maturity, the beak ¼–⅓ the length of the ovoid or suborbicular body .... 16. *C. gravida*

23. Perigynia deep green at maturity, the beak ⅓–½ the length of the ovoid body 24

24. Scales ½ the length of the perigynia, acute to obtuse; leaf sheaths truncate at the mouth, the lowermost cross-puckered ... 18. *C. cephaloidea*

24. Scales about the length of the perigynia, acuminate to aristate; leaf sheath concave at the mouth, none of them cross-puckered .... 17. *C. aggregata*

25. Inflorescence ovoid to oblongoid; spikes in a head 26

25. Inflorescence linear, elongate; spikes interrupted 28

26. Perigynia more than 2 mm wide; scales about as long as or longer than

the perigynia; at least the lowest spike subtended by a conspicuous
filiform bract .... 14. *C. muhlenbergii*
26. Perigynia less than 2 mm wide; scales much shorter than the perigynia;
bracts subtending the spikes inconspicuous or absent 27
27. Perigynia broadest at the truncate-cordate base, flat ventrally, with the
beak entire or nearly so; leaf sheaths thin at the mouth ....
.... 12. *C. leavenworthii*
27. Perigynia broadest below the middle, round-tapering to the base, the
margins raised ventrally, with the beak serrulate; leaf sheaths slightly
thickened at the mouth .... 11. *C. cephalophora*
28. Perigynia 2 times longer than wide, 4.5–5.0 mm long, about 2.25 mm
wide, with the beak at least 1.5 mm long; ligule much longer than wide
.... 13. *C. spicata*
28. Perigynia less than 2 times longer than wide, 3–4 mm long, 2.25–0 mm
wide, with the beak 1.00–1.25 mm long; ligule about as long as wide 29
29. Perigynia sessile, 3.0–3.5 mm long, 2.25–2.50 mm wide, wider than the
scales; bracts slender based; achenes about 2 mm long ....
.... 14. *C. muhlenbergii*
29. Perigynia stipitate, 3.5–4.0 mm long, 2.5–3.0 mm broad, mostly covered
by the scales; bracts dilated at the base; achenes 2.2–2.5 mm long .....
.... 15. *C. austrina*

Group 8

Spikes essentially alike; stigmas 2; achenes lenticular or plano-convex;
staminate flowers (or remnants) at the base of some spikes (gynecandrous).

1. Perigynia wing-margined (at least above), the lower part firm and without spongy thickenings 2
1. Perigynia without wings, at most thin margined, the lower part spongy-
thickened 22
2. Plants mostly solitary, from well-developed rhizomes; lower 1–3 spikes
usually pistillate, the middle spikes staminate, the terminal spikes gynecandrous .... 5. *C. foenea*
2. Plants mostly cespitose, without well-developed rhizomes; most spikes
gynecandrous 3
3. Spikelets over 15 mm long, pointed at both ends; perigynia 6.5–10.0 mm
long .... 36. *C. muskingumensis*
3. Spikelets less than 15 mm long, rounded at the apex or the base or both;
perigynia mostly less than 6 mm long 4
4. Mature perigynia ovoid-lanceoloid to lanceoloid (2–) 3–5 times longer
than wide 5
4. Mature perigynia ovoid to suborbicular, not more than 2½ times longer
than wide, with some as wide as long 8
5. Sterile culms with spreading leaves common; mature perigynia 1.0–1.5
mm wide, the wing narrowed at or below the middle of the body and/or
wingless at the base 6
5. Sterile culms uncommon, if present, the leaves ascending or clustered
at the apex of the culm; mature perigynia 1.25–2.25 mm wide, the wing
not distinctly narrowed at or below the middle, winged to the base .. 7
6. Spikes most often aggregated in an elongated head, ovoid-turbinoid;
leaves 3–7 mm wide, stiff and often ascending ...... 38. *C. tribuloides*
6. Spikes, or at least the lower 3–5, separated from each other, subglobose; leaves 1–3 mm wide, flaccid and spreading ...... 39. *C. projecta*

7. Spikelets clavate at the base; perigynia 3–4 times longer than wide ....
.... 37. *C. scoparia*
7. Spikelets abruptly contracted at the base; perigynia at most 2 times longer than wide .... 42. *C. bebbii*
8. Spikes globose; beaks of mature perigynia spreading to recurved, the wing abruptly narrowed below the middle and wingless at the base; sterile culms with spreading, non-clustered leaves common ....
.... 41. *C. cristatella*
8. Spikes ovoid to lanceoloid-ovoid; beaks of mature perigynia ascending to appressed (somewhat spreading in *C. normalis*), the wing continuing to the base; sterile leafy culms uncommon (except *C. normalis*) 9
9. Mature perigynia not more than 2 mm wide, mostly less than 4 mm long
.... 10
9. Mature perigynia more than 2 mm wide, usually more than 4 mm long
.... 14
10. Body of perigynium orbicular to broadly obovoid, widest near the summit, abruptly beaked .... 46. *C. albolutescens*
10. Body of perigynium narrowly to broadly ovoid or suborbicular, widest near the middle or at the base, tapering to the beak or abruptly beaked
.... 11
11. Perigynia narrowly or broadly ovoid, 2–4 mm long, gradually tapering into the beak 12
11. Body of perigynium suborbicular, 3.5–4.0 mm long, abruptly contracted into the beak .... 45. *C. festucacea*
12. Spikelets (especially the lower ones) well separated; leaves 0.5–2.5 mm wide .... 44. *C. tenera*
12. Spikelets mostly overlapping; leaves 2.0–6.5 mm wide 13
13. Spikelets crowded into a head, abruptly contracted at the base, acute at the apex .... 42. *C. bebbii*
13. Spikelets singly overlapping, clavate at the base, somewhat obtuse at the apex .... 43. *C. normalis*
14. Perigynia widest above the middle, the body broadly obovoid to orbicular 15
14. Perigynia widest at or below the middle, the body broadly ovoid to suborbicular 17
15. Perigynia 2.75–3.75 mm wide, the beaks ascending to appressed; pistillate scales long-acuminate to aristate; spikelets pointed at the apex ..
.... 53. *C. alata*
15. Perigynia 2.5–4.0 mm wide, the beaks spreading; pistillate scales obtuse to acuminate; spikelets rounded at the apex 16
16. Leaf sheaths green-striate ventrally; spikelets densely aggregated; perigynium 3–4 mm long, 2.5 mm wide .... 47. *C. cumulata*
16. Leaf sheaths white-hyaline ventrally; spikelets slightly overlapping; perigynia 4–5 mm long, 3–4 mm wide .... 49 *C. brevior*
17. Leaf sheaths green-striate ventrally to the hyaline band at the mouth; perigynia widest at the middle 18
17. Leaf sheaths narrowly or broadly white-hyaline ventrally to the mouth; body of perigynium usually widest below the middle (except sometimes widest at the middle in *C. reniformis*) 19
18. Spikelets pointed; pistillate scales obtuse to cuspidate; perigynia nerveless or obscurely 1- to 2-nerved ventrally, the body tapering from the middle to the base .... 48. *C. suberecta*
18. Spikelets obtuse; pistillate scales long-acuminate to aristate; perigynia nerved ventrally, the body curved from the beak to the base   40. *C. richii*

19. Perigynia 5.25–6.00 mm long, 3.25–4.25 mm wide 20
19. Perigynia 3–5 mm long, 2.00–3.25 mm wide 21
20. Perigynia nerved dorsally and ventrally, hyaline and smooth, the body orbicular .... 52. *C. bicknellii*
20. Perigynia nerved ventrally, brown with wrinkles, the body broader than long .... 51. *C. reniformis*
21. Spikelets mostly aggregated, the tip broadly obtuse, the base abruptly contracted; perigynia finely nerved dorsally and ventrally, gradually tapering into the beak .... 50. *C. molesta*
21. Spikelets mostly separated, the tip usually pointed, the base gradually tapering; perigynia nerveless or nearly so ventrally, the body abruptly contracted into the beak .... 49. *C. brevior*
22. Perigynia 4–5 mm long, narrowly lanceoloid, appressed at maturity .... 32. *C. bromoides*
22. Perigynia 2–4 mm long, ovoid or ellipsoid, spreading (except *C. trisperma*) at maturity 23
23. Spikes 2- to 4-flowered; perigynia ellipsoid, the beak 0.5 mm long, entire and emarginate, appressed at maturity .... 31. *C. trisperma*
23. Spikes (2-) 4- to 15-flowered; perigynia ovoid, the beak 0.5–1.0 mm long, serrulate, distinctly bidentate, spreading at maturity 24
24. Pistillate scales about ¾ the length of the body of the perigynia; perigynia ellipsoid, nerveless ventrally or only short-nerved at the base, with the beak about ¼ the total length .... 33. *C. interior*
24. Pistillate scales as long as or longer than the length of the body of the perigynia; perigynia ovoid-deltoid, obscurely to strongly nerved ventrally, with the beak about ⅓–½ the total length 25
25. Terminal spike sometimes totally staminate; pistillate scales lanceolate, acuminate, hyaline, with a wide brown center; perigynia deeply bidentate, the teeth usually bent or twisted, the beak about ½ the length of the body, setose-serrate .... 34. *C. sterilis*
25. Terminal spike always gynecandrous; pistillate scales ovate, acute, hyaline, with a narrow green center; perigynia shallowly bidentate, the teeth straight, the beak about ⅓ the length of the body, serrate .... 35. *C. incomperta*

1. C. stenophylla Wahlenb. var. enervis (C. A. Mey.) Kükenth. April–May. Gravel-bluff prairie, very rare; Winnebago Co. *C. eleocharis* Bailey—G.

2. C. praegracilis W. Boott. April–Aug. Low prairies, roadsides, dry sterile soil, rare; DeKalb, Kane, and Winnebago cos.

3. C. chordorrhiza L. f. May–June. Sphagnous swamps, very rare; Lake and McHenry cos.

4. C. sartwellii Dew. April–July. Low wet prairies and meadows, marshes, peaty swamps, open cold bogs; occasional in the n. one-half of the state.

5. C. foenea Willd. May–June. Perigynia nerved on both faces, tapering gradually to the beak. Sandy, usually prairie, soil, not common; confined to the n. one-sixth of the state; also Menard Co. *C. siccata* Dewey—G, J.

var. enervis Evans & Mohlenbr. Perigynia nerveless on the ventral face, tapering abruptly to the beak. "Sandy plains," according to the collector; exact locality unknown.

6. C. retroflexa Muhl. April–June. Dry woods; occasional in the s. two-fifths of the state, rare or absent elsewhere.

7. C. texensis (Torr.) Bailey. April–June. Dry woods, old lawns; occasional in the s. one-half of Ill. *C. retroflexa* Muhl. var. *texensis* (Torr.) Fern.—G.

8. C. convoluta Mack. April–June. Dry or moist woods, not common; scattered throughout the state. *C. rosea* Schk.—G.

9. C. rosea Schk. April–June. Dry or moist woods; common throughout the state.

10. C. socialis Mohlenbr. & Schwegm. April–June. Low woods, rare; Johnson, Massac, Pulaski, and Union cos.

11. C. cephalophora Muhl. April–July. Dry woods, lawns, and fields; common throughout the state.

12. C. leavenworthii Dewey. April–June. Dry open woods; not common but scattered in Ill.

13. C. spicata Huds. June. Native of Europe; adventive along a railroad; Champaign Co.

14. C. muhlenbergii Schk. May–July. Perigynia nerved on both faces. Dry woods and fields; occasional throughout the state.

var. enervis Boott. May–July. Perigynia nerveless on the upper face. Dry woods and fields; occasional throughout the state.

15. C. austrina (Small) Mack. May–July. Prairies, rare; Jackson and Perry cos. *C. muhlenbergii* Schk.—J; *C. muhlenbergii* Schk. var. *australis* Olney—G.

16. C. gravida Bailey. May–July. Perigynia half as wide as long, obscurely nerved on the lower convex face. Prairies and fields; common in the n. three-fourths of the state, rare elsewhere.

var. lunelliana (Mack.) F. J. Herm. May–July. Perigynia $^2/_5$–¾ as wide as long, conspicuously nerved on the lower convex face. Prairies and fields; scattered in the n. half of the state. *C. gravida* Bailey—J.

17. C. aggregata Mack. May–July. Woods, rare; Adams and Madison cos. *C. sparganioides* Muhl. var. *aggregata* (Mack.) Gl.—G.

18. C. cephaloidea Dewey. May–July. Woods; occasional in the n. half of the state, rare in the s. half. *C. sparganioides* Muhl. var. *cephaloidea* (Dewey) Carey—G.

19. C. sparganioides Muhl. May–July. Dry or moist woods; occasional in the n. half of the state, rare in the s. half.

20. C. vulpinoidea Michx. May–Aug. Wet ground, common; probably in every co. (Including *C. setacea* Dewey.)

21. C. annectens Bickn. May–July. Perigynia yellow-brown, prominently nerved on the convex face. Fields, not common; widely scattered in Ill.

var. xanthocarpa (Bickn.) Wieg. May–July. Perigynia dark brown, obscurely nerved on the convex face. Fields, occasional; scattered throughout the state. *C. brachyglossa* Mack.—J.

22. C. decomposita Muhl. June–Aug. Swamps, very rare; Pulaski and Union cos.

23. C. diandra Schrank. May–Aug. Wet meadows, rare; confined to the n. half of the state.

24. C. prairea Dewey. May–Aug. Wet meadows, rare; DeKalb, Stark, and Winnebago cos.

25. C. alopecoidea Tuckerm. June–July. Wet meadows; occasional in the n. half of the state; also Pope Co.

26. C. conjuncta Boott. May–June. Moist woods; occasional throughout the state.

27. C. stipata Muhl. May–Aug. Leaves up to 8 mm. broad. Swamps and moist woods; occasional throughout the state.

var. maxima Chapm. May–Aug. Leaves 8 mm broad or broader. Swamps and moist woods; occasional in the s. one-sixth of the state.

28. C. laevivaginata (Kükenth.) Mack. May–Aug. Swamps and wet woods, rare; scattered in Ill.

29. C. crus-corvi Shuttlew. June–Aug. Swamps and wet woods; occasional in the s. half of the state, becoming less common in the n. half.

30. C. disperma Dewey. May–Aug. Wet meadows; occasional in the n. one-third of the state.

31. C. trisperma Dewey. June–July. Bog, very rare; Lake Co.

32. C. bromoides Schk. May–July. Swamps and moist woods, not common; confined to the ne. corner of Ill.; also Jackson and Pope cos.

33. C. interior Bailey. May–Aug. Swamps and wet woods; occasional in the n. half of the state; also Christian and Pope cos.

34. C. sterilis Willd. May–Aug. Swampy meadows; occasional in the n. half of the state. *C. muricata* L. var. *sterilis* (Carey) Gl.—G.

35. C. incomperta Bickn. May–Aug. Swampy woods, rare; Pope Co.

36. C. muskingumensis Schwein. July–Aug. Swamps and wet woods; occasional throughout the state.

37. C. scoparia Schk. May–Aug. Marshes and wet prairies; occasional throughout the state.

38. C. tribuloides Wahlenb. June–Sept. Wet woods, marshes, wet prairies; occasional to common throughout the state.

39. C. projecta Mack. June–Aug. Wet woods; not common, but scattered in Ill.

40. C. richii (Fern.) Mack. June–Aug. Along a railroad; Menard Co. *C. straminea* Willd.—F, G.

41. C. cristatella Britt. June–Aug. Wet ground; occasional throughout the state. (Including f. *catelliformis* [Farw.] Fern.)

42. C. bebbii Olney. June–Aug. Wet ground; occasional in the n. half of the state, absent elsewhere.

43. C. normalis Mack. May–Aug. Moist or dry woods, prairies; occasional to common throughout the state.

44. C. tenera Dewey. May–Aug. Wet ground; occasional in the n. half of the state, less common in the s. half.

45. C. festucacea Schk. May–July. Wet woods and wet prairies; occasional throughout the state.

46. C. albolutescens Schw. May–Aug. Wet ground; occasional throughout the state.

47. C. cumulata (Bailey) Mack. June–Sept. Moist soil, very rare; Kankakee Co.

48. C. suberecta (Olney) Britt. May–July. Wet meadows and wet prairies; occasional in the n. one-fourth of the state, rare elsewhere.

49. C. brevior (Dewey) Mack. May–July. Prairies and meadows; occasional to common throughout the state.

50. C. molesta Mack. May–July. Moist soil, rare; Jackson and St. Clair cos. *C. brevior* (Dewey) Mack—J.

51. C. reniformis (Bailey) Small. May–June. Wet ground, very rare; Massac Co.

52. C. bicknellii Britt. May–July. Dry or moist prairies and meadows; occasional or common in the n. three-fourths of the state.

53. C. alata Torr. & Gray. May–July. Wet ground, very rare; Massac Co.

54. C. leptalea Wahlenb. June–Aug. Bogs and wet meadows; known from the extreme ne. cos. and Peoria, Tazewell, and Woodford cos.

55. C. jamesii Schwein. May–June. Moist or dry woods; occasional throughout the state.

56. C. pensylvanica Lam. April–June. Perigynia less than 3 mm long, the beak about one-fourth as long as the body. Dry woods; occasional to common in the n. half of Ill., much rarer southward.

var. distans Peck. Perigynia 3–4 mm long, the beak at least one-half as long as the body. Lake Co.

57. C. communis Bailey. May–July. Woods, rare; Peoria Co.

58. C. emmonsii Dewey. April–June. Dry woods, rare; Union Co.

59. C. artitecta Mack. April–June. Dry woods; common in the s. half of the state, becoming less common northward. *C. nigromarginata* Schwein. var. *muhlenbergii* (Gray) Gl.—G.

60. C. physorhyncha Liebm. April–June. Dry, rocky woods, rare; Randolph and Union cos. *C. nigromarginata* Schwein. var. *muhlenbergii* (Gray) Gl.—G.

61. C. nigromarginata Schwein. April–June. Woods, rare; Pope and Wabash cos.

62. C. umbellata Schk. March–June. Dry woods and on sandstone rocks; occasional in the n. and s. cos., rare in the cent. cos.

63. C. abdita Bickn. April–June. Dry woods and prairies, rare; Jackson and Pope cos.

64. C. tonsa (Fern.) Bickn. April–June. Woods, very rare; Pope Co.

65. C. pedunculata Muhl. April–May. Woods, rare; JoDaviess, Kane, and Winnebago cos.

66. C. richardsonii R. Br. May–June. Dry, rocky woods, rare; Hancock, Lake, Menard, Peoria, and Winnebago cos.

67. C. eburnea Boott. May–Aug. Rocky soil, not common; scattered in the n. cos. and the cos. along the Mississippi River.

68. C. hirtifolia Mack. May–June. Woods; occasional throughout the state.

69. C. garberi Fern. June–July. Sandy soil along Lake Michigan; Cook and Lake cos.

70. C. aurea Nutt. June–July. Moist soil, very rare; Lake Co.

71. C. crinita Lam. May–Aug. Awns of lower pistillate scales at least twice as long as the perigynia; perigynia 1–2 mm thick. Low woods, swamps; not common but scattered throughout the state.

var. brevicrinis Fern. May–Aug. Awns of lower pistillate scales up to twice as long as the perigynia; perigynia 2–3 mm thick. Swampy woods, rare; Pope Co.

72. C. substricta (Kükenth.) Mack. May–Aug. Wet ground; occasional in the n. one-sixth of the state, rare or absent elsewhere. *C. aquatilis* Wahl.—G; *C. aquatilis* Wahl. var. *altior* (Rydb.) Fern.—F.

73. C. stricta Lam. May–Aug. Swamps; occasional in the n. three-fourths of the state. (Including var. *strictior* [Dewey] Carey.)

74. C. emoryi Dewey. May–June. Swamps and low woods; not common but scattered in the n. one-third of the state; also Fayette Co. *C. stricta* Lam. var. *elongata* (Boeckl.) Gl.—G.

75. C. haydenii Dewey. May–Aug. Wet ground; occasional in the n. half of the state.

76. C. torta Boott. May–July. In shallow, rocky streams; confined to the s. one-sixth of the state.

77. C. buxbaumii Wahlenb. May–Aug. Wet meadows and bogs; oc-

casional in the n. half of the state; also Clay Co. (Including f. *hetero-stachya* Anderss.)

78. C. limosa L. May–Aug. Bogs, rare; Cook, McHenry, Peoria, Taze-well, and Will cos.

79. C. shortiana Dewey. May–June. Moist woods, wet meadows, along streams; occasional to common throughout the state.

80. C. lasiocarpa Ehrh. May–Aug. Bogs, swamps, wet roadside ditches; occasional in the n. half of the state. *C. lasiocarpa* Ehrh. var. *americana* Fern.—F, G.

81. C. lanuginosa Michx. May–Aug. Wet woods, meadows, and road-side ditches; occasional throughout the state. *C. lasiocarpa* Ehrh. var. *latifolia* (Boeckl.) Gl.—G.

82. C. pallescens L. May–Aug. Wet ground, very rare; "N. Ill.," accord-ing to the collector. *C. pallescens* L. var. *neogaea* Fern.—F.

83. C. hirsutella Mack. May–Aug. Dry woods and fields; common in the s. half of the state, becoming less common northward. *C. com-planata* Torr. & Hook. var. *hirsuta* (Bailey) Gl.—G.

84. C. caroliniana Schwein. May–July. Wet ground; Alexander, Jack-son, McDonough, and Pope cos.

85. C. bushii Mack. May–July. Moist or dry woods and fields; oc-casional to common in the s. half of the state; also McDonough Co.

86. C. virescens Muhl. May–July. Moist or dry woods; not common but scattered in the ne. and s. cos.; also Edgar Co.

87. C. swanii (Fern.) Mack. May–July. Dry woods and fields; not com-mon but scattered throughout the state.

88. C. prasina Wahlenb. May–June. Wet ground, very rare; Adams Co.

89. C. gracillima Schwein. May–Aug. Moist ground; occasional in the n. one-third of the state.

90. C. oxylepis Torr. & Hook. April–June. Swampy woods, rare; Har-din, Johnson, and Union cos.

91. C. davisii Schwein. & Torr. May–June. Moist ground; occasional to common throughout the state.

92. C. debilis Michx. var. rudgei Bailey. May–Aug. Dry woods, very rare; Hardin Co. *C. debilis* Michx.—J.

93. C. sprengelii Dewey. May–Aug. Moist woods and fields; oc-casional in the n. half of the state.

94. C. granularis Muhl. May–July. Perigynia ovoid to subglobose, 1.5–2.5 mm thick. Wet ground; occasional to common throughout the state.

var. haleana (Olney) Porter. May–July. Perigynia oblongoid-ovoid to ellipsoid, up to 1.5 mm thick. Wet ground; rare but scattered in Ill. *C. granularis* Muhl.—G, J.

95. C. crawei Dewey. June–July. Moist soil, rare; Boone, Cook, Grundy, and Lake cos.

96. C. conoidea Schk. May–Aug. Moist soil; occasional in the n. half of the state; also Massac Co.

97. C. amphibola Steud. May–June. Moist woods, not common; Alex-ander, Cook, DuPage, Jackson, Johnson, St. Clair, and Stark cos. (Including var. *rigida* [Bailey] Fern.)

98. C. grisea Wahlenb. May–July. Dry or moist woods; common throughout the state. *C. amphibola* Steud. var. *turgida* Fern.—F.

99. C. flaccosperma Dewey. April–June. Rich woods; occasional in the s. one-fifth of the state; also McDonough Co.

100. C. glaucodea Tuckerm. May–July. Woods; occasional in the s.

half of the state. *C. flaccosperma* Dewey var. *glaucodea* (Tuckerm.) Kükenth.—F; *C. flaccosperma* Dewey—J.

101. C. oligocarpa Schk. May–July. Moist woods; occasional throughout the state.

102. C. hitchcockiana Dewey. May–July. Moist woods, not common; apparently confined to the ne. one-fourth of the state.

103. C. meadii Dewey. May–June. Prairies; occasional in the n. half of the state, becoming much less common southward.

104. C. tetanica Schk. May–July. Bogs, wet prairies, low woods; occasional in the n. half of the state, rare in the s. half.

105. C. woodii Dewey. May–June. Rich woods, rare; Cook, Kankakee, and Will cos. *C. tetanica* Schk. var. *woodii* (Dewey) Wood—G.

106. C. plantaginea Lam. April–June. Rich woods, very rare; Cook Co.

107. C. careyana Torr. May–June. Rich woods; rare and scattered in the s. half of the state.

108. C. platyphylla Carey. April–June. Rich woods, very rare; McHenry Co.

109. C. digitalis Willd. May–July. Dry woods, rare; Hardin, Massac, and Pope cos.

110. C. laxiculmis Schwein. May–July. Rich woods, rare; Fulton, Kane, Peoria, and Will cos.

111. C. albursina Sheldon. April–June. Rich woods; occasional throughout the state. *C. laxiflora* Lam. var. *latifolia* Boott—G.

112. C. laxiflora Lam. April–June. Rich woods, very rare; Pope Co.

113. C. striatula Michx. May–June. Rich woods, very rare; Pope and Union cos. *C. laxiflora* Lam. var. *angustifolia* Dewey—G.

114. C. blanda Dewey. April–June. Rich woods; occasional to common throughout the state. *C. laxiflora* Lam. var. *blanda* (Dewey) Boott—G.

115. C. gracilescens Steud. April–June. Rich woods; occasional throughout the state. *C. laxiflora* Lam. var. *gracillima* (Boott) Robins. & Fern.—G.

116. C. styloflexa Buckl. April–June. Rich woods, very rare; Jackson Co.

117. C. cryptolepis Mack. June–Aug. Moist ground, very rare; Cook and Lake cos. *C. flava* L. var. *fertilis* Peck—F.

118. C. viridula Michx. June–Sept. Wet ground, rare; Cook, Lake, and McHenry cos.

119. C. frankii Kunth. June–Sept. Rich woods, wet ground; occasional in the s. three-fourths of the state; also Kendall and Will cos.

120. C. squarrosa L. June–Sept. Wet woods and ditches; common in the s. half of the state, becoming less common northward.

121. C. typhina Michx. June–Sept. Wet woods and ditches; occasional in the s. half of the state, less common in the n. half.

122. C. lacustris Willd. May–Aug. Swamps and wet ground; occasional throughout the state.

123. C. hyalinolepis Steud. May–July. Wet ground; occasional to common in the s. half of the state, absent elsewhere. *C. lacustris* Willd. var. *laxiflora* Dewey—G.

124. C. atherodes Spreng. June–Aug. Wet ground, very rare; Winnebago Co.

125. C. X fulleri Ahles. June–Aug. Wet ground, very rare; Winnebago Co.

126. C. laeviconica Dewey. June–Aug. Wet ground; occasional in the n. half of the state; also Lawrence and St. Clair cos.

127. C. trichocarpa Muhl. June–Aug. Wet ground; occasional in the n. half of the state.

128. C. X subimpressa Clokey. June–July. Wet ground, rare; Macon, Perry, and St. Clair cos. Reputed to be a hybrid between *C. hyalinolepis* Steud. and *C. lanuginosa* Michx.

129. C. comosa Boott. June–Aug. Swamps and wet ground; occasional in the n. half of the state; also Gallatin and Union cos.

130. C. hystricina Muhl. June–Aug. Swamps and wet ground; occasional in the n. half of the state, rare elsewhere.

131. C. lurida Wahlenb. June–Oct. Swamps and wet ground; occasional throughout the state.

132. C. grayi Carey. June–Oct. Low woods and wet ground; occasional throughout the state. (Including var. *hispidula* Gray.)

133. C. intumescens Rudge. May–Sept. Low woods and wet ground, not common; Cook, Johnson, and Peoria cos.

134. C. louisianica Bailey. June–Sept. Achene ellipsoid-ovoid, widest near the middle, not conspicuously knobbed. Swamps and low woods, not common; confined to the s. one-fourth of the state.

135. C. lupulina Muhl. June–Oct. Wet ground; occasional throughout the state. (Including var. *pedunculata* Gray.)

136. C. lupuliformis Sartwell. June–Oct. Wet ground; not common but scattered throughout the state.

137. C. gigantea Rudge. May–Sept. Wet ground, rare; Johnson and Union cos.

138. C. retrorsa Schw. July–Oct. Wet ground; rare and scattered in Ill.

139. C. rostrata Stokes var. utriculata (Boott) Bailey. June–Oct. Wet ground, very rare; Henderson and Winnebago cos. *C. rostrata* Stokes—J.

140. C. tuckermanii Boott. June–Sept. Low woods and wet ground, rare; Cook, Hancock, Lake, and Winnebago cos.

141. C. oligosperma Michx. June–Aug. Bogs, very rare; Kane and Lake cos.

142. C. vesicaria L. June–Aug. Swamps and wet ground; occasional in the n. half of the state, much less common in the s. half. (Including var. *monile* [Tuckerm.] Fern.)

27. ARACEAE

1. True spathe absent; ovary 2- to 3-celled; fruit dry; leaves linear, the midvein not central .... 1. *Acorus*

1. Spathe present; ovary 1-celled; fruit fleshy; leaves broad, the midvein central 2

2. Flowers perfect; spadix nearly globoid; perianth parts 4; fruit 8–12 cm in diameter; leaves simple, ovate; plants with rhizomes . 2. *Symplocarpus*

2. Flowers unisexual; spadix elongate; perianth absent; fruit to 6 cm long, narrower; leaves simple, arrowhead-shaped, or compound; plants with corms or fleshy roots 3

3. Staminodia present; berries brownish or greenish; leaves simple, arrowhead-shaped 4

3. Staminodia absent; berries red; leaves compound .... 5. *Arisaema*

4. Plants rhizomatous; spathe tubular only at base; flowers confined to the lower half of the spadix .... 3. *Arum*

4. Plants with fleshy roots; spathe tubular at both ends, opening at the middle; flowers covering all or most of the spadix .... 4. *Peltandra*

1. Acorus L.   Sweet Flag

1. A. calamus L. Sweet Flag. May–Aug. Marshes and other low areas; occasional throughout the state.

2. Symplocarpus Salisb.   Skunk Cabbage

1. S. foetidus (L.) Nutt. Skunk Cabbage. Feb.–March. Swamps and other low areas; occasional in the n. three-fifths of the state.

3. Arum L.   Arum

1. A. italicum Mill. Arum. May. Native of w. Europe and N. Af.; rarely escaped in Ill.; Jackson Co.

4. Peltandra Raf.   Arrow Arum

1. P. virginica (L.) Kunth. Arrow Arum. May–June. Swamps, shallow water, wet ditches; occasional in the s. four-fifths of the state; also Cook Co.

5. Arisaema Mart.   Jack-in-the-Pulpit

1. Leaflets (3–) 5–13; spathe convolute below and above, open near the middle; spadix slender-tapering, exserted beyond the spathe .... 1. *A. dracontium*
1. Leaflets 3; spathe convolute below, open and arching above; spadix club-shaped, obtuse, covered by the arching spathe ... 2. *A. triphyllum*

1. A. dracontium (L.) Schott. Green Dragon. May–June. Rich woodlands, common; probably in every co.
2. A. triphyllum (L.) Schott. Jack-in-the-Pulpit. April–May. Lateral leaflets strongly asymmetrical; extended portion of spathe 4–7 cm wide; fruiting head 3–6 cm long. Rich woods, common; in every co. *A. atrorubens* (Ait.) Blume—F.
var. pusillum Peck. Small Jack-in-the-Pulpit. May. Lateral leaflets scarcely asymmetrical; extended portion of spathe to 3 cm wide; fruiting head to 2 cm long. Rich wooldlands; known only from Alexander Co. *A. triphyllum* (c.) Schoot.—F.

28. LEMNACEAE

1. Rootlets 1-several per plant; reproductive pouches lateral, 2 per frond   2
1. Rootlets none; reproductive pouch basal, 1 per plant 3
2. Rootlets (1–) 2 or more per plant .... 1. *Spirodela*
2. Rootlet 1 per plant .... 2. *Lemna*
3. Frond thin, linear .... 3. *Wolffiella*
3. Frond thick, globose or ellipsoid .... 4. *Wolffia*

1. Spirodela Schleiden

1. Fronds orbicular, 5- to 8-nerved; rootlets 2–10 .... 1. *S. polyrhiza*
1. Fronds obovate to reniform, obscurely 3- (5-) nerved; rootlets (1–) 2–5
.... 2. *S. oligorhiza*

    1. S. polyrhiza (L.) Schleiden. Duckweed. Standing water; through-
out the state; probably in every co.
    2. S. oligorhiza (Kurtz) Hegelm. Duckweed. Standing water; Alexan-
der, Johnson, and Union cos.

2. Lemna L. Duckweed

1. Fronds spatulate with long, persistent stipes, often submerged in compact masses; rootlet frequently absent .... 1. *L. trisulca*
1. Fronds orbicular to elliptic, floating; rootlet present 2
2. Fronds 3- to 5-nerved 3
2. Fronds 1-nerved or obscurely nerved 6
3. Root sheaths cylindrical 4
3. Root sheaths winged 5
4. Lower surface of frond flattened or only weakly convex, apex rounded to
acute, frond usually 3-nerved .... 2. *L. minor*
4. Lower surface of frond moderately to strongly convex, apex broadly
rounded or often obtuse, frond 3- to 5-nerved .... 3. *L. gibba*
5. Fronds weakly to strongly asymmetrical, thick, the nerves often inconspicuous .... 4. *L. perpusilla*
5. Fronds symmetrical or nearly so, membranous, the 3 nerves prominent
.... 5. *L. trinervis*
6. Fronds 1-nerved, the lower suface green 7
6. Fronds obscurely nerved, the lower surface frequently reddish-purple ..
.... 8. *L. obscura*
7. Fronds ovate-elliptic to elliptic, weakly to moderately asymmetrical,
often floating in compact masses .... 6. *L. valdiviana*
7. Fronds obovate to orbicular, symmetrical, usually solitary   7. *L. minima*

    1. L. trisulca L. Duckweed. Standing water; local throughout the
state, but rare in the s. cos.
    2. L. minor L. Duckweed. Standing water; rather common through-
out the state.
    3. L. gibba L. Duckweed. Standing water; Knox, Mason, and Will
cos.
    4. L. perpusilla Torr. Duckweed. Standing water; scattered through-
out the state, but not common.
    5. L. trinervis (Austin) Small. Duckweed. Standing water; through-
out the state, but not common.
    6. L. valdiviana Phil. Duckweed. Standing water; in the s. one-third
of the state; also Lake Co.
    7. L. minima Phil. Duckweed. Standing water; Carroll, Madison, and
Will cos.
    8. L. obscura (Austin) Daubs. Duckweed. Standing water; scattered
throughout the state.

3. Wolffiella Hegelm.

1. W. floridana (J. D. Smith) Thompson. Duckweed. Standing water; scattered throughout the state, but not common.

4. Wolffia Horkel

1. Plants punctate, flattened above, with a single papule, or with papule absent 2
1. Plants epunctate, convex above, frequently with a median row of small papules .... 3. *W. columbiana*
2. Upper surface with a single acuminate papule, ovate-elliptic ....
.... 1. *W. papulifera*
2. Upper surface flattened, without a papule, elliptic ...... 2. *W. punctata*

1. W. papulifera Thompson. Water Meal. Standing water; scattered throughout the state, but apparently not common.
2. W. punctata Griseb. Water Meal. Standing water; scattered in Ill., but not common.
3. W. columbiana Karst. Water Meal. Standing water; throughout the state

29. XYRIDACEAE

1. Xyris L.   Yellow-eyed Grass

1. Plants twisted, from a bulbous base; lateral sepals 3–4 mm long, the keel ciliate .... 1. X.*torta*
1. Plants not twisted, from a non-bulbous base; lateral sepals 4–6 mm long, the keel toothed but not ciliate .... .2. X. *jupicai*

1. X. torta Sm. Twisted Yellow-eyed Grass. July–Aug. Low, sandy areas; restricted in the n. half of the state.
2. X. jupicai L. Rich. Yellow-eyed Grass. July–Aug. Low, sandy areas, very rare. X. *caroliniana* Walt.—F, G.

30. COMMELINACEAE

1. Calyx radially symmetrical; corolla radially symmetrical; fertile stamens 6, the filaments villous; ovary 3-celled, with all cells fertile; inflorescence appearing umbellate, from an elongated, leaf-like bract   1. *Tradescantia*
1. Calyx bilaterally symmetrical, with two of the sepals connate at the base; corolla bilaterally symmetrical; fertile and sterile stamens each 3 (rarely 3 plus 2), the filaments glabrous; ovary 3-celled, with one cell aborted; inflorescence cymose, from a spathe .... 2. *Commelina*

1. Tradescantia L. Spiderwort

1. Stems flexuous; leaves usually 2–4 cm broad; cymes several, terminal and lateral; sepals 5–10 mm long .... 1. *T. subaspera*
1. Stems straight; leaves usually 0.5–1.5 cm broad; cyme usually solitary, terminal; sepals 8–15 mm long 2
2. Stems generally 4–10 dm tall; leaves and stems glabrous and glaucous; sepals glabrous or with tuft of eglandular hairs at the tip   2. *T. ohiensis*
2. Stems generally 0.5–4.0 dm tall; leaves and stems glabrous or pubescent, but not glaucous; sepals pubescent throughout with glandular or eglandular hairs 3
3. Pubescence of sepals eglandular .... 3. *T. virginiana*
3. Pubescence of sepals glandular-viscid .... 4. *T. bracteata*

1. T. subaspera Ker. Spiderwort. May–Aug. Woodlands; occasional in the s. two-thirds of the state, rare in the n. one-third.
2. T. ohiensis Raf. Spiderwort. April–Aug. Edges of woodlands, prairies, common; probably in every co.
3. T. virginiana L. Spiderwort. April–June. Woodlands and prairies; common in the s. two-thirds of the state, rare in the n. one-third.
4. T. bracteata Small. Prairie Spiderwort. May–Aug. Prairies, very rare; Jersey, Mason, and Morgan cos.

2. Commelina L. Day Flower

1. Plants rooting at the nodes; margins of spathes free; seeds black, rugose or reticulate 2
1. Plants not rooting at the nodes; margins of spathes united for about one-third the way up from the base; seeds reddish or brown, smooth or puberulent, not rugose or reticulate 3
2. Plants annual; leaf-sheaths glabrous at summit; anthers 6; seeds 4; petals 10–15 mm long, the lower petal white; seeds 3.5–4.0 mm long, rugose .... 1. *C. communis*
2. Plants perennial; leaf-sheaths ciliate at summit; anthers 5; seeds 5; petals 6–8 mm long, the lower petal blue; seeds 2.0–2.5 mm long, reticulate .... 2. *C. diffusa*
3. Plants with rhizomes; lowest petal blue, only slightly smaller than the other two; seeds about 6 mm long, reddish .... 3. *C. virginica*
3. Plants with thick, fleshy roots; lowest petal white, much smaller than the other two; seeds about 3 mm long, brown .... 4. *C. erecta*

1. C. communis L. Common Day Flower. June–Oct. Native of Asia; moist waste ground; occasional throughout the state.
2. C. diffusa Burm. f. Day Flower. July–Oct. Moist woodlands and moist waste ground; occasional in the s. one-third of the state, absent in the n. two-thirds, except for Jersey and Mason cos.
3. C. virginica Ļ. Day Flower. June–Sept. Low woodlands, not common; restricted to the s. one-fourth of the state; also McDonough Co.
4. C. erecta L. Day Flower. June–Sept. Moist or dry sandy soil; occasional in the w. cos.; also Cook, Grundy, and Kankakee cos. (Including var. *angustifolia* [Michx.] Fern., f. *crispa* [Woot.] Fern., and var. *deamiana* Fern.)

31. PONTEDERIACEAE

1. Perianth funnelform, bilaterally symmetrical; stamens 6; ovary 3-celled, with 2 cells abortive; fruit a utricle; inflorescence a spike-like panicle with more than 8 flowers; perianth segments blue, with the uppermost marked with yellow .... 1. *Pontederia*
1. Perianth salverform, radially symmetrical; stamens 3; ovary 1-celled or partially 3-celled; fruit a capsule; inflorescence a spike of 2–8 flowers, or flower solitary; perianth segments blue, white, or pale yellow, but not blue marked with yellow 2
2. Flower pale yellow; stamens uniform; ovary 1-celled; capsule indehiscent, few-seeded; leaves linear, 2–6 mm broad .... 2. *Zosterella*
2. Flowers blue or white; two stamens shorter than the third; ovary partially 3-celled; capsule dehiscent, many-seeded; leaves broader, at least 1 cm broad .... 3. *Heteranthera*

1. Pontederia L.   Pickerelweed

1. P. cordata L. Pickerelweed. May–Sept. Wet situations, sometimes in standing water; occasional in most of Ill., except for the e. cent. cos. (Including var. *angustifolia* [Pursh] Torr.)

2. Zosterella Small   Water Star Grass

1. Z. dubia (Jacq.) Small. Water Star Grass. July–Aug. Shallow water, muddy shores, not common; confined to the n. one-half of the state. *Heteranthera dubia* (Jacq.) MacM.—F, J.

3. Heteranthera R. & P.   Mud Plantain

1. Flower solitary; perianth tube 2–4 cm long; filaments glabrous; stigma 3-lobed; leaves lanceolate to ovate, tapering, rounded, or subcordate at the base .... 1. *H. limosa*
1. Flowers 2–8 in a spike; perianth tube 5–10 cm long; filaments pilose; stigma capitate; leaves orbicular, cordate .... 2. *H. reniformis*

1. H. limosa (Sw.) Willd. Mud Plantain. June–July. Muddy shores; shallow water; Alexander, Hardin, and St. Clair cos.
2. H. reniformis R. & P. Mud Plantain. July–Aug. Muddy shores; shallow water; restricted to the s. one-third of the state.

32. JUNCACEAE

1. Plants hairy; capsule 3-seeded .... 1. *Luzula*
1. Plants glabrous; capsule several-seeded .... 2. *Juncus*

1. Luzula DC.   Wood Rush

1. Flowers solitary (rarely paired) at the tips of the inflorescence rays; seeds over 2 mm long, including the strongly curved caruncle ....
.... 1. *L. acuminata*

1. Flowers crowded in glomerulate spikes; seeds 1.2–2.0 mm long, including the conical caruncle .... 2. *L. multiflora*

> 1. L. acuminata Raf. Wood Rush. May. Open woods, rare; LaSalle and Ogle cos. *L. saltuensis* Fern.—J.
>
> 2. L. multiflora (Retz.) Lejeune. Wood Rush. April–May. Rays of umbel erect to ascending. Woods and shaded cliffs; common in the s. cos., less common northward. (Including *L. bulbosa* [Wood] Rydb.) *L. campestris* (L.) DC.—G.
>
> var. echinata (Small) Mohl. Wood Rush. April–May. Rays of umbel horizontally spreading to reflexed. Woods and cliffs; occasional to common in the s. one-fifth of the state, rare elsewhere. (Including var. *mesochorea* Herman.) *L. echinata* (Small) F. J. Herm.—F, J.

2. Juncus L. Rush

1. Leaf sheaths without blades, apiculate or mucronate; inflorescences appearing lateral on the stems 2
1. Leaf sheaths with definite blades; inflorescences terminal 3
2. Stems densely cespitose; stamens 3, anthers 0.5–0.8 mm long; seeds 0.5 mm long; capsules beakless .... 1. *J. effusus* var. *solutus*
2. Stems single at intervals from elongate rhizomes; stamens 6, anthers 1.5 to nearly 2.0 mm long; seeds 1 mm long; capsules with beaks 0.5–1.0 mm long .... 2. *J. balticus* var. *littoralis*
3. Flowers in heads, not prophyllate (i.e., not with bractlets) 4
3. Flowers borne singly on the inflorescence branches, not in heads, prophyllate 15
4. Leaves flat, not terete and cross-septate; anthers purplish-brown 5
4. Leaves terete, cross-septate; anthers yellow 6
5. Stems solitary, approximately 3.5 cm apart on conspicuous, scaly rhizomes, 4.9–8.8 (–10.2) dm tall; leaves 2.0–6.5 mm wide; heads (13–) 20–135 .... 3. *J. biflorus*
5. Stems cespitose, 0.45–5.90 dm tall; leaves 1.0–2.5 (–2.9) mm wide; heads 3–28 (–32) .... 4. *J. marginatus*
6. Seeds fusiform, 0.8–1.9 mm long, caudate 7
6. Seeds ellipsoid to oblongoid or ovoid, 0.4–0.6 mm long, apiculate ... 8
7. Seeds 1.2–1.9 mm long, the tails comprising (one-third) ½–⅝ the total length of the seeds; heads 5- to 50-flowered; stamens 3 5. *J. canadensis*
7. Seeds 0.8–1.2 mm long, the tails comprising ¼–⅖ the total length of the seeds; heads 2- to 5- (10-) flowered; stamens 3 or 6 6. *J. brachycephalus*
8. Stamens 6 9
8. Stamens 3 11
9. Involucral leaves shorter than the inflorescences; heads hemispherical or ellipsoid, 2–7 mm wide, 2- to 9-flowered; sepals 1.9–3.0 mm long, acuminate to acute or obtuse; capsules oblongoid or ellipsoid, acute or obtuse .... 7. *J. alpinus*
9. Involucral leaves usually exceeding the inflorescences; heads hemispherical or spherical, 8–15 mm wide, 9- to 90-flowered; sepals 2.5–5.0 mm long, subulate; capsules lanceoloid, subulate 10
10. Sepals 2.5–4.0 mm long; petals equalling to exceeding the sepals by 0.8 mm; heads 8–11 (–12) mm wide; stems to 6 dm tall ...... 8. *J. nodosus*
10. Sepals 4–5 mm long; petals 1 mm shorter than to nearly equalling the sepals; heads 10–15 mm wide; stems to 10.7 dm tall .... 9. *J. torreyi*

11. Capsules linear-lanceoloid, acute, exceeding the sepals by at least 1.5 mm (usually by 2.0 mm) .... 10. *J. diffusissimus*
11. Capsules ellipsoid or oblongoid, obtuse to subulate, shorter than to exceeding the sepals by 1 mm 12
12. Capsules oblongoid, subulate, exceeding the sepals by 0.75–1.00 mm; perianth segments subulate .... 11. *J. scirpoides*
12. Capsules ellipsoid, acute or obtuse, shorter than to exceeding the sepals by 0.75 mm; perianth segments subulate or acuminate 13
13. Perianth segments acuminate; heads 2- to 35-flowered, hemispherical to spherical; petals 0.7 mm shorter than to equalling the sepals; capsules slightly shorter than the petals to exceeding the sepals by 0.75 mm   14
13. Perianth segments subulate; heads densely 50- to 80-flowered, spherical; petals distinctly (approximately 1 mm) shorter than the sepals; capsules usually 0.5 mm shorter than the petals ...... 14. *J. brachycarpus*
14. Sepals 2.0–2.5 mm long; heads 150–280, 3–5 mm wide, 2- to 7- (8-) flowered; leaves 1.5–4.5 mm wide .... 12. *J. nodatus*
14. Sepals 3–4 mm long; heads (1–) 2–82, 5–10 mm wide, 5- to 35-flowered; leaves 1–3 mm wide .... 13. *J. acuminatus*
15. Annual; auricles absent; sepals 4–7 mm long; inflorescences comprising ¼–⁴/₅ of the total height of the plants .... 15. *J. bufonius*
15. Perennial; auricles present; sepals 2.3–6.0 mm long; inflorescences comprising less than one-half of the total height of the plants 16
16. Sepals obtuse; anthers 1 mm long, three times longer than the filaments .... 16. *J. gerardii*
16. Sepals acuminate, subulate, or aristate; anthers shorter than to as long as the filaments 17
17. Leaves terete, at least distally; capsules usually exceeding the perianths; seeds 0.5–1.3 mm long 18
17. Leaves flat or involute; capsules shorter than to exceeding the perianths by 0.1 mm; seeds 0.3–0.5 mm long 19
18. Leaves involute near the summit of the sheath, becoming closed and terete above; inflorescences 1.5–6.5 (–8.0) cm long; petals acute or obtuse; capsules exceeding the sepals by (0.75–) 1.0–1.6 mm; seeds 0.5–0.6 mm long, apiculate at both ends .... 17. *J. greenei*
18. Leaves terete throughout; inflorescences (1.0–) 2.0–3.5 cm long; petals acuminate or aristate; capsules slightly shorter than to exceeding the sepals by 1 mm; seeds 1.0–1.3 mm long, caudate at both ends .... 18. *J. vaseyi*
19. Petals 0.2 mm shorter than to exceeding the sepals by 0.5 mm; tips of the inflorescence branches incurved; leaves to 13 cm long .... 19. *J. secundus*
19. Petals 1 mm shorter than to equalling the sepals; tips of the inflorescence branches not incurved; leaves to 30 cm long 20
20. Auricles friable, not firm or rigid, scarious, hyaline, lanceolate, white or brownish, prolonged 1.0–4.5 (–5.0) mm beyond point of insertion; perianth segments spreading .... 20. *J. tenuis*
20. Auricles firm at apex or rigid, cartilaginous or membranceous, occasionally prolonged to 2 mm beyond point of insertion; perianth segments spreading or appressed 21
21. Auricles cartilaginous, opaque, rigid, often slightly flaring, obtuse, yellow or orange-brown, less than 1 mm long and usually 0.75 mm long or less; perianth segments spreading; prophylls obtuse to acute .... 21. *J. dudleyi*
21. Auricles membranous, hyaline, not cartilaginous or rigid, usually firm at

apex, pale or brown, very slightly prolonged to exserted 2 mm beyond point of insertion; perianth segments appressed; prophylls acuminate to aristate .... 22. *J. interior*

1. J. effusus L. var. solutus Fern. & Wieg. Soft Rush. June–Sept. Wet ground of shores, swamps, and ditches; common in the s. half of the state, occasional in the n. half. *J. effusus* L.—J.

2. J. balticus Willd. var. littoralis Engelm. Rush. June–Sept. Wet, sandy shores, swamps, not common; confined to the n. one-fourth of the state. *J. balticus* Willd.—J.

3. J. biflorus Ell. Rush. May–Sept. Heads predominantly 2- to 4- (5-) flowered. Wet ground; occasional to common in the s. half of the state; also Cook, Kankakee, and Winnebago cos.

f. andinus Fern. & Grisc. Rush. May–Sept. Heads predominantly 6- to 10-flowered. Wet ground; scattered in the s. half of the state.

4. J. marginatus Rostk. Rush. June–Sept. Wet, often sandy ground of pond borders, ditches, fields; occasional throughout the state.

5. J. canadensis J. Gay. Rush. July–Oct. Wet ground of bogs, swamps, meadows; occasional in the northern half of the state, becoming less common in the southern half. (Including f. *conglobatus* Fern.).

6. J. brachycephalus (Engelm.) Buch. Rush. July–Sept. Marshes, wet ground along bodies of water; local in the n. half of the state; also Johnson and Montgomery cos.

7. J. alpinus Vill. var. rariflorus Hartm. Rush. July–Aug. One or several of the flowers elevated above the others on slightly elongated pedicels. Wet sandy shores and marshes, rare; confined to extreme ne. Ill. *J. richardsonianus* Schult.—J.

var. fuscescens Fern. Rush. July–Aug. Flowers sessile or equally short-pedicellate. Wet sandy shores and marshes, rare; confined to extreme ne. Ill.

8. J. nodosus L. Rush. July–Aug. Wet, often sandy ground of marshes, swamps, fields, and shores; confined to the n. one-half of the state; also St. Clair and Washington cos.

9. J. torreyi Coville. Rush. July–Aug. Wet ground; common throughout the state. (Including var. *globularis* Farw.)

10. J. diffusissimus Buckl. Rush. July–Aug. Wet ground, ditches; local in the s. one-third of the state.

11. J. scirpoides Lam. Rush. July–Aug. Wet and often sandy ground, rare; Cass, Lawrence, and Menard cos.

12. J. nodatus Coville. Rush. July–Sept. Wet grounds, ditches, edges of water bodies; occasional in the s. one-half of the state; also Hancock Co.

13. J. acuminatus Michx. Rush. July–Sept. Wet, low ground of ditches, ponds; occasional to common throughout the state.

14. J. brachycarpus Engelm. Rush. July–Aug. Low, wet ground of fields, prairies, ditches, roadsides; occasional throughout the state, except for the nw. cos.

15. J. bufonius L. Toad Rush. June–Aug. Terminal flowers and flowers on the inflorescence branches borne singly, separated. Wet roadsides, fields; local and scattered in Ill.

var. congestus Wahlb. Rush. July–Aug. Many of the terminal flowers and flowers on the inflorescence branches closely aggregated into 2- and 4-flowered fascicles. Wet roadsides, rare.

16. J. gerardii Loisel. Black Grass. July–Aug. Adventive in disturbed, marshy area; Cook Co.

17. J. greenei Oakes & Tuckerm. Rush. July–Sept. Wet, usually sandy ground of meadows, prairies, and fields; local in a few n. cos.

18. J. vaseyi Engelm. Rush. July–Sept. Wet meadows, bogs, shores; Cook, McHenry, and Winnebago cos.

19. J. secundus Beauv. Rush. July–Sept. Open sandstone ledges and outcroppings; confined to the s. one-sixth of the state.

20. J. tenuis Willd. Path Rush. June–Oct. Moist or dry ground, often in disturbed areas; in every co.

21. J. dudleyi Wieg. Rush. June–Oct. Moist ground; occasional throughout the state.

22. J. interior Wieg. Rush. June–Oct. Wet ground of fields, prairies, ditches, roadsides; common throughout the state.

33. LILIACEAE

31. Flowers campanulate, with the white perianth segments tipped with green .... 26. *Leucojum*
31. Flowers not campanulate, the perianth parts not green-tipped 32
32. Flowers with a corona, often borne singly or, if in an umbel of 2–6, white 33
32. Flowers without a corona, borne in spikes or, if in an umbel of 2–6, yellow 34
33. Flowers recurved; perianth tube (not corona) up to 2 (–4) cm long; stamens included within the corona; ovules several per cell of ovary; bract one .... 27. *Narcissus*
33. Flowers ascending; perianth tube (not corona) 5–10 cm long; stamens exserted from the corona; ovules 2 per cell of ovary; bracts 2-several .. .. 28. *Hymenocallis*
34. Perianth parts united to form a tube, greenish-yellow, leaves large, glabrous, fleshy .... 29. *Polianthes*
34. Perianth parts free, bright yellow; leaves small, pubescent, not fleshy .. .. 30. *Hypoxis*

1. Tofieldia Huds.   False Asphodel

1. T. glutinosa (Michx.) Pers. False Asphodel. June–July. Bogs, rare; restricted to the extreme ne. cos.

2. Veratrum L.   False Hellebore

1. V. woodii Robbins. False Hellebore. July–Aug. Rich, moist woodlands, rare; apparently confined to a few cent. cos.

3. Melanthium L.   Bunch-flower

1. M. virginicum L. Bunch-flower. June–July. Meadows, wet prairies; occasional in the w. cent. cos., rare in the sw. cos.; absent elsewhere.

4. Zigadenus Michx.   Camass

1. Z. glaucus Nutt. White Camass. July–Sept. Limestone cliffs or low areas near rivers, rare; JoDaviess and Kane cos.

5. Stenanthium Gray   Grass-leaved Lily

1. S. gramineum (Ker) Morong. Grass-leaved Lily. June–Aug. Moist woods and along streams, not common; confined to the s. half of the state. (Including var. *robustum* [S. Wats.] Fern.)

6. Chamaelirium Willd.   Devil's Bit

1. C. luteum (L.) Gray. Fairy Wand. May–June. Low, wooded hillsides, rare; Hardin, Massac, and Pope cos.

7. Lilium L.   Lily

1. All the leaves alternate, the upper with bulblets in the axils; stems scabrous .... 1. *L. lancifolium*

1. Some of the leaves borne in whorls; bulblets absent; stems glabrous   2
2. Flowers erect; perianth parts with claws; only the uppermost group of
leaves in a whorl; plants less than 1 m tall ....
.... 2. *L. philadelphicum* var. *andinum*
2. Flowers nodding; perianth parts without claws; several whorls of leaves
borne on the stem; plants generally over 1 m tall 3
3. Bulbs yellow; margins and nerves of leaves roughened; midvein of outer
perianth parts rounded on the back; anthers 8–15 mm long; filament attached 1–2 mm from end of anther .... 3. *L. michiganense*
3. Bulbs white; margins and nerves of leaves smooth; midvein of outer
perianth parts sharply ridged on the back; anthers 17–25 mm long; filament attached 4–8 mm from end of anther .... 4. *L. superbum*

1. L. lancifolium Thunb. Tiger Lily. July–Aug. Native of e. Asia; escaped from cultivation into waste ground. *L. tigrinum*—F, G, J.

2. L. philadelphicum L. var. andinum (Nutt.) Ker. Wood Lily. June–
July. Dry woodlands, occasional; restricted to the upper half of the
state; also Macoupin Co. *L. umbellatum* Pursh—J.

3. L. michiganense Farw. Turk's-cap Lily. May–July. Moist woodlands and prairies; rather common throughout the state.

4. L. superbum L. Superb Lily. July. Low, moist woodlands, rare;
Jackson and Pope cos. *L. michiganense* Farw.—G.

8. Hemerocallis L.   Day Lily

1. Flowers orange .... 1. *H. fulva*
1. Flowers yellow .... 2. *H. lilio-asphodelus*

1. H. fulva L. Orange Day Lily. June–Aug. Introduced from Eurasia;
roadsides and waste ground, common; probably in every co.

2. H. lilio-asphodelus L. Yellow Day Lily. June–Aug. Introduced
from Asia; rarely escaped into waste ground. *H. flava* L.—F, G, J.

9. Hosta Tratt   Plantain Lily

1. H. lancifolia (Thunb.) Engl. Plantain Lily. Aug. Native of e. Asia;
rarely escaped into disturbed areas. *H. japonica* (Thunb.) Voss—F, G,
J.

10. Ornithogalum L.   Star-of-Bethlehem

1. O. umbellatum L. Star-of-Bethlehem. April–June. Native of
Europe; adventive in fields, open woods, and along roads; common
throughout the state.

11. Aletris L.   Colic Root

1. A. farinosa L. Colic Root. June–Aug. Moist, sandy prairies, sandy
flats; restricted to the n. one-third of the state.

12. Muscari Mill.   Grape Hyacinth

1. Leaves flat, most or all of them over 3 mm broad 2
1. Leaves terete, not more than 3 mm broad .... 4. *M. atlanticum*

2. Raceme up to 6 cm long; perianth to 6 mm long 3
2. Raceme at least 9 cm long; perianth 9–11 mm long .... 3. *M. comosum*
3. All flowers fertile, purple or blue; perianth 3.5–5.0 mm long ....
.... 1. *M. botryoides*
3. Fertile flowers deep violet with white teeth; sterile flowers pale blue; perianth 5–6 mm long .... 2. *M. armeniacum*

    1. M. botryoides (L.) Mill. Grape Hyacinth. April. Native of Europe; escaped to fields, particularly grassy places; occasional in the s. half of the state.
    2. M. armeniacum Leicht. Heavenly Blue. April–May. Native of sw. Asia; escaped from cultivation in Piatt Co.
    3. M. comosum (L.) Mill. Grape Hyacinth. April. Native of s. Europe and Asia; adventive in Jackson Co.
    4. M. atlanticum Boiss. & Reut. Blue Bottle. April–May. Native of Europe; occasionally escaped into waste ground. *M. racemosum* (L.) Mill.—F, G, J.

13. *Camassia Lindl.   Wild Hyacinth*

1. Scape with 0–2 (–3) deciduous bracts; inflorescence at anthesis 3–5 cm broad; capsule as broad as long; perianth parts white to pale blue to pale lilac; plants beginning to flower in early April ..... 1. *C. scilloides*
1. Scape with 3–24 persistent bracts; inflorescence at anthesis 2–3 (–3.5) cm broad; capsule longer than broad; perianth parts deep lavender to pale purple; plants beginning to flower in May .... 2. *C. angusta*

    1. C. scilloides (Raf.) Cory. Wild Hyacinth. April–June. Moist prairies and woodlands; occasional throughout the state.
    2. C. angusta (Engelm. & Gray) Blankinship. Wild Hyacinth. May–July. Prairies and woodlands, very rare; Macon Co.

14. *Erythronium L.   Dog-tooth Violet*

1. Flowers yellow; stigmas united .... 1. *E. americanum*
1. Flowers white; stigmas free .... 2. *E. albidum*

    1. E. americanum Ker. Yellow Dog-tooth Violet; Yellow Trout Lily. April–May. Moist woodlands, frequently on shaded bluffs; occasional in the ne., e. cent., and s. cos.; also Calhoun Co.
    2. E. albidum Nutt. White Dog-tooth Violet; White Trout Lily. April–May. Woods and occasionally fields; common throughout the state.

15. *Uvularia L.   Bellwort*

1. Leaves perfoliate, puberulent beneath; flowers bright yellow, the perianth parts nearly all over 25 mm long .... 1. *U. grandiflora*
1. Leaves sessile, not perfoliate, glabrous; flowers stramineous, the perianth parts nearly all less than 25 mm long .... 2. *U. sessilifolia*

    1. U. grandiflora Sm. Yellow Bellwort. April–May. Rich woodlands; common throughout the state.
    2. U. sessilifolia L. Sessile-leaved Bellwort. April–May. Rich woodlands, not common; confined to the s. one-third of the state.

16. Polygonatum Mill. Solomon's Seal

1. Leaves pilose on the nerves beneath .... 1. *P. pubescens*
1. Leaves glabrous beneath 2
2. Leaves more or less clasping or sheathing at the base, the largest ones
with over 100 nerves; perianth 17–20 mm long, the lobes 5–7 mm long
.... 2. *P. commutatum*
2. Leaves sessile, the largest ones with less than 100 nerves; perianth
10–17 mm long, the lobes 3–4 mm long .... 3. *P. biflorum*

1. P. pubescens (Willd.) Pursh. Small Solomon's Seal. May–June.
Moist, shaded woods; restricted to a few extreme n. cos.
2. P. commutatum (Schult.) A. Dietr. Solomon's Seal. April–June.
Woods, common; in every co.
3. P. biflorum (Walt.) Ell. Small Solomon's Seal. April–June. Dry
woods, sandstone cliffs; restricted to the s. tip of the state.

17. Smilacina Desf. False Solomon's Seal

1. Flowers paniculate; stamens longer than the perianth parts; perianth
parts to 3 mm long; pedicels less than 4 mm long; berry red ....
.... 1. *S. racemosa*
1. Flowers racemose; stamens shorter than or equalling the perianth parts;
perianth parts 4–6 mm long; pedicels at least 4 mm long; berry black or
greenish-black .... 2. *S. stellata*

1. S. racemosa (L.) Desf. False Solomon's Seal; False Spikenard.
April–June. Rich, moist woods, common; in every co.
2. S. stellata (L.) Desf. Small False Solomon's Seal. May–June.
Moist woodlands, prairies; occasional in the n. three-fifths of the
state; also Crawford and Wabash cos.

18. Convallaria L. Lily-of-the-Valley

1. C. majalis L. Lily-of-the-Valley. May–June. Native of Europe; occasionally escaped from cultivation.

19. Asparagus L. Asparagus

1. A. officinalis L. Asparagus. May–June. Native of Europe; commonly escaped from cultivation; probably in every co.

20. Maianthemum Weber False Lily-of-the-Valley

1. M. canadense Desf. False Lily-of-the-Valley. May–June. Leaves
glabrous beneath; transverse veins of leaf conspicuous in transmitted
light. Moist woodlands, rare; Cook Co.
var. interius Fern. False Lily-of-the-Valley. May–June. Leaves pubescent beneath; transverse veins of leaf obscure in transmitted light.
Moist woodlands, occasional; restricted to the n. one-fourth of the
state.

182   LILIACEAE

21. *Allium* L. Onion

1. A. tricoccum Ait. Wild Leek; Ramp. June–July. Petioles and leaf sheaths reddish; blades mostly 2.6–6.0 cm broad, elliptic. Moist, rich woodlands; occasional in the n. half of the state, rare elsewhere.

var. burdickii Hanes. Wild Leek; Ramp. June–July. Petioles and leaf sheaths greenish or whitish; blades mostly 0.8–2.0 cm broad, lanceolate. Low, moist woods; occasional in the n. one-fourth of the state.

2. A. sativum L. Garlic. June–Sept. Native of Europe and w. Asia; occasionally escaped from cultivation.

3. A. ampeloprasum L. var. atroviolaceum (Boiss.) Regel. Wild Leek. July–Sept. Native of se. Europe and Asia; rarely escaped along roads; Pope and Union cos.

4. A. canadense L. Wild Onion. May–July. Dry woodlands, prairies, waste ground, common; in every co.

5. A. porrum L. Leek. May–July. Cultivated plant rarely escaped from cultivation; Johnson Co.

6. A. cernuum Roth. Nodding Onion. July–Sept. Wooded banks, not common; restricted to ne. Ill.

7. A. stellatum Ker. Cliff Onion. July–Sept. Hill prairies in calcareous areas; Jackson, McHenry, Monroe, Randolph, and Union cos.

8. A. mutabile Michx. Wild Onion. June–Aug. Dry areas, not common; DuPage, Jackson, Johnson, Kane, and Will cos.

9. A. vineale L. Field Garlic. May–Aug. Native of Europe; naturalized in waste ground, common; in every co.

10. A. schoenoprasum L. Chives. June–Aug. Bulbs several. Native of Europe; rarely escaped from cultivation; Jackson Co.

var. sibiricum (L.) Hartm. June–Aug. Bulbs 1–2. Native of Europe and Asia; rarely escaped from cultivation; Kankakee Co.

11. A. cepa L. Onion. May–Aug. Native of sw. Asia; rarely escaped from cultivation; Union Co.

22. Nothoscordum Kunth False Garlic

1. N. bivalve (L.) Britt. False Garlic. April–June. Dry woods, bluffs, and prairies; rather common in the s. half of the state, apparently absent elsewhere.

23. Medeola L. Cucumber Root

1. M. virginiana L. Indian Cucumber Root. May–June. Rich, moist woodlands, rare; Cook and LaSalle cos.

24. Trillium L. Wake Robin

1. Flower sessile, basically maroon or green (rarely yellow). 2
1. Flower pedunculate, white or occasionally pink or purple, never green or yellow 5
2. Leaves petiolate; petals abruptly tapering to distinct claws; sepals reflexed .... 1. *T. recurvatum*
2. Leaves sessile; petals scarcely or gradually tapering, usually without distinct claws; sepals spreading or erect, not reflexed 3
3. Stamens one-half as long as petals; connectives 1–3 mm long 4
3. Stamens one-third as long as petals; connectives less than 1 mm long
.... 4. *T. cuneatum*
4. Stems and veins on lower leaf surface glabrous; anthers 5–6 times longer than the filaments; pollen yellow .... 2. *T. sessile*
4. Summit of stems and veins on lower leaf surface minutely scabrous; anthers 4 times longer than the filaments; pollen olive to brownish .....
.... 3. *T. viride*
5. Ovary and fruit 3-angled .... 5. *T. nivale*
5. Ovary and fruit 6-angled 6
6. Ovary purple (in Ill. specimens) .... 6. *T. erectum*
6. Ovary white 7
7. Petals strongly recurved, 1.5–2.5 cm long; peduncle 10–40 mm long; anthers 4–7 mm long .... 7. *T. cernuum* var. *macranthum*
7. Petals spreading, 2–6 cm long; peduncle 25–100 mm long; anthers (6–) 7–15 mm long 8
8. Stigmas erect; anthers slightly longer than the filaments ....
.... 8. *T. grandiflorum*
8. Stigmas spreading; anthers about twice as long as the filaments ....
.... 9. *T. flexipes*

1. T. recurvatum Beck. Red Trillium; Wake Robin. March–May. Rich, moist woodlands, common; in every co. (Including f. *luteum* Clute and f. *shayii* Palmer & Steyerm.)

2. T. sessile L. Sessile Trillium; Sessile Wake Robin. March–April. Rich woodlands, occasional; scattered throughout the state, except in the w. cos. (Including f. *viridiflorum* Beyer.)

3. T. viride Beck. Green Trillium. April–May. Rich woodlands, prairies, not common; Jackson, Macoupin, Pike, and Union cos.

4. T. cuneatum Raf. Trillium. April. Rich, moist woodlands, very rare; Jackson Co.

5. T. nivale Riddell. Snow Trillium. March–April. Rich woodlands, occasional; restricted to the n. three-fifths of the state.

6. T. erectum L. Purple Trillium. May. Rich woods, very rare; Lake and McHenry cos.

7. T. cernuum L. var. macranthum Eames & Wieg. Nodding Trillium. May–June. Moist woodlands, very rare; Cook and McHenry cos.

8. T. grandiflorum (Michx.) Salisb. Large White Trillium. April–May. Rich, moist woodlands; occasional in the n. half of the state, very rare elsewhere.

9. T. flexipes Raf. White Trillium. April–May. Rich woodlands; occasional throughout the state. *T. gleasonii* Fern.—F.

25. Yucca. L. Adam's Needle

1. Y. filamentosa L. var. smalliana (Fern.) Ahles. Yucca; Adam's Needle. May–July. Native of the east coast; escaped from cultivation along roads and in cemeteries. *Y. filamentosa* L.—F, G, J.

26. Leucojum L. Snowflake

1. L. aestivum L. Summer Snowflake. April–May. Native of Europe; rarely escaped to roadside ditches; Pope Co.

27. Narcissus L. Narcissus

1. Perianth yellow; corona yellow, tubular, about as long as the perianth . . . . 1. *N. pseudo-narcissus*
1. Perianth white; corona white, with red margin, cupular, less than one-fourth as long as the perianth . . . . 2. *N. poeticus*

1. N. pseudo-narcissus L. Daffodil. April–May. Native of Europe; occasionally escaped from cultivation; scattered in Ill.

2. N. poeticus L. Poet's Narcissus. April–May. Native of Europe; rarely escaped from cultivation along roads; scattered but not common in Ill.

28. Hymenocallis Salisb. Spider Lily

1. H. occidentalis (LeConte) Kunth. Spider Lily. July–Sept. Low woods, swamps; restricted to the s. one-sixth of the state; also Wabash Co.

29. Polianthes L. Aloe

1. P. virginica (L.) Shinners. American Agave. May–July. Sandstone outcroppings; dry woodlands, occasional; restricted to the s. one-fourth of the state; also Jersey Co. *Agave virginica* L.—F, G, J.

30. *Hypoxis* L. *Star Grass*

1. H. hirsuta (L.) Coville. Yellow Star Grass. April–June. Dry woods, prairies, fields, sandstone outcroppings, calcareous fens; occasional throughout the state.

31. *Scilla* L. *Squill*

1. S. sibirica Andr. April–May. Native of Eurasia; rarely spreading from cultivation; McDonough Co.

34. SMILACACEAE

1. Stems woody, with few to many prickles, rarely without prickles; ovule 1 per cell of ovary 2
1. Stems herbaceous, without prickles; ovules 2 per cell of ovary 5
2. Leaves pale beneath, usually glaucous .... 1. *S. glauca*
2. Leaves green on both sides 3
3. Stems flexuous, with stout spines, or spines absent; leaves subcoriaceous to coriaceous 4
3. Stems not flexuous, with weak spines, or spines absent; leaves membranaceous .... 4. *S. hispida*
4. Leaves thick-margined, coriaceous, often blotched with white, often panduriform; peduncles much longer than the subtending petioles; berries not glaucous .... 2. *S. bona-nox*
4. Leaves thin-margined, subcoriaceous, not blotched with white, never panduriform; peduncles about as long as the subtending petioles; berries glaucous .... 3. *S. rotundifolia*
5. Stems twining or climbing, with numerous tendrils all along the stem; peduncles borne from the axils of developed leaves 6
5. Stems erect, with few (or no) tendrils near the apex; peduncles borne from bladeless sheaths 8
6. Leaves glabrous and glaucous beneath .... 5. *S. herbacea*
6. Leaves pubescent beneath 7
7. Blades light green, not shiny beneath; petiole generally short; berries blue .... 6. *S. lasioneuron*
7. Blades dark green, shiny beneath; petiole long; berries black ....
.... 7. *S. pulverulenta*
8. Basal leaves narrowly ovate or elliptical, the base mostly truncate to subcordate; petioles usually equal to or longer than blade . 8. *S. illinoensis*
8. Basal leaves broadly ovate, the base cordate; petioles generally equal to or shorter than blade .... 9. *S. ecirrata*

1. S. glauca Walt. Catbrier. May–June. Leaves minutely pubescent beneath. Dry woods, edges of fields and bluffs; common in the s. one-third of Ill., absent elsewhere.
var. leurophylla Blake. Catbrier. May–June. Leaves glabrous beneath. Edges of swamps; Johnson, Pope, and Union cos.
2. S. bona-nox L. Catbrier. May–June. Leaves spinulose, with patches of white coloration. Dry woods and fields, on bluffs; confined to the s. one-fourth of the state.

var. hederaefolia (Beyrich) Fern. Catbrier. May–June. Leaves entire or with very weak marginal spinules, green throughout. Woods and fields, not common; restricted to the s. one-sixth of the state.

3. S. rotundifolia L. Catbrier. April–May. Dry woods, edge of fields; common in the s. one-third of the state, absent elsewhere.

4. S. hispida Muhl. Bristly Catbrier. May–June. Usually moist woods; probably in every co. *S. tamnoides* L. var. *hispida* (Muhl.) Fern.—F.

5. S. herbacea L. Carrion Flower. April–June. Moist woods, very rare; Jackson Co.

6. S. lasioneuron Hook. Carrion Flower. April–June. Moist woods; edge of fields; in every co. *S. herbacea* L. var. *lasioneura* (Hook.) A. DC.—F, G.

7. S. pulverulenta Michx. Carrion Flower. April–June. Moist or dry woods; edges of fields; occasional in the s. two-fifths of the state; rare elsewhere. *S. herbacea* L. var. *pulverulenta* (Michx.) Gray—G.

8. S. illinoensis Mangaly. Carrion Flower. May–June. Thickets, often near roads; occasional in the n. three-fourths of the state; absent elsewhere.

9. S. ecirrata (Engelm.) S. Wats. Carrion Flower. May–June. Moist woods; occasional in the n. one-half of Ill., becoming uncommon in the s. cos.

35. DIOSCOREACEAE

1. All leaves (except sometimes the lowermost) alternate; capsule 1.5–2.5 cm long; seeds (including wing) 7–14 mm broad; petiole essentially glabrous at point of attachment of blade .... 1. *D. villosa*
1. Lowest leaves whorled, becoming opposite or alternate on the upper part of the stem; capsule 2.5–3.0 cm long; seeds (including wing) 15–18 mm broad; petiole puberulent at point of attachment of blade ....
.... 2. *D. quaternata*

1. D. villosa L. Wild Yam. June–July. Dry or moist woodlands, common; probably in every co. (Including f. *glabrifolia* [Bartlett] Fern.)

2. D. quaternata [Walt.] J. F. Gmel. Wild Yam. June–July. Dry or moist woodlands, occasional; confined to the s. one-third of the state. (Including var. *glauca* [Muhl.] Fern.)

36. IRIDACEAE

1. Sepals somewhat recurved; petals spreading or erect; styles petal-like; stamens concealed by the arching styles .... 1. *Iris*
1. Sepals and petals similar, spreading or tubular; styles not petal-like; stamens not concealed 2
2. Perianth tubular, forming a short funnel, scarlet; upper three perianth segments longer than the lower three .... 2. *Gladiolus*

2. Perianth spreading and orange, or rotate and blue or white; all perianth segments essentially equal 3
3. Flowers orange, with purplish spots, at least 3 cm broad; filaments distinct; flowers bracteate, the bracts soon withering; capsule at least 2 cm long; seeds attached to a central column, resembling a blackberry; plants rhizomatous .... 3. *Belamcanda*
3. Flowers blue or white, less than 2 cm broad; filaments united to apex; flowers borne from 1 or 2 persistent spathes; capsule up to 6 mm long; seeds not attached to a central column; plants with fibrous roots ......
.... 4. *Sisyrinchium*

1. *Iris* L. Iris

1. Upper surface of sepals with a beard of hairs 2
1. Sepals without a beard of hairs 3
2. Stems very short or none, bearing one flower; leaves to 15 cm long, to 7 mm broad; sepals to 17 mm broad .... 1. *I. pumila*
2. Stems to nearly 1 m tall, bearing several flowers; leaves to 100 cm long, to 3 cm broad; sepals over 20 mm broad .... 2. *I. germanica*
3. Rhizomes stout, at least 1 cm thick; leaves 40–100 cm long; flowering stem usually branched, 20–100 cm long; spathes subtending flower unequal in length; flowers several per stem; sepals 1- to 2-ridged, or without ridges; capsule obtusely 3-angled or 6-angled, 3–9 cm long; plants often of more aquatic situations 4
3. Rhizomes slender, less than I cm thick; leaves (at flowering time) to 25 cm long, at maturity as long as 40 cm; flowering stem unbranched, to 4.5 cm long; spathes subtending flower nearly equal in length; flowers 1–2 per stem; sepals sharply 3-ridged above; capsule sharply 3-angled, about 1 cm long; plants of rich woodlands, usually near streams ....
.... 7. *I. cristata*
4. Ovary and capsule 6-angled; capsule indehiscent; leaves rather soft . 5
4. Ovary and capsule 3-angled (occasionally 6-angled in the yellow-flowered *I. pseudacorus*); capsule dehiscent; leaves firm 6
5. Flowering stem 20–40 cm tall; flowers dark blue; capsule 3–5 cm long; leaves 15–30 mm broad; seeds more or less globoid ... 3. *I. brevicaulis*
5. Flowering stem at least 50 cm tall; flowers coppery; capsule 4.5–7.5 cm long; leaves 10—15 mm broad; seeds flattened .... 4. *I. fulva*
6. Flowers usually some shade of blue or violet; sepals not 2-ridged on upper surface; perianth tube constricted above the ovary . 5. *I. shrevei*
6. Flowers basically yellow; sepals 2-ridged on upper surface; perianth not constricted above the ovary .... 6. *I. pseudacorus*

    1. I. pumila L. Dwarf Iris. March–April. Native of Europe and Asia; escaped into waste ground; DuPage Co.
    2. I. germanica L. Bearded Iris. April–June. Escaped from cultivation into waste land; McLean, Piatt, and Pope cos.
    3. I. brevicaulis Raf. Blue Iris. May–July. Marshes and wet prairies; scattered in the s. two-thirds of the state.
    4. I. fulva Ker. Swamp Red Iris. May–June. Swamps, usually in shallow water; Alexander, Hamilton, Johnson, Massac, Pulaski, and Union cos.
    5. I. shrevei Small. Wild Blue Iris. May–June. Wet situations, common; probably in every co. *I. virginica* L. var. *shrevei* (Small) E. Anders—F.

6. I. pseudacorus L. Yellow Iris. June–Aug. Native of Europe; scattered throughout Ill.

7. I. cristata Ait. Dwarf Crested Iris. April–May. Lowland woods, usually along streams; restricted to the s. tip of Ill.

2. Gladiolus L.   Gladiolus

1. G. X colvillei Sweet. Scarlet Gladiolus. July–Aug. Escaped from cultivation along roads; Johnson and Massac cos. Reputed to be a hybrid between *G. concolor* Salisb. and *G. cardinalis* Curt.

3. Belamcanda Adans.   Blackberry Lily

1. B. chinensis (L.) DC. Blackberry Lily. July–Aug. Native of Asia; adventive along roads; scattered throughout the state.

4. Sisyrinchium L.   Blue-eyed Grass

1. Spathes long-pedunculate from the axils of leaf-like bracts; capsules dark brown 2
1. Spathes sessile, terminal; capsules pale brown, stramineous, or greenish (occasionally becoming dark brown in *S. montanum*) 3
2. Leaves bright green; stems broadly winged, 3–5 mm wide; capsule 4–6 mm long; plants becoming black on drying .... 1. *S. angustifolium*
2. Leaves pale green or glaucous; stems narrowly winged, 1–3 mm wide; capsule 3.0–4.5 mm long; plants generally not becoming black on drying .... 2. *S. atlanticum*
3. Spathes 2 .... 3. *S. albidum*
3. Spathe 1 4
4. Capsules 2–4 mm long; margins of outer bract free to base or united for less than 2 mm 5
4. Capsules 4–6 mm long; margins of outer bract united for 2–6 mm above base .... 6. *S. montanum*
5. Plants pale green or glaucous, drying pale; stems winged; margins of outer bract free to base; flowers pale blue or white ... 4. *S. campestre*
5. Plants dark green, drying blackish; stems marginate; margins of outer bract united for 1–2 mm above base; flowers bright violet ....
.... 5. *S. mucronatum*

1. S. angustifolium Mill. Blue-eyed Grass. May–June. Low woodlands, moist prairies, common; in nearly every co. *S. graminoides* Bickn.—G; *S. bermudiana* L.—J.

2. S. atlanticum Bickn. Blue-eyed Grass. May–June. Wet prairies, very rare; Kankakee Co.

3. S. albidum Raf. Blue-eyed Grass. April–June. Open woodlands, fields, prairies, common; probably in every co.

4. S. campestre Bickn. Blue-eyed Grass. May–June. Prairies, particularly in sandy soil; generally restricted to nw. and w. cent. cos; Pope Co.

5. S. mucronatum Michx. Blue-eyed Grass. May–June. Sandy prairies, rare; Hancock and Mason cos. *S. campestre* Bickn.—J.

6. S. montanum Greene. Blue-eyed Grass. May–June. Plants pale green or glaucous, drying pale; capsule whitish, greenish, or stramineous, even at maturity. Sandy prairies, rare; Cook, Lake, and Winnebago cos. *S. campestre* Bickn.—J.

var. crebrum Fern. Blue-eyed Grass. May–June. Plants bright green, drying blackish; capsule greenish or pale brown at first, becoming dark at maturity. Fields, very rare; Kankakee Co.

37. BURMANNIACEAE

1. Thismia Griff.

1. T. americana N. E. Pfeiffer. Aug. Wet prairies, undoubtedly extinct; Cook Co.

38. ORCHIDACEAE

1. Plants with green leaves at flowering time 2
1. Plants without green leaves, at least at flowering time 15
2. Leaves whorled .... 10. Isotria
2. Leaves alternate or basal 3
3. Lip inflated, sac-like, at least 18 mm long; anthers 2 ... 1. Cypripedium
3. Lip flat or, if inflated, less than 18 mm long; anther 1 4
4. Leaves 1 or 2, basal (leaf 1–2 and cauline in Malaxis) 5
4. Leaves more than 2, basal or cauline or, if only 1–2, then the leaf cauline
.... 8
5. Flowers totally pink, resupinate; leaf generally 1, linear to linear-lanceolate .... 2. Calopogon
5. Flowers greenish-yellow, madder-purple, or white and pink, not resupinate; leaves generally 2, lanceolate to elliptic to ovate to orbicular 6
6. Flowers spurred, usually fragrant; anther persistent, difficult to detach 7
6. Flowers spurless, not fragrant; anther easily and soon detachable ......
.... 5. Liparis
7. Flowers white and pink; sepals and petals united to form a hooded structure (galea) behind the column .... 3. Orchis
7. Flowers greenish-yellow; sepals and petals not forming a galea ....
.... 4. Habenaria
8. Leaves 1–2, cauline .... 6. Malaxis
8. Leaves 3 or more, basal or cauline or both 9
9. All leaves basal 10
9. At least 1 or more cauline leaves present (basal leaves also may be present) 11
10. Leaves green throughout; lip not sac-like .... 7. Spiranthes
10. Leaves conspicuously marked with white; lip sac-like .... 8. Goodyera
11. Flowers spurred; anther persistent, not easily detached .. 4. Habenaria
11. Flowers not spurred; anther easily detached 12
12. Leaves both basal and cauline; flowers 1–2, terminal, or numerous, sessile and spicate 13
12. Leaves all cauline; flowers numerous, pedicellate, racemose 14
13. Cauline leaf 1; sepals and petals free; flowers 1–2, pink, at least 15 mm long .... 9. Pogonia
13. Cauline leaves 2 or more; upper sepal united with petals; flowers numerous, white, at most 10 mm long .... 7. Spiranthes

14. Flowers white (rarely pink), nodding, 1.0–1.5 cm long, the lip not sac-like; sepals and petals lanceolate; leaves weak, oval ...... 11. *Triphora*
14. Flowers green and purple, ascending, well over 1.5 cm long, the lip sac-like at the base; sepals and petals ovate; leaves firm, lanceolate to ovate
.... 12. *Epipactis*
15. Green leaves never produced; stem brown or purple, with colored sheaths; rhizomes coralline 16
15. Green leaves produced, but these usually absent at flowering time; plants with fleshy roots or with corms or tubers connected in a series 17
16. Lip without longitudinal ridges; flowers at most 1 cm long, brown, yellow-green, or white spotted with purple .... 13. *Corallorhiza*
16. Lip with 5–6 longitudinal ridges; flowers 1.7–2.3 cm long, madder-purple
.... 14. *Hexalectris*
17. Leaves more than 1; plants with fleshy roots, but without connected tubers; flowers whitish .... 7. *Spiranthes*
17. Leaf 1; plants with connected tubers; flowers yellow or brown or madder-purple 18
18. Leaves green on both sides; flowers not spurred .... 15. *Aplectrum*
18. Leaves purple beneath; flowers spurred .... 16. *Tipularia*

1. *Cypripedium* L. Lady's-slipper

1. Lip cleft down the middle, pink; leaves 2, basal .... 1. *C. acaule*
1. Lip not cleft, yellow, white, or with pink or purple streaks; leaves more than 2, cauline 2
2. Sepals and petals acuminate, yellow or greenish-yellow, or streaked with purple or brown 3
2. Sepals and petals obtuse, white .... 6. *C. reginae*
3. Lip yellow 4
3. Lip white, occasionally marked with purple streaks 5
4. Lateral petals purple-brown; staminodium ovate .... 2. *C. calceolus*
4. Lateral petals yellow-green striped with brown; staminodium oblong ...
.... 3. *C. calceolus* var. *pubescens* X *C. candidum*
5. Lip white; staminodium narrowly triangular ....
.... 4. *C calceolus* var. *parviflorum* X *C. candidum*
5. Lip white, streaked with purple; staminodium oblong .. 5. *C. candidum*

1. C. acaule Ait. Lady's-slipper Orchid. May–July. Bogs and acid woodlands, very rare; Cook, Lake, McHenry, and Ogle cos.

2. C. calceolus L. var. parviflorum (Salisb.) Fern. Small Yellow Lady's-slipper Orchid. May–June. Lateral petals 3.5–5.0 cm long; sepals usually purplish. Bogs and marshes, rare; restricted to the n. cos. *C. parviflorum* Salisb.—J.

var. pubescens (Willd.) Correll. Yellow Lady's-slipper Orchid. April–May. Lateral petals 5–8 cm long; sepals usually yellow-green. Moist or dry woodlands, not common; scattered throughout the state. *C. parviflorum* Salisb.—J.

3. C. calceolus var. pubescens X C. candidum. June. Shaded ground, very rare; Cook, Henderson, and Woodford cos. Reputed to be a hybrid between *C. calceolus* var. *pubescens* and *c. candidum. C.* X *favillianum* J. T. Curtis.

4. C. calceolus var. parviflorum X C. candidum. Lady's-slipper Orchid. June. Low, springy areas, very rare; McHenry and Cook cos. Reputed to be a hybrid between *C. calceolus* var. *parviflorum* and *C. candidum. C.* X *andrewsii* A. M. Fuller.

5. C. candidum Muhl. White Lady's-slipper Orchid. May–June. Bogs, swamps, and wet prairies, very rare; restricted to the n. half of Ill.; also from Macoupin and St. Clair cos.

6. C. reginae Walt. Showy Lady's-slipper Orchid. June–July. Bogs; low, springy areas, rare; restricted to the n. one-half of the state.

2. Calopogon L. C. Rich. Grass Pink Orchid

1. C. tuberosus (L.) BSP. Grass Pink Orchid. May–July. Wet prairies, not common; confined to the n. half of the state; also Jackson and Macoupin cos. *C. pulchellus* (Salisb.) R. Br.—F, G, J.

3. Orchis L. Orchis

1. O. spectabilis L. Showy Orchis. April–June. Rich woodlands, not common; scattered throughout the state.

4. Habenaria Willd. Fringed Orchis

1. Leaves 2, basal, suborbicular; lip elongated, entire, 8–20 mm long, the spur 13–40 mm long 2
1. Leaves 1-several, cauline, linear to lanceolate to oval; lip lobed, toothed, erose, or fringed, if entire and elongated, then the lip 4–8 mm long and the spur 2–12 mm long 3
2. Lip linear, 12–20 mm long; flowers greenish-white, with pedicels 5–13 mm long; stem bracted .... 1. *H. orbiculata*
2. Lip triangular-elongated, 8–12 mm long; flowers yellow-green, sessile; stem without bracts .... 2. *H. hookeri*
3. Lip entire, erose, 2- to 3-toothed, or very shallowly 3-lobed, with the lobes entire; spur 2–12 mm long; flowers sessile 4
3. Lip fringed or deeply 3-lobed and toothed; spur 12 mm long or longer; flowers pedicellate 8
4. Leaf 1; inflorescence 2–6 cm long; all bracts shorter than the flowers; spur 8–12 mm long; roots slender .... 3. *H. clavellata*
4. Leaves 2-several; inflorescence 5–30 cm long; at least the lower bracts longer than the flowers; spur 2–8 mm long; rootstocks tuberous 5
5. Sepals at least twice as long as the spur, the spur 2–3 mm long; flowers greenish, tinged with purple; lip 2- to 3-toothed ....
.... 4. *H. viridis* var. *bracteata*
5. Sepals shorter than to as long as the spur, the spur 3–8 mm long; flowers greenish-white, greenish-yellow, or white; lip entire or erose . 6
6. Lip erose, with a tubercle borne near the summit .... 5. *H. flava*
6. Lip entire, without a tubercle 7
7. Flowers greenish-white, faintly odorous; lip gradually broadened at the base .... 6. *H. hyperborea*
7. Flowers creamy-white, with a spicy fragrance; lip abruptly broadened at the base .... 7. *H. dilatata*
8. Lip simple, fringed 9
8. Lip deeply 3-lobed, the lobes fringed, long-toothed, or erose 10
9. Flowers orange; leaves lanceolate, some or all over 2 cm broad ....
.... 8. *H. ciliaris*
9. Flowers white; leaves linear to linear-lanceolate, less than 2 cm broad
.... 9. *H. blephariglottis*
10. Flowers yellow-green or white; lobes of the lip fringed 11
10. Flowers reddish-purple; lobes of the lip long-toothed or erose 12

11. Flowers yellow-green or greenish-white; sepals 4.5–7.0 mm long; spur
14–20 mm long .... 10. *H. lacera*
11. Flowers white; sepals 7–12 mm long; spur 20–48 mm long ....
.... 11. *H. leucophaea*
12. Lip shallowly erose, 13–22 mm long, the terminal lobe notched ....
.... 12. *H. peramoena*
12. Lip long-toothed, 9–13 mm long, the terminal lobe not notched ....
.... 13. *H. psycodes*

1. H. orbiculata (Pursh) Torr. Round-leaved Orchid. July–Aug. Moist
woodlands, very rare; Cook and Kane cos.

2. H. hookeri Torr. Hooker's Orchid. June–July. Moist woodlands,
rare; Cook, Hancock, and Lake cos.

3. H. clavellata (Michx.) Spreng. Wood Orchid. July–Aug. Moist,
shaded areas, rare; Cass, Cook, JoDaviess, Kankakee, Lake, Pope,
and Will cos.

4. H. viridis (L.) R. Br. var. bracteata (Muhl.) Gray. Bracted Green
Orchid. May–June. Rich woodlands, rare; confined to the n. one-half
of the state. *H. bracteata* (Muhl.) R. Br.—J.

5. H. flava (L.) R. Br. Tubercled Orchid. June–Aug. Bracts of lowest
flowers equalling the flowers; inflorescence lax; lip suborbicular.
Low, shaded areas, rare; Johnson, Massac, and Wabash cos.

var. herbiola (R. Br.) Ames & Correll. Tubercled Orchid. June–July.
Moist woodlands, rare; restricted to the n. one-half of the state; also
St. Clair and Wabash cos.

6. H. hyperborea (L.) R. Br. var. huronensis (Nutt.) Farw. Green Orchid. June–Aug. Swampy areas, rare; known only from the n. one-half
of the state. *H. huronensis* (Nutt.) Spreng.—J; *H. hyperborea* (L.) R.
Br.—G.

7. H. dilatata (Pursh) Hook. White Orchis. June–July. Springy situations, very rare; Carroll, Kankakee, and McHenry cos.

8. H. ciliaris (L.) R. Br. Yellow Fringed Orchid. June–Aug. Low
ground, rare; Cook Co.; also reported from Kankakee and Union cos.

9. H. blephariglottis (Willd.) Hook. White Fringed Orchid. June–July.
Low areas, very rare; Macon Co.

10. H. lacera (Michx.) Lodd. Green Fringed Orchid. June–July.
Swamps; rare in the n. three-fourths of Ill.; also Massac and Pope
cos.

11. H. leucophaea (Nutt.) Gray. White Fringed Orchid. June–July.
Swampy areas, not common; scattered in the n. two-thirds of the
state.

12. H. peramoena Gray. Purple Fringeless Orchid. June–Sept. Low
woodlands, wet prairies; occasional in the s. two-fifths of the state,
absent elsewhere.

13. H. psycodes (L.) Spreng. Purple Fringed Orchid. June–Aug. Low
woodlands, bogs, rare; Cook, Lake, and Winnebago cos.

5. *Liparis* Rich. Twayblade Orchid

1. Pedicels 5–10 mm long, as long as the flowers; sepals greenish-white,
10–12 mm long; lateral petals 10–12 mm long, madder-purple; lip 8–12
mm long, madder-purple, flat .... 1. *L. liliifolia*
1. Pedicels 4–5 mm long, usually a little shorter than the flowers; sepals
yellow-green, 4–6 mm long; lateral petals 4–6 mm long, yellowish; lip 4–6

mm long, yellow-green, turned up slightly along the margins ....
.... 2. *L. loeselii*

    1. L. liliifolia (L.) Rich. Twayblade Orchid. May–July. Rich wood-
lands, occasional; scattered throughout the state.
    2. L. loeselii (L.) Rich. Lesser Twayblade Orchid. June–July. Low
woodlands, bogs, rare; restricted to the n. half of the state.

6. *Malaxis* Sw. Adder's Mouth Orchid

1. Pedicels 4–9 mm long; lip bilobed, ascending; upper sepal 1.2–1.6 mm
    long .... 1. *M. unifolia*
1. Pedicels 1–3 mm long; lip entire, deflexed; upper sepal 2.0–2.5 mm long
    .... 2. *M. monophylla* var. *brachypoda*

    1. M. unifolia Michx. Adder's Mouth Orchid. June–Aug. Dry or moist
woodlands, rare; Hancock, Henderson, and Menard cos.
    2. M. monophylla (L.) Sw. var. brachypoda (Gray) F. Morris. Adder's
Mouth Orchid. July. Bogs, very rare; Kane and McHenry cos. *M. bra-
chypoda* (Gray) Fern.—F; *M. monophylla* (L.) Sw.—G.

7. *Spiranthes* Rich. Ladies' Tresses

1. Flowers produced in 2–3 rows, the spikes usually crowded; rachis pube-
    scent (see also *S. vernalis* under second no. 1 of couplet) 2
1. Flowers produced in a single, twisted row, often appearing secund, the
    spikes laxly-flowered; rachis glabrous (pubescent in *S. vernalis*) 5
2. Lip yellow; flowers produced from May to July .... 1. *S. lucida*
2. Lip white or cream; flowers produced from August to October 3
3. Sepals and petals 4–5 mm long; lip 4–5 mm long, with two basal, con-
    spicuous, incurved callosities .... 2. *S. ovalis*
3. Sepals and petals 7–12 mm long; lip 7–12 mm long, with two basal,
    rather obscure (sometimes prominent in *S. cernua*), rounded callosities
    4
4. Leaves usually persisting, at least through anthesis; flowers white, not
    strongly scented .... 3. *S. cernua*
4. Leaves usually absent during anthesis; flowers cream, strongly scented
    .... 4. *S. magnicamporum*
5. Rachis pubescent; sepals, petals, and the lip 6–10 mm long, the lip yel-
    lowish; lowest cauline leaves usually resembling the basal leaves,
    present at flowering time .... 5. *S. vernalis*
5. Rachis glabrous; sepals, petals, and the lip 3–4 mm long, the lip white or
    white with a green area; all leaves basal, withered or absent at flowering
    time 6
6. Tuberous roots several; lip white with a green central area; largest
    leaves usually 2–4 cm long 7
6. Tuberous root 1 or occasionally 2 or 3; lip white throughout; largest
    leaves at most 2 cm long .... 8. *S. tuberosa*
7. Leaves present but withered at flowering time, thin; spike little spiraling;
    summit of lip green with a broad white border; plants flowering from
    mid-June to early September, averaging by August 5 ...... 6. *S. lacera*
7. Leaves absent at flowering time, thick; spike strongly spiraling; summit
    of lip green with a narrow white border; plants flowering from late July
    to early October, averaging by September 2 .... 7. *S. gracilis*

1. S. lucida (H. H. Eaton) Ames. Yellow-lipped Ladies' Tresses. May–June. Wet situations, rare; Cook, Hancock, Will, and Woodford cos.

2. S. ovalis Lindl. Ladies' Tresses. Sept.–Oct. Rich woodlands, not common; scattered in the s. three-fourths of the state.

3. S. Cernua (L.) Rich. Nodding Ladies' Tresses. Aug.–Oct. Moist or dry situations in usually open areas; occasional throughout the state.

4. S. magnicamporum Sheviak. Aug.–Oct. Dry prairies; scattered in most of the state, except the southernmost cos.

5. S. vernalis Engelm. & Gray. Ladies' Tresses. July–Aug. Rich woods, prairies, rare; Effingham, Massac, Pope, St. Clair, and Wabash cos.

6. S. lacera (Raf.) Raf. Slender Ladies' Tresses. June–Sept. Mostly dry woodlands, rare; restricted to the n. half of the state. *S. gracilis* (Bigel.) Beck—J.

7. S. gracilis (Bigel.) Beck. Slender Ladies' Tresses. July–Oct. Dry, open woodlands, rare; restricted to the s. one-third of Ill.

8. S. tuberosa Raf. Little Ladies' Tresses. July–Oct. Dry, open woodlands; confined to the s. one-half of the state. *S. grayi* Ames—J.

8. Goodyera R. Br. Rattlesnake Plantain Orchid

1. G. pubescens (Willd.) R. Br. Rattlesnake Plantain Orchid. June–Aug. Rich, moist woodlands; scattered throughout the state.

9. Pogonia Juss. Pogonia

1. P. ophioglossoides (L.) Ker. Snake-mouth. June–July. Low ground, rare; restricted to the n. one-fourth of the state.

10. Isotria Raf. Whorled Pogonia

1. I. verticillata (Willd.) Raf. Whorled Pogonia. May–June. Lowland woods, very rare; Pope Co.

11. Triphora Nutt. Nodding Pogonia

1. T. trianthophora (Sw.) Rydb. Nodding Pogonia. Aug.–Oct. Rich woodlands, not common; scattered throughout the state.

12. Epipactis Sw. Helleborine

1. E. helleborine (L.) Crantz. Helleborine. July–Aug. Native of Europe; adventive in disturbed woodlands; Cook, DuPage, Lake, and McHenry cos. *E. latifolia* (L.) Crantz—G.

13. Corallorhiza Chat. Coral-root Orchid

1. Lip with two lateral lobes or teeth; plants flowering generally from mid-May to mid-August. 2
1. Lip entire or erose, not lobed or toothed; plants flowering generally from March to mid-May or from mid-August to October 3
2. Stems yellowish; sepals and petals yellow-green, rarely spotted with purple, 4–5 mm long, 1-nerved; spur absent; lip 4–5 mm long, with 2 short

lateral lobes; capsule greenish, 6–10 mm long; rhizome white . . . .
. . . . 1. *C. trifida*
2. Stems purplish; sepals and petals white with purple spots, 6–8 mm long,
3-nerved; spur present; lip 6–8 mm long, with 2 well-developed lateral
teeth or lobes; capsule brownish, 10–20 mm long; rhizome brown . . . . .
. . . . 2. *C. maculata*
3. Stems purple; sepals and petals greenish-yellow, spotted with purple,
linear-lanceolate, 6–8 mm long; lip 5–6 mm long, notched at apex, long-
clawed; capsule 8–12 mm long; flowering in early spring 3. *C. wisteriana*
3. Stems purple or brown below, greenish above; sepals and petals pur-
plish-green to purple, oblong, 3–5 mm long; lip 3–4 mm long, undulate
at apex but not notched, short-clawed or clawless; capsule 5–8 mm
long; flowering in autumn . . . . 4. *C. odontorhiza*

    1. C. trifida Chat. Pale Coral-root Orchid. May. Moist woods, very
rare; Cook Co.
    2. C. maculata Raf. Spotted Coral-root Orchid. June–Aug. Woods
rich in humus, rare; confined to the n. two-fifths of the state.
    3. C. wisteriana Conrad. Wister's Coral-root Orchid. March–May.
Rich woodlands; scattered in the s. half of the state; also LaSalle Co.
    4. C. odontorhiza (Willd.) Nutt. Fall Coral-root Orchid. Aug.–Oct.
Woodlands; scattered throughout the state, but not common.

14. *Hexalectris* Raf.   Crested Coral-root Orchid

    1. H. spicata (Walt.) Barnh. Crested Coral-root Orchid. June–Sept.
Dry woodlands; limestone ledges; Jackson, Monroe, Pope, and Randolph cos.

15. *Aplectrum* (Nutt.) Torr.   Putty-root Orchid

    1. A. hyemale (Muhl.) Torr. Putty-root Orchid. May–June. Rich
woods, occasional; scattered throughout the state.

16. *Tipularia* Nutt.   Crane-fly Orchid

    1. T. discolor (Pursh) Nutt. Crane-fly Orchid. July–Aug. Rich woods,
rare; confined to the s. tip of Ill.

39. SAURURACEAE

1. *Saururus* L.   Lizard's-tail

    1. S. cernuus L. Lizard's-tail. May–Sept. Swampy woods, sometimes
in standing water; occasional to common in the two-thirds of Ill.; also
Cook, DeKalb, Grundy, and Kankakee cos.

40. SALICACEAE

1. Leaves twice as long as broad or longer; bud scale one; catkins not
drooping . . . . 1. *Salix*

1. Leaves never twice as long as broad; bud scales several; catkins drooping .... 2. *Populus*

1. Salix L. Willow

1. One or more glands present at upper end of petiole 2
1. Glands absent at upper end of petiole 4
2. Leaves whitish beneath; capsule 7–10 mm long .... 6. *S. serissima*
2. Leaves green beneath; capsule less than 7 mm long 3
3. Leaves with long-tapering, almost tail-like, tip .... 5. *S. lucida*
3. Leaves merely acute to short-acuminate at tip .... 4. *S. pentandra*
4. Leaves purplish, at least some of them opposite .... 24. *S. purpurea*
4. Leaves green or whitish, alternate 5
5. Leaves glabrous on the lower surface 6
5. Leaves pubescent, at least on the lower surface 17
6. Leaves remotely denticulate; petioles less than 3 mm long 10. *S. interior*
6. Leaves closely crenate or crenate-serrate; petioles 3 mm long or longer
.... 7
7. Leaves entire, revolute; some of the stems creeping   17. *S. pedicellaris*
7. Leaves crenate, serrate, or denticulate; all stems upright 8
8. Stipules absent or minute and falling away early on vegetative sprouts
and young branchlets 9
8. Stipules persistent on vegetative sprouts and young branchlets 13
9. Leaves green beneath, coarsely undulate-serrate .... 7. *S. fragilis*
9. Leaves whitish beneath, finely serrulate 10
10. Branchlets "weeping"; capsule 1.0–2.5 mm long ..... 8. *S. babylonica*
10. Branchlets not "weeping"; capsule 3–9 mm long 11
11. Leaves tapering to a long-acuminate, almost tail-like, tip ....
.... 3. *S. amygdaloides*
11. Leaves acute to acuminate, but not tail-like 12
12. Flowers appearing before the leaves; teeth along margin of blade not extending all the way to the base .... 20. *S. petiolaris*
12. Flowers appearing with the leaves; teeth along margin of blade extending all the way to the base .... 9. *S. alba*
13. Leaves green on both sides 14
13. Leaves pale on the lower surface 15
14. Leaves lanceolate, rarely as much as 2 cm broad (except sometimes
those on the sprouts), tapering to base; each staminate flower with two
glands at the base; stamens 3 .... 1. *S. nigra*
14. Leaves oblong-lanceolate, some of them at least 2 cm broad, rounded or
even subcordate at base; each staminate flower with one gland at the
base; stamens 2 .... 11. *S. rigida*
15. Leaves irregularly crenate-serrate; flowers appearing before leaves ex-
pand; capsule puberulent .... 18. *S. discolor*
15. Leaves finely serrulate; flowers appearing with the leaves; capsule glabrous or granular 16
16. Leaves lanceolate, often falcate; each staminate flower with two glands
at base; stamens 4–8; capsule 4–6 mm long, the pedicel at least half as
long .... 2. *S. caroliniana*
16. Leaves oblong to narrowly ovate, not falcate; each staminate flower with
one gland at base; stamens 2; capsule 7–10 mm long, the pedicel much
less than half as long .... 13. *S. glaucophylloides* var. *glaucophylla*
17. Young branchlets and new leaves covered with white wool ....
.... 23. *S. candida*

1. S. nigra Marsh. Black Willow. April–May. Along streams, common; in every co.

2. S. caroliniana Michx. Ward's Willow. May. Wet soil, not common; Jackson, Madison, Perry, St. Clair and Williamson cos.

3. S. amygdaloides Anderss. Peach-leaved Willow. April–May. Along streams; occasional throughout the state.

4. S. pentandra L. Bay-leaved Willow. May. Native of Europe; escaped along creek in Champaign Co.

5. S. lucida Muhl. Shining Willow. May. Bogs and wet sandy areas, not common; confined to the n. one-fourth of the state.

6. S. serissima (Bailey) Fern. Autumn Willow. June. Marshes and bogs, very rare; Lake Co.

7. S. fragilis L. Crack Willow. April–May. Native of Europe; occasionally escaped in all parts of the state.

8. S. babylonica L. Weeping Willow. April–May. Native of Europe and Asia; rarely escaped from cultivation.

9. S. alba L. White Willow. April–May. Leaves silky beneath. Native of Europe; occasionally escaped in most parts of Ill.

var. calva G. F. W. Mey. April. Leaves glabrate, branchlets brown. Native of Europe; occasionally escaped in Ill.

var. vitellina (L.) Stokes. Golden Willow. April. Leaves glabrate, branchlets yellow. Native of Europe; rarely escaped; Champaign, Kankakee, and Vermilion cos. S. vitellina L.—J.

10. S. interior Rowlee. Sandbar Willow. April–May. Leaves glabrate beneath. Sandy shores of rivers and streams; common throughout the state.

f. wheeleri (Rowlee) Rouleau. April–May. Leaves silvery-silky beneath. Sandy shores; occasional in most parts of the state.

11. S. rigida Muhl. Heart-leaved Willow. April–May. Moist ground; occasional throughout the state. (Including var. angustata [Pursh] Fern.)

12. S. X myricoides (Muhl.) Carey. May. Moist soil, rare; Jackson Co. Reputedly a hybrid between S. rigida Muhl. and S. sericea Marsh.

13. S. glaucophylloides Fern. var. glaucophylla (Bebb) Schneid. Blue-leaf Willow. May. Open sand, calcareous pond borders, marsh borders, not common; confined to the n. one-third of Ill. S. glaucophylloides Fern.—G, J.

14. S. eriocephala Michx. April. Low woods, rare; confined to the s. one-third of Ill.

15. S. syrticola Fern. Sand-dune Willow. June. Sand dunes, rare; Cook and Lake cos. S. cordata Michx.—G, J.

16. S. bebbiana Sarg. Bebb Willow. May. Bogs, occasional; confined to the n. one-fourth of Ill.

17. S. pedicellaris Pursh var. hypoglauca Fern. April–May. Bogs, rare; confined to the n. one-third of the state. S. pedicellaris Pursh—J.

18. S. discolor Muhl. Pussy Willow. March–May. Marshes and swamps; occasional throughout the state, but less common in the s. cos. (Including var. latifolia Anderss.)

19. S. humilis Marsh. Prairie Willow. March–May. Branchlets 2 mm or more thick, leaves pilose beneath, the petiole 3 mm long or longer. Prairies; occasional throughout the state.

var. hyporhysa Fern. March–May. Branchlets 2 mm or more thick, leaves glabrate beneath, the petiole 3 mm long or longer. Prairies; occasional in the s. cos.

var. microphylla (Anderss.) Fern. Sage Willow. March–April. Branchlets less than 2 mm thick, leaves with petioles less than 3 mm long. Prairies; Cook, Kankakee, and McDonough cos. S. tristis Ait.—G.

20. S. petiolaris Sm. Petioled Willow. April–June. Low prairies, rich woods, marshes, bogs; confined to the n. one-fourth of the state. S. gracilis Anderss. var. textoris Fern.—F.

21. S. sericea Marsh. Silky Willow. March–May. Low ground, along streams, in bogs; occasional and scattered throughout the state.

22. S. caprea L. Goat Willow. March–April. Native of Europe; frequently planted but rarely escaped from cultivation; Jackson and Wabash cos.

23. S. candida Fluegge. Hoary Willow. April–May. Bogs; confined to the n. one-third of the state.

24. S. purpurea L. Purple Osier. April–May. Native of Europe; occasionally cultivated but rarely escaped in Ill.; Cook and Lake cos.

25. S. X subsericea (Anderss.) Schneid. April–June. Low ground, rare; Cook and Kankakee cos. Reputed to be a hybrid between *S. petiolaris* Sm. and *S. sericea* Marsh.

2. *Populus* L.   Poplar

1. Petioles flat 2
1. Petioles round 6
2. Leaves triangular-ovate to rhombic 3
2. Leaves ovate to suborbicular 4
3. Leaves triangular-ovate; tree with a broad crown, the branches spreading .... 1. *P. deltoides*
3. Leaves rhombic; tree with a narrow crown, the branches strongly ascending .... 2. *P. nigra* var. *italica*
4. Margin of leaf dentate, with 5–25 teeth (averaging 10–20); buds pubescent 5
4. Margin of leaf finely crenate, with 20 or more teeth (averaging 31); buds glabrous or nearly so .... 5. *P. tremuloides*
5. Margin of leaf with 5–15 teeth (averaging 10); petiole 5–10 cm long (averaging 7 cm) .... 3. *P. grandidentata*
5. Margin of leaf with 12–25 teeth (averaging 20); petiole 3–6 cm long (averaging 5.5 cm) .... 4. *P.* X *smithii*
6. Leaves covered with a white felt on the lower surface .... 6. *P. alba*
6. Leaves glabrous or variously pubescent beneath, but not covered with a white felt on the lower surface 7
7. Leaves rounded or truncate at base, if cordate, the buds heavily resinous 8
7. Leaves cordate at base; buds not heavily resinous . 10. *P. heterophylla*
8. Leaves sinuate-dentate; buds to 6 mm long, not resinous .... 7. *P. canescens*
8. Leaves serrulate or crenulate; buds to 25 mm long, resinous 9
9. Twigs usually glabrous; leaves glabrous, or with a few hairs on the midvein beneath .... 8. *P. balsamifera*
9. Twigs usually pubescent; leaves pubescent beneath .. 9. *P. gileadensis*

1. P. deltoides Marsh. Cottonwood. March–May. Along rivers and streams, common; probably in every co. (Including var. *missouriense* [Henry] Rehd.)

2. P. nigra L. var. italica Muenchh. Lombardy Poplar. March–May. First found in Europe; rarely escaped from cultivation.

3. P. grandidentata Michx. Large-toothed Aspen. April. Disturbed areas in and around woodlands; occasional to common in the n. half of the state, extending s. to Marion and Wabash cos.

4. P. X smithii Boivin. Barnes' Aspen. April–May. Low woods; LaSalle and Peoria cos. Reputed to be a hybrid between *P. grandidentata* Michx. and *P. tremuloides* Michx.

5. P. tremuloides Michx. Quaking Aspen. April–May. Low ground of woods, marshes, and bogs; common in the n. cos., absent in the extreme s. cos.

6. P. alba L. White Poplar. April–May. Native of Europe; frequently spreading from cultivation.

7. P. canescens (Ait.) Sm. Gray Poplar. April–May. Native of Europe;

infrequently escaped from cultivation; Adams, Champaign, and Lake cos.

8. P. balsamifera L. Balsam Poplar. April. Dunes near Lake Michigan, rare; Cook and Lake cos. (Including *P.* X *jackii* Sarg.)

9. P. X gileadensis Rouleau. Balm-of-Gilead. April. Rarely escaped from cultivation; Lake Co. Reported to be a hybrid between *P. deltoides* Marsh. and *P. balsamifera* L.

10. P. heterophylla L. Swamp Cottonwood. April–May. Swamps and low woods, often in standing water; occasional in the s. one-third of Ill.

41. MYRICACEAE

*1. Comptonia L'Hé*r.   Sweet-fern

1. C. peregrina (L.) Coult. Sweet-fern. April–May. Sand flats and barrens; confined to the n. one-fourth of Ill. *Myrica asplenifolia* L.—G.

42. JUGLANDACEAE

1. Husk of nuts not splitting; pith chambered .... 1. *Juglans*
1. Husk of nuts splitting (at least partially) at maturity; pith not chambered
.... 2. *Carya*

1. Juglans L.   Walnut

1. Branchlets and husk of nuts downy; fruit ellipsoid .... 1. *J. cinerea*
1. Branchlets and husk of nuts not downy; fruit globose (very rarely ellipsoid) .... 2. *J. nigra*

1. J. cinerea L. Butternut. April–May. Rich woodlands; occasional throughout the state.

2. J. nigra L. Black Walnut. April–May. Rich woodlands; common throughout the state. (Including f. *oblonga* [Marsh.] Fern.)

2. Carya Nutt.   Hickory

1. At least some of the leaves with nine or more leaflets (if only 7 in *C. cordiformis,* then the buds mustard-yellow and very elongated) 2
1. Leaves with (3–) 5–7 leaflets (occasionally 9 in the very tomentose *C. tomentosa*) 4
2. Nut cylindric, sweet; buds with yellow hairs .... 1. *C. illinoensis*
2. Nut usually compressed, bitter; buds with yellow glands or yellow scurfiness 3
3. Buds with yellow glands; leaflets strongly falcate; fruit with wings extending to the base .... 2. *C. aquatica*
3. Buds with yellow scurfiness; leaflets not usually strongly falcate; fruit with wings extending only about half-way to the base   3. *C. cordiformis*
4. Buds yellow-lepidote 5

4. Buds not yellow-lepidote 6
5. Rachis (at least in spring) with reddish hairs .... 4. *C. texana*
5. Rachis without reddish hairs .... 5. *C. pallida*
6. Terminal bud up to 1.2 cm long; leaflets without cilia along the margin 7
6. Terminal bud over 1.2 cm long; leaflets with cilia along the margin .. 8
7. Husk of fruit readily splitting; bark becoming platy or even shaggy at maturity; leaflets mostly 7; outermost bud scales hairy throughout on the
margins .... 6 *C. ovalis*
7. Husk of fruit tardily splitting or not splitting at all; bark tight at maturity;
leaflets mostly 5; outermost bud scales hairy only at the tip 7. *C. glabra*
8. Branchlets densely pubescent; bark tight at maturity; kernel bitter .....
.... 8. *C. tomentosa*
8. Branchlets glabrous or sparsely pubescent; bark shaggy at maturity;
kernel sweet 9
9. Leaflets mostly 5; twigs without conspicuous raised orange lenticels ...
.... 9. *C. ovata*
9. Leaflets mostly 7–9; twigs with conspicuous raised orange lenticels ....
.... 10. *C. laciniosa*

1. C. illinoensis (Wang.) K. Koch. Pecan. April–May. Bottomland
woods; occasional throughout the state except for the ne. one-fourth.
A hybrid with *C. laciniosa* Sarg. (*Carya* X *nussbaumeri* Sarg.) is
known from Ill.

2. C. aquatica (Michx. f.) Nutt. Water Hickory. March–April. Swamps
and wet woods, rare; confined to the s. one-eighth of the state.

3. C. cordiformis (Wang.) K. Koch. Bitternut Hickory. May–June.
Moist or dry woods; common throughout the state; in every co.

4. C. texana Buckl. Black Hickory. April–May. Dry woods and bluffs;
occasional to rare in the s. half of the state; apparently absent elsewhere. (Including var. *arkansana* [Sarg.] Sarg. and var. *villosa* [Sarg.]
Little.)

5. C. pallida (Ashe) Engl. & Graebn. Pale Hickory. April–May.
Wooded slopes, rare; Union Co.

6. C. ovalis (Wang.) Sarg. Sweet Pignut Hickory. April–June. Leaflets not glandular beneath; husk of fruit narrowly winged or wingless;
fruit ellipsoid to nearly spherical. Woodlands; occasional to common
in the s. half of the state. (Including var. *obcordata* [Muhl. & Willd.]
Sarg.)

var. obovalis Sarg. April–May. Leaflets not glandular beneath; husk
of fruit narrowly winged or wingless; fruit obovoid. Woodlands; occasional in the s. half of the state.

var. odorata (Marsh.) Sarg. April–May. Leaflets glandular-viscid beneath; husk of fruit strongly winged. Woodlands; rare in the s. tip of
the state.

7. C. glabra (Mill.) Sweet. Pignut Hickory. April–May. Husk of fruit
up to 2.5 mm thick. Woodlands; occasional to common in the s. half
of the state, becoming rare elsewhere.

var. megacarpa Sarg. April–May. Husk of fruit about 3.5 mm thick.
Woodlands; rare in the s. one-fourth of the state, apparently absent
elsewhere.

8. C. tomentosa (Poir.) Nutt. Mockernut Hickory. May–June. Moist
or dry woods; occasional in the s. three-fourths of the state, rare elsewhere. A hybrid with *C. illinoensis* (*C.* X *schneckii* Sarg.) is known
from Ill.

202 JUGLANDACEAE

9. C. ovata (Mill.) K. Koch. Shagbark Hickory. April–May. Fruit 3.5–6.0 cm long; some or all the leaflets 5 cm wide or wider. Rich woodlands, upland woodlands; occasional to common throughout the state. (Including var. *pubescens* Sarg.)

var. nuttallii Sarg. Small Shagbark Hickory. April. Fruit 1.5–2.0 cm long; some or all the leaflets 5 cm wide or wider. Rich woodlands; rare in the s. tip of Ill.

var. fraxinifolia Sarg. Ash-leaved Shagbark Hickory. April. Fruit 3.5–4.0 cm long; none of the leaflets up to 5 cm wide. Rich woodlands, rare; apparently confined to the s. half of the state.

10. C. laciniosa (Michx.) Loud. Kingnut Hickory. April–May. Bottomland woods; occasional to rare in the s. two-thirds of the state.

43. BETULACEAE

1. Nuts borne in elongated catkins or in woody "cones"; staminate flowers with a calyx 2
1. Nuts subtended or enclosed by large bracts; staminate flowers without a calyx 3
2. Nuts borne in elongated catkins; pistillate scales 3-lobed, herbaceous ..
.... 1. *Betula*
2. Nuts borne in woody "cones"; pistillate scales 5-lobed, woody   2. *Alnus*
3. Leaves broadly ovate, cordate, doubly serrate .... 3. *Corylus*
3. Leaves narrowly ovate or oval, not cordate, once-serrate 4
4. Bracts inflated, enclosing the nut; bark rough; lateral nerves of leaves usually branched near the margin .... 4. *Ostrya*
4. Bracts foliaceous, subtending the nut; bark smooth; lateral nerves of leaves unbranched .... 5. *Carpinus*

1. Betula L. Birch

1. Leaves with 8 or more pairs of lateral veins 2
1. Leaves with up to 7 (–8) pairs of lateral veins 3
2. Bark yellowish- or silvery-gray, peeling off in thin layers; leaves thin, green beneath .... 1. *B. lutea*
2. Bark brownish to pinkish, peeling off in shaggy pieces; leaves firm, pale beneath .... 2. *B. nigra*
3. Bark peeling off in thin layers; small trees 4
3. Bark close, not peeling off in thin layers; shrubs 5
4. Leaves pubescent along the veins; most or all the fruiting spikes 3 cm long or longer .... 3. *B. papyrifera*
4. Leaves glabrous beneath; most or all the fruiting spikes less than 3 cm long .... 4. *B. populifolia*
5. Leaves acute to acuminate, some or all over 3 cm long 6
5. Leaves obtuse to subacute, not more than 3 cm long ..... 7. *B. pumila*
6. Bark dark brown; fruiting spikes less than 1 cm thick   5. *B.* X *sandbergii*
6. Bark gray; fruiting spikes at least 1 cm thick .... 6. *B.* X *purpusii*

1. B. lutea Michx. f. Yellow Birch. May–June. Boggy woods, rare; DuPage, Lake, Lee, Ogle, and Winnebago cos.

2. B. nigra L. River Birch. April–May. Along rivers and streams; occasional in the s. half of the state, becoming less common northward.

3. B. papyrifera Marsh. Paper Birch. May–June. Wooded ravines, low dune ridges, rare; Carroll, Cook, JoDaviess, and Lake cos. (Including var. *cordifolia* [Regel] Fern.)

4. B. populifolia Marsh. Gray Birch. May–June. Thickets, very rare; Cook and Winnebago cos.

5. B. X sandbergii Britt. Sandberg's Birch. May–June. Boggy woods, rare; Kane and Lake cos. Reputed to be a hybrid between *B. lutea* and *B. pumila.*

6. *B.* X purpusii Schneid. Purpus' Birch. May–June. Boggy woods, rare; Lake Co. Reputed to be a hybrid between *B. lutea* and *B. pumila.*

7. *B. pumila* L. Dwarf Birch. May–June. Bogs; confined to Boone, Cook, Kane, Lake, McHenry, and Winnebago cos. (Including var. *glabra* Regel and var. *glandulifera* Regel.)

2. Alnus B. Ehrh. Alder

1. Leaves dentate or denticulate, with up to 7 pairs of lateral veins ....
.... 1. *A. glutinosa*
1. Leaves serrate or serrulate, with 8 or more pairs of lateral veins 2
2. Leaves ovate to elliptic, whitened beneath .... 2. *A. rugosa*
2. Leaves obovate, green beneath .... 3. *A. serrulata*

1. A. glutinosa (L.) Gaertn. Black Alder. April–May. Native of Europe; becoming naturalized along rivers in the n. tip of the state; Cook, DuPage, and Will cos.

2. A. rugosa (DuRoi) Spreng. var. americana (Regel) Fern. Speckled Alder. May–June. Moist thickets, rare; Boone, Cook, Kane, Lake, McHenry, and Winnebago cos. *A. rugosa* (DuRoi) Spreng.—J.

3. A. serrulata (Ait.) Willd. Smooth Alder. April–May. Along rocky streams; occasional in the s. one-third of the state. (Including var. *noveboracensis* [Britt.] Fern.)

3. Corylus L. Hazelnut

1. C. americana Walt. Hazelnut. March–April. Thickets, dry, disturbed woods, common; in every co. (Including f. *missouriensis* [A. DC.] Fern.)

4. Ostrya Scop. Hop Hornbeam

1. O. virginiana (Mill.) K. Koch. April–May. Woodlands; in every co. (Including var. *lasia* Fern. and f. *glandulosa* [Spach] Macbr.)

5. Carpinus L. Ironwood

1. C. caroliniana Walt. Blue Beech. April–May. Moist woods; probably in every co. (Including var. *virginiana* [Marsh.] Fern.)

44. FAGACEAE

1. Leaves unlobed 2
1. Leaves lobed .... 3. *Quercus*

2. Fruits with a prickly covering 3
2. Fruits acorns, not prickly .... 3. *Quercus*
3. Bark smooth; nuts triangular .... 1. *Fagus*
3. Bark rough; nuts flattened .... 2. *Castanea*

1. Fagus L.   Beech

1. F. grandifolia Ehrh. Beech. April–May. Rich woods; occasional in the s. one-fourth of the state, extending along the extreme e. border of Ill. to Vermilion Co.; also Lake Co. (Including var. *caroliniana* [Loud.] Fern. & Rehd. and f. *mollis* Fern. & Rehd.)

2. Castanea Mill.   Chestnut

1. C. dentata (Marsh.) Borkh. Chestnut. June–July. Rocky woods; originally in the s. tip of the state, particularly Pulaski Co., now nearly extinct, except for a few sprouts here and there and a single tree in Williamson Co.

3. Quercus L.   Oak

1. Leaves entire 2
1. Leaves toothed or lobed 3
2. Leaves stellate-pubescent beneath; cup of acorn at least 1.5 cm broad
.... 1. *Q. imbricaria*
2. Leaves glabrous or nearly so beneath; cup of acorn up to 1.2 cm. broad
.... 2. *Q. phellos*
3. Leaves toothed 4
3. Leaves lobed 7
4. Fruits on stalks at least 2 cm long; veins of leaves not reaching the tip of the teeth .... 3. *Q. bicolor*
4. Fruits up to 1 cm long; veins of leaves reaching the tip of the teeth   . 5
5. Leaves velvety-tomentose beneath; cup of acorn at least 2.5 cm broad
.... 14. *Q. michauxii*
5. Leaves glabrous or minutely pubescent, but not velvety 6
6. Leaves with pointed teeth; acorns up to 2 cm long   15. *Q. muhlenbergii*
6. Leaves with rounded teeth; acorns 2 cm long or longer .. 16. *Q. prinus*
7. Lobes of leaf with bristle-tips 8
7. Lobes of leaf without bristle-tips 19
8. Leaves permanently uniformly pubescent throughout on the lower surface 9
8. Leaves glabrous beneath, or with tufts of axillary hairs, or irregularly pubescent 12
9. Leaves much broader in the upper half, with 3 (–5) broad, shallow lobes
.... 3. *Q. marilandica*
9. Leaves broader at or below the middle with (3–) 5–11 narrower, deeper lobes 10
10. Lower leaf surface grayish or yellowish; scales of buds reddish-brown, glabrous or nearly so; terminal winter buds up to 6 mm long 11
10. Lower leaf surface brownish or reddish-brown; scales of buds grayish, tomentose; terminal winter buds about 6–12 mm long ... 6. *Q. velutina*
11. Terminal and usually one or two lateral lobes of leaf curved 4. *Q. falcata*
11. All lobes of leaf straight or nearly so .... 5. *Q. pagodaefolia*
12. Leaves much broader in the upper half, with 3 (–5) broad, shallow lobes
.... 3. *Q. marilandica*

12. Leaves broader at or below the middle, with 5–11 narrower, deeper lobes
.... 13
13. Terminal winter buds 6–12 mm long, gray-tomentose; scales along edge
of acorn cup not appressed; upper leaf surface with pubescent midnerve
.... 6. *Q. velutina*
13. Terminal winter buds up to 4 mm long, not gray-tomentose; scales along
edge of acorn cup appressed; upper leaf surface with glabrous midnerve
.... 14
14. Leaves lobed less than half-way to middle; cup covering less than one-
fourth of acorn .... 7. *Q. ruba*
14. Leaves lobed at least half-way to middle; cup covering at least one-
fourth of acorn 15
15. Acorn cup covering much less than one-half of the acorn 16
15. Acorn cup covering about one-half of the acorn 18
16. Acorn up to 1.5 cm long; cup not exceeding 1.5 cm in diameter ....
.... 8. *Q. palustris*
16. Acorn more than 1.5 cm long; cup more than 1.5 cm in diameter ... 17
17. Buds glabrous; leaves not long-tapering at the tip; acorn cup mostly 2
cm broad or broader .... 9. *Q. shumardii*
17. Buds puberulent, or at least ciliate; leaves tending to be long-tapering at
the tip; acorn cup mostly 1.5–2.0 cm broad .... 10. *Q. nuttallii*
18. Acorn cup up to 1.5 cm in diameter, with pubescent scales ....
.... 11. *Q. ellipsoidalis*
18. Acorn cup at least 1.5 cm in diameter, with glabrous scales............
.... 12. *Q. coccinea*
19. Leaves completely glabrous beneath .... 17. *Q. alba*
19. Leaves pubescent beneath 20
20. Leaves 3- to 5-lobed, the upper three with squarish tips, forming a cross;
acorn up to 1 cm across, the cup unfringed and covering less than half
of the nut .... 18. *Q. stellata*
20. Leaves 5- to 11-lobed, the upper three without squarish tips; acorn 1 cm
broad or broader, the cup either fringed or covering half to nearly all the
nut 21
21. Scales of cup forming a fringe around acorn; cup ½–⁴/₅ covering the
acorn .... 19. *Q. macrocarpa*
21. Scales of cup not forming a fringe; cup nearly completely covering the
acorn .... 20. *Q. lyrata*

1. Q. imbricaria Michx. Shingle Oak. April–May. Moist or dry woods,
edges of fields; occasional to common throughout the state. Hybrids
between this species and *Q. falcata* (*Q. X anceps* E. J. Palmer), *Q.
marilandica* (*Q. X tridentata* [A. DC.] Engelm), *Q. palustris* (*Q. X exacta* Trel.), *Q. rubra* (*Q X runcinata* [A. DC.] Engelm.), and *Q. velutina*
(*Q. X leana* Nutt.) have been reported from the state.
2. Q. phellos L. Willow Oak. April–May. Swampy woods; Alexander,
Massac, Pulaski, and Union cos. Hybrids between this species and *Q.
palustris* (*Q. X schochiana* Dieck) and *Q. veluntina* (*Q. X filialis* Little)
have been reported from the state.
3. Q. marilandica Muenchh. Blackjack Oak. April–May. Upland
woods, bluff-tops; occasional to common in the s. three-fourths of
the state, absent elsewhere. Hybrids between this species and *Q. imbricaria* (*Q. X tridentata* [A. DC.] Engelm.) and *Q. velutina* (*Q. X
bushii* Sarg.) have been reported from the state.
4. Q. falcata Michx. Spanish Oak. April–May. Moist or dry woods;
occasional in the s. one-fifth of the state. Hybrids between this spe-

cies and *Q. imbricaria* (*Q.* X *anceps* E. J. Palmer) and *Q. phellos* (*Q.* X *ludoviciana* Sarg.) have been found in the state.

5. Q. pagodaefolia (Ell.) Ashe. Cherry-bark Oak. April–May. Rich woods; occasional in the s. one-fifth of the state. *Q. falcata* Michx. var. *pagodaefolia* Ell.—F, G; *Q. falcata* Michx.—J.

6. Q. velutina Lam. Black Oak. April–May. Young branchlets and lower surface of leaves glabrate at maturity, except in the leaf axils. Upland woods; common throughout the state; in every co. Hybrids between this species and *Q. imbricaria* (*Q.* X *leana* Nutt.), *Q. marilandica* (*Q.* X *bushii* Sarg.), and *Q. phellos* have been reported from the state.

f. missouriensis (Sarg.) Trel. April–May. Branchlets and lower surface of leaves permanently pubescent. Upland woods; Saline Co.

7. Q. rubra L. Red Oak. April–May. Upland woods; common throughout the state. Including var. *borealis* [Michx. f.] Farw.) Hybrids between this species and *Q. imbricaria* (*Q.* X *runcinata* [A. DC.] Engelm.) have been reported from the state.

8. Q. palustris Muenchh. Pin Oak. April–May. Wet ground; occasional to common in all parts of Ill. except the extreme nw. cos. Hybrids between this species and *Q. imbricaria* (*Q.* X *exacta* Trel.) and *Q. phellos* (*Q.* X *schochiana* Dieck) have been reported from the state.

9. Q. shumardii Buckley. Shumard's Oak. April–May. Low woods; occasional in the s. one-third of the state; also McLean Co. (Including *Q. shumardii* Buckley var. *schneckii* [Britt.] Sarg.)

10. Q. nuttallii E. J. Palmer. Nuttall's Oak. April–May. Wet woods, very rare; Alexander Co.

11. Q. ellipsoidalis E. J. Hill. Hill's Oak. April–May. Upland woods; occasional in the n. one-fourth of the state.

12. Q. coccinea Muenchh. Scarlet Oak. April–May. Upland woods; occasional in the s. one-third of the state.

13. Q. bicolor Willd. Swamp White Oak. April–May. Low woods and swamps; occasional throughout the state. Hybrids between this species and *Q. alba* (*Q.* X *jackiana* Schneider), *Q. lyrata* (*Q.* X *humidicola* E. J. Palmer), and *Q. macrocarpa* (*Q.* X *hillii* Trelease) have been reported from the state.

14. Q. michauxii Nutt. Basket Oak. April–May. Low woods and swamps; occasional in the s. one-third of the state.

15. Q. muhlenbergii Engelm. Yellow Chestnut Oak. April–May. Moist or dry woods; occasional throughout the state. (Including f. *alexanderi* [Britt.] Trel.) *Q. prinoides* Willd. var. *acuminata* (Michx.) Gl.—G. Hybrids between this species and *Q. alba* (*Q.* X *deamii* Trel.) and *Q. macrocarpa* (*Q.* X *fallax* E. J. Palmer) have been reported from the state.

16. Q. prinus L. Rock Chestnut Oak. April–May. Rocky woods, rare; Alexander, Hardin, Saline, and Union cos.

17. Q. alba L. White Oak. April–May. Upland woods; common throughout the state; in every co. (Including f. *repanda* [Michx.] Trel. and f. *latiloba* [Sarg.] Palmer & Steyerm.) Hybrids between this species and *Q. bicolor* (*Q.* X *jackiana* Schneider), *Q. macrocarpa* (*Q.* X *bebbiana* Schneider), *Q. muhlenbergii* (*Q.* X *deamii* Trel.), and *Q. stellata* (*Q.* X *fernowi* Trel.) have been reported from Ill.

18. Q. stellata Wangh. Post Oak. April–May. Upland woods, bluffs; occasional to common in the s. half of the state; also Grundy Co.

Hybrids between this species and *Q. alba* (*Q.* X *fernowi* Trel.) have been reported from Ill.

19. Q. macrocarpa Michx. Bur. Oak. April–May. Bottomland woods; occasional to common throughout the state; in every co. Hybrids between this species and *Q. alba* (*Q.* X *bebbiana* Schneider), *Q. bicolor* (*Q.* X *hillii* Trel.), and *Q. muhlenbergii* (*Q.* X *fallax* E. J. Palmer) have been reported from Ill. (Including f. *olivaeformis* [Michx. f.] Trel.)

20. Q. lyrata Walt. Overcup Oak. April–May. Bottomland woods; occasional in the s. half of the state.

45. ULMACEAE

1. Leaves with strong lateral veins arising from the main vein at a distance above the very base of the blade; pith of branches solid; sepals united; fruit a samara or nut 2
1. Leaves with a pair of strong lateral veins arising from the main vein at the very base of the blade; pith of branches chambered; sepals free; fruit a drupe .... 3. *Celtis*
2. Fruit a samara; all flowers perfect, appearing before the leaves; leaves mostly doubly toothed .... 1. *Ulmus*
2. Fruit a wingless nut; at least some of the flowers unisexual, appearing with the leaves; leaves mostly singly toothed .... 2. *Planera*

1. Ulmus L. Elm

1. Upper surface of leaves harshly pubescent; winter buds rusty-pubescent; samaras eciliate on the margins but pubescent at the center of both sides .... 1. *U. rubra*
1. Upper surface of leaves glabrous or pubescent, but never harshly so; winter buds glabrous or pubescent, but not with rusty pubescence; samaras ciliate or, if eciliate, the center of each side glabrous ...ˌ...... 2
2. None of the branches corky-winged 3
2. Some of the branches corky-winged 4
3. Leaves doubly serrate, usually strongly asymmetrical at the base; samaras ciliate; flowers pendulous from long pedicels .... 2. *U. americana*
3. Leaves mostly singly serrate, usually nearly symmetrical at the base; samaras eciliate; flowers sessile or nearly so .... 3. *U. pumila*
4. Buds glabrous or nearly so; leaves sessile or on petioles up to 3 mm long; samaras (excluding cilia) up to 5 mm broad .... 4. *U. alata*
4. Buds downy-pubescent; leaves on petioles 3 mm long or longer; samaras (excluding cilia) 1.0–1.5 cm broad 5
5. Most of the mature leaves over 8 cm long, glabrous above; flowers pendulous from long pedicels; samaras ciliate .... 5. *U. thomasii*
5. Most of the mature leaves less than 8 cm long, scabrous above; flowers sessile or nearly so; samaras eciliate .... 6. *U. procera*

1. U. rubra Muhl. Slippery Elm. Feb.–April. Woods, often in disturbed areas, common; in every co.

2. U. americana L. American Elm. Feb.–April. Rich woods, wooded flood plains, common; in every co.

3. U. pumila L. Siberian Elm. May. Native of Asia; escaped into waste areas; occasional to common.

4. U. alata Michx. Winged Elm. Feb.–March. Wooded slopes, cliffs, or low woodlands; common in the s. one-fourth of the state; apparently absent elsewhere.

5. U. thomasii Sarg. Rock Elm. March–May. Rich woods, not common in the n. half of Ill. absent elsewhere.

6. U. procera Salisb. English Elm. April. Native of Europe; rarely escaped in Ill.; Jackson Co.

2. Planera J. F. Gmel.   Water Elm

1. P. aquatica [Walt.] J. F. Gmel. Water Elm. March–April. Swampy woods rare; Alexander, Jackson, Johnson, Massac, Pope, Pulaski, Randolph, and Union cos.

3. Celtis L.   Hackberry

1. Leaves conspicuously several-toothed 2
1. Leaves entire or only sparingly toothed 5
2. Leaves harshly scabrous on the upper surface ...... 1. *C. occidentalis*
2. Leaves smooth or nearly so on the upper surface 3
3. Most of the leaves more than half as long as broad, very strongly asymmetrical at the base .... 1. *C. occidentalis*
3. Most of the leaves less than half as long as broad, only slightly asymmetrical at the base 4
4. Drupe 8–11 mm long, dark purple or dark brown .... 1. *C. occidentalis*
4. Drupe 5–7 mm long, orange, pale brown, or red .... 2. *C. laevigata*
5. Leaves harshly scabrous on the upper surface 6
5. Leaves smooth or nearly so on the upper surface 7
6. Leaves mostly over half as broad as long .... 3. *C. tenuifolia*
6. Leaves mostly less than half as broad as long .... 2. *C. laevigata*
7. Leaves more than half as broad as long, acute to short-acuminate at the apex .... 3. *C. tenuifolia*
7. Leaves less than half as broad as long, long-tapering at the apex ......
.... 2. *C. laevigata*

1. C. occidentalis L. Hackberry. April–May. Leaves harshly scabrous on the upper surface. Low woods; common throughout the state.

var. pumila (Pursh) Gray. Hackberry. April–May. Leaves more or less smooth on the upper surface, most of them more than half as broad as long. Low woods, bluffs; occasional throughout the state.

var. canina (Raf.) Sarg. Hackberry. April–May. Leaves more or less smooth on the upper surface, most of them less than half as broad as long. Low woods; occasional throughout the state.

2. C. laevigata Willd. Sugarberry. April–May. Leaves more or less smooth on the upper surface, entire or nearly so, the petioles essentially glabrous. Low woods; occasional to common in the s. half of the state.

var. smallii (Beadle) Sarg. Sugarberry. April–May. Leaves more or less smooth on the upper surface, regularly toothed, the petioles essentially glabrous. Low woods, occasional in the extreme s. cos.

var. texana Sarg. Dwarf Hackberry. April–May. Leaves harshly scabrous on the upper surface, entire or nearly so, the petioles pubescent. Dry woods and cliffs; occasional in the s. one-fifth of the state.

3. C. tenuifolia Nutt. Dwarf Hackberry. April–May. Leaves more or less smooth on the upper surface, membranaceous, the petioles glabrous or nearly so. Dry woods and cliffs; occasional, mostly in the s. one-fifth of the state. *C. pumila* Pursh—J.

var. georgiana (Small) Fern. & Schub. Dwarf Hackberry. April–May. Leaves harshly scabrous on the upper surface, leathery, the petioles densely hairy. Dry woods and cliffs; occasional, mostly in the s. one-fifth of the state. *C. pumila* Pursh—J.

46. MORACEAE

1. Trees or shrubs; latex present 2
1. Herbs or vines; latex absent 4
2. Leaves toothed, often lobed; branches without spines; fruit 1–2 cm in diameter, white, orange, or red 3
2. Leaves entire, unlobed; branches with short spines; fruit 10–20 cm in diameter, greenish-yellow .... 3. *Maclura*
3. Leaves glabrous or short-hairy on the lower surface, not velvety; most or all the petioles less than 3 cm long, glabrous or appressed-pubescent; bark roughened .... 1. *Morus*
3. Leaves velvety on the lower surface; most or all the petioles more than 3 cm long, pilose; bark smooth .... 2. *Broussonetia*
4. Stems climbing or trailing; leaves simple although often lobed ....
.... 4. *Humulus*
4. Stems erect; leaves compound .... 5. *Cannabis*

1. Morus L. Mulberry

1. Lower surface of leaves pubescent on the blade as well as on the veins
.... 1. *M. rubra*
1. Lower surface of leaves glabrous, or pubescent only on the veins, or the hairs confined to axillary tufts .... 2. *M. alba*

1. M. rubra L. Red Mulberry. April–May. Lowland or upland woods, edges of fields; common throughout most of Ill.

2. M. alba L. White Mulberry. April–May. Fruits white or pale pink, rarely dark purple or black, usually over 1 cm long. Native of Asia; naturalized throughout the state. (Including f. *skeletoniana* [Schneider] Rehder.)

var. tatarica (L.) Loudon. Russian Mulberry. Fruits red, up to 1 cm long. Native of Europe and Asia; naturalized along the edges of woods; occasional throughout the state. *M. tatarica* L.—J.

2. Broussonetia L'Hér. Paper Mulberry

1. B. papyrifera (L.) L'Hér. Paper Mulberry. April–May. Native of Asia; occasionally naturalized in the s. one-fourth of the state.

3. Maclura Nutt. Osage Orange

1. M. pomifera (Raf.) Schneider. Osage Orange. May–June. Native of the s. cent. U.S.; apparently naturalized in Ill.; occasional throughout the state.

4. Humulus L.   Hops

1. Most of the leaves 3-lobed, with resinous glands beneath   1. *H. lupulus*
1. Most of the leaves 5- to 7-lobed, without resinous glands beneath . . . . . .
. . . . 2. *H. japonicus*

1. H lupulus L. Common Hops. July–Aug. Fencerows, thickets, dry
wooded slopes; occasional throughout the state. *H. americanus*
Nutt.—J.
2. H. japonicus Sieb. & Zucc. Japanese Hops. July–Sept. Native of
Asia; naturalized in waste areas; occasional and scattered in Ill.

5. Cannabis L.   Hemp

1. C. sativa L. Marijuana. June–Oct. Native of Asia; naturalized and
escaped into waste areas; occasional throughout the state, although
apparently less common in the southernmost cos.

47. URTICACEAE

1. Leaves opposite (a few alternate leaves may occur in *Boehmeria*)   . . . 2
1. Leaves all alternate 4
2. Leaves and usually the stems pubescent 3
2. Leaves and stems glabrous . . . . 3. *Pilea*
3. Pistillate flowers with free sepals; stinging hairs often present   1. *Urtica*
3. Pistillate flowers with united sepals; stinging hairs never present . . . .
. . . . 2. *Boehmeria*
4. Leaves serrate; stinging hairs present . . . . 4. *Laportea*
4. Leaves essentially entire; stinging hairs absent . . . . 5. *Parietaria*

1. Urtica L.   Nettle

1. Perennials often attaining a height of 1 m or more; inflorescences 2 cm
long or longer . . . . 1. *U. dioica*
1. Annuals never attaining a height of 1 m; inflorescences less than 2 cm
long 2
2. Leaves dentate, tapering to the base . . . . 2. *U. urens*
2. Leaves crenate, some of them cordate at the base . . . .
. . . . 3. *U. chamaedryoides*

1. U. dioica L. Stinging Nettle. June–Sept. Rich woods, moist waste
ground; occasional to common in the n. half of the state, rare in the s.
half. (Including *U. gracilis* Ait. and *U. procera* Muhl.)
2. U. urens L. Burning Nettle. June–Sept. Native of Europe and
Asia; rarely escaped into waste areas.
3. U. chamaedryoides Pursh. Nettle. April–June. Swampy woods,
river banks; Alexander and Jackson cos.

2. Boehmeria Jacq.   False   Nettle

1. B. cylindrica (L.) Sw. False Nettle. July–Oct. Leaves smooth or
slightly scabrous above. Moist woods, common; throughout the state.

var. drummondiana Wedd. False Nettle. July–Oct. Leaves harshly scabrous above. Edge of bogs and marshes, rare; Alexander, Dewitt, Jackson, Lake, and Will cos. *B. drummondiana* Wedd.—J.

3. *Pilea Lindl.* Clearweed

1. Achenes green, averaging 1 mm wide .... 1. *P. pumila*
1. Achenes black, averaging 1.5 mm wide 2
2. Plants leafy above the middle; achenes averaging 1.5 mm long ....
.... 2. *P. fontana*
2. Plants leafy from near the base; achenes averaging 2.0 mm long ....
.... 3. *P. opaca*

1. P. pumila (P.) Gray. Clearweed. July–Sept. Leaves mostly tapering to the base, with 3–11 teeth on each margin. Moist, usually shaded, ground, common; probably in every co. (Including var. *deamii* [Lunell] Fern.)
2. P. fontana (Lunell) Rydb. July–Sept. Moist soil; occasional and scattered in Ill.
3. P. opaca (Lunell) Rydb. July–Sept. Moist soil; occasional and scattered in Ill.

4. *Laportea Gaud.* Wood Nettle

1. L. canadensis (L.) Wedd. Wood Nettle. June–Sept. Moist soil; common throughout the state.

5. *Parietaria L.* Pellitory

1. P. pensylvanica Muhl. Pellitory. May–Sept. In shade beneath cliffs and along buildings; common throughout the state.

48. SANTALACEAE

1. *Comandra Nutt.* False Toadflax

1. C. richardsiana Fern. False Toadflax. May–Aug. Woods, prairies, occasional; throughout the state, except for the very southernmost cos. *C. umbellata* (L.) Nutt.—J.

49. LORANTHACEAE

1. *Phoradendron Nutt.* Mistletoe

1. P. flavescens (Pursh) Nutt. Mistletoe. Sept.–Oct. On various deciduous trees, mostly in low areas; occasional in the s. one-sixth of the state and extending along the Wabash River to Clark Co.

50. ARISTOLOCHIACEAE

1. Leaves 2, basal; stamens 12; calyx regular .... 1. *Asarum*
1. Leaves more than 2, cauline; stamens 6; calyx irregular   2. *Aristolochia*

1. *Asarum* L.   Wild Ginger

1. A. canadense L. Wild Ginger. April–May. Tips of calyx lobes extended into a point 5–20 mm long. Rich woodlands; restricted to the n. half of the state. *A. canadense* var. *acuminatum* Ashe—F; *A. acuminatum* (Ashe) Bickn.—J.

var. reflexum (Bickn.) Robins. Wild Ginger. April–May. Tips of calyx lobes merely short-pointed, the point not exceeding 5 mm in length. Rich woodlands; common throughout Ill. *A. reflexum* Bickn.—J.

2. *Aristolochia* L.   Birthwort

1. Erect herb; lower surface of leaves pubescent, but not white-woolly; calyx dark purple, to 15 mm long .... 1. *A. serpentaria*
1. High-climbing woody twiner; lower surface of leaves white-woolly; calyx yellow, with a purplish orifice, over 15 mm long ...... 2. *A. tomentosa*

1. A. serpentaria L. Virginia Snakeroot. May–July. Some or all the leaves over 2 cm across midway from base to apex. Rich woodlands; local in the s. half of the state.

var. hastata (Nutt.) Duchartre. Narrow-leaved Snakeroot. June–July. Leaves never more than 2 cm across midway from base to apex. Rich or swampy woods, rare; Alexander and Johnson cos. *A. nashii* Kearney—J.

2. A. tomentosa Sims. Dutchman's Pipe-vine. May–June. Chiefly calcareous woods, rare; confined to the s. half of Ill.

51. POLYGONACEAE

1. Climbing shrubs with tendrils; pedicels winged on one side by the calyx
.... 1. *Brunnichia*
1. Prostrate or ascending or erect herbs or, if climbing, then without tendrils; pedicels not winged by the calyx 2
2. Leaves needle-like .... 2. *Polygonella*
2. Leaves linear or broader, not needle-like 3
3. Sepals 6, the inner 3 longer than the outer in fruit .... 3. *Rumex*
3. Sepals usually 5, the inner not longer than the outer in fruit 4
4. Stems retrorse-prickly .... 5. *Polygonum*
4. Stems not retrorse-prickly 5
5. Leaves hastate-deltoid .... 4. *Fagopyrum*
5. Leaves not hastate-deltoid .... 5. *Polygonum*

1. *Brunnichia* Banks

1. B. cirrhosa Banks. Ladies' Eardrops. June–Aug. Swampy woods; confined to the s. one-sixth of the state.

2. *Polygonella* Michx. Jointweed

1. P. articulata (L.) Meisn. Jointweed. July–Oct. Sandy soil; restricted to the n. one-third of the state.

3. *Rumex* L. Dock

1. Some of the leaves hastate (rarely entire in one form of *R. acetosella*); plants dioecious 2
1. Leaves not hastate; plants monoecious 3
2. Plants with stolons or slender rhizomes; achenes exserted from calyx .. .. 1. *R. acetosella*
2. Plants with a taproot; achenes enclosed by the calyx . 2. *R. hastatulus*
3. Fruiting sepals with spinulose bristles 4
3. Fruiting sepals entire or only minutely toothed 5
4. Stems hollow; tubercle of fruit long and slender; bristles longer than width of fruiting sepals .... 3. *R. maritimus*
4. Stems firm; tubercle of fruit only slightly longer than broad; bristles not longer than width of fruiting sepals .... 4. *R. obtusifolius*
5. Only one of the 3 fruiting sepals with a tubercle 6
5. Each of the fruiting sepals with a tubercle 8
6. Some of the leaves over 10 cm broad; each fruiting sepal 8–10 mm broad .... 5. *R. patientia*
6. None of the leaves 10 cm broad; each fruiting sepal up to 6 mm broad   7
7. Leaves wavy along the margins .... 6. *R. crispus*
7. Leaves flat .... 7. *R. altissimus*
8. Leaves with conspicuous wavy margins .... 6. *R. crispus*
8. Leaves flat, entire or crenulate 9
9. Leaves crenulate; lateral veins of leaves forming right angles with vertical veins .... 8. *R. orbiculatus*
9. Leaves entire; lateral veins of leaves ascending 10
10. Fruiting pedicels two to five times as long as the calyx .... 9. *R. verticillatus*
10. Fruiting pedicels shorter than or barely longer than the calyx 11
11. Leaves narrowly lanceolate, never more than 3 cm broad .... 10. *R. mexicanus*
11. Leaves broadly lanceolate, at least some of them over 3 cm broad .... 7. *R. altissimus*

1. R. acetosella L. Sour Dock. May–Aug. Leaves hastate. Native of Europe; adventive in fields, roadsides, waste areas; in every co.

f. integrifolius (Wallr.) G. Beck. Leaves unlobed. Native of Europe; rarely adventive in Ill.; Kane Co.

2. R. hastatulus Baldw. Sour Dock. May–Aug. Sandy soil, rare; Madison and St. Clair cos.

3. R. maritimus L. var. fueginus (Phil.) Dusén. July–Oct. Sandy shores; scattered throughout the state. *R. fueginus* Phil.—J.

4. R. obtusifolius L. Bitter Dock. June–Sept. Native of Europe; adventive in waste ground; occasional throughout the state.

5. R. patientia L. Patience Dock. June–July. Native of Europe and Asia; adventive in waste ground; scattered throughout the state.

6. R. crispus L. Curly Dock. May–Aug. Native of Europe; adventive in waste ground; in every co. (Including f. *unicallosus* Peterm.)

7. R. altissimus Wood. Pale Dock. May–Aug. Moist soil; occasional to common throughout the state.

8. R. orbiculatus Gray. Water Dock. June–Sept. Moist soil; restricted to the n. two-thirds of the state.

9. R. verticillatus L. Swamp Dock. June–Sept. Wet ground; common in the n. half of the state, occasional in the s. half.

10. R. mexicanus Meisn. Dock. July–Sept. Moist soil; scattered throughout the state. *R. triangulivalvis* (Danser) Rech. f.—J.

4. *Fagopyrum Mill. Buckwheat*

1. F. esculentum Moench. Buckwheat. June–Sept. Native of Asia; adventive in waste ground; occasional in the n. three-fourths of the state, apparently absent elsewhere. *F. sagittatum* Gilib.—F.

5. *Polygonum L. Smartweed*

18. Peduncles without stipitate glands 19
19. Apex of ocreae with an expanded rim .... 17. *P. orientale*
19. Apex of ocreae rimless 20
20. Calyx punctate or glandular-dotted 21
20. Calyx neither punctate nor glandular-dotted 22
21. Racemes usually interrupted; achenes shining ...... 18. *P. punctatum*
21. Racemes mostly continuous; achenes dull .... 19. *P. hydropiper*
22. Racemes densely crowded; annuals without elongated rootstocks .. 23
22. Racemes more or less interrupted; perennials with elongated rootstocks
.... 24
23. Leaves often with a dark blotch; most or all the spikes at least 7 mm
thick .... 20. *P. persicaria*
23. Leaves without a dark blotch; most of all the spikes less than 7 mm thick
.... 21. *P. cespitosum*
24. Leaves strigose on the upper surface; calyx white .... 22. *P. setaceum*
24. Leaves glabrous or scabrous on the upper surface; calyx mostly pink or
greenish 25
25. Achene completely included; calyx pinkish or rosy ....
.... 23. *P. hydropiperoides*
25. Achene partly exserted; calyx greenish or greenish-purple ....
.... 24. *P. opelousanum*
26. Spikes 1–3; perennials with elongated rootstocks 27
26. Spikes usually more than 3 28
27. Peduncles and ocreolae glabrous .... 25. *P. amphibium*
27. Peduncles and ocreolae pubescent .... 26. *P. coccineum*
28. Some or all the flowers with long-exserted stamens and styles ....
.... 27. *P. longistylum*
28. None of the flowers with exserted stamens and styles 29
29. Spikes greenish .... 28. *P. scabrum*
29. Spikes pink or rose or white 30
30. Spikes arching or pendulous; achenes up to 2 mm broad..............
.... 29. *P. lapathifolium*
30. Spikes erect; achenes broader than 2 mm .... 30. *P. pensylvanicum*

1. P. sagittatum L. Tear Thumb. July–Oct. Swampy woods, wet ground; occasional throughout the state.

2. P. arifolium L. var. pubescens (Keller) Fern. Tear Thumb. July–Oct. Wet ground, very rare; McHenry and Macon cos. *P. arifolium* L.—J.

3. P. convolvulus L. Black Bindweed. May–Oct. Native of Europe; fields and edges of woods; occasional to common throughout the state.

4. P. cristatum Engelm. & Gray. Crested Bindweed. Aug.–Sept. Woods; occasional in the s. four-fifths of the state, rare elsewhere. (Including *P. dumetorum* L.) *P. scandens* L. var. *cristatum* (Engelm. & Gray) Gl.—G.

5. P. scandens L. False Buckwheat. Aug.–Oct. Woods and thickets; occasional to common throughout the state

6. P. tenue Michx. Slender Knotweed. July–Sept. Dry, sandy soil; crevices of sandstone blufftops; occasional throughout the state.

7. P. aviculare L. Knotweed. June–Oct. Leaves thin, linear to lanceolate, more or less acute. Native of Europe; waste ground; common throughout the state. (Including var. *vegetum* Ledeb.)

var. littorale (Link) W. D. J. Koch. Knotweed. June–Oct. Leaves thick, oblong to elliptic, obtuse. Waste ground, rare; scattered in Ill. (Including *P. buxiforme* Small.)

8. P. prolificum (Small) Robins. Knotweed. July–Oct. Waste ground, rare; Johnson Co.

9. P. exsertum Small. Long-fruited Knotweed. Aug.–Oct. Banks and shores, usually in sand; occasional throughout the state. *P. ramosissimum* Michx.—G, J.

10. P. ramosissimum Michx. Knotweed. July–Oct. Sandy soil; occasional throughout the state.

11. P. erectum L. Knotweed. Aug.–Oct. Waste ground; occasional to common throughout the state.

12. P. achoreum Blake. Knotweed. Aug.–Oct. Waste ground, rare; Champaign and JoDaviess cos. *P. erectum* L.—J.

13. P. sachalinense F. Schmidt. Giant Knotweed. Aug.–Oct. Native of e. Asia; rarely adventive in Ill.; Jackson Co.

14. P. cuspidatum Sieb. & Zucc. Japanese Knotweed. Aug.–Sept. Native of e. Asia; occasionally spreading from cultivation in Ill.

15. P. virginianum L. Virginia Knotweed. July–Sept. Woods; common throughout the state. *Tovara virginiana* (L.) Raf.—F.

16. P. careyi Olney. Carey's Smartweed. July–Sept. Sandy soil, rare; Kankakee Co.

17. P. orientale L. Prince's Feather. June–Oct. Native of Eurasia; escaped into waste ground; occasional and scattered throughout the state.

18. P. punctatum Ell. Smartweed. July–Oct. Wet ground; common throughout the state. (Including var. *leptostachyum* [Meisn.] Small.)

19. P. hydropiper L. Smartweed. June–Oct. Native of Europe; adventive in wet ground; occasional throughout the state.

20. P. persicaria L. Lady's Thumb. May–Sept. Native of Europe; adventive in waste ground; common throughout the state.

21. P. cespitosum Blum var. longisetum (DeBruyn) Steward. Creeping Smartweed. June–Oct. Native of se. Asia; adventive in shaded ground and in lawns; scattered throughout the state. *P. longisetum* DeBruyn—J.

22. P. setaceum Baldw. var. interjectum Fern. Smartweed. July–Oct. Wet ground, not common; confined to the s. one-third of the state. *P. hydropiperoides* Michx.—J; *P. hydropiperoides* Michx. var. *setaceum* (Baldw.) Gl.—G.

23. P. hydropiperoides Michx. Mild Water Pepper. June–Oct. Wet ground, sometimes in shallow water; occasional to common throughout the state.

24. P. opelousanum Riddell. Water Pepper. July–Oct. Wet ground, not common; confined to the s. one-fourth of the state. *P. hydropiperoides* Michx.—J; *P. hydropiperoides* Michx. var. *opelousanum* (Riddell) Stone—G.

25. P. amphibium L. var. stipulaceum (Coleman) Fern. Water Smartweed. June–Sept. Ocreae usually with bristles. Wet ground; occasional in the n. half of the state, rare elsewhere.*P. fluitans* Eaton f. *hartwrightii* (Gray) G. N. Jones—J; *P. natans* Eaton f. *hartwrightii* (Gray) Stanford—G.

f. fluitans (Eat.) Fern. Water Smartweed. June–Sept. Ocreae without bristles. In water; occasional in the n. half of the state, rare elsewhere. *P. fluitans* Eaton—J; *P. natans* Eaton—G.

26. P. coccineum Muhl. Water Smartweed. July–Oct. Wet ground;

occasional throughout the state. (Including var. *pratincola* [Greene] Stanford.)

27. P. longistylum Small. Smartweed. July–Oct. Wet ground, rare; Alexander, St. Clair, and Union cos. *P. pennsylvanicum* L.—J.

28. P. scabrum Moench. Smartweed. June–Aug. Native of Europe; rarely adventive along railroads; Champaign and Lee cos. (Including *P. tomentosum* Schrank.) *P. lapathifolium* L.—G, J.

29. P. lapathifolium L. Pale Smartweed. July–Oct. Wet ground; common throughout the state.

30. P. pensylvanicum L. Common Smartweed. July–Oct. Leaves strigose; peduncles with spreading hairs. Wet ground; occasional throughout the state.

var. durum Stanford. July–Sept. Leaves strigose; peduncles strigose. Wet ground, rare; Pope Co.

var. laevigatum Fern. July–Oct. Leaves glabrous. Wet ground, common; in every co.

52. CHENOPODIACEAE

1. Leaves reduced to small, opposite scales; stems jointed, fleshy . . . . 1. *Salicornia*
1. Leaves linear-subulate or broader; stems not conspicuously jointed or fleshy (sometimes succulent in *Salsola*) . . . .'. 2
2. Leaves linear-subulate to linear, up to 3 mm broad 3
2. Leaves linear-lanceolate or broader, over 3 mm broad. 7
3. Leaves at most about 1 cm long . . . . 2. *Polycnemum*
3. Leaves 2 cm long or longer 4
4. Uppermost leaves spinose . . . . 3. *Salsola*
4. Uppermost leaves not spinose 5
5. Sepals 5 6
5. Sepal 1 . . . . 6. *Corispermum*
6. Stems villous above; leaves more or less pubescent; all sepals winged in fruit . . . . 4. *Kochia*
6. Stems glabrous; leaves glabrous; only 1 or 3 of the sepals winged in fruit . . . . 5. *Suaeda*
7. Fruiting calyx horizontally winged . . . . 7. *Cycloloma*
7. Fruiting calyx sometimes variously keeled or ridged, but not horizontally winged 8
8. Flowers unisexual; fruit enclosed by 2 bracteoles . . . . 8. *Atriplex*
8. Flowers perfect; fruit enclosed by the persistent calyx 9. *Chenopodium*

1. Salicornia L.   Glasswort

1. S. europaea L. Glasswort. Aug.–Oct. Native of Europe, Asia, and Africa; rarely adventive in soils with a high mineral content; Cook Co.

2. Polycnemum L.

1. P. majus A. Br. July–Sept. Native of Europe; rarely adventive along a railroad; Monroe Co.

3. Salsola L. Saltwort

1. S. kali L. var. tenuifolia Tausch. Russian Thistle. July–Sept. Native of Europe and Asia; adventive along railroads and on sand beaches; occasional in the n. half of the state, rare in the s. half. *S. pestifer* Nels.—J.

4. Kochia Roth Burning-bush

1. K. scoparia (L.) Roth. Summer Cypress. July–Sept. Native of Europe and Asia; naturalized in waste areas; common in the n. half of the state, occasional to rare elsewhere. (Including var. *culta* Farw.)

5. Suaeda Forsk. Sea-Blite

1. S. depressa (Pursh) S. Wats. Sea-Blite. Aug.–Oct. Native of the w. US.; adventive along a highway, apparently in response to salting; Cook Co.

6. Corispermum L. Bugseed

1. C. hyssopifolium L. Common Bugseed. July–Sept. Sandy soil, rare; Cook, Lake, Mason, Menard, and Whiteside cos.

7. Cycloloma Moq. Winged Pigweed

1. C. atriplicifolium (Spreng.) Coult. Winged Pigweed. July–Aug. Sandy soil; occasional throughout the state.

8. Atriplex L. Orach

1. Most or all the lowermost leaves opposite or subopposite; stems and leaves not mealy or scurfy (rarely mealy in young specimens of *A. hortensis*) 2
1. All the leaves alternate; stems and leaves mealy or scurfy 3
2. Bracts subtending fruits 10–15 mm broad; seeds 2–4 mm across ....
.... 1. *A. hortensis*
2. Bracts subtending fruits up to 5 mm broad; seeds 1–2 mm across ......
.... 2. *A. patula*
3. Stems and leaves mealy; bracts subtending fruits entire or dentate .....
.... 3. *A. rosea*
3. Stems and leaves scurfy; bracts subtending fruits deeply cleft ....
.... 4. *A. argentea*

1. A. hortensis L. Garden Orach. Aug.–Sept. Native of Asia; rarely escaped from cultivation; DuPage, Lake, and Vermilion cos.
2. A. patula L. Spear Scale. July–Sept. Waste ground, often in saline areas; occasional in the n. half of the state, uncommon elsewhere. (Including *A. hastata* L. and *A. patula* L. var. *hastata* [L.] Gray.)
3. A. rosea L. Red Scale. Aug.–Oct. Native of Europe and Asia; rarely adventive in alkaline waste ground; Cook Co.
4. A. argentea Nutt. Silver Scale. July–Oct. Native of the w. U.S.; rarely adventive in Ill.; Cook, Hancock, and Menard cos.

9. *Chenopodium L. Goosefoot*

1. Plants with glandular hairs or resinous sessile glands on the leaves and/or the stems; plants strongly aromatic 2
1. Plants without glandular-pubescence or without sessile glands; plants usually not aromatic 3
2. Calyx lobes glandular-pubescent; stems with spreading, glandular hairs .... 1. *C. botrys*
2. Calyx lobes not glandular-pubescent; stems glabrous or with appressed, glandless hairs .... 2. *C. ambrosioides*
3. Lower surface of leaves densely white-mealy 4
3. Lower surface of leaves not white-mealy 12
4. Leaves entire, mostly linear, oblong, or narrowly lanceolate 5
4. Leaves lobed or toothed, mostly lanceolate, ovate, deltoid, or rhombic 6
5. Leaves 1-nerved, linear; pericarp somewhat firmly attached to seed ....
.... 3. *C. pallescens*
5. Leaves 3-nerved, oblong or narrowly lanceolate; pericarp readily removed from seed .... 4. *C. desiccatum*
6. Most of the seeds vertical, about 0.5 mm broad, free from the pericarp; calyx lobes mostly 3–4 .... 5. *C. glaucum*
6. Most of the seeds horizontal, 0.8–2.0 mm broad, firmly attached to the pericarp (easily removed in *C. desiccatum*); calyx lobes 5 7
7. Seeds honeycombed on the surface 8
7. Seeds smooth, or at least not honeycombed on the surface 9
8. Calyx lobes conspicuously keeled; seeds 0.8–1.5 mm broad ....
.... 6. *C. berlandieri*
8. Calyx lobes slightly or not at all keeled; seeds 1.5–2.0 mm broad ....
.... 7. *C. bushianum*
9. Pericarp easily removed from the seed .... 4. *C. desiccatum*
9. Pericarp firmly attached to the seed 10
10. Seed 1.1–1.5 mm broad .... 8. *C. album*
10. Seeds 0.8–1.2 mm broad 11
11. Lowermost leaves serrate; middle leaves entire; calyx lobes not covering the fruit. .... 9. *C. strictum*
11. Lowermost and middle leaves coarsely toothed; calyx lobes covering the fruit .... 10. *C. missouriense*
12. Calyx mostly 3-lobed; seeds nearly all vertical 13
12. Calyx 5-lobed; seeds horizontal 15
13. Mature flower-heads 5–10 mm in diameter; fruit berry-like ....
.... 11. *C. capitatum*
13. Mature flower-heads less than 5 mm in diameter; fruit not berry-like ...
.... 14
14. Seeds 0.8–1.0 mm broad; annuals; calyx fleshy .... 12. *C. rubrum*
14. Seeds 1.5 mm broad; perennials; calyx herbaceous ....
.... 13. *C. bonus-henricus*
15. Seeds 1.5–2.5 mm broad 16
15. Seeds less than 1.5 mm broad 17
16. Seeds honeycombed, the pericarp closely adherent to the seeds; leaves less than 5 cm long .... 7. *C. bushianum*
16. Seeds smooth, the pericarp easily removed from the seeds; most or all the leaves over 5 cm long .... 14. *C. gigantospermum*
17. Seeds honeycombed .... 6. *C. berlandieri*
17. Seeds smooth 18
18. All leaves entire .... 15. *C. polyspermum*

18. Some or all of the leaves toothed 19
19. Pericarp easy to remove from the seed .... 16. *C. standleyanum*
19. Pericarp closely adherent to the seed 20
20. Seeds 1.2–1.5 mm broad, dull .... 17. *C. murale*
20. Seeds up to 1.2 mm broad, lustrous 21
21. Leaves thin, lustrous on the upper surface .... 18. *C. urbicum*
21. Leaves thickish, not lustrous on the upper surface 22
22. Lowermost leaves serrate; middle leaves entire; calyx lobes not covering the fruit .... 9. *C. strictum*
22. Lowermost and middle leaves coarsely toothed; calyx lobes covering the fruit .... 10. *C. missouriense*

1. C. botrys L. Jerusalem Oak. July–Oct. Native of Europe and Asia; adventive in waste ground; occasional throughout the state.

2. C. ambrosioides L. Mexican Tea. Aug.–Nov. Inflorescence interspersed with leafy bracts; lobes of calyx minutely keeled. Native of trop. Am.; adventive in waste ground; common throughout the state, except for the extreme n. cos.

var. anthelminticum (L.) Gray. Mexican Tea. Aug.–Nov. Inflorescence without leafy bracts; lobes of calyx not keeled. Native of trop. Am.; adventive in waste ground; occasional in the s. half of the state.

3. C. pallescens Standl. Narrow-leaved Goosefoot. June–Sept. Rocky ground, waste areas; occasional in the n.-cent. counties. (Included in *C. leptophyllum* Nutt. by F and G and in *C. pratericola* Rydb. by J.)

4. C. desiccatum A. Nels. var. leptophylloides (Murr.) Wahl. Narrow-leaved Goosefoot. June–Oct. Waste ground, occasional; mostly in the n. half of the state. (Included in *C. leptophyllum* Nutt. by F and G and in *C. pratericola* Rydb. by J.)

5. C. glaucum L. Oak-leaved Goosefoot. July–Sept. Native of Europe; adventive in waste ground; occasional throughout the state.

6. C. berlandieri Moq. var. zschackei (Murr.) June–Oct. Waste ground; occasional throughout the state. *C. berlandieri* Moq.—J; *C. album* L.—G.

7. C. bushianum Aellen. Goosefoot. Aug.–Oct. Fields, woods, waste ground; occasional throughout the state. *C. paganum* Reich.—G, J.

8. C. album L. Lamb's Quarters. May–Oct. Flower clusters continuous; margin of calyx yellow. Waste ground; common throughout the state; in every co.

var. lanceolatum (Muhl.) Coss. & Germ. Lamb's Quarters. May–Oct. Flower clusters interrupted; margin of calyx white. Waste ground; occasional throughout the state. *C. lanceolatum* Muhl.—F.

9. C. strictum Roth var. glaucophyllum (Aellen) Wahl. Goosefoot. July–Sept. Waste ground, rare.

10. C. missouriense Aellen. Goosefoot. Sept.–Oct. Fields, waste ground; occasional throughout the state. *C. berlandieri* Moq.—F; *C. album* L.—G.

11. C. capitatum (L.) Aschers. Strawberry Blite. May–Aug. Native of Europe and Asia; rarely escaped in Ill.; McHenry and Peoria cos.

12. C. rubrum L. Coast Blite. July–Oct. Native of the w. U.S.; rarely adventive in Ill.; Cook Co.

13. C. bonus-henricus L. Good King Henry. July–Oct. Native of Europe; rarely adventive in Ill.; Cook Co.

14. C. gigantospermum Aellen. Maple-leaved Goosefoot. June–Oct.

Shaded ledges, rocky woods; occasional throughout the state. *C. hybridum* L.—G, J; *C. hybridum* var. *gigantospermum* (Aellen) Rousseau—F.

15. C. polyspermum L. Many-seeded Goosefoot. June–Oct. Native of Europe; rarely adventive in waste ground; Jackson Co.

16. C. standleyanum Aellen. Goosefoot. June–Oct. Woodlands; occasional throughout the state. C. boscianum Moq.—F, J.

17. C. murale L. Nettle-leaved Goosefoot. June–Nov. Native of Europe; adventive in waste areas; scattered throughout the state.

18. C. urbicum L. City Goosefoot. July–Oct. Native of Europe; adventive in waste areas; scattered throughout the state, except for the s. one-fifth.

53. AMARANTHACEAE

1. Leaves alternate .... 1. *Amaranthus*
1. Leaves opposite 2
2. Pubescence stellate; flowers axillary .... 2. *Tidestromia*
2. Pubescence simple; flowers terminal 3
3. Leaves ovate, glabrous or sparsely hairy beneath; inflorescence paniculate .... 3. *Iresine*
3. Leaves linear to elliptic, densely pubescent beneath; inflorescence spicate .... 4. *Froelichia*

1. *Amaranthus* L.   Amaranth

1. Leaves with a pair of spines in the axils .... 1. *A. spinosus*
1. Leaves without axillary spines 2
2. Plants monoecious, with both staminate and pistillate flowers on the same plant. 3
2. Plants dioecious, with staminate and pistillate flowers on separate plants 9
3. Flowers in short axillary clusters; plants mostly low and spreading (sometimes erect in *A. albus*) 4
3. Flowers in elongated terminal and axillary spikes; plants erect 5
4. Seeds less than 1 mm long; bracts subulate, much longer than the calyx .... 2. *A. albus*
4. Seeds more than 1 mm long; bracts narrowly ovate, about equalling the calyx or a little longer .... 3. *A. graecizans*
5. Calyx of pistillate flowers 3.0–3.3 mm long, up to half as long as the bracts 6
5. Calyx of pistillate flowers 1.5–2.0 mm long, more than half as long as the bracts 7
6. Sepals of pistillate flowers acute; stamens 3 .... 4. *A. powellii*
6. Sepals of pistillate flower obtuse to emarginate ...... 5. *A. retroflexus*
7. Sepals of pistillate flowers acute; calyx usually longer than the fruit .... 6. *A. hybridus*
7. Sepals of pistillate flowers obtuse; calyx shorter than the fruit 8
8. Sepals of pistillate flowers overlapping at their margins; terminal spike arching .... 7. *A. caudatus*
8. Sepals of pistillate flowers not overlapping at their margins; terminal spike erect .... 8. *A. cruentus*

9. Pistillate flowers with 5 sepals 10
9. Pistillate flowers lacking a calyx, or sometimes with 1 or 2 sepals only
.... 12
10. Bracts of pistillate flowers longer than the calyx; leaves acute; seeds
1.3–1.4 mm across .... 9. *A. palmeri*
10. Bracts of pistillate flowers shorter than the calyx; leaves obtuse; seeds
1.0–1.1 mm across 11
11. Fruits dehiscent .... 10. *A. torreyi*
11. Fruits indehiscent .... 11. *A. ambigens*
12. Fruit indehiscent or splitting irregularly; bracts without a conspicuous
midnerve; outer sepals as long as inner sepals .... 12. *A. tuberculatus*
12. Fruit circumscissile; bracts with a conspicuous midnerve; outer sepals
longer than inner sepals .... 13. *A. tamariscinus*

1. A. spinosus L. Spiny Pigweed. July–Oct. Native of trop. Am.;
waste ground; common in the s. two-thirds of the state, rare elsewhere.

2. A. albus L. Tumbleweed. July–Sept. Fields and waste areas; common throughout the state.

3. A. graecizans L. Prostrate Amaranth. July–Sept. Native of the w.
U.S.; waste ground; common throughout the state.

4. A. powellii S. Wats. Smooth Pigweed. July–Oct. Native of the sw.
U.S.; occasional throughout the state, becoming very common northward.

5. A. retroflexus L. Rough Pigweed. Aug.–Oct. Fields and waste
areas; in every co.

6. A. hybridus L. Green Amaranth. Aug.–Oct. Fields and waste
ground; common throughout the state.

7. A. caudatus L. Purple Amaranth. July–Oct. Native of trop. Am.;
rarely escaped from cultivation; Jackson Co.

8. A. cruentus L. Purple Amaranth. July–Oct. Native of Asia; rarely
escaped from cultivation; occasional in the ne. corner of the state.

9. A. palmeri S. Wats. Palmer's Amaranth. Aug.–Oct. Native of the
sw. U.S.; rarely adventive in Ill.; Cook, DeKalb, and Kane cos.

10. A. torreyi (Gray) Benth. Torrey's Amaranth. July–Oct. Native of
the w. U.S.; rarely adventive in Ill.; Crawford Co. *A. arenicola* I. M.
Johnston—F, J.

11. A. ambigens Standl. Water Hemp. July–Oct. Moist soil, rare; Winnebago Co. The type was collected by M. S. Bebb from Winnebago
Co. *Acnida altissima* (Riddell) Riddell—J.

12. A. tuberculatus (Moq.) Sauer. Water Hemp. Aug.–Oct. Wet
ground, often on sand bars; scattered throughout the state. *Acnida
altissima* Riddell—G, J; *Acnida altissima* var. *subnuda* (S. Wats.)
Fern.—F; *Acnida subnuda* (S. Wats.) Standl.—G, J; *Acnida altissima*
var. *prostrata* (Uline & Bray) Fern.—F.

13. A. tamariscinus Nutt. Water Hemp. July–Oct. Wet soil; scattered
throughout the state. *Acnida tamariscina* (Nutt.) Wood—F, G, J.

2. *Tidestromia Standl.*

1. T. lanuginosa (Nutt.) Standl. July–Oct. Native of w. U.S.; rarely
adventive along a railroad; Cook Co.

3. *Iresine* P. Br. Bloodleaf

1. I. rhizomatosa Standl. Aug.–Oct. Wet woodlands, rare; confined to the s. one-fourth of the state.

4. *Froelichia* Moench Cottonweed

1. Flowering spikes over 1 cm thick; leaves oblanceolate to elliptic; stem usually unbranched from near the base .... 1. *F. floridana*
1. Flowering spikes less than 1 cm thick; leaves linear to lanceolate; stems usually branched from near the base .... 2. *F. gracilis*

    1. F. floridana (Nutt.) Moq. var. campestris (Small) Fern. Cottonweed. May–Sept. Sandy soil; occasional in the n. two-thirds of the state, apparently absent elsewhere. *F. campestris* Small—J.
    2. F. gracilis (Hook.) Moq. Cottonweed. May–Sept. Native of w. U.S.; adventive, particularly along railroads; occasional throughout the state.

54. NYCTAGINACEAE

1. *Mirabilis* L. Four-o'clock

1. Leaves linear, at least 10 times as long as broad, glaucous   1. *M. linearis*
1. Leaves linear-lanceolate to ovate, at most about 5 times as long as broad, not glaucous 2
2. Leaves more or less ovate, cordate or truncate at base; stems glabrous or nearly so .... 2. *M. nyctaginea*
2. Leaves linear-lanceolate to elliptic, tapering or somewhat rounded at the base; stems hirsute or with short, incurved hairs in two longitudinal lines 3
3. Stems hirsute; fruit rugose .... 3. *M. hirsuta*
3. Stems with short, incurved hairs in two longitudinal lines; fruit tuberculate (at least on the angles) .... 4. *M. albida*

    1. M. linearis (Pursh) Heimerl. Narrow-leaved Umbrella-wort. June–Aug. Native to the w. U.S.; rarely adventive in waste areas; Cook, St. Clair, and Will cos. *Oxybaphus linearis* (Pursh) Robins.—G.
    2. M. nyctaginea (Michx.) MacM. Wild Four-o'clock. May–Aug. Native to the w. U.S.; naturalized in waste areas, particularly along railroads; common throughout the state. *Oxybaphus nyctagineus* (Michx.) Sweet—G.
    3. M. hirsuta (Pursh) MacM. Hairy Umbrella-wort. July–Aug. Native to the w. U.S.; rarely adventive in waste areas; Cook, DuPage, and JoDaviess cos. *Oxybaphus hirsutus* (Pursh) Sweet—G.
    4. M. albida (Walt.) Heimerl. Pale Umbrella-wort. July–Aug. Native to the s. cent. U.S.; rarely adventive in waste areas; Cook, Grundy, and Logan cos. *Oxybaphus albidus* (Walt.) Sweet—G.

55. PHYTOLACCACEAE

1. *Phytolacca L. Pokeweed*

1. P. americana L. Pokeweed. July–Oct. Fields, woods, waste areas; occasional to common throughout the state.

56. AIZOACEAE

1. *Mollugo L. Indian Chickweed*

1. M. verticillatus L. Carpetweed. June–Oct. Native of trop. Am.; naturalized in waste areas; common throughout the state.

57. PORTULACACEAE

1. Flowers yellow .... 1. *Portulaca*
1. Flowers not yellow 2
2. All leaves from base of plant .... 2. *Talinum*
2. Leaves cauline 3
3. Only one pair of flat leaves on each stem .... 3. *Claytonia*
3. Several alternate terete leaves on each stem .... 1. *Portulaca*

1. *Portulaca L. Purslane*

1. Leaves spatulate, flat; plants glabrous; flowers up to 1 cm across ......
.... 1. *P. oleracea*
1. Leaves linear, terete; plants hairy at the nodes; flowers at least 2 cm across .... 2. *P. grandiflora*

    1. P. oleracea L. Purslane. June–Oct. Native of Europe; naturalized in waste areas; occasional to common throughout the state.
    2. P. grandiflora Hook. Rose Moss. June–Oct. Native of S. Am.; occasionally escaped from cultivation.

2. *Talinum Adans. Fameflower*

1. Flowers pale pink; stamens usually 5, never more than 8 ....
.... 1. *T. parviflorum*
1. Flowers rose; stamens 10 or more 2
2. Petals up to 8 mm long; stamens 10–25; capsule up to 5 mm long
.... 2. *T. rugospermum*
2. Petals 10 mm long or longer; stamens usually more than 25; capsule 6–8 mm long .... 3. *T. calycinum*

    1. T. parviflorum Nutt. Small Flower-of-an-hour. June–July. Exposed sandstone cliffs, rare; Johnson, Pope, and Union cos.

2. T. rugospermum Holz. Flower-of-an-hour. June–Aug. Sandy soil, cliffs; occasional to rare in the n. one-third of the state.

3. T. calycinum Engelm. Large Flower-of-an-hour. June–Sept. Edge of sandstone cliff, very rare; Randolph Co.

3. *Claytonia* L. *Spring Beauty*

1. C. virginica L. Spring Beauty. March–May. Woods, fields; common throughout the state. (Including f. *robusta* [Somes] Palmer & Steyerm.)

58. CARYOPHYLLACEAE

1. Leaves with scarious stipules at their base 2
1. Leaves without stipules 4
2. Petals absent; leaves elliptic to oblanceolate; styles 2-cleft ....
.... 1. *Paronychia*
2. Petals 5; leaves linear-filiform; styles 3 or 5 3
3. Leaves opposite; petals pink or white; styles 3 .... 2. *Spergularia*
3. Leaves whorled; petals white; styles 5 .... 3. *Spergula*
4. Petals absent; flowers green 5
4. Petals present; flowers variously colored, not green 6
5. Fruit with 1 seed; styles 2; sepals united into a cup .... 4. *Scleranthus*
5. Fruit with several seeds; styles 5; sepals free .... 5. *Sagina*
6. Sepals free, or united only at the very base 7
6. Sepals united into a tube 12
7. Petals emarginate, bifid, or jagged toothed along the margins 8
7. Petals entire 11
8. Petals jagged toothed along the margins .... 6. *Holosteum*
8. Petals emarginate or bifid 9
9. Styles 3 .... 7. *Stellaria*
9. Styles 5 10
10. Styles alternate with the sepals; capsule ovoid .... 8. *Myosoton*
10. Styles opposite the sepals; capsule cylindric .... 9. *Cerastium*
11. Stamens 5; styles 5 .... 5. *Sagina*
11. Stamens 10; styles 3 .... 10. *Arenaria*
12. Calyx subtended by 2–6 bracts .... 11. *Dianthus*
12. Calyx not subtended by bracts 13
13. Petals bifid .... 12. *Tunica*
13. Petals entire, or fringed at the tip 14
14. Styles 5 15
14. Styles 2 or 3 16
15. Calyx lobes longer than calyx tube .... 13. *Agrostemma*
15. Calyx lobes shorter than calyx tube .... 14. *Lychnis*
16. Styles 3; calyx obviously 10-nerved .... 15. *Silene*
16. Styles 2; calyx obviously 5-nerved, or the nerves obscure 17
17. Calyx up to 5 mm long, conspicuously nerved .... 16. *Gypsophila*
17. Calyx over 8 mm long, obscurely nerved 17.*Saponaria*

1. *Paronychia* Mill. *Forked Chickweed*

1. Stems glabrous; sepals obtuse .... 1. *P. canadensis*
1. Stems pubescent; sepals mucronulate .... 2. *P. fastigiata*

1. P. canadensis (L.) Wood. Forked Chickweed. June–Oct. Dry woods; occasional throughout the state.
2. P. fastigiata (Raf.) Fern. Forked Chickweed. June–Oct. Dry woods; occasional in all but the northernmost cos. (Including var. *paleacea* Fern.)

2. Spergularia J. & C. Presl

1. Capsule 5 mm long or longer; seeds 0.6 mm long or longer, narrowly winged .... 1. *S. media*
1. Capsule up to 5 mm long; seeds up to 0.6 mm long, wingless 2. *S. rubra*

1. S. media (L.) C. Presl. July–Sept. Native of Europe; rarely adventive along a highway, in response to salting; Cook, DuPage, and Kane cos.
2. S. rubra (L.) J. & C. Presl. July–Sept. Native of Europe and Asia; rarely adventive in Ill.; Cook Co.

3. Spergula L.

1. S. arvensis L. June–Sept. Native of Europe; rarely adventive in Ill.; Champaign, Cook, and Marion cos.

4. Scleranthus L.

1. S. annuus L. April–Aug. Native of Europe; rarely adventive in Ill.; Kankakee and Winnebago cos.

5. Sagina L. Pearlwort

1. S. decumbens (Ell.) Torr. & Gray. Pearlwort. April–May. Moist, open areas; occasional in the s. half of the state, rare in the n. half.

6. Holosteum L. Jagged Chickweed

1. H. umbellatum L. Jagged Chickweed. April–May. Native of Europe; rapidly becoming common throughout the s. half of the state, rare elsewhere.

7. Stellaria L. Chickweed

1. Median leaves ovate, petiolate; petals shorter than the sepals .... 1. *S. media*
1. Median leaves linear or lanceolate or elliptic, sessile; petals longer than the sepals 2
2. Stems puberulent in lines, at least in the upper half of the stem; leaves elliptic; capsule shorter than the sepals .... 2. *S. pubera*
2. Stems glabrous or essentially so; leaves lanceolate or linear; capsule longer than the sepals 3
3. Leaves lanceolate; seeds roughened; sepals strongly nerved .... 3. *S. graminea*
3. Leaves linear; seeds smooth; sepals nerveless or essentially so .... 4. *S. longifolia*

1. S. media (L.) Cyrillo. Common Chickweed. Jan.–Dec. Native of Europe and Asia; in every co.

2. S. pubera Michx. Great Chickweed. March–May. Wooded bluffs, rare; Cook and Pope cos.

3. S. graminea L. Common Stitchwort. May–June. Native of Europe; adventive in moist areas; scattered throughout the state.

4. S. longifolia Muhl. Chickweed. May–July. Moist ground; scattered throughout the state.

8. Myosoton Moench

1. M. aquaticum (L.) Moench. Giant Chickweed. June–Sept. Native of Europe; naturalized in moist waste ground; occasional in the n. one-sixth of the state; also Clark, Cumberland, and Sangamon cos. *Stellaria aquatica* (L.) Scop.—G, J.

9. Cerastium L. Mouse-ear Chickweed

1. Margins of bracts scarious 2
1. Bracts herbaceous throughout 4
2. Petals about as long as the sepals; stems more or less viscid 3
2. Petals at least twice as long as the sepals; stems not viscid .... 3. *C. velutinum*
3. Stamens 10; plants perennial, with basal offshoots .... 1. *C. vulgatum*
3. Stamens 5; plants annual, without basal offshoots ...... 2. *C. pumilum*
4. Pedicels at least twice as long as the capsules 5
4. Pedicels up to twice as long as the capsules or shorter 7
5. Stamens 4 or 5 .... 4. *C. tetrandrum*
5. Stamens 10 6
6. Filaments glabrous; stems viscid throughout; sepals pubescent but not distinctly bearded at the tip; capsule curved .... 5. *C. nutans*
6. Filaments ciliate; stems viscid only near apex; sepals long-bearded at the tip; capsule straight .... 6. *C. brachypetalum*
7. Sepals acute; capsule about twice as long as the sepals 7. *C. viscorum*
7. Sepals obtuse to subacute; capsule about three times as long as the sepals .... 8. *C. brachypodum*

1. C. vulgatum L. Common Mouse-ear Chickweed. May–Sept. Native of Europe and Asia; naturalized in waste ground; in every co.

2. C. pumilum Curtis. Mouse-eared Chickweed. April–May. Native of Europe; rarely found in waste ground; Jackson Co.

3. C. velutinum Raf. Field Mouse-ear Chickweed. May–June. Rocky woods; scattered throughout the state. *C. arvense* L.—G; *C. arvense* L. var. *villosum* (Muhl.) Hollick & Britt.—F. (Including *C. arvense* L. f. *oblongifolium* [Torr.] Pennell.)

4. C. tetrandrum Curtis. March–April. Native of Europe; rarely adventive along roadsides; Champaign and Union cos.

5. C. nutans Raf. Nodding Mouse-ear Chickweed. March–June. Moist ground; common throughout the state, except for the extreme nw. cos.

6. C. brachypetalum Pers. April–May. Native of Europe; adventive along roadsides; Pulaski and Union cos.

7. C. viscosum L. Mouse-ear Chickweed. April–May. Native of

Europe; adventive in moist soils; occasional in the s. half of the state; absent elsewhere except for Boone, DeKalb, and Kane cos.

8. C. brachypodum (Engelm.) B. L. Robins. Mouse-ear Chickweed. April–May. Native of the w. U.S.; occasional in the s. half of the state; also McLean and Winnebago cos.

10. *Arenaria L.   Sandwort*

1. Leaves subulate, setaceous, filiform, or linear 2
1. Leaves oval, oblong, or ovate 3
2. Leaves often in fascicles; seeds about 1 mm long .... 1. *A. stricta*
2. Leaves opposite; seeds 0.5–0.6 mm long .... 2. *A. patula*
3. Perennial with slender rhizomes; leaves obtuse to subacute, most or all of them over 7 mm long; sepals obtuse, 2–3 mm long; seeds 1 mm long or longer .... 3. *A. lateriflora*
3. Annual; leaves acuminate, rarely over 7 mm long; sepals acuminate, 3–4 mm long; seeds 0.5–0.6 mm long .... 4. *A. serpyllifolia*

1. A. stricta Michx. Stiff Sandwort. May–July. Sand ridges, hill prairies, limestone glade; occasional or rare in the n. one-fourth of the state; also Tazewell Co.

2. A. patula Michx. Slender Sandwort. May–June. Wooded slopes, rare; Cook, Grundy, Kankakee, St. Clair, and Will cos.

3. A. lateriflora L. Sandwort. May–June. Woodlands; occasional in the n. half of the state; apparently absent elsewhere.

4. A. serpyllifolia L. Thyme-leaved Sandwort. April–June. Native of Europe; adventive in sandy waste ground; scattered throughout the state.

11. *Dianthus L.   Pink*

1. Flowers solitary, on long pedicels; leaves less than 2 cm long; bracts about half as long as the calyx .... 1. *D. deltoides*
1. Flowers clustered, sessile or nearly so; leaves (or most of them) more than 2 cm long; bracts about as long as the calyx 2
2. Pubescent annual; leaves linear, or linear-lanceolate, to 4 mm broad; calyx pubescent .... 2. *D. armeria*
2. Glabrous perennial; leaves lanceolate to ovate-lanceolate, mostly at least 4 mm broad; calyx glabrous .... 3. *D. barbatus*

1. D. deltoides L. Maiden Pink. June–Aug. Native of Europe; rarely escaped from cultivation; Kane and Vermilion cos.

2. D. armeria L. Deptford Pink. May–Aug. Native of Europe; adventive in fields and along roads; occasional throughout the state.

3. D. barbatus L. Sweet William. June–Aug. Native of Europe; rarely adventive in Ill.; Hancock and Jackson cos.

12. *Tunica Scop.*

1. T. saxifraga (L.) Scop. Saxifrage Pink. June–July. Native of Europe; rarely adventive in Ill.; Champaign Co.

13. Agrostemma L.

1. A. githago L. Corn Cockle. May–July. Native of Europe; adventive in fields and waste ground; occasional throughout the state.

14. Lychnis L.   Campion

1. Plants viscid-pubescent; teeth of calyx not twisted; petals bifid 2
1. Plants white-woolly; teeth of calyx twisted; petals emarginate ....
.... 3. *L. coronaria*
2. Flowers white, opening at dusk; teeth of calyx narrowly lanceolate; capsule ovoid .... 1. *L. alba*
2. Flowers pink or red, opening in the morning; teeth of calyx deltoid; capsule globose .... 2. *L. dioica*

    1. L. alba Mill. Evening Campion. May–Aug. Native of Europe; adventive in waste ground; common in the n. two-thirds of the state, rare elsewhere.
    2. L. dioica L. Red Campion. June–Aug. Native of Europe; rarely adventive in Ill.; Winnebago Co.
    3. L. coronaria (L.) Desr. Mullein Pink. June–Aug. Native of Europe; rarely escaped from cultivation; Cook Co.

15. Silene L.   Catchfly

1. Petals deep red, crimson, or scarlet 2
1. Petals white or pink or purplish 3
2. Stem with 15 or more pairs of leaves; petals entire or emarginate ......
.... 1. *S. regia*
2. Stem with up to 8 pairs of leaves; petals bifid .... 2. *S. virginica*
3. Leaves whorled; petals fringed .... 3. *S. stellata*
3. Leaves opposite; petals bifid 4
4. Flowers up to 4 mm across .... 4. *S. antirrhina*
4. Flowers 1 cm or more across 5
5. Stems glutinous below each node; flowers pink or purple 5. *S. armeria*
5. Stems not glutinous below each node; flowers white or only pink-based
.... 6
6. Stems viscid-pubescent or hirsute; calyx pubescent 7
6. Stems and calyx glabrous 8
7. Flowers nodding, opening at dusk; viscid-pubescent annual ....
.... 6. *S. noctiflora*
7. Flowers ascending, opening during the day; hirsute biennial ....
.... 7. *S. dichotoma*
8. Plants green; leaves acuminate; flowers solitary in the upper axils .....
.... 8. *S. nivea*
8. Plants glaucous; leaves acute; flowers in cymose panicles 9
9. Calyx up to 15 mm long in fruit, scarcely inflated; vertical nerves of calyx sparsely or not at all connected by cross-veins .... 9. *S. cserei*
9. Calyx up to 20 mm long in fruit, inflated; vertical nerves of calyx regularly connected by cross-veins .... 10. *S. cucubalus*

    1. S. regia Sims. Royal Catchfly. July–Aug. Dry soil, rare; Clark, Cook, Lawrence, St. Clair, Wabash, Will, and Winnebago cos.

2. S. virginica L. Firepink. April–July. Dry woods; occasional in the ne. quarter of the state, rare or absent elsewhere.

3. S. stellata (L.) Ait. f. Starry Campion. June–Aug. Woods; occasional throughout the state.

4. S. antirrhina L. Sleepy Catchfly. May–July. Fields, waste ground; occasional to common throughout the state.

5. S. armeria L. Sweet William Catchfly. June–July. Native of Europe; rarely escaped from gardens; Cook, Kane, and Will cos.

6. S. noctiflora L. Night-flowering Catchfly. June–July. Native of Europe; adventive in waste ground; occasional in the n. three-fifths of the state, absent elsewhere.

7. S. dichotoma Ehrh. Forked Catchfly. June–July. Native of Europe; rarely adventive in waste ground; Champaign, Cook, Douglas, McLean, and Winnebago cos.

8. S. nivea (Nutt.) Otth. Snow Campion. June–July. Woods, calcareous fens; occasional in the n. two-thirds of the state, absent elsewhere.

9. S. cserei Baumg. Glaucous Campion. June–July. Native of Europe; adventive along railroads and in waste ground; occasional to common in the n. one-third of the state; also Coles, Lawrence, St. Clair, and Vermilion cos.

10. S. cucubalus Wibel. Bladder Catchfly. May–July. Native of Europe; adventive in waste ground; occasional in the n. two-thirds of the state.

16. Gypsophila L.

1. Petals at least twice as long as the calyx; calyx 3–5 mm long .... 1. *G. elegans*
1. Petals about as long as the calyx; calyx up to 3 mm long 2
2. Leaves cuneate, 1-nerved; teeth of calyx suborbicular, with a narrow green center .... 2. *G. paniculata*
2. Leaves subcordate, 3- to 5-nerved; teeth of calyx ovate, with a broad green center .... 3. *G. perfoliata*

1. G. elegans Bieb. Baby's Breath. May–June. Native of Europe and Asia; rarely escaped from cultivation; Champaign Co.

2. G. paniculata L. Baby's Breath. June–Aug. Native of Europe and Asia; rarely escaped from cultivation; Cook, Mason, and Winnebago cos.

3. G. perfoliata L. Baby's Breath. June–Oct. Native of Europe and Asia; rarely escaped from cultivation; Cook and McHenry cos.

17. Saponaria L.

1. Each petal with a scale at the base; flowers 1.8–2.5 cm broad, usually pink or white .... 1. *S. officinalis*
1. Petals without scales at the base; flowers less than 1.5 cm broad, usually red .... 2. *S. vaccaria*

1. S. officinalis L. Bouncing Bet. June–Sept. Native of Europe; adventive in waste ground; common throughout the state.

2. S. vaccaria L. Cow Herb. June–Aug. Native of Europe; adventive in waste ground; scattered in the n. half of the state; also Jackson Co.

59. CERATOPHYLLACEAE

1. Ceratophyllum L. Hornwort

1. Achenes with 2 basal spines; ultimate leaf segments toothed on the margins .... 1. *C. demersum*
1. Achenes with several spines both lateral and basal; ultimate leaf segments not toothed on the margins .... 2. *C. echinatum*

    1. C. demersum L. Coontail. July–Sept. Quiet waters; common throughout Ill.
    2. C. echinatum Gray. Coontail. July–Sept. Quiet waters, not common; scattered in Ill.

60. NYMPHAEACEAE

1. Flowers yellow; sepals 5–6; leaves oval .... 1. *Nuphar*
1. Flowers white or pinkish; sepals 4; leaves orbicular ...... 2. *Nymphaea*

1. Nuphar Smith Pond Lily

    1. N. luteum L. ssp. variegatum (Engelm.) Beal. Bullhead Lily. May–Aug. Petiole conspicuously flattened; sepals usually red-tinged. Ponds, rare; Cook and Lake cos. *N. variegatum* Engelm.—F, G, J.
    ssp. macrophyllum (Small) Beal. Yellow Pond Lily. May–Aug. Petiole terete to more or less flattened; sepals green. Ponds; occasional throughout the state. *N. advena* (Ait.) Ait. f.—F, G, J.

2. Nymphaea L. Water Lily

1. Petals subacute at tip; flowers fragrant; seeds 2 mm long 1. *N. odorata*
1. Petals rounded at tip; flowers not fragrant; seeds 3–4 mm long ....
.... 2. *N. tuberosa*

    1. N. odorata Ait. Fragrant Water Lily. June–Sept. Lakes and ponds, rare; Cook, Franklin, Johnson, McHenry, Perry, and Union cos. (Including var. *gigantea* Tricker.)
    2. N. tuberosa Paine. White Water Lily. June–Aug. Ponds and streams; occasional in the n. two-thirds of Ill.

61. NELUMBONACEAE

1. Nelumbo Adans. Lotus

    1. N. lutea (Willd.) Pers. American Lotus. July–Aug. Lakes and ponds; occasional throughout Ill.

62. CABOMBACEAE

1. Leaves uniform; stamens 12–18; carpels 4 or more .... 1. *Brasenia*
1. Leaves dimorphic; stamens 3–6; carpels 2–4 .... 2. *Cabomba*

1. *Brasenia Schreb. Watershield*

1. B. schreberi Gmel. Watershield. June–Sept. Ponds and quiet streams, rare; scattered in Ill.

2. *Cabomba Aubl.*

1. C. caroliniana Gray. Carolina Watershield. May–Sept. Ponds, rare; Franklin, Massac, Union, and Wabash cos.

63. RANUNCULACEAE

1. Flowers yellow or white or occasionally pinkish 2
1. Flowers red, blue, purple, or green 15
2. Stems climbing; leaves opposite .... 17. *Clematis*
2. Stems erect or creeping or floating; leaves basal or alternate 3
3. Flowers yellow 4
3. Flowers white or occasionally pinkish 5
4. Sepals and petals present and differentiated; leaves variously toothed or lobed, but not crenate; fruit an achene .... 1. *Ranunculus*
4. Sepals yellow, petal-like; petals absent; leaves crenate; fruit a follicle ..
.... 2. *Caltha*
5. Plants aquatic; sepals and petals each 5, differentiated   1. *Ranunculus*
5. Plants not true aquatics; petals absent or, if present, either stamen-like or with a spur 6
6. Flowers spurred .... 3. *Delphinium*
6. Flowers without a spur 7
7. Flowers numerous in racemes, panicles, or corymbs; sepals inconspic-uous, falling away as the flower opens 8
7. Flowers 1–4, never in a raceme; sepals showy, petal-like, persistent   11
8. Leaves simple, although deeply lobed .... 4. *Trautvetteria*
8. Leaves variously compound 9
9. Flowers unisexual, arranged in much-branched panicles . 5. *Thalictrum*
9. Flowers perfect, arranged in racemes or corymbs 10
10. Fruit fleshy, berry-like; raceme simple .... 6. *Actaea*
10. Fruit dry, follicular; raceme sparingly branched .... 7. *Cimicifuga*
11. Leaves 3-lobed, all basal .... 8. *Hepatica*
11. Leaves either not 3-lobed or not all basal 12
12. Leaves on the stem alternate 13
12. Leaves on the stem opposite or whorled 14
13. Leaves ternately compound; sepals 5; pedicels glabrous . 10. *Isopyrum*
13. Leaves simple, palmately lobed; sepals 3; pedicels pubescent..........
.... 9. *Hydrastis*
14. Leaves ternately compound; roots tuberous-thickened   11. *Anemonella*

14. Leaves deeply to shallowly palmately lobed, but not ternately divided; plants rhizomatous, or with a woody caudex .... 12. *Anemone*
15. Leaves all basal 16
15. At least some of the leaves cauline 17
16. Leaves linear, entire; sepals spurred; flowers greenish; receptacle elongated .... 13. *Myosurus*
16. Leaves palmately lobed; sepals not spurred; flowers pinkish or light purplish; receptacle not elongated .... 8. *Hepatica*
17. One or more petals or sepals prolonged backward into a spur 18
17. None of the perianth parts spurred 19
18. Flowers red and yellow, with each of the five petals prolonged backward into a long spur .... 14. *Aquilegia*
18. Flowers purple, blue, or green; one of the petal-like sepals prolonged backward into a spur .... 3. *Delphinium*
19. Cauline leaves opposite or whorled 20
19. Cauline leaves alternate 21
20. Sepals 4, thick and fleshy; leaves opposite .... 17. *Clematis*
20. Sepals usually 5–20, thin; leaves whorled .... 12. *Anemone*
21. Flowers green .... 15. *Helleborus*
21. Flowers purplish or blue 22
22. Flowers blue; inflorescence subtended by a deeply divided involucre; flowers perfect .... 16. *Nigella*
22. Flowers purplish; inflorescence without an involucre; flowers unisexual .... 5. *Thalictrum*

1. *Ranunculus* L.   Buttercup

1. Petals white; achenes covered by horizontal wrinkles 2
1. Petals yellow; achenes smooth or variously marked, but not with horizontal wrinkles 3
2. Leaves becoming limp after removal from the water; beak of achene less than 1 mm long, or absent .... 1. *R. trichophyllus*
2. Leaves remaining firm after removal from the water; beak of achene about 1 mm long .... 2. *R. longirostris*
3. At least some of the leaves simple and unlobed 4
3. None of the leaves simple and unlobed 12
4. All leaves simple and unlobed 5
4. At least some of the leaves lobed or divided 8
5. Leaves reniform, cordate; petals often slightly shorter than the sepals .. .. 3. *R. cymbalaria*
5. Leaves linear to lanceolate, tapering to base; petals equalling or slightly longer than the sepals 6
6. Petals 5–7 in number, 3–9 mm long; stamens 20–50 7
6. Petals 1–3 in number, 1.0–2.5 mm long; stamens 3–10 ... 6. *R. pusillus*
7. Perennial; stamens 25–50; achenes flattened, about 2 mm long .... 4. *R. ambigens*
7. Annual; stamens 12–25; achenes plump, about 1 mm long .... 5. *R. laxicaulis*
8. Petals longer than the sepals 9
8. Petals equalling or shorter than the sepals 10
9. Stamens in one or two series; fruiting head less than 6 mm thick; sepals without long white hairs .... 7. *R. harveyi*
9. Stamens in 3–5 series; fruiting head 6–10 mm thick; sepals with long white hairs .... 8. *R. rhomboideus*

10. Plants more or less fleshy; achenes with corky thickenings at base; stems often hollow .... 9. *R. sceleratus*
10. Plants not fleshy; achenes without corky thickenings at base; stems not hollow 11
11. Achenes shiny; receptacle pubescent; roots slender .. 10. *R. abortivus*
11. Achenes dull; receptacle glabrous, except sometimes near the tip; roots thickened .... 11. *R. micranthus*
12. Plants aquatic or, if creeping in mud, some of the leaves finely dissected 13
12. Plants not truly aquatics; leaves not finely divided 14
13. Achene rugose on the sides, corky-thickened at the base; beak of achene about 1.5 mm long .... 12. *R. flabellaris*
13. Achene smooth on the sides, not corky-thickened at the base; beak of achene up to 0.8 mm long .... 13. *R. gmelinii*
14. Petals up to 6 mm long 15
14. Petals 6 mm long or longer 17
15. Achenes with short, hooked spines; petals 1–2 mm long .... 14. *R. parviflorus*
15. Achenes smooth; petals 2–6 mm long 16
16. Petals about equalling the sepals; achenes flat, with strongly recurved beaks; terminal lobe of leaves not stalked .... 15. *R. recurvatus*
16. Petals distinctly shorter than the sepals; achenes not flat, with nearly straight beaks; terminal lobe of leaves stalked .... 16. *R. pensylvanicus*
17. Achenes smooth on the flattened sides 18
17. Achenes papillate on the sides .... 24. *R. sardous*
18. Petals at least half as broad as long 19
18. Petals less than half as broad as long 24
19. Terminal segment of leaves not stalked .... 17. *R. acris*
19. Terminal segment of leaves stalked 20
20. Stems erect or ascending, not rooting at the nodes 21
20. Stems creeping and rooting at the nodes, at least at maturity 22
21. Beak of achene less than 1 mm long, usually curved .. 18. *R. bulbosus*
21. Beak of achene more than 1 mm long, more or less straight .... 19. *R. hispidus*
22. Achene plump, the beak up to 1.5 mm long .... 22. *R. repens*
22. Achene flattened, the beak 1.5–3.0 mm long 23
23. Achene up to 3.5 (–4.5) mm long, with a low narrow keel near the margin .... 20. *R. septentrionalis*
23. Achene 3.5–5.0 mm long, with a high broad keel near the margin .... 21. *R. carolinianus*
24. Some of the roots tuberous-thickened .... 23. *R. fascicularis*
24. All roots fibrous .... 21. *R. carolinianus*

1. R. trichophyllus Chaix. White Water-crowfoot. May–Aug. Ponds and slow streams; mostly in the n. cos. *R. aquatilis* L. var. *capillaceus* (Thuill.) DC.—G.

2. R. longirostris Godr. White Water-crowfoot. May–Aug. Ponds and slow streams; mostly in the n. cos. *R. circinatus* Sibth.—G.

3. R. cymbalaria Pursh. Seaside Crowfoot. May–Aug. Wet soil, rare; restricted to the n. one-fourth of the state.

4. R. ambigens Wats. Spearwort. June–Sept. Swampy woods and ditches, rare; Fulton, St. Clair, and Wabash cos.

5. R. laxicaulis (Torr. & Gray) Darby. Spearwort. May–July. Wet woods and ditches; generally in s. Ill., n. to Fulton Co.

6. R. pusillus Poir. Small Spearwort. May–June. Swamps, wet woods, ditches; confined to the s. half of Ill.

7. R. harveyi (A. Gray) Britt. Harvey's Buttercup. April–May. Leaves and stems glabrous. Sandstone ravines, rare; Effingham, Jackson, and Randolph cos.

f. pilosus (Benke) Palmer & Steyerm. May. Leaves and stems pilose. Sandstone ravines, rare; Randolph Co.

8. R. rhomboideus Goldie. Prairie Buttercup. May. Prairies, rare; restricted to the extreme n. cos.

9. R. sceleratus L. Cursed Crowfoot. May–Aug. Wet meadows, ditches, and river banks; occasional throughout Ill., although rare in the s. cos.

10. R. abortivus L. Small-flowered Crowfoot. April–June. Leaves, stems, and peduncles glabrous. Fields, moist woods; common throughout Ill.

var. acrolasius Fern. April–June. Some of the leaves, stems, and peduncles pilose. Fields, moist woods; much less common than the preceding.

11. R. micranthus Nutt. Small-flowered Buttercup. March–May. Moist or dry woods; occasional in the s. half of Ill. (Including var. delitescens [Greene] Fern.)

12. R. flabellaris Raf. Yellow Water-crowfoot. April–June. Swamps and ponds; occasional throughout the state.

13. R. gmelinii DC. var. hookeri (D. Don) L. Benson. Small Yellow Water-crowfoot. July–Aug. Ponds; Cook Co. R. purshii Richards.—J.

14. R. parviflorus L. April–June. Naturalized from Europe; waste ground; Jackson Co.

15. R. recurvatus Poir. April–June. Wet woods; occasional throughout the state.

16. R. pensylvanicus L. f. Bristly Crowfoot. July–Sept. Wet ground, rare; scattered throughout the state.

17. R. acris L. Tall Buttercup. May–Aug. Naturalized from Europe; roadsides and fields; occasional in the n. half of Ill.

18. R. bulbosus L. Bulbous Buttercup. April–July. Naturalized from Europe, apparently rare; Henderson and Johnson cos.

19. R. hispidus Michx. Bristly Buttercup. April–May. Pubescence spreading; achenes 3.0–3.5 mm long (excluding the beak). Woods; occasional in the s. three-fifths of Ill.

var. marilandicus (Poir.) L. Benson. Bristly Buttercup. April–May. Pubescence generally appressed; achenes 2.0–2.5 mm long (excluding the beak). Dry woods; rare in s. Ill. Var. falsus Fern.—F, G.

20. R. septentrionalis Poir. Swamp Buttercup. April–July. Petioles and lower part of stems glabrous or appressed-pubescent. Low woods and ditches; common throughout Ill.

var. caricetorum (Greene) Fern. April–July. Petioles and lower part of stems with retrorse pubescence; much less common than the preceding.

21. R. carolinianus DC. April–May. Lowland woods, rare; Champaign, Hancock, and Union cos.

22. R. repens L. Creeping Buttercup. April–Aug. Petals 10 or fewer. Native of Europe; roadsides and fields; rare in the n. half of Ill.; also Jersey Co.

var. pleniflorus Fern. Double-flowered Creeping Buttercup. Apr.– Aug. Petals more than 10. Planted in gardens, rarely escaped into waste ground; Jackson Co.

23. R. fascicularis Muhl. Early Buttercup. April–May. Open woods and meadows; throughout the state. (Including var. *apricus* [Greene] Fern.)

24. R. sardous Crantz. May–July. Native of Europe; low fields and waste places; confined to a few cos. in the southernmost part of Ill.

2. Caltha L. Marsh Marigold

1. C. palustris L. Marsh Marigold. April–June. Wet meadows; confined to the n. two-thirds of Ill.

3. Delphinium L. Larkspur

1. Pistil 1; follicle 1; cultivated annual .... 1. *D. ajacis*
1. Pistils 3–5; follicles 3–5; native perennial 2
2. Stems glabrous; follicles spreading at maturity; seeds smooth ....
.... 2. *D. tricorne*
2. Stems pubescent; follicles erect at maturity; seeds roughened 3
3. Seeds winged, with appressed scales .... 3. *D. carolinianum*
3. Seeds unwinged, with projecting scales .... 4. *D. virescens*

1. D. ajacis L. Rocket Larkspur. June–Aug. Escaped from gardens; occasional throughout the state.

2. D. tricorne Michx. Dwarf Larkspur. April–June. Rich woods; occasional to common in the s. four-fifths of the state, absent elsewhere. Occasional white-flowered specimens (f. *albiflora* Millsp.) have been observed.

3. D. carolinianum Walt. Wild Blue Larkspur. May–June. Dry, often sandy soil, rare; Adams, Henderson, Macon, Mercer, Moultrie, and Pike cos. (Our plants belong to var. *crispum* Perry.)

4. D. virescens Nutt. Prairie Larkspur. May–July. Prairies, very rare; Hancock Co.

4. Trautvetteria Fisch. & Mey. False Bugbane

1. T. caroliniensis (Walt.) Vail. False Bugbane. June–July. Moist ground along a stream, very rare; Cass Co.

5. Thalictrum L. Meadow Rue

1. Middle and upper leaves sessile 2
1. Middle and upper leaves petiolate .... 3. *T. dioicum*
2. Lower surface of leaflets eglandular and glabrous 4
2. Lower surface of leaflets glandular or pubescent 3
3. Lower surface of leaflets glandular .... 1. *T. revolutum*
3. Lower surface of leaflets pubescent but eglandular .. 2. *T. dasycarpum*
4. Margins of leaflets revolute, the lower surface of the leaflets conspicuously reticulate .... 1. *T. revolutum*
4. Margins of leaflets flat, the lower surface of the leaflets conspicuously reticulate .... 2. *T. dasycarpum*

1. T. revolutum DC. Waxy Meadow Rue. Late May–June. Lower surface of leaflets glandular. Prairies and open woods; occasional throughout the state.

f. glabrum Pennell. Waxy Meadow Rue. Late May–June. Lower surface of leaflets glabrous and eglandular. Prairies and open woods; not as common as the preceding.

2. T. dasycarpum Fisch. & Lall. Purple Meadow Rue. May–June. Leaflets firm, pubescent but eglandular beneath. Moist, wooded ravines, occasional; scattered throughout Ill.

var. hypoglaucum (Rydb.) Boivin. Meadow Rue. June–July. Leaflets thin, glabrous and eglandular beneath. Moist, wooded ravines, occasional; scattered throughout Ill. *T. hypoglaucum* Rydb.—J.

3. T. dioicum L. Early Meadow Rue. April–early May. Rich woodlands and prairies; occasional throughout Ill.

6. *Actaea* L. *Baneberry*

1. Pedicels much narrower than the peduncles in fruit; seeds 10 or more per berry, up to 4 mm long .... 1. *A. rubra*
1. Pedicels about as thick as the peduncles in fruit; seeds 3–9 (–10) per berry, over 4 mm long .... 2. *A. pachypoda*

1. A. rubra (Ait.) Willd. Red Baneberry. April–July (fruiting from Aug.–Oct.). Berries red. Rich woods, not common; confined to the n. cos. of Ill.

f. neglecta (Gillman) Robins. April–July (fruiting from Aug.–Oct.). Berries white. Rich woods, apparently not as common in Ill. as the preceding.

2. A. pachypoda Ell. Doll's-eyes. April–June (fruiting from July–Oct.). Rich woods; occasional throughout most of Ill. *A. alba* (L.) Mill.—G, J.

7. *Cimicifuga* L. *Bugbane*

1. At least the terminal leaflet cordate .... 1. *C. rubifolia*
1. All leaflets truncate or subcordate .... 2. *C. racemosa*

1. C. rubifolia Kearney. Black Cohosh. July–Sept. Rich woods, rare; Hardin, Massac, and Pope cos. *C. racemosa* (L.) Nutt.—G; *C. racemosa* var. *cordifolia* (Pursh) Gray—F; *C. cordifolia* Pursh—J.

2. C. racemosa (L.) Nutt. Black Cohosh. June–July. Woods, rare; St. Clair and Wabash cos.

8. *Hepatica* Mill. *Hepatica*

1. H. nobilis Schreb. var. obtusa (Pursh) Steyerm. Round-lobed Liverleaf. March–April. Lobes of the leaves, as well as the bracts, rounded at the tip. Rich woods, not common; confined to the ne. cos. of Ill. *H. americana* (DC.) Ker—F, G, J.

var. acuta (Pursh) Steyerm. Sharp-lobed Liverleaf. March–May. Lobes of the leaves, as well as the bracts, acute at the tip. Rich woods; occasional throughout the state. *H. acutiloba* DC.—F, G, J.

9. *Hydrastis* Ellis *Goldenseal*

1. H. canadensis L. Goldenseal. April–May. Rich woods, occasional; scattered throughout Ill.

10. *Isopyrum* L.

1. I. biternatum (Raf.) Torr. & Gray. False Rue Anemone. March–May. Rich woods; common throughout the state.

11. *Anemonella Spach*

1. A. thalictroides (L.) Spach. Rue Anemone. April–May. Dry or moist open woods; occasional throughout the state. (Including f. *favilliana* Bergseng, a form with petaloid stamens.)

12. *Anemone L. Anemone*

1. Styles 2–4 cm long, plumose; staminodia present .... 1. *A. patens*
1. Styles up to 4 mm long, usually pubescent but not plumose; staminodia absent 2
2. Plants arising from a tuber; sepals 10–20 .... 2. *A. caroliniana*
2. Plants arising from rhizomes; sepals 5 (–6) 3
3. Leaves of the involucre sessile; beak of achene 2–5 mm long ....
.... 3. *A. canadensis*
3. Leaves of the involucre petiolate; beak of achene less than 2 mm long 4
4. Basal leaf solitary; plants at maturity less than 30 cm tall; achenes hirsutulous, but not woolly .... 4. *A. quinquefolia*
4. Basal leaves 2-several; plants at maturity more than 30 cm tall; achenes woolly 5
5. Leaves of the involucre 5–9; fruiting heads more than twice as long as wide; styles less than 1 mm long .... 5. *A. cylindrica*
5. Leaves of the involucre 3; fruiting heads less than twice as long as wide; styles 1 mm long or longer .... 6. *A. virginiana*

1. A. patens L. Pasque-flower. March–April. Prairies; confined to the extreme n. cos. of the state. *A. ludoviciana* Nutt.—J; *A. patens* var. *wolfgangiana* (Bess.) Koch—G.

2. A. caroliniana Walt. Carolina Anemone. April–May. Prairie soil, bluffs, not common; confined to the n. half of Ill.

3. A. canadensis L. Meadow Anemone. May–July. Open woodlands, moist prairies; occasional in the n. two-thirds of Ill.; also Gallatin and Jackson cos.

4. A. quinquefolia L. Wood Anemone. April–May. Rich woods, not common; confined to the n. one-fourth of Ill. (Including var. *interior* Fern.)

5. A. cylindrica Gray. Thimbleweed. May–Aug. Open woods, prairies, occasional; limited to the n. two-thirds of Ill.; also Jackson Co.

6. A virginiana L. Tall Anemone. June–Aug. Open, usually dry woods; common throughout Ill.

13. *Myosurus L. Mousetail*

1. M. minimus L. Mousetail. April–June. Moist ground in woods and fields; occasional to common in the s. cos., rare northward.

14. Aquilegia L. Columbine

1. Spurs of flower straight; flowers red or yellow .... 1. *A. canadensis*
1. Spurs of flower hooked; flowers blue, purple, pink, or white ....
.... 2. *A. vulgaris*

    1. A. canadensis L. Columbine. April–July. Rocky woods; occasional to common throughout the state. (Including var. *coccinea* [Small] Munz.)
    2. A. vulgaris L. Garden Columbine. May–July. Escaped from cultivation; rarely observed outside of gardens.

15. Helleborus L. Hellebore

    1. H. viridis L. Green Hellebore. March–April. Escaped from cultivation; rarely observed outside of gardens.

16. Nigella L.

    1. N. damascena L. Love-in-a-mist. June–Aug. Escaped from cultivation; rarely observed outside of gardens; Jackson Co.

17. Clematis L. Clematis

1. Inflorescence paniculate; sepals thin, white; anthers blunt 2
1. Flower solitary; sepals thick, bluish; anthers with an attenuated tip .. 3
2. Leaves primarily 3-foliolate; anthers up to 1.5 mm long; achenes with spreading hairs .... 1. *C. virginiana*
2. Leaves primarily 5-foliolate; anthers at least 2 mm long; achenes with appressed, silky hairs .... 2. *C. dioscoreifolia*
3. Tails of fruits densely plumose; only the tips of the sepals recurved ....
.... 3. *C. viorna*
3. Tails of fruits glabrous or pubescent, but not plumose; upper half of the sepals recurved 4
4. Leaves thick, conspicuously reticulate beneath; sepals over 25 mm long, the margins crisped .... 5. *C. pitcheri*
4. Leaves thin, not conspicuously reticulate beneath; sepals less than 25 mm long, the margins not crisped .... 4. *C. crispa*

    1. C. virginiana L. Virgin's Bower. July–Sept. Moist soil, particularly at the edges of woods; occasional to common throughout the state.
    2. C. dioscoreifolia Levl. & Vaniot. Virgin's Bower. July–Oct. Naturalized from Japan and Korea; roadsides and borders of woods, occasional.
    3. C. viorna L. Leatherflower. May–July. Along streams, very rare; Johnson and Richland cos.
    4. C. crispa L. Blue Jasmine. April–July. Swampy woods, wet ditches, not common; confined to s. Ill.
    5. C. pitcheri Torr. & Gray. Leatherflower. May–Sept. Woods and thickets, occasional; throughout Ill. except for the northernmost counties.

64. BERBERIDACEAE

1. Prickly shrubs; fruit a red berry .... 1. *Berberis*
1. Unarmed herbs; fruit a capsule or a blue or yellow berry 2
2. Flower solitary on each plant, white; leaves simple 3
2. Flowers in panicles, yellow-green; leaves ternately compound ....
.... 4. *Caulophyllum*
3. Leaves 7- to 9-lobed; sepals usually 6; stamens twice as many as the petals; fruit a berry .... 2. *Podophyllum*
3. Leaves 2-lobed; sepals usually 4; stamens the same number as the petals; fruit a capsule .... 3. *Jeffersonia*

1. Berberis L.   Barberry

1. Leaves entire; flowers solitary or in clusters of 2–4; prickles unbranched;
berries dry .... 1. *B. thunbergii*
1. Leaves toothed; flowers in racemes; prickles mostly forked; berries
fleshy 2
2. Leaves with up to 20 teeth on the margin; branchlets brown; petals
notched; berries ovoid .... 2. *B. canadensis*
2. Leaves with 25 or more teeth on the margin; branchlets gray; petals not
notced; berries ellipsoid .... 3. *B. vulgaris*

   1. B. thunbergii DC. Japanese Barberry. April–May. Native of Asia;
spreading from cultivation, occasional.
   2. B. canadensis Mill. American Barberry. May. Dry woodlands and
sandstone cliffs, very rare; Jackson and Tazewell cos.
   3. B. vulgaris L. Common Barberry. May–June. Native of Europe;
rarely escaped from cultivation.

2. Podophyllum L.   Mayapple

   1. P. peltatum L. Mayapple. April–June. Mostly open woods; common throughout the state; in every co. (Including f. *polycarpum* Clute,
f. *biltmoreanum* Steyerm., f. *aphyllum* Plitt, and f. *deamii* Raymond.)

3. Jeffersonia Bart.   Twinleaf

   1. J. diphylla (L.) Pers. Twinleaf. April–May. Rich woods, not common; scattered throughout Ill. although rare in the s. one-third of the
state.

4. Caulophyllum Michx.   Blue Cohosh

   1. C. thalictroides (L.) Michx. Blue Cohosh. April–May. Rich woods,
occasional; throughout Ill.

65. MENISPERMACEAE

1. Leaves deeply 3- to 7-lobed; petals none; drupe 15–25 mm long ....
.... 1. *Calycocarpum*

1. Leaves entire or with 3–7 angles or shallow lobes; petals 6–8; drupe 5–10 mm long 2
2. Drupe blue-black; stamens 12–24 .... 2. *Menispermum*
2. Drupe red; stamens 6 .... 3. *Cocculus*

1. Calycocarpum Nutt. Cupseed

1. C. lyonii (Pursh) Gray. Cupseed. May–June. Swampy woodlands, rare; confined to s. Ill.

2. Menispermum L. Moonseed

1. M. canadense L. Moonseed. May–July. Moist woodlands, thickets; common throughout the state.

3. Cocculus DC. Snailseed

1. C. carolinus (L.) DC. Snailseed. July–Aug. Moist woods, thickets; occasional in s. Ill.

66. MAGNOLIACEAE

1. Leaves entire; petals without an orange blotch at the base within; seeds unwinged .... 1. *Magnolia*
1. Leaves 4-lobed; petals with an orange blotch at the base within; seeds winged .... 2. *Liriodendron*

1. Magnolia L. Magnolia

1. M. acuminata L. Cucumber Magnolia. April–May. Rich woodlands; restricted to the s. three tiers of cos.

2. Liriodendron L. Tulip Tree

1. L. tulipifera L. Tulip Tree. April–May. Rich woodlands, generally common; confined to the s. two-thirds of the state.

67. ANNONACEAE

1. Asimina Adans. Pawpaw

1. A. triloba (L.) Dunal. Pawpaw. April–May. Low woods and wooded slopes; common in the s. cos., becoming rare northward.

68. LAURACEAE

1. Some of the leaves lobed; flowers appearing as the leaves unfold, in modified racemes; fruits blue .... 1. *Sassafras*

1. None of the leaves lobed; flowers appearing before the leaves, in axillary clusters; fruits red .... 2. *Lindera*

1. Sassafras Nees Sassafras

1. S. albidum (Nutt.) Nees. Sassafras. April–May. Leaves glabrous or nearly so on the lower surface. Dry or moist woodlands, thickets, roadsides; common in the s. three-fourths of Ill., rare or absent elsewhere.

var. molle (Raf.) Fern. Red Sassafras. April–May. Leaves permanently pubescent on the lower surface. Woodlands, thickets, roadsides; occasional in the s. three-fourths of Ill.

2. Lindera Thunb. Spicebush

1. L. benzoin (L.) Blume. Spicebush. March–May. Leaves glabrous. Rich woodlands; common throughout the state, except the nw. portion.

var. pubescens (Palmer & Steyerm.) Rehd. Spicebush. March–May. Leaves permanently short-hairy on both surfaces. Swampy woods, rare; Johnson and Union cos.

69. PAPAVERACEAE

1. Flowers actinomorphic; stamens numerous; sap milky or variously colored 2
1. Flowers zygomorphic; stamens 6; sap clear 10
2. Flowers basically white (pink in a rare form of *Sanguinaria*) 3
2. Flowers not white 6
3. Perianth parts 6 or more 4
3. Perianth parts 2 .... 3. *Macleaya*
4. Plants without stems; petals 6 or more .... 1. *Sanguinaria*
4. Plants with stems; petals 4–6 5
5. Leaves prickly .... 2. *Argemone*
5. Leaves not prickly .... 4. *Papaver*
6. Leaves pinnatifid 7
6. Leaves ternately divided .... 7. *Eschscholtzia*
7. Leaves prickly .... 2. *Argemone*
7. Leaves not prickly 8
8. Flowers red or orange; sap milky; capsule opening by pores along the edge .... 4. *Papaver*
8. Flowers yellow; sap yellow; capsule opening from the bottom upward   9
9. Capsule bristly; style distinct; petals 2–3 cm long ...... 5. *Stylophorum*
9. Capsule smooth; style inconspicuous or absent; petals about 1 cm long .... 6. *Chelidonium*
10. Corolla with two spurs, or subcordate 11
10. Corolla with a single spur 12
11. Leaves all from the base; plants erect .... 8. *Dicentra*
11. Leaves cauline and alternate; plants climbing .... 9. *Adlumia*
12. Flowers yellow or pink; capsule elongated, several-seeded 10. *Corydalis*
12. Flowers purplish, tipped with red; fruit nearly round, 1-seeded ....
.... 11. *Fumaria*

1. Sanguinaria L.   Bloodroot

1. S. canadensis L. Bloodroot. March–April. Rich woods; common throughout the state. (Including f. *colbyorum* Benke, with pink petals, and var. *rotundifolium* [Greene] Fedde, with flowers elevated above the leaves.)

2. Argemone L.   Prickly Poppy

1. Flowers white or pink; leaves of a uniform color .... 1. *A. albiflora*
1. Flowers yellow or orange or cream; leaves with patches of pale green . . .. 2. *A. mexicana*

1. A. albiflora Hornem. Prickly Poppy. May–Sept. Naturalized from the w. U.S.; occasional; a Morgan Co. collection may represent a native population (Including *A. intermedia* Sweet.)
2. A. mexicana L. Mexican Poppy. May–Sept. Introduced from Mexico; rarely escaped; Henderson, Mason, Menard, and Stephenson cos.

3. Macleaya R. Br.   Plume Poppy

1. M. cordata (Willd.) R. Br. Plume Poppy. June–July. Introduced from e. Asia; rarely escaped; Cook and Henry cos.

4. Papaver L.   Poppy

1. Stems glabrous; cauline leaves cordate-clasping; plants glaucous; capsule nearly spherical .... 1. *P. somniferum*
1. Stems hirsute; cauline leaves not cordate-clasping; plants not glaucous; capsule longer than broad 2
2. Capsule narrowly obovoid .... 2. *P. dubium*
2. Capsule broadly obovoid .... 3. *P. rhoeas*

1. P. somniferum L. Opium Poppy. June–Aug. Introduced from Europe; rarely escaped from cultivation; Crawford Co.
2. P. dubium L. Poppy. May–Aug. Introduced from Europe; rarely escaped from cultivation; Wabash Co.
3. P. rhoeas L. Corn Poppy. May–Sept. Introduced from Europe; occasionally escaped from cultivation.

5. Stylophorum Nutt.

1. S. diphyllum (Michx.) Nutt. Celandine Poppy. March–May. Rich woods, not common; restricted to the s. counties; also Cook Co.

6. Chelidonium L.   Celandine

1. C. majus L. Celandine. May–Aug. Native of Europe; occasionally escaped from cultivation.

7. Eschscholtzia Cham.   California Poppy

1. E. californica Cham. California Poppy. June–Sept. Escaped from cultivation; rare; Kane and Wabash cos.

8. Dicentra Bernh.

1. Spurs of corolla spreading, subacute; flowers without an odor; plants from granular, white tubers .... 1. *D. cucullaria*
1. Spurs of corolla not spreading, rounded; flowers with a sweet odor; plants from yellow corn-like tubers .... 2. *D. canadensis*

    1. D. cucullaria (L.) Bernh. Dutchman's-breeches. April–May. Rich woods; common throughout Ill.
    2. D. canadensis (Goldie) Walp. Squirrel-corn. April–May. Rich woods, occasional; scattered throughout the state.

9. Adlumia Raf.

    1. A. fungosa (Ait.) Greene. Climbing Fumitory. June–Sept. Escaped from cultivation; rare; Kankakee and Ogle cos.

10. Corydalis Medic. Corydalis

1. Flowers pink, with yellow tips .... 1. *C. sempervirens*
1. Flowers wholly yellow 2
2. Outer petals with a winged crest down the back 3
2. Outer petals not crested down the back 5
3. Winged crest of outer petals with 3–4 teeth; stalk of capsule 1 cm long or longer; seeds rugulose .... 2. *C. flavula*
3. Winged crest of outer petals entire; stalk of capsule up to 5 mm long; seeds smooth 4
4. Capsules scarcely torulose, up to 15 mm long; uppermost raceme not overtopping subtending leaves .... 3. *C. micrantha*
4. Capsules strongly torulose, 15–25 mm long; uppermost raceme overtopping subtending leaves .... 4. *C. halei*
5. Spur ⅓–²/₅ the length of the corolla; sepals up to 1 mm long; seeds 1.3–1.7 mm long .... 5. *C. campestris*
5. Spur less than one-third the length of the corolla; sepals 1.5–2.0 mm long; seeds 2.0–2.5 mm long 6
6. Capsules pendulous or widely spreading; margins of seeds rounded ...
.... 6. *C. aurea*
6. Capsules ascending; margins of seeds sharp .... 7. *C. montana*

    1. C. sempervirens (L.) Pers. Pink Corydalis. May–Aug. Rocky woods, rare; Cook, LaSalle, and Ogle cos.
    2. C. flavula (Raf.) DC. Pale Corydalis. April–May. Moist woods; common in the s. cos., becoming less common northward.
    3. C. micrantha (Engelm.) Gray. Slender Corydalis. May–July. Rocky woods, not common; scattered throughout the state.
    4. C. halei (Small) Fern. & Schub. April–May. Moist soil at base of cliffs, very rare; Monroe Co.
    5. C. campestris (Britt.) Buchholz & Palmer. Plains Corydalis. May–July. Prairies, rare; restricted to w. cent. Ill.; also Grundy Co.
    6. C. aurea Willd. Golden Corydalis. May–Aug. Rocky soil, very rare; Cook, LaSalle, Ogle, and Winnebago cos.
    7. C. montana (Engelm.) Gray. April–May. Rocky woods, rare; confined to the n. one-third of Ill. *C. aurea* Willd. var. *occidentalis* Engelm.—F.

11. *Fumaria* L. *Fumitory*

1. F. officinalis L. Fumitory. May–Aug. Adventive from Europe; occasional in Ill.

70. CAPPARIDACEAE

1. Stamens (6–) 7–32 .... 1. *Polanisia*
1. Stamens 6 .... 2. *Cleome*

1. *Polanisia* Raf. *Clammyweed*

1. Leaflets obovate to lance-elliptic, all or most of them over 6 mm broad; fruits 5–10 mm broad .... 1. *P. dodecandra*
1. Leaflets linear, up to 4 mm broad; fruits 3–4 mm broad ... 2. *P. jamesii*

1. P. dodecandra (L.) DC. Clammyweed. June–Sept. Largest petals 3.5–6.5 (–8.0) mm long; longest stamens 4–10 (–14) mm long, scarcely exceeding the petals. Gravelly soil, often along railroads, river banks; scattered throughout the state. *P. graveolens* Raf.—F, G, J.
ssp. trachysperma (Torr. & Gray) Iltis. Clammyweed. June–Sept. Largest petals (7–) 8–16 mm long; longest stamens (9–) 12–30 mm long, usually much exceeding the petals. Gravelly soil, sandy soil; scattered in all parts of the state, but not common. *P. trachysperma* Torr. & Gray—F, G, J.
2. P. jamesii (Torr. & Gray) Iltis. June–Aug. Sandy soil, rare; Carroll, JoDaviess, and Mason cos. *Cristatella jamesii* Torr. & Gray— F, G, J.

2. *Cleome* L.

1. Plants essentially glabrous; leaves 3-foliolate .... 1. *C. serrulata*
1. Plants pubescent, minutely spinose at the nodes; leaves 5- to 7-foliolate
.... 2. *C. hassleriana*

1. C. serrulata Pursh. July–Sept. Adventive from the w. U.S.; occasional in waste areas.
2. C. hassleriana Chod. Spider Flower. June–Sept. Rarely escaped from cultivation; Coles, Jackson, JoDaviess, Kendall, and Washington cos. *C. speciosissima* Deppe—J.

71. CRUCIFERAE

1. Flowers white or purple 2
1. Flowers yellow 38
2. Leaves deeply palmately divided .... 1. *Dentaria*
2. Leaves simple, pinnately divided, or pinnatifid 3
3. Flowers purplish 4
3. Flowers white 10

4. Plants pubescent 5
4. Plants glabrous 8
5. Some or all the leaves pinnatifid 6
5. None of the leaves pinnatifid 7
6. Plants hispid; roots thickened; petals 1.5 cm long or longer; silique 5–10 mm in diameter .... 2. *Raphanus*
6. Plants more or less glandular-hirtellous; roots not conspicuously thickened; petals less than 1.5 cm long; silique up to 2 mm in diameter .....
.... 35. *Chorispora*
7. Basal leaves cordate, on petioles longer than the blades; cauline leaves obtuse to subacute; siliques up to 4 cm long .... 10. *Cardamine*
7. Basal leaves tapering to base, sessile or on petioles much shorter than the blades; cauline leaves acuminate; siliques 5–14 cm long   3. *Hesperis*
8. At least some of the upper leaves auriculate at the base . 4. *Iodanthus*
8. None of the leaves auriculate at the base 9
9. Plants fleshy; leaves sinuate-dentate; flowers 4–6 mm across . 5. *Cakile*
9. Plants not fleshy; some or all the leaves pinnatifid; flowers 10–15 mm across .... 6. *Eruca*
10. Basal rosette present (although sometimes withered at flowering time) 11
10. Basal rosette absent 20
11. Some of the leaves pinnate or pinnatifid 12
11. None of the leaves pinnate or pinnatifid 18
12. Some or all the cauline leaves sagittate or auriculate at the base ... 13
12. None of the cauline leaves sagittate or auriculate at the base 14
13. Ovary and fruit triangular, broader than long .... 7. *Capsella*
13. Ovary and fruit not triangular, much longer than broad ...... 8. *Arabis*
14. Leaves bi- or tri-pinnate; pubescence stellate .... 9. *Descurainia*
14. Leaves once-pinnate or merely pinnatifid; pubescence simple (hairs bifurcate in *A. lyrata*) 15
15. Petals 5–8 mm long .... 8. *Arabis*
15. Petals up to 4 mm long 16
16. Basal leaves pinnate .... 10. *Cardamine*
16. Basal leaves pinnatifid 17
17. Silique beaked; seeds wingless .... 10. *Cardamine*
17. Silique nearly beakless; seeds winged .... 11. *Sibara*
18. Ovary and fruit only 2–7 times longer than broad .... 12. *Draba*
18. Ovary and fruit much greater than seven times as long as broad ... 19
19. Leaves of basal rosette never exceeding 1 cm in width   13. *Arabidopsis*
19. Most leaves of basal rosette 1 cm broad or broader .... 8. *Arabis*
20. Some or all the leaves pinnate or pinnatifid 21
20. None of the leaves pinnate or pinnatifid 26
21. Some or all the cauline leaves auriculate or sagittate .... 14. *Lepidium*
21. None of the cauline leaves auriculate or sagittate 22
22. Petals to 2 mm long, or absent 23
22. Petals 4–20 mm long 24
23. Plants erect; silicle flattened .... 14. *Lepidium*
23. Plants matted; silicle subglobose .... 36. *Coronopus*
24. Petals 4–8 mm long; plants glabrous 25
24. Petals 10–20 mm long; plants more or less hispid .... 2. *Raphanus*
25. Some or all the cauline leaves simple; fruits up to three times as long as broad .... 15. *Armoracia*
25. All cauline leaves pinnately compound; fruits more than three times as long as broad .... 16. *Nasturtium*
26. Some or all the cauline leaves clasping or sagittate 27

47. Petals up to 4 mm long; silicle 4-seeded; leaves up to 3 cm long ......
.... 26. *Alyssum*
47. Petals 4–7 mm long; silicle several-seeded; most of the leaves over 3 cm
long .... 23. *Lesquerella*
48. Plants glabrous 49
48. Plants pubescent 52
49. Style absent; fruits borne on pendulous pedicels .... 27. *Isatis*
49. Style elongate; fruits borne on ascending pedicels 50
50. None of the leaves on the stem entire .... 22. *Barbarea*
50. Some or all of the leaves on the stem entire 51
51. Petals 7–10 mm long; cauline leaves obtuse at apex ....
.... 28. *Conringia*
51. Petals up to 6 mm long; cauline leaves acute at apex .... 29. *Camelina*
52. Petals 2–3 mm long; fruit 1-seeded .... 30. *Neslia*
52. Petals 4–6 mm long; fruit several-seeded .... 29. *Camelina*
53. Some of the leaves auriculate at the base 54
53. None of the leaves auriculate at the base 56
54. Ovary and fruit several times longer than broad 55
54. Ovary and fruit at most only about three times longer than broad ......
.... 34. *Rorippa*
55. Fruit tapering to a slender beak; seeds globose .... 31. *Brassica*
55. Fruit tapering to a short, thick style; seeds not globose .. 22. *Barbarea*
56. Plants with stellate pubescence .... 9. *Descurainia*
56. Plants glabrous or with pubescence of simple hairs 57
57. Racemes subtended by pinnatifid bracts .... 32. *Erucastrum*
57. Racemes without pinnatifid bracts subtending them (except rarely the
lowest raceme) 58
58. Petals up to 1 cm long 59
58. Petals 1–2 cm long 62
59. Ovary and fruit conspicuously beaked 60
59. Ovary and fruit beakless 61
60. Seeds in two rows .... 24. *Diplotaxis*
60. Seeds in one row .... 31. *Brassica*
61. Ovary and fruit at least 15 times longer than broad; seeds in one row ..
.... 33. *Sisymbrium*
61. Ovary and fruit up to 7 times longer than broad; seeds in two rows ....
.... 34. *Rorippa*
62. Petals with brown or purple veins 63
62. Petals without brown or purple veins .... 31. *Brassica*
63. Pedicels up to 6 mm long; fruit not constricted between the seeds .....
.... 6. *Eruca*
63. Pedicels over 6 mm long; fruit more or less constricted between the
seeds .... 2. *Raphanus*

1. *Dentaria* L. Toothwort

1. D. laciniata Muhl. Toothwort. Feb.–May. Woods, common;
throughout the state.

2. *Raphanus* L. Radish

1. Petals yellowish or white; fruit up to 6 mm thick .... 1. *R. raphanistrum*
1. Petals pink or pale purple; fruit 6–10 mm thick .... 2. *R. sativus*

1. R. raphanistrum L. Wild Radish. June–Aug. Native of Europe and Asia; occasional in waste places; Champaign, DeKalb, McHenry, Peoria, and St. Clair cos.

2. R. sativus L. Radish. May–Sept. Native of Europe and Asia; occasionally escaped from cultivation.

3. Hesperis L.   Rocket

1. H. matronalis L. Rocket. May–Aug. Native of Europe and Asia; occasionally escaped from cultivation.

4. Iodanthus Torr. & Gray

1. I. pinnatifidus (Michx.) Steud. Purple Rocket. May–July. Low woods; occasional throughout the state.

5. Cakile Hill   Sea Rocket

1. C. edentula (Bigel.) Hook. var. lacustris Fern. Sea Rocket. July–Sept. Shores of Lake Michigan, rare; Cook and Lake cos. C. edentula (Bigel.) Hook.—J.

6. Eruca Mill.   Garden Rocket

1. E. sativa Mill. Garden Rocket. June–Sept. Native of Europe; rarely escaped from cultivation; Peoria Co.

7. Capsella Medic.   Shepherd's-purse

1. C. bursa-pastoris (L.) Medic. Shepherd's-purse. Jan.–Dec. Native of Europe; naturalized in waste areas; known from every co.

8. Arabis L.   Rockcress

1. None of the leaves sagittate nor auriculate at the base 2
1. Some or all of the cauline leaves sagittate or auriculate at the base   . 3
2. Fruits strongly ascending at maturity; leaves on stem up to 5 mm broad
. . . . 1. A. lyrata
2. Fruits arching and usually pointing downward at maturity; leaves on stem 5 mm broad or broader . . . . 2. A. canadensis
3. Stems glabrous (except sometimes at the very base in A. glabra and A. drummondii) 4
3. Stems hairy to summit 6
4. At least some of the fruits over 1 mm wide, arched-recurving at maturity
. . . . 5. A. laevigata
4. Fruit never exceeding 1 mm in width, strongly erect at maturity 5
5. Petals creamy-yellow, about as long as the sepals; fruits terete . . . .
. . . . 3. A. glabra
5. Petals white, about twice as long as the sepals; fruits flat . . . .
. . . . 4. A. drummondii
6. Leaves and stems with simple and stellate hairs; petals 2–3 mm long; seeds wingless . . . . 6. A. shortii
6. Leaves and stems with simple hairs; petals 4–6 mm long; seeds narrowly winged . . . . 7. A. hirsuta

1. A. lyrata L. Sand Cress. May–July. Dunes, sandy woods, gravel prairies; restricted to the n. one-third of Ill.

2. A canadensis L. Sicklepod. May–July. Wooded slopes, moist woods; occasional throughout the state although rather rare in the se. cos.

3. A. glabra (L.) Bernh. Tower Mustard. May–July. Moist prairies, limestone woods, not common; throughout the state, but very rare in the s. cos.

4. A. drummondii Gray. Rock Cress. May–July. Gravelly soil, very rare; Cook and Kane cos. A. confinis S. Wats.—J.

5. A. laevigata (Muhl.) Poir. Smooth Rock Cress. March–June. Moist woods; occasional to common throughout Ill.

6. A. shortii (Fern.) Gl. Rock Cress. April–May. Fruits stellate-pubescent. Moist woods; occasional in the n. half of the state, rare in the s. half. A. perstellata E. L. Br. var. shortii Fern.—F.

var. phalacrocarpa (M. Hopkins) Steyerm. Rock Cress. May. Fruits glabrous. Moist woods, rare; Lake Co. A. perstellata E. L. Br. var. phalacrocarpa (M. Hopkins) Fern.—F.

7. A. hirsuta (L.) Scop. var. pycnocarpa (M. Hopkins) Rollins. Hairy Rock Cress. May–June. Stems hirsute to summit. Limestone cliffs and woods; not common in the n. half of the state; also Jackson Co. A. pycnocarpa Hopkins—J.

var. adpressipilis (M. Hopkins) Rollins. Rock Cress. May–June. Stem with appressed hairs to summit. Limestone woods, rare; confined to ne. Ill.; also St. Clair Co.

var. glabrata Torr. & Gray. Rock Cress. June. Stem glabrous above the middle. Woods, very rare; Kankakee Co.

9. *Descurainia* Webb & Berthelot   Tansy Mustard

1. Fruits up to 1 mm thick; stems and leaves grayish, not glandular-hairy
.... 1. *D. sophia*
1. Fruits 1–2 mm thick; stems and leaves green, somewhat glandular-hairy
.... 2. *D. pinnata* var. *brachycarpa*

1. D. sophia (L.) Webb. Tansy Mustard. June–Aug. Native of Europe; escaped into waste areas; rare and scattered.

2. D. pinnata (Walt.) Britt. var. brachycarpa (Richards.) Fern. Tansy Mustard. April–June. Sandstone and limestone cliffs, along railroads; occasional to common throughout the state. D. brachycarpa (Richards.) O. E. Schulz—J.

10. *Cardamine* L.   Bitter Cress

1. None of the leaves pinnate or pinnatifid 2
1. Some or all the leaves pinnate or pinnatifid 3
2. Stems glabrous; sepals green .... 1. *C. bulbosa*
2. Stems hirsute; sepals purplish .... 2. *C. douglassii*
3. Petioles of cauline leaves ciliate at base .... 3. *C. hirsuta*
3. Petioles of cauline leaves not ciliate at base 4
4. Terminal leaflet broader than the lateral leaflets; bases of leaflets decurrent along the rachis .... 4. *C. pensylvanica*
4. Terminal leaflet about the same width as the lateral leaflets; bases of leaflets not decurrent along the rachis ... 5. *C. parviflora* var. *arenicola*

1. C. bulbosa (Schreb.) BSP. Spring Cress. March–June. Wet soil in woods and along streams; occasional to common throughout the state.

2. C. douglassii (Torr.) Britt. Purple Cress. April–May. Low woods; occasional in the n. two-thirds of Ill.; also Johnson Co.

3. C. hirsuta L. Spring Cress. March–April. Native of Europe and Asia; occasionally naturalized in wet areas throughout the state.

4. C. pensylvanica Muhl. Bitter Cress. March–July. Low woods, damp fields, base of moist cliffs; occasional throughout the state.

5. C. parviflora L. var. arenicola (Britt.) O. E. Schulz. Small-flowered Bitter Cress. March–July. Dry woods, wet ledges; occasional throughout the state. *C. arenicola* Britt.—J.; *C. parviflora* L.—G.

11. Sibara Greene

1. S. virginica (L.) Rollins. March–May. Open woods, waste ground; occasional in the s. three-fifths of Ill.; also Kankakee Co. *Arabis virginica* (L.) Poir.—G, J.

12. Draba L. Whitlow Grass

1. No leaves present on flowering scape; petals deeply 2-cleft   1. *D. verna*
1. One or more leaves present on flowering stem 2
2. Leaves confined to lower part of stem 3
2. Leaves all along stem 4
3. Leaves dentate; pedicels pubescent .... 2. *D. cuneifolia*
3. Leaves entire; pedicels glabrous .... 3. *D. reptans*
4. Fruits to 5 mm long, glabrous, with up to 15 seeds . 4. *D. brachycarpa*
4. Fruits at least 6 mm long, pubescent, with 20 or more seeds ....
.... 2. *D. cuneifolia*

1. D. verna L. Vernal Whitlow Grass. Feb.–April. Seeds 40 or more; fruits more than twice as long as broad. Native of Europe and Asia; naturalized in waste areas; occasional in the s. cos., rare in the n. cos.

var. boerhaavii Van Hall. Seeds less than 40; fruits never more than twice as long as broad. Native of Europe; Jackson Co.

2. D. cuneifolia Nutt. Whitlow Grass. Feb.–May. Leaves basal and sometimes from the lower nodes. Limestone ledges, rare; Monroe and Randolph cos.

var. foliosa Mohlenbr. Leaves cauline as well as basal. Limestone ledge; Monroe Co.

3. D. reptans (Lam.) Fern. Whitlow Grass. Feb.–May. Fruits glabrous. Sand prairies, limestone ledges; occasional in the n. cos., rare in the s. cos.

var. micrantha (Nutt.) Fern. Fruits hispidulous; occasional in the n. cos.

4. D. brachycarpa Nutt. Short-fruited Whitlow Grass. March–May. Fields, lawns, prairies, woods, along railroads; common in the s. two-fifths of Ill.; also DeKalb Co., where it may be adventive.

13. Arabidopsis Heynh. Mouse-ear Cress

1. A. thaliana (L.) Heynh. Mouse-ear Cress. March–May. Native of Europe and Asia; naturalized in waste ground; occasional in the s. half of the state; absent from the n. half.

14. Lepidium L.   Peppergrass

1. Plants densely hairy .... 1. *L. campestre*
1. Plants glabrous or puberulent 2
2. Cauline leaves auriculate at base .... 2. *L. perfoliatum*
2. Cauline leaves tapering to base .... 3.
3. Fruits winged their entire length, 5–6 mm long; stamens 6 ....
.... 3. *L. sativum*
3. Fruits unwinged, or winged only at tip, 2–4 mm long; stamens 2 (–4)   4
4. Petals about as long as the sepals .... 4. *L. virginicum*
4. Petals absent 5
5. Plants with a fetid odor; basal leaves bipinnatifid .... 5. *L. ruderale*
5. Plants without a fetid odor; basal leaves once-pinnatifid or coarsely
toothed .... 6. *L. densiflorum*

1. L. campestre (L.) R. Br. Field Peppergrass. April–June. Native of Europe; naturalized in waste areas; occasional throughout the state.

2. L. perfoliatum L. Perfoliate Peppergrass. April–June. Native of Europe; naturalized in waste areas; occasional, primarily in the n. cos.

3. L. sativum L. Garden Peppergrass. May–June. Native of Europe; rarely escaped from cultivation; Cook Co.

4. L. virginicum L. Common Peppergrass. Feb.–Dec. Waste grounds, prairies, edge of woods, fields; in every co.

5. L. ruderale L. Stinking Peppergrass. May–June. Native of Europe; rarely naturalized; DeKalb Co.

6. L. densiflorum Schrad. Peppergrass. April–Oct. Native of Europe and Asia; naturalized in waste areas; occasional throughout the state.

15. Armoracia Gaertn., Meyer, & Scherb.

1. Basal leaves divided into filiform divisions; aquatic plants   1. *A. aquatica*
1. None of the leaves divided into filiform divisions; terrestrial plants .....
.... 2. *A. lapathifolia*

1. A. aquatica (Eat.) Wieg. Lake Cress. May–Aug. Swamps and quiet streams; occasional throughout the state. *Neobeckia aquatica* (Eat.) Greene—J.

2. A. lapathifolia Gilib. Horseradish. May–July. Native of Europe and Asia; escaped from cultivation; occasional throughout Ill. *A. rusticana* (Lam.) Gaertn.—G, J.

16. Nasturtium R. Br.   Water Cress

1. N. officinale R. Br. Water Cress. April–Oct. Cool springs and branches; occasional throughout the state. (Including var. *siifolium* [Reichenb.] Koch.)

17. Thlaspi L.   Penny Cress

1. T. arvense L. Field Penny Cress. April–June. Native of Europe and Asia; naturalized into waste areas; occasional throughout the state.

18. Cardaria Desv.

1. C. draba (L.) Desv. Hoary Cress. April–June. Native of Europe and Asia; naturalized in waste areas; occasional in the n. half of the state, rare in the s. half.

19. Alliaria Scop.   Garlic Mustard

1. A. officinalis Andrz. Garlic Mustard. May–June. Native of Europe and Asia; naturalized into woods and waste places; occasional in the n. half of Ill.; also Coles Co.

20. Berteroa DC.

1. B. incana (L.) DC. Hoary Alyssum. May–Sept. Native of Europe; naturalized in waste areas; not common, mostly found in the n. cos.

21. Lobularia Desv.   Sweet Alyssum

1. L. maritima (L.) Desv. Sweet Alyssum. June–Aug. Native of Europe; rarely escaped from cultivation; Champaign, Cook, Hancock, and Lake cos.

22. Barbarea R. Br.   Winter Cress

1. All leaves pinnatifid; beak of fruit up to 1 mm long . . . . 1. *B. verna*
1. Uppermost leaves merely dentate; beak of fruit 1.5–3.0 mm long . . . .
. . . . 2. *B. vulgaris*

1. B. verna (Mill.) Aschers. Early Winter Cress. April–June. Native of Europe; rarely naturalized in Ill.; Johnson Co.
2. B. vulgaris R. Br. Yellow Rocket. April–June. Fruits strongly ascending. Native of Europe; naturalized in waste ground; rare and scattered in Ill.
var. arcuata (Opiz) Fries. Yellow Rocket. April–June. Fruits spreading. Native of Europe; naturalized in waste ground; common throughout the state. *B. vulgaris* R. Br.—J.

23. Lesquerella S. Wats.

1. Plants perennial; fruits densely stellate-pubescent . . . . 1. *L. ludoviciana*
1. Plants annual; fruits glabrous or nearly so . . . . 2. *L. gracilis*

1. L. ludoviciana (Nutt.) S. Wats. Silvery Bladder Pod. May–Aug. Sandy soil, very rare; Mason Co.
2. L. gracilis (Hook.) S. Wats. Slender Bladder Pod. May–Aug. Native of the w. U.S.; adventive along a railroad in Cook Co.

24. Diplotaxis DC.

1. Sepals 5–8 mm long; stems leafy throughout; fruits stipitate . . . .
. . . . 1. *D. tenuifolia*
1. Sepals 2.5–5.0 mm long; stems leafy only in lower half; fruits sessile . . .
. . . . 2. *D. muralis*

1. D. tenuifolia (L.) DC. Sand Rocket. July–Sept. Native of Europe; rarely adventive in Ill.; Cook Co.

2. D. muralis (L.) DC. Wall Rocket. June–Aug. Native of Europe; naturalized in waste areas; occasional in the n. one-fourth of Ill.

25. Erysimum L.   Treacle Mustard

1. Petals well over 1 cm long, orange-yellow . . . . 1. *E. capitatum*
1. Petals up to 1 cm long, yellow 2
2. Petals up to 5 mm long . . . . 2. *E. cheiranthoides*
2. Petals 5–10 mm long 3
3. Annual; fruit more than 5 cm long; plants pale green . . 3. *E. repandum*
3. Perennial; fruit up to 5 cm long; plants gray-green   4. *E. inconspicuum*

1. E. capitatum (Dougl.) Greene. Western Wallflower. May–July. Sandy bluffs, rare; w. cent. cos.; also Kendall and LaSalle cos. *E. asperum* (Nutt.) DC.—G, J; *E. arkansanum* Nutt.—F.

2. E. cheiranthoides L. Wormseed Mustard. May–Sept. Native of Europe and Asia; naturalized in the n. half of Ill.

3. E. repandum L. Treacle Mustard. April–June. Native of Europe and Asia; naturalized throughout the state.

4. E. inconspicuum (S. Wats.) MacM. May–June. Native of the w. U.S.; occasionally adventive in waste areas in the n. half of Ill.

26. Alyssum L.   Alyssum

1. A. alyssoides L. Pale Alyssum. May–June. Native of Europe; naturalized in waste areas; occasional in the n. half of the state.

27. Isatis L.   Woad

1. I. tinctoria L. Dyer's Woad. May–June. Native of Europe; rarely escaped from cultivation; Cook Co.

28. Conringia Link   Hare's-ear Mustard

1. C. orientalis (L.) Dumort. Hare's-ear Mustard. May–Aug. Native of Europe; naturalized in waste places; occasional throughout the state.

29. Camelina Crantz   False Flax

1. Stems and leaves glabrous or with appressed pubescence; fruits 7–9 mm long . . . . 1. *C. sativa*
1. Stems and leaves hirsutulous; fruits 4–5 (–7) mm long   2. *C. microcarpa*

1. C. sativa (L.) Crantz. False Flax. April–Aug. Native of Europe; naturalized in waste areas; rare in the n. cos.; also Moultrie Co.

2. C. microcarpa Andrz. False Flax. April–Aug. Native of Europe; naturalized in waste areas; occasional throughout the state.

30. Neslia Desv.   Ball Mustard

1. N. paniculata (L.) Desv. Ball Mustard. May–Sept. Native of Europe; naturalized in Ill.; DuPage Co.

31. *Brassica* L. Mustard

1. None of the leaves auriculate and clasping at the base 2
1. Some of the cauline leaves auriculate and clasping at the base 5
2. Leaves all pinnatifid; fruits densely white-bristly .... 1. *B. hirta*
2. At least the uppermost leaves merely toothed, not pinnatifid; fruits glabrous or nearly so 3
3. Plants more or less hispid or hirsute, green; fruiting pedicels 3–7 mm long 4
3. Plant glabrous or nearly so, glaucous; fruiting pedicels 7–10 mm long .. .. 4. *B. juncea*
4. Fruits 2.5–5.5 cm long, ascending, the 1-seeded beak 10–15 mm long; upper leaves sessile .... 2. *B. kaber*
4. Fruits 1–2 cm long, erect, the seedless beak 1–3 mm long; upper leaves short-petiolate .... 3. *B. nigra*
5. Plants glaucous 6
5. Plants green .... 7. *B. napus*
6. Leaves not thick; petals 7–10 mm long; young leaves setose   5. *B. rapa*
6. Leaves thick; petals 15 mm long or longer; young leaves glabrous .... 6. *B. oleracea*

1. B. hirta Moench. White Mustard. April–July. Native of Europe and Asia; rarely naturalized in Ill.; Adams, Boone, Cook, DeKalb, Henry, Kane, Kankakee, Lake, McHenry, and Will cos. *B. alba* (L.) Rabenh.—J.

2. B. kaber (DC.) L. C. Wheeler var. pinnatifida (Stokes) L. C. Wheeler. Charlock. April–July. Fruits slightly torulose, 3–4 mm thick. Native of Europe and Asia; naturalized in waste areas; occasional throughout the state. *B. kaber* (DC.) L. C. Wheeler—G, J.

var. schkuhriana (Reichenb.) L. C. Wheeler. Field Mustard. April–July. Fruit strongly torulose, 1.5–2.0 mm thick. Native of Europe and Asia; naturalized in waste areas; rare and scattered in Ill.

3. B. nigra (L.) Koch. Black Mustard. April–Oct. Native of Europe and Asia; naturalized in waste areas; occasional throughout the state.

4. B. juncea (L.) Coss. Indian Mustard. April–Sept. Native of Europe and Asia; naturalized in waste areas; occasional throughout the state.

5. B. rapa L. Field Mustard. April–Sept. Native of Europe and Asia; naturalized in waste areas; occasional throughout the state. *B. campestris* L.—J.

6. B. oleracea L. Cabbage. April–Sept. Native of Europe and Asia; rarely escaped from cultivation; Peoria Co.

7. B. napus L. Turnip. May–Sept. Native of Europe and Asia; rarely escaped from cultivation; Champaign and Hancock cos. *B. rapa* L.—J.

32. *Erucastrum* Presl

1. E. gallicum (Willd.) O. E. Schulz. May–Sept. Native of Europe; naturalized in waste areas; occasional and scattered throughout the state.

33. *Sisymbrium* L.

1. Petals up to 3 mm long; fruits up to 2 cm long .... 1. *S. officinale*
1. Petals 5–6 mm long; fruits 2–10 cm long 2

2. At least the upper leaves divided into threadlike divisions; fruits 5–10 cm
long .... 2. *S. altissimum*
2. Leaves divided into triangular lobes; fruits 2–4 cm long ... 3. *S. loeselii*

1. S. officinale (L.) Scop. Hedge Mustard. May–Oct. Fruits pubescent. Native of Europe and Asia; naturalized in waste areas; not common, but scattered throughout the state.
var. leiocarpum DC. May–Oct. Fruits glabrous or nearly so. Native
of Europe and Asia; naturalized into waste areas; occasional throughout the state.
2. S. altissimum L. Tumble Mustard. May–Aug. Native of Europe; naturalized in waste areas; occasional throughout the state.
3. S. loeselii L. May–Oct. Native of Europe; naturalized in waste
areas; rare and apparently confined to the n. cos.

34. *Rorippa* Scop. *Yellow Cress*

1. Petals over 4 mm long, longer than the sepals 2
1. Petals absent, or up to 2 mm long, never longer than the sepals 3
2. Leaves without auricles at base; seeds up to 0.8 mm long 1. *R. sylvestris*
2. Leaves with auricles at base; seeds about 1 mm long .... 2. *R. sinuata*
3. Petals up to 0.5 mm long; stamens 4; fruits 6–10 times longer than the
pedicels; style nearly absent; seeds 175 or more per fruit ....
.... 3. *R. sessiliflora*
3. Petals 1–2 mm long; stamens 6; fruits at most only 4 times longer than
the pedicels; style present; seeds less than 75 per fruit 4
4. Petals 1.0–1.2 mm long; fruits 2–4 times longer than the pedicels ......
.... 4. *R. truncata*
4. Petals 1.7–2.0 mm long; fruits shorter than to up to 2 times longer than
the pedicels .... 5. *R. islandica*

1. R. sylvestris (L.) Bess. Creeping Yellow Cress. May–Sept. Native
of Europe; naturalized in waste areas; occasional throughout the
state.
2. R. sinuata (Nutt.) Hitchc. Spreading Yellow Cress. April–July. Wet
soil, particularly along rivers; occasional throughout the state.
3. R. sessiliflora (Nutt.) Hitchc. Sessile-flowered Yellow Cress.
April–Nov. Along rivers and streams; occasional to common in the s.
four-fifths of Ill.; also Grundy and Kane cos.
4. R. truncata (Jepson) Stuckey. May–Sept. Along rivers, rare; St.
Clair Co. *R. obtusa* (Nutt.) Britt.—F, G.
5. R. islandica (Oeder) Borbas. Marsh Yellow Cress. July–Sept.
stems and leaves glabrous or nearly so; leaves pinnate or deeply pinnatifid, membranaceous. Mud and sand flats, rare; Jackson Co.
var. fernaldiana Butt. & Abbe. Marsh Yellow Cress. May–Nov. Stems
and leaves glabrous or nearly so; leaves coarsely or shallowly
toothed, firm. Wet soil; common throughout the state. *R. islandica*
(Oeder) Borbás—J.
var. hispida (Desv.) Butt. & Abbe. Hairy Marsh Yellow Cress. May–
Oct. Stems and leaves hirsutulous. Wet ground, rare; Cook and
McHenry cos. *R. hispida* (Desv.) Britt.—J.

35. Chorispora DC.

1. C. tenella (Willd.) DC. May–July. Native of Asia; adventive along railroads; Cook, Greene, Kane, Kankakee, and Lake cos.

36. Coronopus Trew. Wart Cress

1. C. didymus (L.) Sm. Wart Cress. April–Oct. Native of Europe; adventive in waste ground; Cook Co.

72. RESEDACEAE

1. Reseda L. Mignonette

1. R. alba L. Mignonette. June–Oct. Rarely escaped from cultivation; Cook and DuPage cos.

73. SARRACENIACEAE

1. Sarracenia L. Pitcher-plant

1. S. purpurea L. Pitcher-plant. May–June. Bogs, rare; Cook, Lake, and McHenry cos.

74. DROSERACEAE

1. Drosera L. Sundew

1. Leaves spatulate; seeds obovoid, with a tight, papillate testa . . . .
. . . . 1. D. intermedia
1. Leaves suborbicular; seeds fusiform, with a loose, striate testa . . . .
. . . . 2. D. rotundifolia

   1. D. intermedia Hayne. Narrow-leaved Sundew. July–Sept. Bogs, rare; Cook, Kane, and Kankakee cos.
   2. D. rotundifolia L. Round-leaved Sundew. July–Sept. Bogs, rare; Cook, Lake, and Ogle cos.

75. CRASSULACEAE

1. Sedum L. Stonecrop

1. Flowers yellow 2
1. Flowers white, pink, or purplish 3

2. Leaves linear, to 6 mm long .... 1. *S. acre*
2. Leaves oblanceolate, 1–2 cm long .... 2. *S. sarmentosum*
3. Leaves terete, linear to linear-spatulate; petals up to 8 mm long ....
.... 3. *S. pulchellum*
3. Leaves flat, elliptic, ovate, or obovate; petals 8 mm long or longer ... 4
4. Flowers white; leaves entire, up to 2 cm long .... 4. *S. ternatum*
4. Flowers pink or purple; leaves dentate, at least the larger more than 2
cm long 5
5. Flowers pink; petals more than twice as long as the sepals ....
.... 5. *S. telephioides*
5. Flowers purple; petals at most only twice as long as the sepals ....
.... 6. *S. purpureum*

    1. S. acre L. Mossy Stonecrop. June–Aug. Native of Europe; oc-
casionally escaped from cultivation.
    2. S. sarmentosum Bunge. Yellow Stonecrop. June. Native of
e. Asia; occasionally escaped from cultivation.
    3. S. pulchellum Michx. Widow's-cross. May–July. Dry or moist
sandstone; known only from the s. one-sixth of the state.
    4. S. ternatum Michx. Three-leaved Stonecrop. May–June. Lime-
stone cliffs; moist wooded ravines; occasional and scattered throughout Ill.
    5. S. telephioides Michx. American Orpine. Aug.–Sept. Sandstone
cliffs, rare; apparently only in the se. one-fourth of the state.
    6. S. purpureum (L.) Link. Live-forever. Aug.–Sept. Native of
Europe; occasionally escaped from cultivation. *S. triphyllum* (Haw.)
S. F. Gray—J; *S. telephium* L. ssp. *purpureum* (Link) Schinz & Keller—
G.

76. SAXIFRAGACEAE

1. Trees or shrubs 2
1. Herbs 6
2. Leaves opposite 3
2. Leaves alternate 5
3. Leaves elliptic to oblong to narrowly ovate; flowers solitary or in cymes
or racemes; stamens 10 or more 4
3. Leaves broadly ovate; flowers in flat-topped corymbs; stamens ten or
less .... 3. *Hydrangea*
4. Stamens 15 or more; pubescence not stellate .... 1. *Philadelphus*
4. Stamens 10; pubescence stellate .... 2. *Deutzia*
5. Leaves palmately lobed; fruit a berry .... 4. *Ribes*
5. Leaves serrulate; fruit a capsule .... 5. *Itea*
6. Leaves essentially all basal, or sometimes with 1–2 leaves on the stem 7
6. Leaves cauline .... 11. *Penthorum*
7. Flower solitary, with gland-tipped staminodia .... 6. *Parnassia*
7. Flowers several, without gland-tipped staminodia 8
8. Stem with a pair of opposite leaves; petals deeply divided .. 7. *Mitella*
8. Stem with a single leaf, or leafless 9
9. Leaves longer than broad; stamens 10 .... 8. *Saxifraga*
9. Leaves as broad as long; stamens 5 10

10. Leaves pubescent; ovary 1-locular; seeds unwinged ...... 9. *Heuchera*
10. Leaves glabrous or nearly so; ovary 2-locular; seeds winged ....
.... 10. *Sullivantia*

1. Philadelphus L. Mock Orange

1. Flowers 1–4 in a terminal inflorescence .... 1. *P. inodorus*
1. Flowers 5–7 in both terminal and lateral racemes 2
2. Flowers faintly aromatic; calyx pubescent; lower surface of leaves
pubescent throughout .... 2. *P. pubescens*
2. Flowers strongly fragrant; calyx glabrous; lower surface of leaves pubescent only on the nerves .... 3. *P. coronarius*

1. P. inodorus L. Scentless Mock Orange. May–June. Native of the
se. U.S.; rarely escaped from cultivation; Union Co.
2. P. pubescens Loisel. Mock Orange. May–June. Wooded bluffs
along Ohio River, very rare; Pope Co.
3. P. coronarius L. Sweet Mock Orange. May–June. Native of
Europe; rarely escaped from cultivation; Cook Co.

2. Deutzia Thunb. Deutzia

1. D. scabra Thunb. Pride-of-Rochester. May–June. Native of China
and Japan; rarely escaped from cultivation.

3. Hydrangea L. Hydrangea

1. H. arborescens L. Wild Hydrangea. June–Aug. Woods; common
in the s. four-fifths of the state, apparently absent elsewhere. (Including var. *oblonga* Torr. & Gray and var. *deamii* St. John.)

4. Ribes L. Gooseberry. Currant

1. Plants with spines or prickles 2
1. Plants lacking spines and prickles 4
2. Berries prickly; lobes of calyx shorter than the tube ... 1. *R. cynosbati*
2. Berries lacking prickles; lobes of calyx as long as or longer than the
tube 3
3. Flowers white to greenish-white; berries at least 10–15 mm in diameter
.... 2. *R. missouriense*
3. Flowers greenish-yellow or purplish; berries 8–10 (–12) mm in diameter
.... 3. *R. hirtellum*
4. Flowers golden-yellow, very fragrant; tube of calyx more than twice as
long as the lobes .... 4. *R. odoratum*
4. Flowers yellow-green, yellow and white, or greenish-purple; tube of
calyx up to twice as long as the lobes 5
5. Leaves and fruits with resinous dots; berries black 6
5. Leaves and fruits lacking resinous dots; berries red ..... 7. *R. sativum*
6. Flowers broadly campanulate; calyx 5–6 mm long, pubescent ....
.... 5. *R. nigrum*
6. Flowers tubular-campanulate; calyx 8–10 mm long, glabrous ....
.... 6. *R. americanum*

1. R. cynosbati L. Prickly Gooseberry. April–May. Moist, often rocky, woods; occasional in the n. half and the s. one-fourth of the state, rare or absent elsewhere. (Including var. *glabratum* Fern.)

2. R. missouriense Nutt. Missouri Gooseberry. April–May. Woods; common in the n. two-thirds of the state, occasional elsewhere.

3. R. hirtellum Michx. Northern Gooseberry. April–June. Moist woods, bogs, rare; confined to the ne. cos., except for Menard Co.

4. R. odoratum Wendl. Buffalo Currant. April–June. Native of the w. U.S.; occasionally escaped from cultivation.

5. R. nigrum L. Black Currant. May–June. Native of Europe; rarely escaped from cultivation; Lake Co.

6. R. americanum Mill. Wild Black Currant. May–June. Moist woods; common in the n. three-fifths of the state, absent elsewhere.

7. R. sativum (Reichenb.) Syme. Red Currant. May–June. Native of Europe; occasionally escaped from cultivation.

5. *Itea* L.

1. I. virginica L. Virginia Willow. May–June. Swampy woods; confined to the s. one-sixth of the state.

6. *Parnassia* L.

1. P. glauca Raf. Grass-of-Parnassus. July–Sept. Calcareous springs, dune flats; restricted to the n. one-half of the state.

7. *Mitella* L. *Miterwort*

1. M. diphylla L. Bishop's-cap. May. Rich, often rocky, woods; occasional in the n. half of the state, rare in the s. half.

8. *Saxifraga* L. *Saxifrage*

1. None of the leaves more than 8 cm long; calyx lobes about as long as calyx tube; petals white, at least 4 mm long .... 1. *S. virginiensis*
1. Some or all the leaves more than 8 cm long; calyx lobes at least twice as long as calyx tube; petals greenish-white or greenish-yellow, 2–3 mm long 2
2. Leaves pilose on the lower surface; petals longer than the sepals .... 2. *S. forbesii*
2. Leaves glabrous or nearly so on the lower surface; petals about as long as the sepals .... 3. *S. pensylvanica*

1. S. virginiensis Michx. Early Saxifrage. April–June. Rocky wooded ravines, rare; Hardin Co.

2. S. forbesii Vasey. Forbes' Saxifrage. May. Moist sandstone cliffs, rare; known from seven cos. in extreme s. Ill. and three cos. in extreme n. Ill.; also Monroe Co.

3. S. pensylvanica L. Swamp Saxifrage. May–June. Springs, bogs, moist meadows; occasional in the n. half of the state, absent elsewhere; also Wabash Co. (Including ssp. *interior* Burns.)

9. *Heuchera L. Alumroot*

1. Calyx to 2 mm long; leaves and stems villous .... 1. *H. parviflora*
1. Calyx 2.5 mm long or longer; leaves and stems hispid or strigose   ... 2
2. Leaves strigose or nearly glabrous; calyx symmetrical or nearly so .....
.... 2. *H. hirsuticaulis*
2. Leaves hispid beneath; calyx asymmetrical .... 3. *H. richardsonii*

1. H. parviflora Bartl. var. rugelii (Shuttlw.) Rosend., Butt. & Lak. Late Alumroot. July–Nov. Moist sandstone cliffs; common in the s. one-sixth of the state, absent elsewhere. *H. parviflora* Bartl.—J.

2. H. hirsuticaulis (Wheelock) Rydb. Tall Alumroot. May–June. Dry woods; occasional in the s. half of the state, absent elsewhere. (Including var. *interior* Rosend., Butt. & Lak.) *H. americana* L. var. *hirsuticaulis* (Wheelock) Rosend., Butt. & Lak.—F.

3. H. richardsonii R. Br. var. grayana Rosend., Butt. & Lak. Prairie Alumroot. May–July. Prairies; occasional in the n. three-fourths of the state, apparently absent elsewhere. (Including var. *affinis* Rosend., Butt. & Lak.)

10. *Sullivantia Torr. & Gray*

1. S. renifolia Rosend. Sullivantia. June–July. Moist shaded cliffs, rare; confined to the extreme nw. corner of Ill.

11. *Penthorum L.*

1. P. sedoides L. Ditch Stonecrop. July–Oct. Wet ground, common; in every co.

77. HAMAMELIDACEAE

1. Leaves star-shaped; flowers without petals; fruit a globose, echinate head .... 1. *Liquidambar*
1. Leaves oval to obovate to suborbicular; flowers with 4 yellow petals; fruit an obovoid capsule .... 2. *Hamamelis*

1. *Liquidambar L. Sweet Gum*

1. L. styraciflua L. Sweet Gum. April–May. Low woods; occasional to common in the s. one-third of the state.

2. *Hamamelis L. Witch-hazel*

1. H. virginiana L. Witch-hazel. Sept.–Nov. Woods; occasional in the n. half of the state; also Richland, Wabash, and White cos.

78. PLATANACEAE

1. *Platanus L. Sycamore*

1. P. occidentalis L. Sycamore. April–May. Moist woods and along streams; common in the s. two-thirds of the state, occasional elsewhere; in every co.

79. ROSACEAE

18. Flowers white; ovaries more than 50 per flower; receptacle bearing the achenes fleshy .... 16. *Fragaria*
19. Leaves 2- to 3-pinnate; flowers dioecious .... 17. *Aruncus*
19. Leaves once-pinnate or palmate; flowers perfect 20
20. Flowers pink, rose, or purple 21
20. Flowers white, cream, or yellow 22
21. Plants glabrous; flowers up to 8 mm across .... 18. *Filipendula*
21. Plants pubescent; flowers at least 1 cm across .... 21. *Geum*
22. Petals absent; calyx 4-parted .... 19. *Sanguisorba*
22. Petals present; calyx 5-parted 23
23. Petals white 24
23. Petals yellow 27
24. Petals strap-shaped, usually 2–3 times longer than broad; each leaf subtended by a pair of foliaceous stipules .... 20. *Gillenia*
24. Petals mostly obovate, usually less than twice as long as broad; each leaf generally not subtended by a pair of foliaceous stipules 25
25. Trailing or ascending, nearly glabrous plants; leaves pedately 3- to 5-foliolate .... 12. *Rubus*
25. Erect, pubescent plants; leaves pinnately 3- to 11-foliolate 26
26. Basal leaves 3- to 5-foliolate; petals as long as or shorter than the sepals .... 21. *Geum*
26. Basal leaves 7- to 11-foliolate; petals longer than the sepals .... 14. *Potentilla*
27. All leaves trifoliolate 28
27. At least some of the leaves 5-foliolate or more 29
28. Receptacle in fruit spongy, greatly enlarged; bractlets between the calyx lobes longer than the lobes .... 22. *Duchesnea*
28. Receptacle in fruit dry, not enlarged; bractlets between the calyx lobes about the same size as the lobes .... 14. *Potentilla*
29. Leaves pinnately compound 30
29. Leaves palmately compound 32
30. Bractlets present between calyx lobes; ovaries at least 30 per flower 31
30. Bractlets between the calyx lobes absent; ovaries 1–4 .. 23. *Agrimonia*
31. Style deciduous from the achene .... 14. *Potentilla*
31. Style persistent on the achene .... 21. *Geum*
32. Style deciduous from the achene .... 14. *Potentilla*
32. Style persistent on the achene .... 21. *Geum*

1. *Physocarpus* Maxim. Ninebark

    1. P. opulifolius (L.) Maxim. Ninebark. May–June. Rocky slopes, rocky banks, moist swales; occasional in the n. half of the state; also Jackson, Pope, and Wabash cos. (Including var. *intermedius* [Rydb.] Robins.)

2. *Spiraea* L. Spiraea

1. Flowers in elongate panicles 2
1. Flowers in corymbs .... 4. *S. prunifolia*
2. Leaves woolly beneath; flowers usually rose .... 1. *S. tomentosa*
2. Leaves glabrous or nearly so beneath; flowers white 3
3. Leaves serrulate; branches yellow-brown .... 2. *S. alba*
3. Leaves coarsely toothed; branches reddish- or purplish-brown .... 3. *S. latifolia*

1. S. tomentosa L. Hardhack. July–Aug. Bogs, moist thickets, rare; Cook, DuPage, Iroquois, Kankakee, Lake, and Stephenson cos. (Including var. *rosea* [Raf.] Fern.)

2. S. albà DuRoi. Meadow-sweet. July–Aug. Wet soil; occasional in the n. four-fifths of the state.

3. S. latifolia (Ait.) Borkh. Meadow-sweet. June–July. Native of ne. N.A.; rarely escaped from cultivation; Champaign and Hancock cos.

4. S. prunifolia Sieb. & Zucc. Bridal-wreath. May–June. Native of the Old World; rarely escaped from cultivation.

3. *Prunus* L.   Plum.   Cherry

14. Leaves firm, crenulate-serrate; lobes of calyx acute; fruit black-purple . .
. . . . 15. *P. serotina*

1. P. persica (L.) Batsch. Peach. April. Introduced from Asia; occasionally escaped along roadsides.

2. P. armeniaca L. Apricot. May. Introduced from Europe; rarely escaped from cultivation; Champaign Co.

3. P. nigra Ait. Canada Plum. May. Mostly along rivers and streams; occasionally in the n. half of the state

4. P. hortulana Bailey. Wild Goose Plum. March–April. Thickets; occasional in the s. two-thirds of the state; rare and probably introduced in Cook, JoDaviess, LaSalle, and Stephenson cos.

5. P. munsoniana Wight & Hedrick. Wild Goose Plum. March–April. Thickets; occasional in the s. one-third of the state.

6. P. angustifolia Marsh. Chickasaw Plum. Thickets; occasional in the s. half of the state, uncommon and probably adventive in the n. half of the state.

7. P. mexicana S. Wats. Big Tree Plum. April. Base of cherty cliff; known only from Jackson Co.

8. P. americana Marsh. Wild Plum. April–May. Leaves glabrous or sparsely pubescent beneath. Thickets, woodlands; occasional throughout the state.

var. lanata Sudw. April–May. Leaves soft-pubescent beneath. Thickets, woodlands; occasional throughout the state. *P. lanata* (Sudw.) Mack. & Bush—J.

9. P. susquehanae Willd. Sand Cherry. May–June. Sandy areas, particularly along Lake Michigan; confined to the n. one-sixth of the state. *P. pumila* L.—G, J.

10. P. pensylvanica L. f. Pin Cherry. April. Sandy soil, particularly along Lake Michigan; confined to the n. one-third of the state.

11. P. cerasus L. Pie Cherry. April–May. Native of Europe; rarely escaped form cultivation; Champaign and Lawrence cos.

12. P. avium L. Sweet Cherry. April–May. Native of Europe and Asia; rarely escaped from cultivation.

13. P. mahaleb L. Mahaleb Cherry. May. Native of Europe; rarely escaped from cultivation; Kane, Sangamon, Union, and Will cos.

14. P. virginiana L. Common Chokecherry. May. Woods and thickets; occasional in the n. half of the state, rare in the s. half.

15. P. serotina Ehrh. Wild Black Cherry. May. Woods; common throughout the state; in every co.

4. Cydonia Mill.   Quince

1. C. oblonga Mill. Common Quince. May–June. Native of Asia; established at Atwood Ridge, Union Co.

5. Amelanchier Medic.   Shadbush

1. Stoloniferous shrub; leaves usually dentate-serrate, at least above the middle; sepals revolute at maturity; petals up to 1 cm long 1. *A. humilis*

1. Non-stoloniferous shrubs or small trees; leaves finely or sharply serrate their total length; sepals reflexed at maturity; petals 1 cm long or longer
. . . . 2

2. Leaves white-tomentose beneath at least until anthesis; pedicels of lowermost flowers in each raceme less than 2 cm long ...... 2. *A. arborea*
2. Leaves glabrous or sparsely pubescent beneath; pedicels of lowermost flowers in each raceme usually at least 2 cm long 3
3. Leaves with up to 10 pairs of veins; ovary and usually the young fruit pubescent .... 3. *A. interior*
3. Leaves with more than 10 pairs of veins; ovary and young fruit glabrous .... 4. *A. laevis*

1. A. humilis Wieg. Low Shadbush. May–June. Rocky or sandy soil; restricted to the n. one-fourth of the state; also Adams Co. *A. spicata* K. Koch—G, J.

2. A. arborea (Michx. f.) Fern. Shadbush. March–May. Wooded slopes and bluffs; occasional throughout the state.

3. A. interior Nielsen. Shadbush. June. Bogs and wet woods, rare; Lake and Winnebago cos.

4. A. laevis Wieg. Shadbush. March–June. Wooded slopes; confined to the n. one-fourth of the state and Vermilion Co.

6. Chaenomeles L. Japanese Quince

1. C. japonica L. Japanese Quince. March–April. Native of Europe; occasionally escaped from cultivation and persistent.

7. Pyrus L. Pear

1. Leaves crenate, glabrous or glabrate; fruit more or less pyriform ....
.... 1. *P. communis*
1. Leaves serrate, tomentose when young; fruit subglobose 2. *P. pyrifolia*

1. P. communis L. Pear. April–May. Introduced from Europe and Asia; escaped from cultivation throughout Ill.

2. P. pyrifolia (Burm. f.) Nakai. Chinese Pear. April. Native of Asia; escaped along a stream in Union Co.

8. Malus Mill. Apple

1. Anthers yellow; most or all leaves unlobed .... 1. *M. pumila*
1. Anthers red; some of the leaves (at least those on the vegetative branches) lobed or deeply toothed 2
2. Calyx glabrous or sparsely pubescent 3
2. Calyx densely pubescent 4
3. Leaves of fruiting branches mostly obtuse at tip ..... 2. *M. angustifolia*
3. Leaves of fruiting branches mostly acute at tip .... 3. *M. coronaria*
4. Fruit up to 3.5 cm in diameter .... 4. *M. ioensis*
4. Fruit up to 5 cm in diameter or broader .... 5. *M.* X *soulardii*

1. M. pumila Mill. Apple. May. Introduced from Europe; occasionally escaped in Ill. *Pyrus malus* L.—F, G.

2. M. angustifolia (Ait.) Michx. Narrow-leaved Crab Apple. May. Low woods, rare; Hardin and Jackson cos. *Pyrus angustifolia* Ait.—F, G.

3. M. coronaria (L.) Mill. Wild Sweet Crab Apple. April–May. Leaves at least half as broad as long, sometimes lobulate. Woods; occasional in the s. half of the state, less common in the n. half. *Pyrus coronaria* L.—F, G.

var. lancifolia Rehder. Narrow-leaved Crab Apple. April–May. Leaves less than half as broad as long, not lobulate. Woods; Gallatin, Jackson, and Pope cos. *Pyrus coronaria* var. *lancifolia* (Rehd.) Fern.—F.

4. M. ioensis (Wood) Britt. Iowa Crab Apple. May. Woods and thickets; occasional or common throughout the state; in every co. *Pyrus ioensis* (Wood) Bailey—F, G.

5. M. X soulardii (Bailey) Britt. Soulard Crab. May. Reputed hybrid between *M. pumila* and *M. ioensis;* originally cultivated at Galena, Ill.; rarely escaped in Ill. *Pyrus X soulardii* Bailey—F, G.

9. Aronia Medic. Chokeberry

1. Axis of inflorescence and lower surface of leaves pubescent . . . .
. . . . 1. *A. prunifolia*
1. Axis of inflorescence and lower surface of leaves glabrous or nearly so
. . . . 1. *A. melanocarpa*

1. A. prunifolia (Marsh.) Rehd. Purple Chokeberry. May–June. Occasional in bogs, otherwise rare; restricted to a few cos. in the ne. corner of the state; also Winnebago Co. *Pyrus floribunda* Lindl.—F.

2. A. melanocarpa (Michx.) Ell. Black Chokeberry. May–June. Bogs, moist woods, sandstone ledge (in Saline Co.); occasional in the ne. one-fourth of the state; also Saline Co. *Pyrus melanocarpa* (Michx.) Willd.—F.

10. Sorbus [Tourn.] L. Mountain Ash

1. Axis of inflorescence, leaflets, and outer bud scales glabrous or nearly
so . . . . 1. *S. americana*
1. Axis of inflorescence, leaflets, and outer bud scales pubescent . . . .
. . . . 2. *S. aucuparia*

1. S. americana Marsh. Mountain Ash. May–July. Rocky woods; known only from Ogle Co. *Pyrus americana* (Marsh.) DC.—F.

2. S. aucuparia L. European Mountain Ash. May–July. Native of Europe and Asia; escaped to bogs and swamps in the n. one-fifth of the state. *Pyrus aucuparia* (L.) Gaertn.—F.

11. Crataegus L. Hawthorn

1. Veins of larger leaves running to the sinuses as well as to the points of
the lobes. 2
1. Veins of the leaves running only to the points of the lobes 4
2. Leaves thin, early deciduous; thorns stout, some or all over 2 cm long. 3
2. Leaves thick, persisting late into the autumn; thorns slender, up to 2 cm
long . . . . 3. *C. monogyna*
3. Leaves of the flowering branchlets deltoid-cordate, usually trilobate, glabrous at maturity; stamens 20; anthers yellow; fruit subglobose, with dry flesh . . . . 1. *C. phaenopyrum*
3. Leaves of the flowering branchlets not deltoid-cordate, usually deeply cut, pubescent along the veins beneath at maturity; stamens 10; anthers red; fruit oblong, with succulent flesh . . . . 2. *C. marshallii*
4. Leaves cuneate, broadest at middle or apex 5
4. Leaves broadest at the usually rounded or truncate or cordate base 21

5. Leaves broadest toward apex 6
5. Leaves broadest at the middle 15
6. Leaves impressed-veined above 7
6. Leaves not impressed-veined above 11
7. Leaves pubescent beneath at maturity, at least on the veins, dull gray-green; fruits 12–20 mm thick .... 4. *C. punctata*
7. Leaves glabrous or nearly so at maturity, dark green or yellow-green; fruits up to 12 mm thick 8
8. Some of the leaves of the flowering branches lobed 9
8. None of the leaves of the flowering branches lobed 10
9. Leaves with obscure lobes near the apex; stamens 10 or 15; anthers red; leaves usually yellow-green .... 5. *C. cuneiformis*
9. Leaves with conspicuous lobes above the middle; stamens 20; anthers white or pale yellow; leaves dark green .... 6. *C. margaretta*
10. Leaves dull green, pubescent when young; corymbs pubescent; stamens 20; fruit pubescent at first, 8–10 mm thick .... 7. *C. collina*
10. Leaves yellow-green, glabrous; corymbs glabrous; stamens 10; fruit glabrous, 7–8 mm thick .... 8. *C. hannibalensis*
11. None of the leaves lobed 12
11. Some of the leaves, particularly the terminal shoot leaves, lobed (sometimes rarely lobed in *C. fecunda*) 13
12. Leaves glabrous, not reticulate-veined; petioles glabrous 9. *C. crus-galli*
12. Leaves pubescent, reticulate-veined; petioles villous 10. *C. engelmannii*
13. Leaves pubescent (at least when young); corymbs villous 11. *C. fecunda*
13. Leaves glabrous; corymbs glabrous 14
14. Terminal shoot leaves at least twice as broad as those of the flowering branches; stamens 10 or 15; anthers pale yellow ..... 12. *C. acutifolia*
14. Terminal shoot leaves usually less than twice as broad as those of the flowering branches; stamens 20; anthers pink .... 13. *C. permixta*
15. Leaves impressed-veined above 16
15. Leaves not impressed-veined above 17
16. Leaves, corymbs, and fruit glabrous .... 14. *C. succulenta*
16. Leaves, corymbs, and fruit pubescent .... 15. *C. calpodendron*
17. Petioles glandular .... 16. *C. neobushii*
17. Petioles glandless 18
18. Leaves at maturity with tufts of tomentum in the axils beneath; corymbs glabrous 19
18. Leaves at maturity glabrous beneath or at least without tufts of tomentum in the axils; corymbs pubescent 20
19. Leaves thin, scarcely lustrous above; fruit 5–8 mm thick . 17. *C. viridis*
19. Leaves thick, lustrous above; fruit 8–10 mm thick .... 18. *C. nitida*
20. Leaves glabrous; anthers pink; fruit succulent .... 19. *C. lucorum*
20. Leaves when young densely covered on the upper surface with short, appressed hairs; anthers pale yellow; fruit dry .... 20. *C. faxonii*
21. Leaves completely glabrous, bluish-green; stamens usually 20 22
21. Leaves pubescent above and/or below, at least on the veins, or scabrous above, yellow-green or dark green; stamens 10 or 20. 23
22. Leaves thin; fruit succulent, not pruinose .... 21. *C. tortilis*
22. Leaves thick; fruit dry and often mealy, pruinose ...... 22. *C. pruinosa*
23. Leaves pubescent above at maturity .... 23. *C. mollis*
23. Leaves glabrous or nearly so above at maturity or scabrous above . 24
24. Leaves scabrous above, otherwise usually glabrous 25
24. Leaves usually smooth above, but with some pubescence beneath, at least on the veins 26

25. Leaves mostly broadly ovate; corymbs glabrous; fruits 10–17 mm thick
.... 24. *C. holmesiana*
25. Leaves mostly elliptic-ovate to oval; corymbs villous; fruits 7–10 mm
thick .... 25. *C. pedicellata*
26. Leaves bright yellow-green; corymbs usually pubescent (sometimes glabrous in *C. putnamiana*) 27
26. Leaves dark green or dark yellow-green; corymbs glabrous 28
27. Branchlets villous when young; stamens 5 or 10; corymbs villous; fruit
oblong, 8–12 mm thick .... 26. *C. pringlei*
27. Branchlets usually glabrous; stamens 20; corymbs usually glabrous;
fruit usually subglobose, 12–15 mm thick .... 27. *C. corusca*
28. Edge of leaf often crisped at maturity; flowers 2.4–2.6 cm wide; stamens 20; fruit subglobose .... 28. *C. coccinioides*
28. Edge of leaf not crisped; flowers 1.3–1.8 cm wide; stamens 10 or fewer;
fruit obovoid or oblongoid .... 29. *C. macrosperma*

1. C. phaenopyrum (L. f.) Medic. Washington Thorn. May–June. Open woods; occasional in the s. one-third of the state.

2. C. marshallii Egglest. Parsley Haw. April. Wet, pin oak woods, rare; Jackson Co.

3. C. monogyna Jacq. English Hawthorn. May. Native of Europe and Asia; rarely escaped in Ill.; Cook Co.

4. C. punctata Jacq. Dotted Thorn. May–June. Open woods, pastures; occasional throughout the state. (Including var. *aurea* Ait., *C. mortonis* Laughlin, and *C. sucida* Sarg.)

5. C. cuneiformis (Marsh.) Egglest. Hawthorn. May. Open woods; occasional in the n. half of the state. (Including *C. pratensis* Sarg., *C. peoriensis* Sarg., and *C. disperma* Ashe.) *C. disperma* Ashe—F, G.

6. C. margaretta Ashe. Hawthorn. April–May. Open woods, thickets; occasional in the n. three-fifths of the state.

7. C. collina Chapm. Hawthorn. April–May. Low woods, rare; St. Clair and Union cos. (Including *C. lettermanii* Sarg.)

8. C. hannibalensis Palmer. Hawthorn. May. Thickets, rare; Adams Co. *C. cuneiformis* (Marsh.) Egglest.—J.

9. C. crus-galli L. Cock-spur Thorn. May–June. Leaves thick. Open woods and thickets; occasional throughout the state. (Including *C. attenuata* Ashe, *C. arduennae* Sarg., *C. farwellii* Sarg., *C. acantha-colonensis* Laughlin, and *C. pachyphylla* Sarg.)
var. barrettiana (Sarg.) Palmer. Barrett's Thorn. May–June. Leaves thin. Thickets, rare; St. Clair Co.

10. C. engelmannii Sarg. Barberry-leaved Hawthorn. May. Open woods, bluffs, pastures; occasional in the s. one-fifth of the state.

11. C. fecunda Sarg. Fruitful Thorn. May. Moist woods, rare; Gallatin and St. Clair cos. (Including *C. pilifera* Sarg.) *C. engelmanni* Sarg.—J.

12. C. acutifolia Sarg. Hawthorn. May. Low woods, rare; Richland and St. Clair cos. (Including *C. erecta* Sarg. and var. *insignis* [Sarg.] Palmer.) *C. viridis* L.—J.

13. C. permixta Palmer. Hawthorn. May. Open woods; Peoria Co.

14. C. succulenta Link. Hawthorn. May. Dry or moist woods and thickets; occasional in the n. half of the state, rare in the s. half. (Including *C. macracantha* Lodd.; *C. illinoiensis* Ashe; *C. gemmosa* Sarg.; *C. vegeta* Sarg.; *C. gaultii* Sarg.; *C. longispina* Sarg.; *C. rutila* Sarg.; *C. laxiflora* Sarg.; *C. corporea* Sarg.; *C. leucantha* Laughlin.)

15. C. calpodendron (Ehrh.) Medic. Hawthorn. May. Woods and

thickets; occasional throughout the state. (Including *C. tomentosa* DuRoi; *C. chapmani* [Beadle] Ashe; *C. structilis* Ashe; *C. mollicula* Sarg.; *C. hispidula* Sarg.; *C. pertomentosa* Ashe; *C. chrysocarpa* Ashe; *C. whitakeri* Sarg.)

16. C. neobushii Sarg. Hawthorn. May. Rocky woods; confined to the s. one-fourth of the state.

17. C. viridis L. Green Thorn. May. Low woods; occasional in the s. half of the state. (Including *C. schneckii* Ashe; *C. durifolia* Ashe; *C. ovata* Sarg.; *C. mitis* Sarg.; *C. lanceolata* Sarg.; *C. dawsoniana* Sarg.; *C. pechiana* Sarg.)

18. C. nitida (Engelm.) Sarg. Hawthorn. May. Low woods; Gallatin, Henderson, and St. Clair cos. *C. viridis* L.—J.

19. C. lucorum Sarg. Hawthorn. May. Woods and stream banks; Cook, DuPage, Lake, and Will cos. (Including *C. apiomorpha* Sarg.) *C. macrosperma* Ashe—J.

20. C. faxonii Sarg. Hawthorn. May. Rocky woods and thickets; St. Clair Co. *C. margaretta* Ashe—J.

21. C. tortilis Ashe. Hawthorn. May. Woods and thickets; Cook Co. *C. macrosperma* Ashe—J.

22. C. pruinosa (Wendl.) K. Koch. Hawthorn. May. Thickets and rocky woods; occasional throughout the state. (Including *C. gattingeri* Ashe; *C. virella* Ashe; *C. conjuncta* Sarg.; *C. dissona* Sarg.; *C. platycarpa* Sarg.)

23. C. mollis (Torr. & Gray) Scheele. Red Haw. May. Woods; occasional throughout the state; common northward. (Including *C. sera* Sarg.; *C. altrix* Ashe; *C. valens* Ashe; *C. venosa* Ashe; *C. nupera* Ashe; *C. verna* Ashe; *C. declivitatis* Sarg.; *C. umbrosa* Sarg.; *C. lanigera* Sarg.; *C. ridgwayi* Sarg.; *C. pachyphylla* Sarg.)

24. C. holmesiana Ashe. Hawthorn. May. Woods and thickets; confined to the n. one-half of the state. (Including *C. amicta* Ashe; *C. biltmoreana* Beadle; *C. intricata* Lange; *C. magniflora* Sarg.) *C. coccinea* L.—J.

25. C. pedicellata Sarg. Hawthorn. May. Woods and thickets; scattered in Ill. (Including *C. albicans* Ashe; *C. paucispina* Sarg.; *C. robesoniana* Sarg.) *C. coccinea* L.—J.

26. C. pringlei Sarg. Hawthorn. May. Thickets and woods; Lake Co. *C. coccinea* L.—J.

27. C. corusca Sarg. Hawthorn. May. Woods and thickets; Lake Co. (Including *C. putnamiana* Sarg.) *C. coccinea* L.—J; *C. putnamiana* Sarg.—G.

28. C. coccinioides Ashe. Hawthorn. May. Thickets and rocky woods; confined to the s. one-third of the state.

29. C. macrosperma Ashe. Hawthorn. May. Thickets and woods; occasional in the ne. one-fourth of the state. (Including *C. egani* Ashe; *C. ferrissii* Ashe; *C. demissa* Sarg.; *C. cyanophylla* Sarg.; *C. sextilis* Sarg.; *C. tarda* Sarg.; *C. depilis* Sarg.; *C. hillii* Sarg.; *C. blothra* Laughlin; *C. taetrica* Sarg.)

12. *Rubus* L.   Blackberry

1. Leaves simple; flowers purple .... 1. *R. odoratus*
1. Leaves compound; flowers white 2
2. Stems without prickles or bristles, herbaceous .... 2. *R. pubescens*

1. R. odoratus L. Purple Flowering Raspberry. May–June. Woods; Carroll, Cook, DeKalb, and LaSalle cos.

2. R. pubescens Raf. Dwarf Raspberry. May–June. Bogs and springy places; Cook, DeKalb, Lake, and Winnebago cos.

3. R. occidentalis L. Black Raspberry. May–June. Edge of woods, thickets, roadsides, bluffs; common throughout the state.

4. R. phoenicolasius Maxim. Wineberry. May–July. Native of Asia; escaped in Jersey, Lake, and Morgan cos.

5. R. idaeus L. Cultivated Raspberry. May–June. Native of Europe; occasionally escaped and persistent from cultivation.

6. R. strigosus Michx. Red Raspberry. May–June. Bogs and swampy woods; confined to the n. one-fourth of the state. *R. idaeus* L. var. *strigosus* (Michx.) Maxim.—F.

7. R. procerus P. J. Muell. Himalaya-berry. May–July. Native of Europe; escaped along a road in Randolph Co.

8. R. laciniatus Willd. Evergreen Blackberry. June–July. Native of Europe; scattered in Ill., but not common.

9. R. hispidus L. Swampy Dewberry. June–July. Bog, black oak woods, marshes, thickets; restricted to the n. one-half of the state. (Including *R. signatus* Bailey.) *R. hispidus* var. *obovalis* (Michx.) Fern.—F.

10. R. trivialis Michx. Southern Dewberry. April–May. Fields, waste ground; confined to the s. one-fourth of the state.

11. R. occidualis Bailey. Velvet-leaved Dewberry. May–June. Edge of woods; Pope Co. *R. flagellaris* L.—G.

12. R. flagellaris Willd. Dewberry. April–June. Fields, edge of woods, roadsides; common to occasional throughout most of the state. (Including *R. baileyanus* Britt.)

13. R. enslenii Tratt. Arching Dewberry. May–June. Rocky woods; Randolph Co. (Including *R. mundus* Bailey.)

14. R. schneideri Bailey. Bristly Blackberry. June. Sandy swales; Kankakee Co. (Including *R. offectus* Bailey and *R. wheeleri* Bailey.) *R. setosus* Bigel.—G, J.

15. R. allegheniensis Porter. Common Blackberry. May–June. Roadsides, thickets, woods; common throughout the state. (Including *R. impos* Bailey.)

16. R. alumnus Bailey. Blackberry. May–June. Along railroad; Jackson and Wabash cos.

17. R. argutus Link. Highbush Blackberry. May–June. Thickets, woods; common throughout the state. (Including *R. blakei* Bailey, *R. abactus* Bailey, *R. schneckii* Bailey, and *R. virilis* Bailey.) *R. ostryifolius* Rydb.—G, J.

18. R. avipes Bailey. Blackberry. May–June. Thickets, rare; confined to n. Ill.

19. R. pensylvanicus Poir. Blackberry. May–June. Fields, thickets, woods; occasional to common throughout the state. (Including *R. recurvans* Blanch. and *R. bellobatus* Bailey.)

20. R. frondosus Bigel. Blackberry. May–June. Woods and thickets; occasional throughout the state. *R. pensilvanicus* Poir.—G.

13. *Rosa* L.   Rose

1. Styles united into a column 2
1. Styles free 3
2. Flowers white (rarely pinkish), up to 2.5 cm across; leaflets (5–) 7–9; stipules pectinate .... 1. *R. multiflora*
2. Flowers pink or rose, 3 cm broad or broader; leaflets 3 (–5); stipules entire or toothed .... 2. *R. setigera*
3. Achenes lining only the base of the receptacle 4

3. Achenes lining the walls and the base of the receptacle 10
4. Sepals reflexed after flowering, rarely persisting on the mature fruit . 5
4. Sepals not reflexed after flowering, usually persisting on the mature fruit
.... 6
5. Mature plants over 1 m tall; leaves finely toothed .... 3. *R. palustris*
5. Mature plants less than 1 m tall; leaves coarsely toothed . 4. *R. carolina*
6. Flowers borne singly 7
6. Flowers clustered in corymbs 8
7. Stems glabrous, except for the numerous spines .. 5. *R. pimpinellifolia*
7. Stems tomentose as well as spiny .... 6. *R. rugosa*
8. Leaflets glabrous, rarely more than 2 cm long .... 7. *R. lunellii*
8. Leaflets pubescent, usually at least 2 cm long 9
9. Prickles absent on flowering branches, although often present near base
of plant; leaflets 5–7 (–9) .... 8. *R. blanda*
9. Prickles present on flowering branches as well as on the lower part of
the stem; leaflets 9–11 .... 9. *R. suffulta*
10. Rusty resinous glands present on at least the lower surface of the leaflets 11
10. Rusty resinous glands absent from the surfaces of the leaflets (glands
sometimes present on lower surface of leaflets in *R. canina*) 12
11. Leaflets with many rusty glandular dots on both surfaces; styles pubescent; calyx persistent on the fruit .... 10. *R. eglanteria*
11. Leaflets with few rusty glandular dots on the surfaces; styles glabrous or
nearly so; calyx falling from the fruit .... 11. *R. micrantha*
12. Receptacle and pedicels glabrous or nearly so .... 12. *R. canina*
12. Receptacle pubescent .... 13. *R. gallica*

1. R. multiflora Thunb. Multiflora Rose. May–June. Native of China and Japan; frequently planted and often escaped throughout the state.

2. R. setigera Michx. Prairie Rose. June–July. Leaflets glabrous be-neath, or pilose on the nerves. Woods, thickets, clearings; occasional to common throughout the state.

var. tomentosa Torr. & Gray. Leaflets tomentose beneath. Woods, thickets, clearings; occasional throughout the state.

3. R. palustris Marsh. Swampy Rose. June–Aug. Swamps, bogs, moist thickets; occasional throughout the state.

4. R. carolina L. Pasture Rose. May–July. Lower surface of leaflets glabrous or sparsely pubescent. Prairies, fields, dry woods; common throughout the state. (Including var. *grandiflora* [Baker] Rehd.)

var. villosa (Best) Rehd. Lower surface of leaflets softly pubescent beneath. Prairies, fields; occasional throughout the state.

5. R. pimpinellifolia L. Burnet Rose. May–July. Native of Europe; oc-casionally escaped from cultivation. *R. spinosissima* L.—F, G, J.

6. R. rugosa Thunb. Rugose Rose. June–Sept. Native of Asia; rarely escaped from cultivation; Kane Co.

7. R. lunellii Greene. Lunell's Rose. June–July. Native of the w. U.S.; along a railroad; JoDaviess Co.

8. R. blanda Ait. Meadow Rose. May–June. Thickets, woods, open areas; occasional in the n. one-third of the state, very rare elsewhere.

9. R. suffulta Greene. Sunshine Rose. June–July. Thickets, woods; occasional in the n. two-thirds of the state, apparently absent else-where. (Including *R. relicta* Erlanson.) *R. arkansana* Porter var. *suf-fulta* (Greene) Cockerell—F.

10. R. eglanteria L. Sweet-brier. May–July. Native of Europe; occasionally escaped from cultivation.

11. R. micrantha Sm. Small Sweet-brier. May–July. Native of Europe; rarely escaped from cultivation; Winnebago Co.

12. R. canina L. Dog Rose. May–July. Native of Europe; occasionally escaped from cultivation.

13. R. gallica L. French Rose. June–July. Native of Europe; rarely escaped from cultivation; Piatt Co.

14. *Potentilla L.* Cinquefoil

1. Petals purple .... 1. *P. palustris*
1. Petals yellow or white 2
2. Petals white 3
2. Petals yellow 4
3. Leaflets 3, usually entire, evergreen; achenes hairy .... 2. *P. tridentata*
3. Leaflets 7–11, serrate, not evergreen; achenes glabrous ... 3. *P. arguta*
4. Shrub; achenes pubescent; style lateral .... 4. *P. fruticosa*
4. Herbs; achenes glabrous; style terminal 5
5. Flowers solitary 6
5. Flowers in cymes 7
6. Leaves palmately compound, with usually 5 leaflets ...... 5. *P. simplex*
6. Leaves pinnately compound, with 7–25 leaflets .... 6. *P. anserina*
7. Lower leaves ternate or digitate 8
7. Lower leaves pinnate 12
8. Lower leaves with 5–7 leaflets 9
8. Lower leaves with 3 leaflets 11
9. Achenes rugose; flowers 1.5–2.5 cm across, the petals emarginate; stems and leaves hirsute .... 7. *P. recta*
9. Achenes smooth; flowers up to 1.5 cm across, the petals entire; stem and leaves white-woolly or gray-tomentulose 10
10. Stem and leaves white-woolly; leaflets sometimes nearly pinnatifid .... 8. *P. argentea*
10. Stem and leaves gray-tomentulose; leaflets at most coarsely dentate .... 9. *P. intermedia*
11. Plants softly villous; flowers up to 4 mm across; stamens about 10 .... 10. *P. millegrana*
11. Plants rough-hirsute; flowers usually at least 6 mm across; stamens 15–20 .... 11. *P. norvegica*
12. Leaflets of the lower leaves approximate; achenes smooth .... 12. *P. rivalis*
12. Leaflets of the lower leaves more distant; achenes longitudinally ribbed .... 13. *P. paradoxa*

1. P. palustris (L.) Scop. Marsh Cinquefoil. June–July. Bogs, rare; Cook, Lake, and McHenry cos.

2. P. tridentata Ait. Three-toothed Cinquefoil. June–Sept. Gravel ridge, very rare; Cook Co.

3. P. arguta Pursh. Prairie Cinquefoil. June–July. Prairies; occasional in the n. half of the state.

4. P. fruticosa L. Shrubby Cinquefoil. June–Aug. Interdunal ponds, hill prairies, boggy fens, rare; Cook, JoDaviess, Kane, Kendall, and Lake cos.

5. P. simplex Michx. Common Cinquefoil. May–June. Stems villous or hirsute; lower surface of leaflets strigose. Dry woods, prairies, fields, common; in every co.

var. calvescens Fern. Common Cinquefoil. May–June. Stems and lower surface of leaflets glabrous or strigillose. Similar habitats; occasional throughout the state.

var. argyrisma Fern. Common Cinquefoil. May–June. Stems spreading-villous; lower surface of leaflets densely silvery-silky. Dry woods; rare in the s. cos.

6. P. anserina L. Silverweed. May–Aug. Sandy beaches, interdunal ponds, meadows, gravel bars; restricted to the ne. cos. of Cook, De-Kalb, Lake, McHenry, and Will.

7. P. recta L. Sulfur Cinquefoil. May–July. Native of Europe. Fields and roadsides; occasional to common throughout the state.

8. P. argentea L. Silvery Cinquefoil. May–Sept. Native of Europe; disturbed sandy soil; occasional in the n. half of the state.

9. P. intermedia L. Intermediate Cinquefoil. June–Aug. Native of Europe; adventive along railroad; Champaign Co.

10. P. millegrana Engelm. Cinquefoil. June–July. Moist soil, rare; Johnson and St. Clair cos.

11. P. norvegica L. Rough Cinquefoil. June–July. Native of Europe; disturbed soils; occasional throughout the state. *P. monspeliensis* L.—J.

12. P. rivalis Nutt. Brook Cinquefoil. June–Aug. Native of the w. U.S.; adventive once in Cook Co.

13. P. paradoxa Nutt. Cinquefoil. May–Sept. Wet soil, usually along rivers; Jackson, Randolph, and St. Clair cos.

15. *Waldsteinia Willd.*

1. W. fragarioides (Michx.) Tratt. Barren Strawberry. April–May. Sandstone ledge; Pope Co.

16. *Fragaria L.   Strawberry*

1. Most or all petals less than 1 cm long; fruits up to 1.5 cm across 2
1. Most or all petals over 1 cm long; fruits over 1.5 cm across ....
.... 3. *F.* X *ananassa*

1. F. virginiana Duchesne. Wild Strawberry. April–July. Woodlands, prairies, fields; occasional to common throughout the state. (Including var. *illinoensis* [Prince] Gray.)

2. F. americana (Porter) Britt. Hillside Strawberry. May–Aug. Wooded slopes, occasional to rare; confined to the n. half of the state. *F. vesca* L. var. *americana* Porter—F, G.

3. F. X ananassa Duchesne. Cultivated Strawberry. April–May. Rarely escaped from cultivation. Reputed to be a hybrid between *F. chiloensis* (L.) Duchesne and *F. virginiana* Duchesne. *F. chiloensis* Duch. var. *ananassa* Bailey—G.

17. *Aruncus Adans.   Goat's-beard*

1. A. dioicus (Walt.) Fern. Goat's-beard. May–June. Rich woods; occasional throughout the state; absent in the ne. cos., except Grundy Co. (Including var. *pubescens* [Rydb.] Fern.)

18. *Filipendula Mill.*

1. F. rubra (Hill) Robins. Queen-of-the-prairie. June–July. Springy fens, occasional; confined to the n. half of the state.

19. *Sanguisorba L. Burnet*

1. Leaflets up to 2.5 cm long; stamens 12 or more, drooping . 1. *S. minor*
1. Most or all the leaflets 2.5 cm long or longer; stamens 4, erect . . . .
. . . . 2. *S. canadensis*

    1. S. minor Scop. Garden Burnet. May–July. Native of Europe and Asia; adventive in a limestone barren; Will Co.
    2. S. canadensis L. American Burnet. July–Oct. Moist prairies, very rare; Cass, LaSalle, and Will cos.

20. *Gillenia Moench*

    1. G. stipulata (Muhl.) Baill. Indian Physic. May–July. Woods; occasional in the s. half of the state; also La Salle Co.

21. *Geum L. Avens*

1. Flowers purple 2
1. Flowers white, cream, or yellow 3
2. Styles not jointed; flowers mostly less than 2 cm across; style 2 cm long or longer . . . . 1. *G. triflorum*
2. Styles jointed; flowers mostly at least 2 cm across; style up to 1 cm long . . . . 2. *G. rivale*
3. Petals white 4
3. Petals yellow or cream 5
4. Most of the petals 5 mm long or longer; peduncles puberulent; fruiting receptacle densely white-villous . . . . 3. *G. canadense*
4. Petals up to 5 mm long; peduncles hirsute; fruiting receptacle glabrous or nearly so . . . . 4. *G. laciniatum*
5. Receptacle stalked in the calyx; calyx not subtended by bractlets; petals about 2 mm long, yellow . . . . 5. *G. vernum*
5. Receptacle sessile; calyx subtended by bractlets; petals (or some of them) more than 2 mm long, cream, orange, or deep yellow 6
6. Petals 2–4 mm long, shorter than the calyx, cream . . . 6. *G. virginianum*
6. Petals 5–10 mm long, as long as the calyx, orange or deep yellow . . . . . .
. . . . 7. *G. strictum*

    1. G. triflorum Pursh. Prairie Avens. May–June. Dry prairies; occasional in the n. one-sixth of the state; absent elsewhere.
    2. G. rivale L. Purple Avens. May–Aug. Moist soil, very rare; Kane, McHenry, and Winnebago cos.
    3. G. canadense Jacq. White Avens. June–Aug. Fruiting carpel pubescent throughout. Woodlands, common; in every co. (Including var. *camporum* [Rydb.] Fern. & Weath.)
    var. grimesii Fern. & Weath. White Avens. June–Aug. Fruiting carpel hispid above, glabrous below. Woodlands, rare; confined to the s. one-fifth of the state.
    4. G. laciniatum Murr. Rough Avens. June–July. Achenes glabrous.

Meadows and thickets, not common; restricted to the n. three-fourths of the state.

var. trichocarpum Fern. Rough Avens. June–July. Achenes bristly at summit. Meadows and thickets; occasional in the n. three-fourths of the state.

5. G. vernum (Raf.) Torr. & Gray. Spring Avens. April–May. Moist woods; common in the s. half of the state, occasional or rare in the n. half.

6. G. virginianum L. Pale Avens. June–July. Dry woods, rare; restricted to the s. one-fourth of the state.

7. G. strictum Ait. Yellow Avens. June–July. Bogs, moist thickets; occasional in the n. one-half of the state; also St. Clair Co. *G. aleppicum* Jacq. var. *strictum* (Ait.) Fern.—F; *G. aleppicum* Jacq.—G.

22. Duchesnea Sm.

1. D. indica (Andr.) Focke. Indian Strawberry. April–June. Native of Asia; adventive in waste ground; Alexander, DuPage, and Johnson cos.

23. Agrimonia L. Agrimony

1. A. parviflora Ait. Swamp Agrimony. July–Sept. Low ground; occasional throughout the state.

2. A. microcarpa Wallr. Small-fruited Agrimony. July–Aug. Dry woods, very rare; St. Clair Co.

3. A. pubescens Wallr. Soft Agrimony. July–Sept. Dry or moist woods; occasional throughout the state. (Including *A. mollis* [Torr. & Gray] Britt.)

4. A. gryposepala Wallr. Tall Agrimony. June–Aug. Woods and thickets; occasional in the n. two-thirds of the state, absent elsewhere.

5. A. rostellata Wallr. Woodland Agrimony. July–Sept. Woods; occasional in the s. two-thirds of the state; also Cook and Will cos.

80. LEGUMINOSAE

3. Leaves twice-pinnate 4
3. Leaves once-pinnate 6
4. Some or all the leaflets over 1.5 cm broad .... 3. *Gymnocladus*
4. None of the leaflets 1.5 cm broad 5
5. Leaflets 1 cm long or longer; plants usually with large thorns; flowers in elongated spike-like racemes .... 4. *Gleditsia*
5. Leaflets (or most of them) up to 8 mm long; plants without thorns; flowers in spherical heads .... 6. *Albizia*
6. Woody vines .... 7. *Wisteria*
6. Trees or upright shrubs 7
7. Leaflets three .... 8. *Ononis*
7. Leaflets five or more 8
8. Tip of rachis developed into a spine, but soon deciduous . 2. *Caragana*
8. Tip of rachis not spinescent 9
9. Leaves even-pinnate .... 4. *Gleditsia*
9. Leaves odd-pinnate 10
10. Lower surface of leaflets glandular-punctate or canescent; flowers less than 1 cm long; shrubs to 7 mm tall .... 9. *Amorpha*
10. Lower surface of leaflets neither glandular-punctate nor canescent; flowers 1.4 cm long or longer; trees up to 15 m tall 11
11. Stipular spines usually present; stems sometimes glandular or hispid .... 10. *Robinia*
11. Stipular spines absent; stems glabrous .... 11. *Cladrastis*
12. Leaves simple .... 12. *Crotalaria*
12. Leaves compound 13
13. Leaves palmately compound, not trifoliolate 14
13. Leaves pinnately compound or trifoliolate 15
14. Leaflets 7–11 .... 13. *Lupinus*
14. Leaflets 5 .... 14. *Psoralea*
15. Leaves twice-pinnate .... 5. *Desmanthus*
15. Leaves once-pinnate 16
16. Leaves even-pinnate 17
16. Leaves odd-pinnate 21
17. Some or all the leaf rachises ending in a tendril 18
17. None of the leaf rachises prolonged into a tendril 20
18. Calyx lobes leaf-like; lowermost stipules larger than the leaflets they subtend .... 15. *Pisum*
18. Calyx lobes not leaf-like; none of the stipules larger than the leaflets 19
19. Wings and keel petal united; style with a terminal tuft of hairs 16. *Vicia*
19. Wings free from the keel petal; style pubescent only along one side .... 17. *Lathyrus*
20. Petioles with one or more conspicuous glands; corolla nearly regular .. .. 18. *Cassia*
20. Petioles without glands; corolla pea-shaped .... 19. *Sesbania*
21. Leaflets 5 or more 22
21. Leaflets 3 32
22. Stems, petioles, and peduncles with hooked prickles ... 20. *Schrankia*
22. Stems, petioles, and peduncles without hooked prickles 23
23. Leaflets 5–9 24
23. Leaflets 11-many 26
24. Climbing vines; flowers brownish-purple or greenish-white, in rather loose racemes .... 21. *Apios*
24. Upright herbs; flowers bright yellow, bright rose, or white, in umbels or spikes 25

25. Flowers bright yellow, in umbels; stamens 10 .... 22. *Lotus*
25. Flowers bright rose or white, in spikes; stamens 5 .. 23. *Petalostemum*
26. Plants glandular-viscid; fruit with hooked bristles ...... 24. *Glycyrrhiza*
26. Plants not glandular-viscid; fruit without hooked bristles 27
27. Flowers in umbels .... 25. *Coronilla*
27. Flowers in spikes or elongated or shortened racemes 28
28. Flowers in dense, elongated heads, the rachis not exposed; fruits enclosed by the calyx, 1- to 2-seeded 29
28. Flowers in elongated or shortened racemes; fruits exserted, several-
seeded (1-seeded in *Onobrychis*) 30
29. Flowers pink or white; stamens 10; bracts about equalling the calyx;
leaflets obtuse .... 26. *Dalea*
29. Flowers rose, fading to white; stamens 5; bracts longer than the calyx;
leaflets mostly acute .... 23. *Petalostemum*
30. Inflorescences terminal; flowers cream, tipped with purple ....
.... 27. *Tephrosia*
30. Inflorescences axillary; flowers purple, cream, or yellow-green 31
31. Fruit strongly reticulate and occasionally toothed along the margins; corolla pale pink .... 28. *Onobrychis*
31. Fruit neither strongly reticulate nor toothed; corolla yellow-green,
cream, or purple .... 29. *Astragalus*
32. Leaflets toothed 33
32. Leaflets entire 36
33. Plants spiny .... 8. *Ononis*
33. Plants not spiny 34
34. Flowers in elongated racemes .... 30. *Melilotus*
34. Flowers in heads or spikes 35
35. Stipules entire; fruit usually straight, the petals mostly persistent ....
.... 31. *Trifolium*
35. Stipules toothed (rarely entire); fruit coiled or twisted, the petals not persistent .... 32. *Medicago*
36. Foliage with a few glandular dots .... 14. *Psoralea*
36. Foliage without glandular dots 37
37. Flowers 2 cm long or longer 38
37. Flowers up to 1.6 cm long 39
38. Flower 2–3 cm long, in racemes .... 33. *Baptisia*
38. Flowers 4–6 cm long, solitary or only 1– together .... 34. *Clitoria*
39. Leaflets without stipels 40
39. Leaflets with stipels 45
40. Flowers yellow 41
40. Flowers purple, violet, rose, or white with purple markings 43
41. Flowers 1.3–1.6 cm long .... 33. *Baptisia*
41. Flowers less than 1 cm long 42
42. Stipules united to petiole, forming a sheath which encircles the stem ..
.... 35. *Stylosanthes*
42. Stipules free from petiole .... 36. *Lespedeza*
43. Uppermost leaves reduced to one leaflet; fruit a legume 1.5–2.5 cm long
.... 37. *Hosackia*
43. Uppermost leaves trifoliolate; fruit a loment less than 1 cm long or a
legume opening by a lid 44
44. Fruit a legume opening by a lid; flowers in a globose head ....
.... 31. *Trifolium*
44. Fruit a loment; flowers variously arranged, but not in a globose head
.... 36. *Lespedeza*

1. Cercis L. Redbud

1. C. canadensis L. Redbud. April–May. Woods; common in the s. cos., becoming less abundant northward. (Including f. *glabrifolia* L.)

2. Caragana Lam. Pea-tree

1. C. arborescens Lam. Pea-tree. May. Native of Asia; escaped from cultivation and spreading in Winnebago Co.

3. Gymnocladus Lam. Coffee-tree

1. G. dioica (L.) K. Koch. Kentucky Coffee-tree. May–June. Mostly low woods; occasional throughout the state.

4. Gleditsia L. Honey Locust

1. Legume 15 cm long or longer, with sweet pulp between the several seeds .... 1. *G. triacanthos*
1. Legume to 5 cm long, with no pulp surrounding the 1 or 2 seeds ......
.... 2. *G. aquatica*

1. G. triacanthos L. Honey Locust. May–June. Trunk beset by stout thorns. Woods; common in the s. cos., becoming less common northward.

f. inermis (Pursh) Schneid. Thornless Honey Locust. May–June. Trunks without thorns. Occasionally found with the typical form.

2. G. aquatica Marsh. Water Locust. May–June. Swamps, rare; confined to s. Ill., extending northward to Calhoun Co. in the w. and Lawrence Co. in the e.; also Henderson Co.

5. *Desmanthus Willd.*

1. D. illinoensis (Michx.) MacM. Illinois Mimosa. June–Aug. Prairies; along levees; occasional in the s. and n. cent. cos., uncommon or absent elsewhere.

6. *Albizia Duraz.*

1. A. julibrissin Duraz. Mimosa. June–Aug. Native of Asia and Africa; occasionally escaped from cultivation in the s. cos.

7. *Wisteria Nutt.*

1. Leaflets mostly 15–19 .... 1. *W. floribunda*
1. Leaflets mostly 7–13 2
2. Pedicels 15 mm long or longer; ovary and legume velvety 2. *W. sinensis*
2. Pedicels up to 10 mm long; ovary and legume glabrous ....
.... 3. *W. macrostachya*

1. W. floribunda DC. Japanese Wisteria. April–June. Native of Japan; rarely escaped from cultivation; Jackson Co.
2. W. sinensis Sweet. Chinese Wisteria. April–May. Native of Asia; rarely escaped from cultivation; Jackson, Piatt, and Richland cos.
3. W. macrostachya Nutt. Wisteria. June–July. Swampy woods; generally confined to the s. one-fifth of the state, but extending n. to Clark and Richland cos.; adventive in Peoria and Washington cos.

8. *Ononis L. Rest Harrow*

1. O. spinosa L. Rest Harrow. June–Aug. Native of Europe; rarely persisting from cultivation; DuPage Co.

9. *Amorpha L.*

1. Some or all leaflets 1 cm broad or broader, short-petiolulate 2
1. None of the leaflets 1 cm broad, sessile or nearly so .. 3. *A. canescens*
2. Leaflets and branchlets glabrous or nearly so; legume without resinous
dots .... 1. *A. nitens*
2. Leaflets and branchlets pubescent; legume with resinous dots ....
.... 2. *A. fruticosa*

1. A. nitens Boynton. Smooth False Indigo. May–June. Woods along rivers and streams, very rare; Pope Co.
2. A. fruticosa L. False Indigo. May–June. Leaflets mostly less than twice as long as broad, grayish-pubescent. Moist soil; occasional throughout the state.
    var. angustifolia Pursh. False Indigo. May–June. Leaflets at least twice as long as broad, grayish-pubescent. Moist soil; occasional throughout the state. (Including var. *tennesseenis* [Shuttlew.] Palmer; var. *oblongifolia* Palmer.)
    var. croceolanata (P.W. Wats.) Schneid. False Indigo. May–June. Pubescence softly tawny-villous; low woods, rare; Pope Co.
3. A. canescens Pursh. Leadplant. May–Aug. Prairies; common in the n. two-thirds of the state, less common elsewhere.

10. *Robinia* L.   Locust

1. Branchlets neither glandular-viscid nor bristly; flowers white . . . .
. . . . 1. *R. pseudoacacia*
1. Branchlets glandular-viscid or bristly; flowers pale pink, rose, or purple
. . . . 2
2. Branchlets glandular-viscid . . . . 2. *R. viscosa*
2. Branchlets bristly . . . . 3. *R. hispida*

1. R. pseudoacacia L. May–June. Black Locust. Woods and thickets; native in s. Ill., commonly planted elsewhere.
2. R. viscosa Vent. Clammy Locust. May–June. Native of the se.
U.S.; rarely escaped from cultivation; JoDaviess Co.
3. R. hispida L. Bristly Locust. May–June. Native of the se. U.S.; occasionally escaped from cultivation, chiefly in the s. half of the state.

11. *Cladrastis* Raf.

1. C. lutea (Michx. f.) K. Koch. Yellow-wood. May. Wooded slopes,
rare; Alexander and Gallatin cos.

12. *Crotalaria* L.   Rattlebox

1. Leaves acute at apex; flowers up to 1.5 cm long; legume less than 3 cm
long . . . . 1. *C. sagittalis*
1. Leaves obtuse at apex; flowers over 1.5 cm long; legume 3 cm long or
longer . . . . 2. *C. spectabilis*

1. C. sagittalis L. Rattlebox. June–Sept. Fields; edge of woods; occasional in the s. half of the state, becoming less common northward.
2. C. spectabilis Roth. Showy Rattlebox. Aug.–Oct. Native of the
Tropics; rarely adventive in Ill.; Alexander Co.

13. *Lupinus* L.   Lupine

1. L. perennis L. Wild Lupine. May–June. Sandy woods; occasional
in the n. one-fourth of the state, absent elsewhere. *L. perennis* L. var.
occidentalis S. Wats.—F.

14. *Psoralea* L.   Scurf-pea

1. Leaflets always 3; fruits variously wrinkled, often glandular-dotted . . . 2
1. Leaflets sometimes 5; fruits glandular-dotted, but not wrinkled . . . .
. . . . 3. *P. tenuiflora*
2. Some or all the leaflets at least 2 cm broad; peduncles shorter than the
subtending leaves . . . . 1. *P. onobrychis*
2. None of the leaflets 2 cm broad; peduncles exceeding the subtending
leaves . . . . 2. *P. psoralioides* var. *eglandulosa*

1. P. onobrychis Nutt. French Grass. June–July. Thickets, prairies,
not common; in most sections of the state, except for the nw. cos.
2. P. psoralioides (Walt.) Cory var. eglandulosa (Ell.) Freeman.
Sampson's Snakeroot. June–July. Dry woods, prairies; occasional in
the s. one-third of the state, apparently absent elsewhere. *P. psoralioides* (Walt.) Cory—F, G.

3. P. tenuiflora Pursh. Scurf-pea. June–Sept. Prairies; occasional in the n. two-thirds of the state, absent elsewhere. (Including var. *floribunda* [Nutt.] Rydb.)

15. *Pisum* L.   Garden Pea

1. P. sativum L. Garden Pea. May–June. Escaped from cultivation; Adams and Jackson cos.

16. *Vicia* L.   Vetch

1. Peduncles as long as the leaflets or longer 2
1. Peduncles much shorter than the leaves or nearly absent 6
2. Pubescence on stems and axis of inflorescence spreading   1. *V. villosa*
2. Pubescence on stems and axis of the inflorescence appressed or nearly absent 3
3. Flowers white, tipped with purple; racemes rather laxly flowered ....
.... 2. *V. caroliniana*
3. Flowers purple; racemes densely flowered 4
4. Flowers less than 10 per raceme; entire inflorescence shorter than the subtending leaves .... 3. *V. americana*
4. Flowers 10 or more per raceme; entire inflorescence longer than the subtending leaves 5
5. Calyx gibbous at the base on the upper side .... 4. *V. dasycarpa*
5. Calyx merely rounded at the base on the upper side ...... 5. *V. cracca*
6. Stems more or less pubescent; leaflets all similar in shape; most or all the flowers 18 mm long or longer; legume torulose ....
.... 6. *V. sativa*
6. Stems glabrous or nearly so; upper leaflets narrower than lower ones; all of the flowers less than 18 mm long; legume flat ....
.... 7. *V. angustifolia*

1. V. villosa Roth. Winter Vetch. June–Aug. Native of Europe; in fields and along roads; common throughout the state.
2. V. caroliniana Walt. Wood Vetch. April–June. Dry, wooded slopes; occasional in the upper one-fourth of the state, absent elsewhere.
3. V. americana Muhl. American Vetch. July–Aug. Disturbed woods and disturbed prairies; occasional in the n. one-third of the state, absent elsewhere.
4. V. dasycarpa Ten. Hairy-fruited Vetch. May–Sept. Native of Europe; scattered in the state, but not common.
5. V. cracca L. Cow Vetch. May–Aug. Native of Europe; roadsides and fields; occasional throughout the state.
6. V. sativa L. Common Vetch. July–Sept. Native of Europe; roadsides and fields; scattered in Ill., but not common.
7. V. angustifolia Reich. Narrow-leaved Vetch. Native of Europe; waste ground; occasional in the extreme n. and extreme s. cos. (Including var. *segetalis* [Thuill.] W. D. J. Koch.)

17. *Lathyrus* L.   Wild Pea

1. Leaflets 2 2
1. Leaflets 4–12 4
2. Corolla pink, purple, or white 3

2. Corolla bright yellow .... 3. *L. pratensis*
3. Perennial; peduncles 4- to 10-flowered; flowers odorless   1. *L. latifolius*
3. Annual; peduncles 1- to 3-flowered; flowers fragrant .... 2. *L. odoratus*
4. Leaflets fleshy; stipules with 2 basal lobes .... 4. *L. maritimus*
4. Leaflets not fleshy; stipules with one basal lobe 5
5. Corolla yellowish-white; petiolules glabrous .... 5. *L. ochroleucus*
5. Corolla violet or purple; petiolules pubescent 6
6. Leaflets and stems glabrous 7
6. Leaflets hirtellous .... 8. *L. venosus*
7. Stem distinctly winged .... 6. *L. palustris*
7. Stem unwinged .... 7. *L. myrtifolius*

1. L. latifolius L. Everlasting Pea. June–Sept. Native of Europe; occasionally escaped from cultivation.

2. L. odoratus L. Sweet Pea. June–Sept. Native of Europe; escaped to roadsides; occasional in the s. one-fourth of the state.

3. L. pratensis L. Yellow Vetchling. June–Aug. Native of Europe; rarely escaped from cultivation.

4. L. maritimus (L.) Bigel. Beach Pea. June–Aug. Beaches along Lake Michigan; Cook and Lake cos. *L. japonicus* Willd. var. *glaber* (Ser.) Fern.—F; *L. maritimus* (L.) Bigel. var. *glaber* Ser.—G.

5. L. ochroleucus Hook. Pale Vetchling. May–July. Woods; rare in the n. one-sixth of the state; also Gallatin Co.

6. L. palustris L. Marsh Vetchling. May–Sept. Moist, open areas; occasional in the n. half of the state; also Massac, Pope, and Wabash cos.

7. L. myrtifolius Muhl. Vetchling. June–July. Moist soil; occasional in the n. four-fifths of the state.

8. L. venosus Muhl. var. intonsus Butt. & St. John. Veiny Pea. May–June. Dry prairies, woods; occasional in the n. one-fourth of the state. *L. venosus* Muhl.—J.

18. Cassia L.   Senna

1. Leaflets 4–18, at least 7 mm broad 2
1. Leaflets 20 or more, less than 7 mm broad 5
2. Leaflets 4–6; gland midway between lowest pair of leaflets .. 1. *C. tora*
2. Leaflets 8–18; gland near base of petiole 3
3. Leaflets obtuse; stems somewhat villous above .... 2. *C. hebecarpa*
3. Leaflets acute; stems glabrous 4
4. Ovary and fruits pubescent; legume up to 10 cm long 3. *C. marilandica*
4. Ovary and fruits glabrous; legume 10–12 cm long ... 4. *C. occidentalis*
5. Calyx 9–10 mm long; corolla 10–20 mm long; stamens 10; pedicels 10 mm long or longer .... 5. *C. fasciculata*
5. Calyx 3–4 mm long; corolla less than 10 mm long; stamens 5; pedicels 2–4 mm long .... 6. *C. nictitans*

1. C. tora L. Sicklepod. July–Sept. Along streams and railroads, very rare; Jackson, Lake, Pulaski, and Wabash cos.

2. C. hebecarpa Fern. Wild Senna. July–Aug. Along streams, open woods; occasional throughout the state.

3. C. marilandica L. Maryland Senna. July–Aug. Roadsides, thickets; occasional throughout the state.

4. C. occidentalis L. Coffee Senna. Aug.–Sept. Native of the Tropics; rarely adventive in Ill.; Cook Co.

5. C. fasciculata Michx. Partridge Pea. July–Sept. Stems with appressed pubescence. Open areas, prairies, fields; common throughout the state.

var. robusta (Pollard) Macbr. Partridge Pea. July–Sept. Stems hirsute. Fields; rare in Ill.

6. C. nictitans L. Wild Sensitive Plant. July–Sept. Fields, roadsides, edge of woods; occasional in the s. one-fifth of the state, uncommon elsewhere.

19. Sesbania Scop.

1. S. exaltata (Raf.) Cory. July–Oct. Low ground, rare; Alexander and Pulaski cos.

20. Schrankia Willd. Sensitive Brier

1. S. uncinata Willd. Cat-claw. June–Sept. Prairies, rare; Peoria and Winnebago cos. S. nuttallii (DC.) Standl.—F.

21. Apios Medic. Groundnut

1. Flowers brownish-purple; standard without a spongy knob at the apex .... 1. A. americana
1. Flowers greenish-white, tinged with purple; standard with a spongy knob at the apex .... 2. A. priceana

1. A. americana Medic. Groundnut. July–Sept. Thickets, prairies; common throughout the state. (Including var. turrigera Fern.)

2. A. priceana Robins. Price's Groundnut. July–Sept. Woods, very rare; Union Co.

22. Lotus L.

1. L. corniculatus L. Birdsfoot-trefoil. June–Sept. Native of Europe; waste areas; occasional throughout the state.

23. Petalostemum Michx. Prairie Clover

1. Leaflets 5–7 2
1. Leaflets 19–27 .... 3. P. foliosum
2. Flowers white; calyx glabrous .... 1. P. candidum
2. Flowers purple or rose; calyx densely pubescent ..... 2. P. purpureum

1. P. candidum (Willd.) Michx. White Prairie Clover. June–July. Prairies; occasional throughout the state.

2. P. purpureum (Vent.) Rydb. Purple Prairie Clover. June–Aug. Prairies; occasional throughout the state.

3. P. foliosum Gray. Leafy Prairie Clover. July–Sept. Dry soil above streams; once known from a few counties in the ne. one-fourth of the state, now known only from Will Co.

24. Glycyrrhiza L.

1. G. lepidota Pursh. Wild Licorice. June–Aug. Native of w. U.S.; waste ground; occasional in the ne. corner of the state; also St. Clair Co.

25. Coronilla L.   Crown Vetch

1. C. varia L. Crown Vetch. June–Aug. Native of Europe; often planted and frequently escaped.

26. Dalea Juss.

1. D. alopecuroides Willd. Foxtail Dalea. July–Sept. Fields, roadsides; occasional in most sections of the state.

27. Tephrosia Pers.   Hoary Pea

1. T. virginiana (L.) Pers. Goat's-rue. May–Aug. Dry woods; occasional throughout the state.

28. Onobrychis Gaertn.

1. O. viciaefolia Scop. Sainfoin. June–July. Native of Europe and Asia; rarely escaped from cultivation.

29. Astragalus L.   Milk Vetch

1. Calyx tube 2–6 mm long; fruit dry, dehiscent 2
1. Calyx tube 7–10 mm long; fruit fleshy, indehiscent 4
2. Calyx tube hirsute; corolla 1.5–2.0 cm long; fruit up to 1 cm long, long-villous .... 1. *A. goniatus*
2. Calyx tube glabrous or sparsely pubescent, not hirsute; corolla 1.0–1.5 cm long; fruit 1–3 cm long, glabrous 3
3. Fruits lunate, 2–3 cm long; corolla lilac, purple, or white, about 1 cm long; calyx tube 2–3 mm long .... 2. *A. distortus*
3. Fruits more or less straight, 1–2 cm long; corolla greenish-yellow, 1.2–1.5 cm long; calyx tube 4.5–6.0 mm long .... 3. *A. canadensis*
4. Fruits glabrous .... 4. *A. trichocalyx*
4. Fruits pubescent .... 5. *A. tennesseensis*

1. A. goniatus Nutt. Field Milk Vetch. May–July. Native of the w. U.S.; rarely adventive; Boone Co. *A. agrestis* L.—G.
2. A. distortus Torr. & Gray. Bent Milk Vetch. May–June. Rocky prairies; rare in w. Ill.; adventive in Cook Co.
3. A. canadensis L. Canadian Milk Vetch. June–Aug. Prairies, thickets; occasional throughout the state. (Including var. *longilobus* Fassett.)
4. A. trichocalyx Nutt. Large Ground Plum. April–May. Rocky prairies, rare; Macoupin, Madison, and St. Clair cos.; adventive in Will Co. *A. mexicanus* A. DC. var. *trichocalyx* (Nutt.) Fern.—F; *A. crassicarpus* Nutt. var. *trichocalyx* (Nutt.) Barneby—G.
5. A. tennesseensis Gray. Ground Plum. May–June. Dry prairies, rare; confined to the n. one-fourth of the state; also Mason Co.

30. Melilotus Mill. Sweet Clover

1. Flowers white .... 1. *M. alba*
1. Flowers yellow 2
2. Fruit 2.5–3.5 mm long, glabrous or nearly so, prominently reticulate ....
.... 2. *M. officinalis*
2. Fruit 4.5–6.0 mm long, pubescent, obscurely reticulate .. 3. *M. altissima*

    1. M. alba Desr. White Sweet Clover. May–Oct. Native of Europe and Asia; naturalized in waste ground; in every co.
    2. M. officinalis (L.) Lam. Yellow Sweet Clover. June–Sept. Native of Europe; naturalized in waste ground; in every co.
    3. M. altissima Thuill. Tall Yellow Sweet Clover. June–Sept. Native of Europe; naturalized in waste ground; Champaign and Lake cos.

31. Trifolium L. Clover

1. Flowers yellow 2
1. Flowers white, pink, purple, or scarlet 4
2. Petiolule of middle leaflet longer than the petiolules of lateral leaflets 3
2. Petiolule of middle leaflet as long as or shorter than the petiolules of lateral leaflets .... 3. *T. aureum*
3. Flowers 15 or more per head, the head 8–15 mm thick . 1. *T. campestre*
3. Flowers up to 15 per head, the head less than 8 mm thick 2. *T. dubium*
4. Stems pubescent 5
4. Stems glabrous or nearly so 9
5. Flowers sessile or nearly so in the head 6
5. Flowers on pedicels 2 mm long or longer .... 8. *T. reflexum*
6. Flowering heads subglobose, about as broad as long 7
6. Flowering heads elongated, cylindric, longer than broad 8
7. Heads subtended by a pair of leaves; corolla 12–15 mm long; heads usually 2–3 cm in diameter .... 4. *T. pratense*
7. Heads not subtended by a pair of leaves; corolla 5–7 mm long; heads usually less than 2 cm in diameter .... 5. *T. fragiferum*
8. Flowers pale rose, shorter than the calyx; heads covered by silky hairs
.... 6. *T. arvense*
8. Flowers scarlet, longer than the calyx; heads not covered by silky hairs
.... 7. *T. incarnatum*
9. Flowers sessile or nearly so in the head 10
9. Flowers on pedicels 2 mm long or longer 12
10. Heads not subtended by a pair of leaves; corolla 4–6 mm long 11
10. Heads subtended by a pair of leaves; corolla 12–15 mm long ....
.... 4. *T. pratense*
11. Stems repent; bracteoles as long as the calyx; corolla not resupinate ..
.... 5. *T. fragiferum*
11. Stems ascending; bracteoles shorter than the calyx; corolla resupinate
.... 9. *T. resupinatum*
12. Stems creeping, with basal runners; calyx teeth usually shorter than the tube .... 10. *T. repens*
12. Stems ascending, without basal runners; calyx teeth as long as the tube or longer 13
13. Calyx up to 5 mm long, the teeth less than twice as long as the tube; stipules lanceolate .... 11. *T. hybridum*
13. Calyx 6–8 mm long, the teeth at least twice as long as the tube; stipules ovate .... 8. *T. reflexum*

1. T. campestre Schreb. Low Hop Clover. May–Sept. Native of Europe; waste ground; occasional throughout the state. *T. procumbens* L.—F, G, J.

2. T. dubium Sibth. Little Hop Clover. May–Sept. Native of Europe; waste ground, not common; known only from the so. one-fourth of the state.

3. T. aureum Pollich. Yellow Hop Clover. June–Sept. Native of Europe; waste ground; Henry, Peoria, Union, and Winnebago cos. *T. agrarium* L.—F, G, J.

4. T. pratense L. Red Clover. May–Sept. Native of Europe; naturalized in waste ground; in every co. (Including var. *sativum* [Mill.] Schreb.) The white-flowered f. *leucochraceum* Aschers. & Prantl has been collected in Ill.

5. T. fragiferum L. Strawberry Clover. June–Sept. Native of Europe; rarely adventive in Ill.; Cook and Jackson cos.

6. T. arvense L. Rabbit-foot Clover. May–Oct. Native of Europe; occasionally adventive in waste ground in most regions of the state.

7. T. incarnatum L. Crimson Clover. May–July. Native of Europe; occasionally adventive in the s. three-fourths of the state.

8. T. reflexum L. Buffalo Clover. May–June. Stems villous. Woods; not common, but scattered in the state.

var. glabrum Lojacono. Buffalo Clover. May–June. Stems glabrous. Woods; not common, but scattered in the state.

9. T. resupinatum L. Persian Clover. June–Sept. Native of Europe; rarely adventive in Ill.; Cook Co.

10. T. repens L. White Clover. May–Oct. Native of Europe; adventive in lawns and waste ground; in every co.

11. T. hybridum L. Alsike Clover. June–Sept. Native of Europe; adventive in waste ground; in every co. (Including var. *elegans* [Savi] Boiss.)

32. Medicago

1. Flowers blue or purple .... 1. *M. sativa*
1. Flowers yellow 2
2. Fruits spirally coiled, glabrous or spiny 3
2. Fruits not spirally coiled, pubescent .... 4. *M. lupulina*
3. Fruits glabrous; flowers about 3 mm long .... 2. *M. orbicularis*
3. Fruits spiny; flowers 4–6 mm long .... 3. *M. arabica*

1. M. sativa L. Alfalfa. May–Oct. Native of Asia; commonly cultivated and often escaped; in every co. Light-flowered specimens, the result of a cross between *M. sativa* and *M. falcata,* is sometimes found.

2. M. orbicularis (L.) Bartal. Round Medic. May–July. Native of Europe; rarely adventive in waste ground; Jackson Co.

3. M. arabica (L.) Huds. Spotted Medic. May–Sept. Native of Europe; rarely adventive in Ill.; Jackson Co.

4. M. lupulina L. Black Medic. May–July. Native of Europe; adventive in waste ground; in every co. (Including var. *glandulosa* Neilr.)

33. Baptisia Vent.   Wild Indigo

1. Flowers yellow, 1.3–1.6 cm long .... 1. *B. tinctoria*
1. Flowers blue, white, or cream, 2 cm long or longer 2

2. Flowers blue; stipules longer than the petioles 3
2. Flowers white or cream; stipules shorter than the petioles 4
3. Stipe of legume no longer than the calyx; petioles 5 mm long or longer; keel petal up to 2.5 cm long .... 2. *B. australis*
3. Stipe of legume about twice as long as the calyx; petioles less than 5 mm long; keel petal 2.7–3.0 cm long .... 3. *B. minor*
4. Plants glabrous; stipules deciduous; legumes glabrous, up to 3 cm long .... 4. *B. leucantha*
4. Plants usually pubescent; stipules persistent; legumes pubescent, usually at least 4 cm long .... 5. *B. leucophaea*

1. B. tinctoria (L.) R. Br. var. crebra Fern. Yellow Wild Indigo. June–Sept. Sandy woods, very rare and probably extinct in Ill.; Cook Co. *B. tinctoria* (L.) R. Br—G, J.

2. B. australis (L.) R. Br. Blue Wild Indigo. May–June. Woods, very rare; Kane Co.

3. B. minor Lehm. Blue Wild Indigo. June–July. Native of the sw. U.S.; rarely adventive in Ill.; Cook Co.

4. B. leucantha Torr. & Gray. White Wild Indigo. May–July. Prairies, woods; occasional throughout the state.

5. B. leucophaea Nutt. Cream Wild Indigo. May–June. Stems and leaves pubescent. Moist prairies and open woods; occasional in the n. four-fifths of the state.

var. glabrescens Larisey. May–June. Stems and leaves glabrous. Prairies; rare in w. cent. Ill.

34. *Clitoria* L. Butterfly Pea

1. C. mariana L. Butterfly Pea. June–Aug. Dry woods, not common; confined to the s. one-sixth of the state.

35. *Stylosanthes* Sw. Pencil-flower

1. S. biflora (L.) BSP. Pencil-flower. June–Sept. Dry woods, fields; common in the s. one-fourth of the state, rare or absent elsewhere. (Including var. *hispidissima* [Michx.] Pollard & Ball; *S. riparia* Kearney.)

36. *Lespedeza* Michx.

1. Plants annuals with persistent, brown, glabrous stipules 2
1. Plants perennials with deciduous, greenish, usually pubescent stipules 3
2. Petioles 1–3 (–5) mm long; pubescence of stems retrorse .. 1. *L. striata*
2. Petioles 4 mm long or longer; pubescence of stems antrorse .... 2. *L. stipulacea*
3. Stems lying flat on the ground 4
3. Stems erect or ascending 5
4. Pubescence of stems spreading; flowers usually 6 or more in a cluster .... 3. *L. procumbens*
4. Pubescence of stems appressed or absent; flowers usually up to 6 in a cluster .... 4. *L. repens*
5. Pubescence of stems spreading 6
5. Pubescence of stems appressed 9

6. Peduncles longer than the subtending leaves .... 5. *L. hirta*
6. Peduncles shorter than the subtending leaves 7
7. Flowers creamy-white, with a purple spot at base .... 6. *L. capitata*
7. Flowers purplish 8
8. Fruit 3–5 mm long, longer than the calyx; stipules without any apparent
     veins .... 7. *L. stuevei*
8. Fruit 6 mm long or longer, as long as the calyx or a little longer;
     stipules 3-nerved .... 8. *L.* X *simulata*
9. Flowers 12 mm long or longer; plants to 3 m tall ..... 9. *L. thunbergii*
9. Flowers up to 10 mm long; plants at most about 1 m tall 10
10. Flowers white or yellowish, with a purple spot at base 11
10. Flowers purplish 12
11. Leaflets linear, tapering to the apex; calyx pilose .. 10. *L. leptostachya*
11. Leaflets oblanceolate, truncate or retuse at the apex; calyx sericeous or
     nearly glabrous .... 11. *L. cuneata*
12. Peduncles longer than the subtending leaves .... 12. *L. violacea*
12. Peduncles (or most of them) shorter than the subtending leaves ... 13
13. Leaflets oval to elliptic .... 13. *L. intermedia*
13. Leaflets linear to oblong .... 14. *L. virginica*

1. L. striata (Thunb.) Hook. & Arn. Japanese Lespedeza. Aug.–Oct. Native of Asia; roadsides, fields; common in the s. one-third of the state, rare or absent elsewhere.

2. L. stipulacea Maxim. Korean Lespedeza. Aug.–Oct. Native of Asia; roadsides, fields; common in the s. two-thirds of the state, rare or absent elsewhere.

3. L. procumbens Michx. Trailing Bush Clover. July–Sept. Dry woods; occasional in the s. half of the state, apparently absent elsewhere.

4. L. repens (L.) Bart. Creeping Bush Clover. June–Sept. Dry woods; occasional in the s. one-fifth of the state, absent elsewhere.

5. L. hirta (L.) Hornem. Hairy Bush Clover. Aug.–Sept. Dry woods; occasional in the extreme n. and s. cos., but virtually absent elsewhere.

6. L. capitata Michx. Round-headed Bush Clover. Aug.–Sept. Prairies; occasional to common throughout the state. (Including var. *vulgaris* Torr. & Gray; var. *stenophylla* Bissell & Fern.; f. *argentea* Fern.)

7. L. stuevei Nutt. Bush Clover. Aug.–Sept. Dry woods; occasional in the s. half of the state, rare or absent elsewhere.

8. L. X simulata Mack. & Bush. Bush Clover. July–Sept. Prairie, very rare; Crawford Co. Reputed to be a hybrid between *L. capitata* and *L. virginica*.

9. L. thunbergii (DC.) Nakai. Tall Bush Clover. Aug.–Oct. Native of Asia; rarely adventive in Ill.; Coles, Jackson, and Wayne cos.

10. L. leptostachya Engelm. Prairie Bush Clover. Aug.–Sept. Prairies, very rare; Cook, McHenry, and Winnebago cos.

11. L. cuneata (Dum.-Cours.) G. Don. Sericea Lespedeza. Sept.–Oct. Frequently planted in Ill. and often escaped in the s. one-half of the state.

12. L. violacea (L.) Pers. Violet Bush Clover. July–Sept. Dry woods; occasional in the s. half of the state, becoming less common northward.

13. L. intermedia (S. Wats.) Britt. Bush Clover. Aug.–Sept. Dry woods; occasional in the s. half of the state, absent elsewhere.

14. L. virginica (L.) Britt. Slender Bush Clover. Aug.–Sept. Dry woods; occasional throughout the state, but more common in the s. cos.

37. Hosackia Dougl.

1. H. americana (Nutt.) Piper. Deer Vetch. June–Aug. Native of the w. U.S., rarely adventive along railroads; Cook and Greene cos. *Lotus americanus* (Nutt.) Bisch.—F.; *Lotus purshianus* Clem. & Clem.—G.

38. Phaseolus L.   Kidney Bean

1. P. polystachios (L.) BSP. Wild Kidney Bean. July–Sept. Dry woods; occasional in the s. one-third of the state, absent elsewhere.

39. Dioclea

1. D. multiflora (Torr. & Gray) C. Mohr. Aug.–Sept. Moist thickets, very rare; Massac Co.

40. Desmodium Desv.   Tick Trefoil

1. Flowers borne on a long, leafless peduncle .... 1. *D. nudiflorum*
1. Flowers borne on leafy stems 2
2. Leaves clustered in a whorl midway on the flowering stem.............
.... 2. *D. glutinosum*
2. Leaves scattered all along the flowering stem 3
3. Flowers white; stipe of fruit at least twice as long as the calyx ....
.... 3. *D. pauciflorum*
3. Flowers pink or purple; stipe of fruit not longer than the calyx 4
4. Petioles up to 3 mm long, or absent .... 4. *D. sessilifolium*
4. Petioles 4 mm long or longer 5
5. Stipules persistent, conspicuous 6
5. Stipules deciduous, inconspicuous 9
6. Plants lying on the ground; leaflets mostly suborbircular ....
.... 5. *D. rotundifolium*
6. Plants erect or ascending; leaflets variously shaped 7
7. Pubescence on lower surface of leaflets at least of some hooked hairs 8
7. Pubescence on lower surface of leaflets not of hooked hairs, or lower surface glabrous .... 8. *D. cuspidatum*
8. Leaflets obscurely reticulate beneath; joints of fruit angular along the lower margin; rachis of inflorescence with spreading hairs ....
.... 7. *D. canescens*
8. Leaflets conspicuously reticulate beneath; joints of fruit rounded along both margins; rachis of inflorescence with some hooked hairs .... 6. *D. illinoense*
9. Leaflets glaucous beneath .... 9. *D. laevigatum*
9. Leaflets not glaucous beneath 10
10. Leaflets to 3 cm long, to 2 cm broad 11
10. Leaflets, or most of them, longer than 3 cm, broader than 2 cm 12
11. Stems and upper surface of leaflets glabrous or nearly so; pedicels (8–) 10–20 mm long .... 10. *D. marilandicum*
11. Stems and upper surface of leaflets pilose; pedicels 4–8 (–9) mm long .... 11. *D. ciliare*

12. Fruit with 1–2 segments; calyx 2–3 mm long .... 12. *D. rigidum*
12. Fruit with 3 or more segments; calyx 3 mm long or longer 13
13. Most of the flowers at least 9 mm long; primary bracts 4–10 mm long ..
.... 13. *D. canadense*
13. Flowers up to 8 mm long; primary bracts 2–3 mm long 14
14. Leaflets velvety on the lower surface; stipules ovate-deltoid, at most
about 2½ times longer than broad .... 14. *D. nuttallii*
14. Leaflets variously pubescent to nearly glabrous on the lower surface, but
not velvety; stipules filiform or subulate, at least 7 times longer than
broad 15
15. Leaflets rather thick, conspicuously reticulate; terminal leaflet less than
three times longer than broad .... 15. *D. dillenii*
15. Leaflets rather thin, faintly reticulate; terminal leaflet at least three times
longer than broad .... 16. *D. paniculatum*

1. D. nudiflorum (L.) DC. Bare-stemmed Tick Trefoil. July–Aug.
Woods; occasional throughout the state.

2. D. glutinosum (Muhl.) Wood. Pointed Tick Trefoil. June–Aug.
Usually rich woods; occasional throughout the state.

3. D. pauciflorum (Nutt.) DC. White-flowered Tick Trefoil. July–Sept.
Woods; occasional in the s. one-fourth of the stage, absent elsewhere.

4. D. sessilifolium (Torr.) Torr. & Gray. Sessile-leaved Tick Trefoil.
July–Sept. Open woods, sand prairies; occasional throughout the
state, except for the nw. cos.

5. D. rotundifolium DC. Round-leaved Tick Trefoil. Aug.–Sept. Dry
woods, not common; mostly in the s. half of the state.

6. D. illinoense Gray. Illinois Tick Trefoil. July–Aug. Prairies, roadsides; occasional throughout the state.

7. D. canescens (L.) DC. Hoary Tick Trefoil. July–Aug. Dry woods;
occasional in the s. half of the state, less common elsewhere.

8. D. cuspidatum (Muhl.) Loud. Tick Trefoil. July–Sept. Stems, leaflets, and bracts glabrous or nearly so. Dry woods; occasional
throughout the state, except for the extreme n. cos.

var. longifolium (Torr. & Gray) Schub. Tick Trefoil. Stems, leaflets,
and bracts pubescent. July–Sept. Dry woods; occasional throughout
the state. *D. longifolium* (Torr. & Gray) Smythe—J.

9. D. laevigatum (Nutt.) DC. Glaucous Tick Trefoil. July–Sept. Dry
woods; not common; apparently confined to the s. three-fourths of
the state.

10. D. marialandicum (L.) DC. Small-leaved Tick Trefoil. July–Sept.
Dry woods; not common; confined to the s. two-thirds of the state.

11. D. ciliare (Muhl.) DC. Hairy Tick Trefoil. Aug.–Oct. Dry woods; occasional in the s. half of the state.

12. D. rigidum (Ell.) DC. Stiff Tick Trefoil. July–Sept. Dry woods; not
common; restricted to the s. two-thirds of the state.

13. D. canadense (L.) DC. Showy Tick Trefoil. July–Sept. Prairies; occasional throughout the state.

14. D. nuttallii (Schindl.) Schub. Nuttall's Tick Trefoil. Aug.–Sept. Dry
woods; not common; confined to the s. half of the state.

15. D. dillenii Darl. Tick Trefoil. July–Sept. Dry woods; occasional
throughout the state. *D. glabellum* (Michx.) DC.—J; *D. perplexum*
Schub.—G.

16. D. paniculatum (L.) DC. Panicled Tick Trefoil. July–Sept. Dry
woods; occasional throughout the state.

41. Glycine L. Soybean

1. G. max (L.) Merr. Soybean. June–Aug. Native of Asia; frequently cultivated and occasionally escaped throughout the state.

42. Vigna Savi

1. V. sinensis (L.) Endl. Cow Pea. July–Sept. Native of Asia; occasionally cultivated in Ill.; rarely escaped from cultivation; Mason Co.

43. Pueraria DC.

1. P. lobata (Willd.) Ohwi. Kudzu-vine. Aug.–Sept. Native of Japan; occasionally cultivated and sometimes escaping in the s. cos.

44. Strophostyles Ell. Wild Bean

1. Legume up to 3.5 cm long, with spreading pubescence; leaflets silky-pubescent on both surfaces; calyx tube densely pubescent, up to 1.5 mm long .... 1. *S. leiosperma*
1. Legume 4 cm long or longer, with appressed pubescence; leaflets glabrous or sparsely pubescent, at least on the upper surface; calyx tube glabrous or sparsely pubescent, 1.5–2.5 mm long 2
2. Bracteoles as long as the calyx tube; larger leaflets up to 6 cm broad; seeds 6–12 mm long .... 2. *S. helvola*
2. Bracteoles about half as long as the calyx tube; larger leaflets up to 2 cm broad; seeds 3–6 mm long .... 3. *S. umbellata*

1. S. leiosperma (Torr. & Gray) Piper. Wild Bean. July–Sept. Prairies, dry woods, fields, roadsides; occasional in the s. half of the state, becoming less common northward.

2. S. helvola (L.) Ell. Wild Bean. June–Oct. Leaflets, or some of them, lobed. Rocky woods, sand bars, roadsides, fields; occasional throughout the state.

var. missouriensis (S. Wats.) Britt. Wild Bean. June–Oct. Leaflets unlobed. Dry woods; occasional in cos. along the Mississippi River.

3. S. umbellata (Muhl.) Britt. Wild Bean. July–Oct. Dry woods, streambanks; occasional in the s. two-fifths of the state, absent elsewhere.

45. Galactia P. Br. Milk Pea

1. G. volubilis (L.) Britt. var. mississippiensis Vail. Milk Pea. July–Sept. Rocky woods, not common; confined to the s. one-sixth of the state.*G. volubilis* (L.) Britt.—J.

46. Amphicarpa Ell. Hog Peanut

1. A. bracteata (L.) Fern. Hog Peanut. Aug.–Sept. Stem with ascending white hairs; sides of legume glabrous. Woods and thickets; occasional throughout the state.

var. comosa (L.) Fern. Aug.–Sept. Stem with appressed tawny hairs; sides of legume pubescent. Woods and thickets; occasional throughout the state. *A. comosa* (L.) G. Don—J.

81. LINACEAE

1. *Linum* L. *Flax*

1. Flowers blue 2
1. Flowers yellow 3
2. Sepals acuminate; annuals .... 1. *L. usitatissimum*
2. Sepals mostly rounded; perennials .... 2. *L. perenne*
3. Petals 8–10 mm long; outer sepals glandular along the margin; leaves
  with a pair of stipular glands at the base .... 3. *L. sulcatum*
3. Petals up to 8 mm long; outer sepals eglandular (inner sepals glandular
  along the margin in *L. medium*); leaves without a pair of stipular glands
  at the base 4
4. Leaves mostly alternate; outer sepals lanceolate, stiff 5
4. Leaves mostly opposite; outer sepals elliptic, membranaceous ....
  .... 6. *L. striatum*
5. Inner sepals with conspicuous glandular cilia; leaves stiff ....
  .... 4. *L. medium*
5. Inner sepals eglandular or nearly so; leaves rather thin 5. *L. virginianum*

    1. *L. usitatissimum* L. Common Flax. May–Aug. Native of Europe;
adventive in waste areas; occasional throughout the state.
    2. *L. perenne* L. Flax. May–Aug. Native of Europe; rarely escaped
from cultivation; Coles Grundy, and Winnebago cos.
    3. *L. sulcatum* Riddell. Wild Flax. May–Sept. Dry soil, hill prairies;
occasional throughout the state.
    4. *L. medium* (Planch.) Britt. var. *texanum* (Planch.) Fern. Wild Flax.
May–Aug. Mostly dry soil; occasional in the s. half of the state, rare in
the n. half where it occurs on calcareous pond shores. *L. medium*
(Planch.)—J.
    5. *L. virginianum* L. Wild Flax. June–Aug. Dry woods; occasional in
the s. half of the state, rare in the n. half.
    6. *L. striatum* Walt. Wild Flax. June–Sept. Moist soil, sometimes in
woods; occasional in the s. one-fourth of the state.

82. OXALIDACEAE

1. *Oxalis* L. *Wood Sorrel*

1. Leaves all from the base of the plants; flowers purple to lavender (rarely
  white); plants with bulbs .... 1. *O. violacea*
1. Leaves borne along the stem; flowers yellow; plants without bulbs ... 2
2. Plants creeping, rooting at the nodes .... 2. *O. corniculata*
2. Stems erect or ascending, not rooting at the nodes 3
3. Flowers up to 12 mm long; seeds 1.0–1.6 mm long 4
3. Flowers 12–18 mm long; seeds 1.8–2.2 mm long .... 5. *O. grandis*
4. Pedicels of fruits deflexed; pedicels of flowers all arising from same
  point at tip of main axis; stipules oblong .... 3. *O. dillenii*
4. Pedicels of fruits erect or ascending; pedicels of flowers arising both
  from the tip of the main axis as well as along the branches of the inflorescence; stipules linear .... 4. *O. stricta*

1. O. violacea L. Purple Oxalis. April–June. Woods, bluffs, prairies; in every co.

2. O. corniculata L. Creeping Wood Sorrel. April–Oct. Native of the Tropics; in waste areas, often in greenhouses; occasional in the state.

3. O. dillenii Jacq. Yellow Wood Sorrel. May–Nov. Fields, prairies, roadsides; common in every co. *O. stricta* L.—F, G, J.

4. O. stricta L. Yellow Wood Sorrel. May–Oct. Woods, fields, road-sides; in every co. *O. europaea* Jord.—F, G; *O. cymosa* Small—J.

5. O. grandis Small. Large Wood Sorrel. June–Sept. Rich wood-lands, very rare; Pope and Wabash cos.

83. GERANIACEAE

1. Leaves pinnatifid or pinnately compound .... 1. *Erodium*
1. Leaves palmately cleft or divided .... 2. *Geranium*

1. Erodium L'Hér.   Storksbill

1. E. cicutarium (L.) L'Hér. Pin Clover. April–Oct. Native of Europe; naturalized in waste areas; not common.

2. Geranium L.   Cranesbill

1. Petals more than 1 cm long .... 1. *G. maculatum*
1. Petals less than 1 cm long 2
2. Leaves palmately compound .... 2. *G. robertianum*
2. Leaves palmately lobed, often deeply so 3
3. Sepals awnless; stamens 5; seeds without reticulations or pits ....
.... 3. *G. pusillum*
3. Sepals awned or sharp-tipped; stamens 10; seeds reticulate or pitted 4
4. Peduncles 1-flowered ....4. *G. sibiricum*
4. Peduncles 2-flowered 5
5. Flowers purple; most or all the leaves less than 3.5 cm broad ....
.... 5. *G. dissectum*
5. Flowers rose to pale pink to nearly white; leaves 3.5 cm broad or broader 6
6. Beak of mature style column 3–5 mm long; ultimate lobes of leaves acute .... 6. *G. bicknellii*
6. Beak of mature style column 1–2 mm long; ultimate lobes of leaves ob-tuse .... 7. *G. carolinianum*

1. G. maculatum L. Wild Geranium. April–June. Rich woods, com-mon; in every co.

2. G. robertianum L. Herb Robert. July. Woods, fields, rare; Cook, DuPage, and Wabash cos.

3. G. pusillum Burm. f. Small Cranesbill. June–July. Native of Europe; adventive in waste areas, rare; Cook, Kankakee, Peoria, and Piatt cos.

4. G. sibiricum L. Siberian Cranesbill. Aug.–Sept. Native of Europe and Asia; naturalized into waste ground, rare; Champaign, Ogle, and Winnebago cos.

5. G. dissectum L. June. Native of Europe; rarely escaped into edge of prairie; Lake Co.

6. G. bicknellii Britt. Northern Cranesbill. June–Aug. Sandy woods, fields, rare; Cook and Lake cos.

7. G. carolinianum L. Wild Cranesbill. May–Aug. Woods, fields, roadsides; common in the s. half of the state, occasional in the n. half. (Including var. *confertiflorum* Fern.)

84. ZYGOPHYLLACEAE

1. Fruits with 2–4 sharp prickles .... 1. *Tribulus*
1. Fruits without prickles .... 2. *Kallstroemia*

1. Tribulus L. Caltrop

1. T. terrestris L. Puncture-weed. June–Sept. Native of Europe; occasional throughout the state.

2. Kallstroemia Scop.

1. K. intermedia Rydb. June–Sept. Native of the w. U.S.; rarely adventive in Ill.; Cook and St. Clair cos.

85. RUTACEAE

1. Leaflets 5–9; flowers sessile in axillary clusters; sepals absent; fruit an ellipsoid follicle .... 1. *Xanthoxylum*
1. Leaflets 3; flowers in terminal cymes; sepals 3–5; fruit an orbicular samara .... 2. *Ptelea*

1. Xanthoxylum Gmel. Prickly Ash

1. X. americanum Mill. Prickly Ash. April–May. Woodlands; occasional in the n. half of the state, rare in the s. half. *Zanthoxylum americanum* Mill.—G, J.

2. Ptelea L. Hop-tree

1. P. trifoliata L. Wafer Ash. May–July. Branchlets glabrous. Open woods, occasional throughout the state, although less common in the s. cos.

var. mollis Torr. & Gray. May–July. Branchlets densely pubescent. Sand dunes, rare; Cook and Lake cos. *P. trifoliata* L.—J; var. *deamii* Nieuwl.—G.

86. SIMAROUBACEAE

1. *Ailanthus Desf.*   Tree-of-heaven

1. A. altissima (Mill.) Swingle. Tree-of-heaven. June–July. Native of Asia; naturalized in woods fields, and cities; occasional throughout most of Ill.

87. POLYGALACEAE

1. *Polygala L.*   Milkwort

1. Flowers 1.5 cm long or longer .... 1. *P. paucifolia*
1. Flowers up to 1 cm long 2
2. At least the lowermost leaves in whorls 3
2. All leaves alternate 4
3. Leaves mostly obtuse; spikes at least 1 cm thick .... 2. *P. cruciata*
3. Leaves mostly acute; spikes up to 7 mm thick .... 3. *P. verticillata*
4. Petals conspicuously united into a cleft tube; stems glaucous ....
.... 4. *P. incarnata*
4. Petals not united into a cleft tube; stems not glaucous 5
5. Flowers in racemes, the pedicles at least 2 mm long ... 5. *P. polygama*
5. Flowers in spikes, the pedicels absent or up to 1 mm long 6
6. Leaves serrulate; wings 3.0–3.7 mm long; plants with several stems from a thick perennial crown .... 6. *P. senega*
6. Leaves entire or nearly so; wings 4.5 mm long or longer; annuals with solitary stem .... 7. *P. sanguinea*

1. P. paucifolia Willd. Flowering Wintergreen. May–June. Moist woods, very rare; Cook Co.

2. P. cruciata L. var. aquilonia Fern. Cross Milkwort. July–Sept. Acid, sandy soils, not common; mostly in the ne. one-fourth of Ill.; also Clark and Menard cos. *P. cruciata* L.—G, J.

3. P. verticillata L. Whorled Milkwort. July–Sept. All or nearly all leaves (except near inflorescence) whorled; racemes lax; pedicels at least 0.5 mm long. Dry, often sandy soil, not common; mostly in the s. half of the state.

var. isocycla Fern. Whorled Milkwort. July–Sept. All or nearly all leaves (except near the inflorescence) whorled; racemes dense; pedicels at most 0.3 mm long. Dry, acid soil, occasional; throughout most of the state. *P. verticillata* L.—J. (Including var. *sphenostachya* Pennell.)

var. ambigua (Nutt.) Wood. Whorled Milkwort. June–Sept. Only the lowest 1–3 nodes with whorled leaves. Fields, dry woods, occasional; apparently confined to the s. half of Ill. *P. ambigua* Nutt.—G, J.

4. P. incarnata L. Pink Milkwort. July–Sept. Prairies and gravel hills, not common; in the n. half of the state; also Pope Co.

5. P. polygama Walt. var. obtusata Chod. Purple Milkwort. June–Aug. Sandy waste ground and open woods, occasional; confined to the n. half of the state. *P. polygama* Walt.—J.

6. P. senega L. Seneca Snakeroot. May–Sept. Prairies and dry

woods; occasional in the n. three-fourths of Ill., apparently absent in the s. one-fourth. (Including var. *latifolia* Torr. & Gray.)

7. P. sanguinea L. Field Milkwort. July–Sept. Fields, woods, prairies, occasional; throughout the state. (Including f. *viridescens* [L.] Farw. and f. *albiflora* [Wheelock] Millspaugh.)

88. EUPHORBIACEAE

1. Leaves palmately 5- 11-lobed; plants 3 m or more tall .... 1. *Ricinus*
1. Leaves unlobed (if 3- to 5-lobed, then the plants pubescent); plants less than 1 m tall 2
2. Lower cauline leaves alternate, the upper opposite (*Euphorbia corollata,* with five white petal-like structures per flower, may be sought here) ....
.... 8. *Poinsettia*
2. Cauline leaves either all opposite or all alternate, or sometimes whorled
.... 3
3. Leaves alternate along the stem 4
3. Leaves opposite along the stem .... 9. *Chamaesyce*
4. Uppermost leaves subtending the inflorescence whorled; latex present
.... 7. *Euphorbia*
4. None of the leaves at the tip of the stem whorled; latex absent 5
5. Leaves sessile or nearly so; plants glabrous .... 2. *Phyllanthus*
5. Leaves petiolate; plants variously pubescent 6
6. Leaves entire 7
6. Leaves variously toothed 8
7. Some of the petioles at least half as long as the blades; capsule 2- to 3-celled (rarely 1-celled by abortion) .... 3. *Croton*
7. Petioles much less than half as long as the blades; capsule 1-celled ...
.... 4. *Crotonopsis*
8. Plants twining; leaves cordate .... 5. *Tragia*
8. Plants erect; leaves not cordate (except in *Acalypha ostryaefolia,* a plant with a 3-celled ovary) 9
9. Plants with stellate pubescence; leaves with 2 glands at base of blade; pistillate flowers not subtended by cleft bracts .... 3. *Croton*
9. Plants pubescent, but not stellate; leaves without 2 glands at base of blade; pistillate flowers subtended by cleft bracts .... 6. *Acalypha*

1. Ricinus L.   Castor Bean

1. R. communis L. Castor Bean. June–Sept. Native of the Tropics; occasionally escaped from cultivation.

2. Phyllanthus L.

1. P. caroliniensis Walt. June–Oct. Mostly sandy soil; occasional or rare in Ill.

3. Croton L.   Croton

1. Leaves toothed; calyx of staminate flowers 4-parted; rudimentary petals in pistillate flower 5 .... 1. *C. glandulosus*

1. Leaves entire; calyx of staminate flowers 5-parted; petals in pistillate
   flowers absent 2
2. Calyx of pistillate flowers 7- to 12-parted; some or all of the leaves cordate at the base .... 2. *C. capitatus*
2. Calyx of pistillate flowers 5-parted; leaves tapering, rounded, or barely
   subcordate at the base 3
3. Leaves ovate to oblong; stamens 3–8; plants monoecious; seeds pitted;
   capsule 7–9 mm in diameter .... 3. *C. monanthogynus*
3. Leaves linear to linear-oblong; stamens (8–) 10 (–12); plants dioecious;
   seeds reticulate; capsule 4–6 mm in diameter .... 4. *C. texensis*

   1. C. glandulosus L. var. septentrionalis Muell.-Arg. Sand Croton.
July–Oct. Sandy, often disturbed, soil; common in the s. cos., occasional in the n. cos., where it probably is adventive. *C. glandulosus*
L.—J.
   2. C. capitatus Michx. Capitate Croton. July–Sept. Sandy soil, limestone glades; occasional in the s. cos., less common in the cos.,
where it is probably adventive.
   3. C. monanthogynus Michx. Croton. July–Oct. Dry fields, bluffs;
occasional in the s. half of the state; much rarer in the n. half of the
state.
   4. C. texensis (Klotzsch) Muell.-Arg. Texas Croton. June–Sept. Dry
soil, very rare; Menard co.

4. *Crotonopsis* Michx. Rushfoil

1. Fruit spiny at tip; leaves linear-lanceolate .... 1. *C. linearis*
1. Fruit without spines; leaves elliptic to lanceolate .... 2. *C. elliptica*

   1. C. linearis Michx. Rushfoil. July–Sept. Sandy soil, rare; Cass,
LaSalle, Mason, Menard, Mercer, Scott, and Whiteside cos.
   2. C. elliptica Willd. Rushfoil. July–Sept. Dry fields, dry woods,
bluffs; occasional in the s. one-third of the state.

5. *Tragia* L.

   1. T. cordata Michx. July–Sept. Dry woods, bluffs, rare; Hardin,
Johnson, and Pope cos.

6. *Acalypha* L. Three-seeded Mercury

1. Leaves cordate; fruit soft-echinate .... 1. *A. ostryaefolia*
1. Leaves not cordate; fruit not soft-echinate 2
2. Some or all the petioles at least ⅓ as long as the blades 3
2. Petioles at most only ¼ as long as the blades .... 5. *A. gracilens*
3. Bracts subtending pistillate flowers 5- to 9-lobed, often glandular; stems
   glabrous or with incurved hairs 4
3. Bracts subtending pistillate flowers 9- to 15-lobed, usually glandless;
   stems with spreading hairs .... 4. *A. virginica*
4. Capsule 3-seeded; seeds 1.2–2.0 mm long .... 2. *A. rhomboidea*
4. Capsule 2-seeded; seeds 2.2–3.2 mm long .... 3. *A. deamii*

   1. A. ostryaefolia Riddell. Three-seeded Mercury. July–Oct. River
banks, fields, bluffs, roadsides; occasional in the s. half of the state.

2. A. rhomboidea Raf. Three-seeded Mercury. July–Oct. Woods, fields, bluffs, roadsides; in every co.

3. A. deamii (Weatherby) Ahles. Large-seeded Mercury. July–Oct. Wooded river bottoms, rare; Coles and Vermilion cos. *A. rhomboidea* Raf. var. *deamii* Weatherby—F, G.

4. A. virginica L. Three-seeded Mercury. July–Oct. Fields, roadsides, bluffs, woods; occasional to common in the s. three-fourths of the state, rare elsewhere.

5. A. gracilens Gray. Slender Three-seeded Mercury. June–Sept. Leaves oblong to oblong-lanceolate, the petioles more than $^1/_{10}$ as long as the blades; capsule 3-seeded. Woods, fields, roadsides; occasional in the s. half of the state, rare and possibly adventive elsewhere; Grundy and Will cos.

ssp. monococca (Engelm.) Webster. Slender Three-seeded Mercury. June–Sept. Leaves linear to linear-lanceolate, the petioles usually not more than $^1/_{10}$ as long as the blades; capsule 1-seeded. Fields and woods, not common; scattered in the s. two-thirds of the state. *A. gracilens* Gray—J; *A. gracilens* Gray var. *monococca* Engelm—F, G.

7. *Euphorbia* L. Spurge

1. E. marginata Pursh. Snow-on-the-mountain. July–Oct. Native of the w. U.S.; naturalized in waste areas; occasional throughout the state.

2. E. corollata L. Flowering Spurge. May–Oct. Stem and lower surface of the leaves glabrous or nearly so. Prairies, woods, fields, roadsides, common; in every co. (Including var. *paniculata* [Ell.] Boiss.)

var. mollis Millsp. May–Oct. Stem and lower surface of leaves softly hairy. Prairies, woods, fields; occasional in the s. tip of the state.

3. E. cyparissias L. Cypress Spurge. May–Sept. Native of Europe; naturalized along roadsides and in cemeteries; occasional throughout the state.

4. E. esula L. Leafy Spurge. June–Sept. Native of Europe; naturalized in fields and pastures; occasional in the n. half of the state.

5. E. commutata Engelm. Wood Spurge. May–June. Wooded slopes, along streams, and in gravelly soils, rare; mostly in the n. half of the state.

6. E. peplus L. Petty Spurge. June–Oct. Native of Europe; rarely adventive in Ill.; Menard Co.

7. E. helioscopia L. Wart Spurge. June–Oct. Native of Europe; adventive in waste areas; Cass and Menard cos.

8. E. obtusata Pursh. Blunt-leaved Spurge. May–July. Rich woods, wooded slopes; occasional throughout the state.

9. E. spathulata Lam. Spurge. May–July. Limestone ledge at edge of hill prairie, rare; Monroe Co. *E. dictyosperma* Fisch. & Mey.—F, G.

8. *Poinsettia* Graham   Poinsettia

1. Floral bracts usually basally red; seeds not angular, sharply tuberculate; leaves glossy green .... 1. *P. cyathophora*
1. Floral bracts green or basally pale, never red; seeds angular, bluntly tuberculate; leaves dull green .... 2. *P. dentata*

1. P. cyanthophora (Murr.) Kl. & Garcke. Wild Poinsettia. July–Oct. Leaves oval to lanceolate or even pandurate. Fields, roadsides; occasional throughout the state. *Euphorbia heterophylla* L.—F, G; *P. heterophylla* (L.) Kl. & Garcke—J.

var. graminifolia (Michx.) Mohlenbr. July–Oct. Leaves linear to narrowly lanceolate. Moist soil along stream; Pope and Union cos. *Euphorbia heterophylla* var. *graminifolia* (Michx.) Engelm.—F.

2. P. dentata (Michx.) Kl. & Garcke. Wild Poinsettia. June–Sept. Leaves narrowly ovate to rhombic. Fields, roadsides, prairies; occasional throughout the state. *Euphorbia dentata* Michx.—F, G.

var. cuphosperma (Engelm.) Mohlenbr. June–Sept. Leaves linear to narrowly lanceolate. Fields and roadsides; Cook and Hancock cos. *Euphorbia dentata* f. *cuphosperma* (Engelm.) Fern.—F.

9. *Chamaesyce* S. F. Gray

1. Leaves entire 2
1. Leaves toothed, at least near the apex 4
2. Leaves about as broad as long; style 0.2 mm long; seeds about 1 mm long .... 1. *C. serpens*
2. Leaves considerably longer than broad; style 0.7–1.0 mm long; seeds 1.3–2.6 mm long. 3
3. Seeds 2.0–2.6 mm long; capsules 3.0–3.5 mm long . 2. *C. polygonifolia*
3. Seeds 1.3–1.6 mm long; capsules 2.0–2.5 mm long .... 3. *C. geyeri*
4. Ovary and capsule strigose 5
4. Ovary and capsule glabrous 6
5. Seeds with cross-ridges; stems mostly reddish-brown; style 0.3–0.5 mm long, cleft less than halfway to the base .... 4. *C. supina*
5. Seeds smooth or minutely granular; stems mostly green; style 0.7 mm long, cleft halfway to the base .... 5. *C. humistrata*

6. Stems pubescent, at least near the tip; leaves toothed from tip to base   7
6. Stems glabrous; leaves toothed only at apex and occasionally at base   8
7. Stems prostrate; capsules up to 1.9 mm long .... 6. *C. vermiculata*
7. Stems ascending to erect; capsules 1.9 mm long or longer ....
.... 7. *C. maculata*
8. Leaves broadly oblong to ovate; seeds pitted and with short cross-ridges
.... 8. *C. serpyllifolia*
8. Leaves linear-oblong; seeds with 3–6 long cross-ridges ....
.... 9. *C. glyptosperma*

1. C. serpens (HBK.) Small. Round-leaved Spurge. July–Sept. Moist, sandy soil; scattered but not common. *Euphorbia serpens* HBK.—F, G.

2. C. polygonifolia (L.) Small. Seaside Spurge. July–Sept. Dunes and beaches, rare; Cook, Fulton, and Lake cos. *Euphorbia polygonifolia* L.—F, G.

3. C. geyeri (Engelm.) Small. Spurge. June–Sept. Sandy soil; occasional or rare in the w. half of the state. *Euphorbia geyeri* Engelm.—F, G.

4. C. supina (Raf.) Moldenke. Milk Spurge. July–Oct. Disturbed or cultivated habitats; common throughout the state. *Euphorbia supina* Raf.—F, G.

5. C. humistrata (Engelm.) Small. Milk Spurge. July–Sept. Sandy soil, river banks; occasional throughout the state. *Euphorbia humistrata* Engelm.—F, G.

6. C. vermiculata (Raf.) House. Spurge. July–Sept. Rich soil, very rare; Lake Co. *Euphorbia vermiculata* Raf.—F, G.

7. C. maculata (L.) Small. Nodding Spurge. July–Sept. Disturbed areas; common throughout the state. *Euphorbia maculata* L.—F, G.

8. C. serpyllifolia (Pers.) Small. Spurge. July–Oct. Sandy areas, rare; Cook Co. *Euphorbia serpyllifolia* Pers.—F, G.

9. C. glyptosperma (Engelm.) Small. Spurge. June–Sept. Sandy or gravelly soil; occasional in the n. half of the state; also Monroe Co. *Euphorbia glyptosperma* Engelm.—F, G.

89. CALLITRICHACEAE

1. *Callitriche* L. Water Starwort

1. Plants growing in water; flowers each with 2 bracts at the base 2
1. Plants growing on soil; flowers bractless .... 3. *C. terrestris*
2. Fruit as broad as long, rounded at the base .... 1.*C. heterophylla*
2. Fruit longer than broad, narrowed to the base .... 2. *C. palustris*

1. C. heterophylla Pursh. Large Water Starwort. April–Oct. Shallow water, occasional; scattered throughout the state.

2. C. palustris L. Common Water Starwort. April–Oct. Shallow water, not common; scattered throughout the state. *C. verna* L.—G.

3. C. terrestris Raf. Terrestrial Starwort. April–July. Moist soil; occasional in the s. two-thirds of the state, absent elsewhere. *C. deflexa* A. Br. var. *austini* (Engelm.) Hegelm.—F.

90. LIMNANTHACEAE

1. Floerkea Willd.   False Mermaid

1. F. proserpinacoides Willd. False Mermaid. April–June. Rich woodlands; occasional in the n. half of the state, extending southward to Crawford co.

91. ANACARDIACEAE

1. Leaflets 5 or more 2
1. Leaflets 3 3
2. Plants poisonous to touch; drupes whitish; panicles axillary ....
.... 1. *Toxicodendron*
2. Plants not poisonous; drupes reddish; panicles terminal ...... 2. *Rhus*
3. Flowers appearing before or with the leaves in short spikes; drupes red, pubescent; plants not poisonous .... 2. *Rhus*
3. Flowers appearing after the leaves in axillary panicles; drupes white, glabrous or nearly so; plants poisonous to touch ... 1. *Toxicodendron*

1. Toxicodendron Mill

1. Leaflets 7 or more .... 1. *T. vernix*
1. Leaflets 3 .... 2. *T. radicans*

    1. T. vernix (L.) Kuntze. Poison Sumac. May–July. Bogs and marshes; occasional to rare in the ne. one-fourth of the state; also Coles and Woodford cos. *Rhus vernix* L.—F, G, J.
    2. T. radicans (L.) Kuntze. Poison Ivy. May–July. Fields, woods, bluffs, waste areas, common; in every co. *R. radicans* L.—F, G, J. (Including many varieties and forms.)

2. Rhus L.   Sumac

1. Leaflets 5 or more 2
1. Leaflets 3 .... 4. *R. aromatica*
2. Rachis winged; leaves entire or nearly so .... 1. *R. copallina*
2. Rachis unwinged; leaves sharply serrate 3
3. Branches and petioles velvety-pubescent .... 2. *R. typhina*
3. Branches and petioles glabrous .... 3. *R. glabra*

    1. R. copallina L. Dwarf Sumac. July–Sept. Woodlands, fields; common in the s. one-fourth of the state, occasional or rare elsewhere. (Including var. *latifolia* Engl.)
    2. R. typhina L. Staghorn Sumac. June–July. Woods; occasional in the n. half of the state, rare and apparently adventive in the s. half.
    3. R. glabra L. Smooth Sumac. June–July. Woods, fields, roadsides, common; in every co.
    4. R. aromatica Ait. Fragrant Sumac. March–May. Flowers appearing before the leaves. Woods, bluffs, and dunes; occasional throughout the state. (Including var. *illinoensis* [Greene] Rehder.)

var. serotina (Greene) Rehder. Fragrant Sumac. April–May. Flowers appearing when the leaves are partly grown; upright shrubs; leaflets pubescent or nearly glabrous beneath, not velvety. Sandy woods, rare; Union Co.

var. arenaria (Greene) Fern. Fragrant Sumac. April–May. Flowers appearing when the leaves are partly grown; low, sprawling shrub; leaflets velvety-pubescent beneath. Sandy areas, including dunes; apparently restricted to the n. half of the state. *R. arenaria* (Greene) G. N. Jones—J; *R. trilobata* Nutt. var. *arenaria* (Greene) Barkley—G.

92. AQUIFOLIACEAE

1. Ilex L. Holly

1. Leaves coriaceous, evergreen .... 1. *I. opaca*
1. Leaves not coriaceous, deciduous 2
2. Teeth of calyx acute, glabrous .... 2. *I. decidua*
2. Teeth of calyx obtuse, ciliate .... 3. *I. verticillata*

1. I. opaca Ait. American Holly. May–June. Rocky, wooded slope, very rare; Union Co.

2. I. decidua Walt. Swamp Holly. April–May. Swamps, wooded slopes, and bluffs; occasional in the s. counties, becoming rare northward; absent in the upper two-fifths of Ill.

3. I. verticillata (L.) Gray. Winterberry. May–July. Swamps, edge of streams, wooded slopes, and bluffs; occasional in the n. cos., rare in the s. cos.

93. CELASTRACEAE

1. Leaves opposite .... 1. *Euonymus*
1. Leaves alternate .... 2. *Celastrus*

1. Euonymus L.   Spindle-tree

1. Plants trailing or climbing, rooting at the nodes 2
1. Plants erect or ascending, not rooting at the nodes 3
2. Leaves membranaceous, obovate to oblong; sepals and petals each 5; capsule warty; plants trailing .... 1. *E. obovatus*
2. Leaves coriaceous, elliptic; sepals and petals each 4; capsule smooth; plants climbing .... 2. *E. fortunei*
3. Stems unwinged 4
3. Stems winged .... 6. *E. alatus*
4. Leaves petiolate; sepals and petals each 4; capsule smooth; shrubs to 7 meters tall 5
4. Leaves sessile; sepals and petals each 5; capsule warty; shrub to 2.5 meters tall .... 5. *E. americanus*
5. Leaves pubescent on the lower surface; flowers purple ....
.... 3. *E. atropurpureus*

5. Leaves glabrous on the lower surface; flowers yellowish-green . . . .
. . . . 4. *E. europaeus*

    1. E. obovatus Nutt. Running Strawberry-bush. April–June. Rich woods, not common; mostly in the e. half of the state.
    2. E. fortunei (Turcz.) Hand.-Maz. July–Aug. Native of Asia; adventive in low woods; Jackson Co.
    3. E. atropurpureus Jacq. Wahoo. June–July. Woods; occasional throughout the state.
    4. E. europaeus L. European Spindle-tree. May–June. Native of Europe; adventive in a floodplain; Kane Co.
    5. E. americanus L. Strawberry-bush. May–June. Rich woods, rare; Pulaski and Wabash cos.
    6. E. alatus (Thunb.) Sieb. Winged Euonymus. June–July. Native of e. Asia; rarely escaped from cultivation; Coles and DuPage cos.

2. *Celastrus* L. Bittersweet

1. Leaves ovate to ovate-oblong, finely serrate; flowers in terminal panicles or racemes . . . . 1. *C. scandens*
1. Leaves suborbicular, crenate; flowers in axillary cymes 2. *C. orbiculatus*

    1. C. scandens L. Bittersweet. May–June. Woods, occasional; throughout the state.
    2. C. orbiculatus Thunb. Round-leaved Bittersweet. May–June. Native of Asia; rarely adventive in Ill.

94. STAPHYLEACEAE

1. *Staphylea* L. Bladdernut

    1. S. trifolia L. Bladdernut. April–May. Moist woods, common; in every co.

95. ACERACEAE

1. *Acer* L. Maple

1. Leaves pinnately compound . . . . 1. *A. negundo*
1. Leaves simple 2
2. Sinuses between adjacent lobes of the leaves rounded; flowers yellow or yellow-green, appearing at the same time as the leaves; each pair of samaras with a U-shaped sinus 3
2. Sinuses between adjacent lobes of the leaves angular; flowers red or bright yellow, appearing before the leaves; each pair of samaras more or less with a V-shaped sinus 6
3. Latex present; inflorescence pedunculate; samaras horizontally spreading . . . . 2. *A. platanoides*

3. Latex absent; inflorescence sessile; samaras upwardly curved 4
4. Ovary and young fruit pubescent; calyx at most only 3 mm long, pilose
.... 3. *A. barbatum*
4. Ovary and young fruit glabrous; calyx 4–6 mm long, glabrous or nearly
so 5
5. Blades of leaves flat, not drooping along the margins; bark usually gray
.... 4. *A. saccharum*
5. Blades of leaves drooping along the margins; bark usually gray-black to
black .... 5. *A. nigrum*
6. Leaves usually cleft more than halfway to the base; petals absent;
ovaries pubescent .... 6. *A. saccharinum*
6. Leaves usually cleft less than halfway to the base; petals present;
ovaries glabrous .... 7. *A. rubrum*

1. A. negundo L. Box Elder. April–May. Moist soil; common throughout the state. (Including var. *violaceum* [Kirsch.] Jaeg. and var. *texanum* Pax.)

2. A. platanoides L. Norway Maple. May–July. Native of Europe; occasionally planted, but rarely escaped from cultivation.

3. A. barbatum Michx. Southern Sugar Maple. April. Woods, occasional; confined to the s. one-sixth of the state.

4. A. saccharum Marsh. Sugar Maple. April–May. Leaves glabrous or nearly so beneath. Rich woodlands; common throughout the state; in every co. (Including var. *rugelii* [Pax] Rehder and f. *rugelii* [Pax] Palmer & Steyerm.)

var. schneckii Rehder. Schneck's Sugar Maple. April–May. Leaves densely pubescent beneath. Rich woodlands; occasional in the s. one-fourth of the state.

5. A. nigrum Michx. f. Black Maple. May–June. Rich woodlands; occasional in the n. two-thirds of the state, apparently absent (although confused with *A. barbatum*) elsewhere.

6. A. saccharinum L. Silver Maple. Feb.–April. Bottomland woods and along streams; common throughout the state; in every co.

7. A. rubrum L. Red Maple. March–April. Leaves sparsely pubescent or glabrous beneath. Rocky woods, slopes, moist woods; occasional in the s. one-third of the state, rare in the ne. corner of the state, apparently absent elsewhere.

var. rubrum L. f. tomentosum (Desf.) Dansereau. March–April. Leaves permanently tomentose beneath; samaras up to 2.5 cm long. Upland slopes, rare; Union Co.

var. drummondii (H. & A.) Sarg. Drummond's Red Maple. March–April. Leaves permanently tomentose beneath; samaras 3 cm long or longer. Wooded swamps; confined to the s. one-fourth of Ill. *A. drummondii* H. & A.—J.

96. HIPPOCASTANACEAE

1. Aesculus L. Buckeye

1. Petals red, glandular along the margin; capsule smooth   . 1. *A. discolor*
1. Petals greenish-yellow or white marked with red; capsule prickly or covered with scales 2

2. Petals all similar in size and shape; stamens exserted; capsule prickly   3
2. Two of the petals very unlike the others in shape and size; stamens included; capsule covered with scales .... 4. *A. octandra*
3. Petals greenish-yellow; leaflets mostly 5 .... 2. *A. glabra*
3. Petals white marked with red; leaflets mostly 7 ... 3. *A. hippocastanum*

    1. A. discolor Pursh. Red Buckeye. April–May. Rich woodlands, rare; confined to the s. one-sixth of the state; also Richland Co. *A. pavia* L.—G.

    2. A. glabra Willd. Ohio Buckeye. April–May. Bark brown; leaflets green beneath. Rich woodlands; occasional throughout most of the state, except for the extreme n. cos. (Including var. *sargentii* Rehder.)

    var. leucodermis Sarg. Bark whitish; leaflets whitened beneath. Rich woods, very rare; Jackson Co.

    3. A. hippocastanum L. Horse Chestnut. May–June. Native of Europe and Asia; occasionally planted but rarely escaped from cultivation.

    4. A. octandra Marsh. Sweet Buckeye. May–June. Rich woodlands, very rare; Gallatin Co.

97. SAPINDACEAE

1. Annual climber with tendrils; petals white; leaves biternate ....
.... 1. *Cardiospermum*
1. Tree without tendrils; petals yellow; leaves bipinnate .. 2. *Koelreuteria*

1. Cardiospermum L.

    1. C. halicacabum L. Balloon-vine. July–Sept. Native of trop. Am.; rarely adventive in thickets near streams and rivers; Jackson, Monroe, Randolph, and St. Clair cos.

2. Koelreuteria Laxm.   China-tree

    1. K. paniculata Laxm. Golden-rain Tree. July–Sept. Native of Asia; sometimes planted and rarely escaped into waste areas.

98. BALSAMINACEAE

1. Impatiens L.   Jewel-weed

1. Flowers orange with reddish spots .... 1. *I. biflora*
1. Flowers pale yellow .... 2. *I. pallida*

    1. I. biflora Walt. Spotted Touch-me-not. June–Sept. Wet soil; common throughout the state. *I. capensis* Meerb.—F.

    2. I. pallida Nutt. Pale Touch-me-not. July–Sept. Wet soil, often in woodlands; occasional throughout the state.

99. RHAMNACEAE

1. Woody vines .... 1. *Berchemia*
1. Trees or shrubs 2
2. Leaves 3-veined from the base; fruit a capsule .... 2. *Ceanothus*
2. Leaves pinnately veined; fruit a drupe .... 3. *Rhamnus*

1. Berchemia Neck. Supple-jack

1. B. scandens (Hill) K. Koch. Supple-jack. April–June. Edge of pine plantation; Pope Co.

2. Ceanothus L.

1. Leaves elliptic to elliptic-lanceolate, obtuse to subacute .. 1. *C. ovatus*
1. Leaves ovate to ovate-oblong, acute to acuminate (obtuse in var. *pitcheri*) .... 2. *C. americanus*

1. C. ovatus Desf. May–June. Low dunes, sandy soil, rare; Cook, JoDaviess, Lake, Whiteside, and Winnebago cos.
2. C. americanus L. New Jersey Tea. June–Aug. Leaves acute to acuminate, glabrous or nearly so on the upper surface. Woods; occasional throughout the state. (Including var. *intermedius* [Pursh] K. Koch.)
var. pitcheri Torr. & Gray. June–Aug. Leaves obtuse at the tip, pilose on the upper surface. Woods; not common, but throughout the state.

3. Rhamnus L. Buckthorn

1. Low shrub less than 1 m tall; petals absent .... 1. *R. alnifolia*
1. Shrubs or trees, at maturity well over 1 m tall; petals present 2
2. Some of the branchlets ending in a spine; nutlets with a deep, narrow groove. .... 2. *R. cathartica*
2. None of the branchlets ending in a spine; nutlets without a groove, or the groove broad and open 3
3. Winter buds with scales; flowers appearing with the leaves; nutlets with a broad, open groove .... 3. *R. lanceolata*
3. Winter buds without scales; flowers appearing after the leaves; nutlets without a furrow 4
4. Umbels peduncled; leaves more or less acute; most or all the petioles 5 mm long or longer; pedicels pubescent .... 4. *R. caroliniana*
4. Umbels sessile; leaves more or less obtuse; most or all the petioles less than 5 mm long; pedicels glabrous or nearly so .... 5. *R. frangula*

1. R. alnifolia L'Hér. Alder Buckthorn. May–July. Bogs and wooded swamps, rare; confined to the n. half of Ill.
2. R. cathartica L. Common Buckthorn. May–June. Native of Europe; naturalized into waste areas and woodlands; occasional to common in the n. half of the state.
3. R. lanceolata Pursh. Lance-leaved Buckthorn. April–June. River banks, bluffs, calcareous fens; rare to occasional in the n. two-thirds of the state, rare or absent elsewhere. (Including var. *glabrata* Gl.)
4. R. caroliniana Walt. Carolina Buckthorn. May–June. Leaves gla-

brate beneath at maturity. Woods, not common; in the s. one-fifth of the state. (Including var. *mollis* Fern.)

5. R. frangula L. Glossy Buckthorn. May–July. Leaves obovate. Native of Europe; naturalized in woods and bogs; occasional in the n. half of the state.

var. angustifolia Loud. June–July. Leaves narrowly lanceolate. Native of Europe; rarely naturalized in Ill.; Cook Co.

100. VITACEAE

1. Leaves compound 2
1. Leaves simple 3
2. Leaves palmately compound, with (3–) 5 leaflets .... 1. *Parthenocissus*
2. Leaves ternately or pinnately compound, with 11 or more leaflets ......
.... 2. *Ampelopsis*
3. Pith continuous, white; flowering and fruiting clusters broader than long
.... 2. *Ampelopsis*
3. Pith interrupted at the nodes, brown; flowering and fruiting clusters longer than broad .... 3. *Vitis*

1. Parthenocissus Planch. Virginia Creeper

1. Leaflets three .... 1. *P. tricuspidata*
1. Leaflets 5 2
2. Each tendril ending in an adhesive disk; leaves dull on the upper surface; berries 5–7 mm in diameter .... 2. *P. quinquefolia*
2. Tendrils without adhesive disks; leaves shiny on the upper surface; berries 8–10 mm in diameter .... 3. *P. inserta*

1. P. tricuspidata (Sieb. & Zucc.) Planch. Boston Ivy. June–Aug. Native of Asia; rarely escaped from cultivation.

2. P. quinquefolia (L.) Planch. Virginia Creeper. June–July. Woods, along fences, common; in every co.

3. P. vitacea (Knerr) Hitchc. Virginia Creeper. May–July. Woods, thickets; occasional in the n. one-fourth of the state, absent elsewhere. *P. inserta* (Kerner) K. Fritsch—G.

2. Ampelopsis Michx.

1. Leaves simple .... 1. *A. cordata*
1. Leaves compound .... 2. *A. arborea*

1. A. cordata Michx. Raccoon Grape. May–July. Moist woods and thickets; common to occasional in the s. half of the state, becoming rare or absent elsewhere.

2. A. arborea (L.) Koehne. Pepper-vine. June–Aug. Moist woods and thickets, rare; Alexander, Jackson, Pulaski, Randolph, and Union cos.

3. Vitis L. Grape

1. Leaves felted on the lower surface or covered with cobwebby hairs or merely silvery-glaucous on the lower surface 2

310 VITACEAE

1. Leaves glabrous or glabrate on the lower surface, green 4
2. Leaves felted on the lower surface, but not cobwebby or merely silvery-glaucous .... 1. *V. labruscana*
2. Leaves cobwebby on the lower surface or merely silvery-glaucous on the lower surface 3
3. Seeds 5–8 mm long; branchlets not angular .... 2. *V. aestivalis*
3. Seeds 4–5 mm long; branchlets angular .... 3. *V. cinerea*
4. Leaves broader than long; tendrils lacking, or merely opposite the uppermost leaves .... 4. *V. rupestris*
4. Leaves longer than broad; tendrils opposite all leaves except for every third one 5
5. Leaves unlobed or shallowly 3-lobed .... 5. *V. vulpina*
5. Leaves, or at least those on the fertile branches, sharply 3- to 5-lobed . 6
6. Berries black; partitions in pith at the nodes 4–5 mm thick; leaves scarcely or not ciliate on the margins .... 6. *V. palmata*
6. Berries bluish, glaucous; partitions in pith at the nodes up to 2 mm thick; leaves ciliate on the margins .... 7. *V. riparia*

1. V. labruscana Bailey. Labruscan Vineyard Grape. May–June. Cultivated vineyard species, rarely escaped into waste ground; Jackson Co.

2. V. aestivalis Michx. Summer Grape. May–July. Petiole tomentose or rusty-pubescent. Rocky woods, blufftops; occasional to common throughout the state. (Including *V. lincecumii* Buckl.)

var. argentifolia (Munson) Fern. Silver-leaved Grape. Petioles glabrous or nearly so. Rocky woods, occasional throughout the state.

3. V. cinerea Engelm. Winter Grape. May–July. Low woods, thickets; occasional in the s. three-fourths of the state, rare elsewhere.

4. V. rupestris Scheele. River-bank Grape. May–June. Rocky banks, very rare; Union Co.

5. V. vulpina L. Frost Grape. May–June. Moist soil; occasional throughout the state.

6. V. palmata Vahl. Catbird Grape. May–July. Swamps, low woods; confined to the s. one-sixth of the state.

7. V. riparia Michx. Riverbank Grape. May–June. Petioles glabrous; berries 8–12 mm in diameter. Low woods; occasional to common throughout the state.

var. praecox Engelm. April. Petioles glabrous; berries 6–7 mm in diameter. Low woods; occasional in the sw. cos. of the state.

var. syrticola (Fern. & Wieg.) Fern. May–June. Petioles densely hairy. Low woods, rare; scattered throughout the state.

101. TILIACEAE

1. Tilia L. Basswood

1. Leaves glabrous beneath, except in the axils of the veins .... 1. *T. americana*
1. Leaves pubescent beneath with stellate hairs 2
2. Leaves green beneath; peduncles and pedicels glabrous 1. *T. americana*
2. Leaves white beneath; peduncles and pedicels pubescent .... 2. *T. heterophylla*

1. T. americana L. Basswood. May–July. Leaves glabrous beneath, except in the axils of the veins. Rich woods; occasional to common throughout the state.

var. neglecta (Spach) Fosberg. Basswood. May–July. Leaves pubescent beneath with stellate hairs. Rich woods, rare; in some of the n. cos. *T. neglecta* Spach—F.

2. T. heterophylla Vent. White Basswood. May–July. Low rocky woods, rare; Hardin, Massac, and Pope cos.

102. STERCULIACEAE

1. Melochia L.

1. M. corchorifolia L. Chocolate-weed. Aug.–Sept. Native of the s. states; adventive in Ill. along a road; Randolph Co.

103. MALVACEAE

1. Some or all the leaves lobed 2
1. None of the leaves lobed 13
2. Calyx subtended by an involucre of bracts at the base 3
2. Calyx without an involucre of bracts at the base 11
3. Bracts at base of calyx 3 4
3. Bracts at base of calyx 6 or more 10
4. Bracts laciniate .... 4. *Gossypium*
4. Bracts unlobed, or at least not laciniate 5
5. All or most of the leaves divided nearly to the base 6
5. None of the leaves divided nearly to the base 7
6. Plants hispid; stems more or less procumbent .... 2. *Callirhoë*
6. Plants with scattered pubescence, but not hispid; stems ascending ....
.... 1. *Malva*
7. Flowers over 2 cm across 8
7. Flowers less than 2 cm across .... 1. *Malva*
8. Lower leaves triangular, unlobed .... 2. *Callirhoë*
8. Lower leaves shallowly lobed or, if unlobed, not triangular 9
9. Bracts at base of calyx lanceolate to ovate-lanceolate; calyx lobes acute; carpels 1-seeded .... 1. *Malva*
9. Bracts at base of calyx linear; calyx lobes acuminate; carpels 2- to 4-seeded .... 3. *Iliamna*
10. Bracts triangular; fruit separating at maturity into 15–20 carpels ....
.... 7. *Althaea*
10. Bracts linear; fruit a 5-celled capsule .... 6. *Hibiscus*
11. Flowers unisexual; flowers white, up to 2 cm across .... 8. *Napaea*
11. Flowers bisexual; flowers blue or, if white, over 2 cm across 12
12. Flowers over 4 cm across .... 2. *Callirhoë*
12. Flowers up to 4 cm across .... 9. *Anoda*
13. Leaves as broad as long, some of them at least 3 cm broad 14
13. Leaves longer than broad, never 3 cm broad 17

14. Calyx not subtended by bracts .... 10. *Abutilon*
14. Calyx subtended by bracts 15
15. Bracts subtending the calyx 3 16
15. Bracts subtending the calyx 6 or more .... 6. *Hibiscus*
16. Flowers over 2 cm broad, deep purple .... 2. *Callirhoë*
16. Flowers much less than 2 cm broad, pale lilac or white ...... 1. *Malva*
17. Calyx subtended by 2–3 setaceous bracts .... 5. *Sphaeralcea*
17. Calyx not subtended by bracts .... 11. *Sida*

1. *Malva* L. Mallow

1. Flowers at least 2.5 cm across, usually broader; petals more than twice as long as the calyx 2
1. Flowers up to 1.5 cm across; petals at most only twice as long as the calyx 3
2. Petals up to four times as long as the calyx; leaves shallowly lobed, the tips of the lobes triangular or broadly rounded; carpels about 10, minutely pubescent to nearly glabrous .... 1. *M. sylvestris*
2. Petals 5–8 times as long as the calyx; leaves deeply divided into narrow divisions, the tips slender and not triangular; carpels usually 15–20, densely pubescent .... 2. *M. moschata*
3. Petals about twice as long as the calyx; leaves with rounded lobes or sometimes without lobes; carpels rounded on the margins 4
3. Petals about as long as the calyx; leaves with angular lobes; carpels with sharp margins .... 5. *M. rotundifolia*
4. Leaves scarcely lobed, flat on the margins; stems procumbent .... 3. *M. neglecta*
4. Leaves with 5–7 rounded lobes, crisped on the margins; stems erect ... . 4. *M. verticillata* var. *crispa*

    1. M. sylvestris L. High Mallow. Aug.–Sept. Lobes of leaf triangular; stems and leaves hirsute. Native of Europe; occasionally escaped from cultivation; scattered in Ill.
    var. mauritiana (L.) Boiss. High Mallow. Aug.–Sept. Lobes of leaf broadly rounded; stems and leaves glabrous or nearly so. Native of Europe; escaped from cultivation; known from the ne. cos.
    2. M. moschata L. Musk Mallow. June–Aug. Native of Europe; rarely escaped from cultivation; Cook, Lawrence, and LaSalle cos.
    3. M. neglecta Wallr. Common Mallow. June–Sept. Native of Europe; naturalized in waste areas, particularly farm lots; occasional to common throughout Ill.
    4. M. verticillata L. var. crispa L. Curled Mallow. July–Sept. Native of Europe; rarely escaped from cultivation in a few of the n. cos. *M. crispa* L.—J.
    5. M. rotundifolia L. Mallow. June–Sept. Native of Europe; naturalized in waste areas; occasional and scattered in Ill.

2. *Callirhoë* Nutt. Poppy Mallow

1. Each flower subtended by three bracts 2
1. Flowers without subtending bracts 3
2. Some of the leaves triangular; flowers in panicles or appearing umbellate; carpels not rugose .... 1. *C. triangulata*
2. None of the leaves triangular; flowers solitary; carpels rugose .... 2. *C. involucrata*

3. Basal leaves more or less triangular; top of carpels pubescent . . . .
. . . . 3. *C. alcaeoides*
3. Basal leaves rounded; top of carpels glabrous . . . . 4. *C. digitata*

1. C. triangulata (Leavenw.) Gray. Poppy Mallow. June–Sept. Sandy soil; mostly in the n. cos.

2. C. involucrata (Torr. & Gray) Gray. Poppy Mallow. June–Aug. Native of the w. U.S.; naturalized in a few mostly cent. cos.

3. C. alcaeoides (Michx.) Gray. Poppy Mallow. May–Aug. Dry, gravelly areas, rare; Christian, Henry, and Peoria cos.

4. C. digitata Nutt. Poppy Mallow May–July. Native of the w. U.S.; rarely adventive in n. Ill.

3. *Iliamna Greene*

1. I. remota Greene. Kankakee Mallow. June–Aug. Dry banks, rare; Kankakee Co.

4. *Gossypium L. Cotton*

1. G. hirsutum L. Cotton. May–June. Native of trop. Am.; escaped from cultivation in scattered Ill. areas.

5. *Sphaeralcea St. Hil. Globe Mallow*

1. S. angusta (Gray) Fern. Globe Mallow. July–Aug. Dry soil, very rare; Grundy, LaSalle, and Rock Island cos. *Malvastrum angustum* Gray—G, J.

6. *Hibiscus L. Rose Mallow*

1. Annual herbs; flowers yellow or whitish with a black or purple center 2
1. Perennial herbs, shrubs, or small trees; flowers not yellow 3
2. Calyx inflated, 5-winged; seeds verrucose . . . . 1. *H. trionum*
2. Calyx like a spathe, splitting down one side; seeds mucilaginous . . . . . .
. . . . 2. *H. esculentus*
3. Stems and leaves glabrous 4
3. Stems and at least the lower surface of the leaves pubescent 5
4. Herb 1.0–2.5 m tall; calyx glabrous or with simple hairs . . 3. *H. militaris*
4. Shrub or small tree 3 m tall or taller; calyx stellate-pubescent . . . .
. . . . 4. *H. syriacus*
5. Leaves glabrous or nearly so above; some of the leaves often 3-lobed; capsules glabrous . . . . 5. *H. palustris*
5. Leaves soft-hairy on both surfaces; none of the leaves lobed; capsules densely hairy . . . . 6. *H. lasiocarpos*

1. H. trionum L. Flower-of-an-hour. July–Oct. Native of Europe; adventive throughout the state, although rare in the extreme s. tip.

2. H. esculentus L. Okra. July–Sept. Native of Africa; rarely escaped from cultivation; Champaign Co.

3. H. militaris Cav. Halberd-leaved Rose Mallow. July–Oct. Wet areas; occasional throughout the state, although rare in the extreme n. cos.

4. H. syriacus L. Rose-of-Sharon. July–Sept. Native of Asia; occasionally persisting from cultivation.

5. H. palustris L. Swamp Rose Mallow. July–Sept. Wet soil; Cook and LaSalle cos. *H. moscheutos* L.—J.

6. H. lasiocarpos Cav. Hairy Rose Mallow. July–Oct. Low, wet soil; occasional to common in the s. half of Illinois, rare or absent in most of the n. cos.

7. *Althaea* L.

1. A. rosea (L.) Cav. Hollyhock. July–Oct. Native of Europe and Asia; occasionally escaped from cultivation.

8. *Napaea* L.   Glade Mallow

1. N. dioica L. Glade Mallow. June–July. Alluvial soil along streams and rivers; occasional in the n. half of the state, apparently absent from the s. cos.

9. *Anoda* Cav.

1. A. cristata (L.) Schlecht. Aug.–Oct. Native of the sw. U.S.; rarely escaped from cultivation; Hancock and Massac cos.

10. *Abutilon* Mill.   Indian Mallow

1. A. theophrastii Medic. Velvet-leaf. Aug.–Oct. Native of India; naturalized in waste areas; throughout the state.

11. *Sida* L.

1. Leaves oblong to lance-ovate, all or most of them over 8 mm broad; carpels 5 per flower .... 1. *S. spinosa*
1. Leaves linear, rarely over 7 mm broad; carpels 8 or more per flower ....
.... 2. *S. elliottii*

1. S. spinosa L. Prickly Sida. June–Oct. Native of trop. Am.; naturalized into waste areas; occasional to common in the s. four-fifths of the state, rare or absent in most of the extreme n. cos.

2. S. elliottii Torr. & Gray. Aug.–Oct. Farm lots, very rare; Alexander Co.

104. HYPERICACEAE

1. Sepals 4; petals 4 .... 1. *Ascyrum*
1. Sepals 5; petals 5 2
2. Petals yellow or orange, convolute in bud; stamens 5 to many, without glands at the base .... 2. *Hypericum*
2. Petals pinkish or flesh-colored, imbricate in bud; stamens 9, with 3 large orange glands .... 3. *Triadenum*

1. Ascyrum L. St. Andrew's Cross

1. A. hypericoides L. St. Andrew's Cross. July–Aug. Ascending shrub with linear-oblanceolate leaves 2–4 (–5) mm broad. Dry soil, very rare; Hancock Co.

var. multicaule (Michx.) Fern. St. Andrew's Cross. July–Aug. Sprawling shrub with oblong-oblanceolate leaves 4 mm broad or broader. Dry woods, on slopes and ridges; occasional in the s. tip of Ill., absent elsewhere. *A. hypericoides* L.—G, J.

2. Hypericum L. St. Johns-wort

1. Either the margins or the surface of the petals streaked or dotted with black 2
1. Petals not dotted or streaked with black 3
2. Only the margins of the petals streaked with black; seeds rough; stems repeatedly branched .... 1. *H. perforatum*
2. Petals black-dotted over entire surface; seeds smooth or nearly so; stems mostly unbranched .... 2. *H. punctatum*
3. Leaves more than 2 mm in width (if less than 2 mm in width, then the leaves 3-nerved) 4
3. None of the leaves exceeding 2 mm in width 17
4. Petals 8 mm long or longer; stamens 20 or more 5
4. Petals at most only 6 mm long; stamens 5–12 (rarely 20) 13
5. Styles 5 (rarely 4) 6
5. Styles 3, often united to appear as a single beak 8
6. Flowers 4–7 cm across; capsule 2–3 cm long; some of the leaves partly clasping .... 3. *H. pyramidatum*
6. Flowers 1.5–3.0 cm across; capsule up to 1 cm long; leaves not clasping 7
7. Capsule 7–10 mm long; stem with papery whitish bark; sepals 5–15 mm long .... 4. *H. kalmianum*
7. Capsule 3.0–6.5 mm long; stem without papery whitish bark; sepals 2–5 mm long .... 5. *H. lobocarpum*
8. Stems arising from a creeping stoloniferous base 9
8. Plants without a creeping stoloniferous base, although underground rhizomes may be present 10
9. Leaves linear-lanceolate; stamens persistent .... 6. *H. adpressum*
9. Leaves elliptic-oblong; stamens deciduous .... 7. *H. ellipticum*
10. Ovary and capsule 3-locular; stems woody nearly throughout 11
10. Ovary and capsule 1-locular; stems herbaceous or slightly woody only at the base 12
11. Capsule 8–15 mm long; some or all the sepals 5 mm long or longer .... 8. *H. spathulatum*
11. Capsule 3–7 mm long; all of the sepals less than 5 mm long .... 9. *H. densiflorum*
12. Styles free; plants virgate .... 10. *H. denticulatum*
12. Styles united into a beak; plants not virgate .... 11. *H. sphaerocarpum*
13. Bracts foliaceous, resembling the foliage leaves .... 12. *H. boreale*
13. Bracts linear-setaceous, much reduced from the foliage leaves 14
14. Leaves ovate to orbicular 15
14. Leaves linear to lanceolate 16
15. Capsules 3.0–3.5 mm long; plants much branched, not virgate .... 13. *H. mutilum*

15. Capsules 4 mm long or longer; plants sparingly branched, virgate . . . . .
. . . . 14. *H. gymnanthum*
16. Leaves linear, 1- to 3-nerved; some or all the sepals less than 4 mm long
. . . . 15. *H. canadense*
16. Leaves lanceolate, 5- to 7-nerved; some or all the sepals 4 mm long or
longer . . . . 16. *H. majus*
17. Leaves scale-like, at most only 3 mm long; capsule at least twice as long
as the calyx . . . . 17. *H. gentianoides*
17. Leaves linear, mostly 6–20 mm long; capsule about as long as the calyx
. . . . 18. *H. drummondii*

1. H. perforatum L. Common St. Johns-wort. June–Sept. Naturalized from Europe; roadsides, dry pastures, and fields; scattered in all parts of Ill.

2. H. punctatum Lam. Spotted St. Johns-wort. July–Aug. Petals 5–7 mm long; uppermost leaves obtuse at the tip. Woods and roadsides; in every co.

var. pseudomaculatum (Bush) Fern. Large Spotted St. Johns-wort. May–July. Petals 8–12 mm long; uppermost leaves acute at the tip. Dry woods, rare; Jackson Co. *H. pseudomaculatum* Bush—G, J.

3. H. pyramidatum Ait. Giant St. Johns-wort. July–Aug. Banks of rivers and streams; occasional in the n. half of the state; also Macoupin and St. Clair cos. *H. ascyron* L.—J.

4. H. kalmianum L. Kalm's St. Johns-wort. June–Aug. Calcareous sand and small bogs, locally common; Cook and Lake cos.

5. H. lobocarpum Gattinger. St. Johns-wort. June–Aug. Low woods, rare; Massac and Pope cos. *H. densiflorum* Pursh var. *lobocarpum* (Gattinger) Svenson—F, G.

6. H. adpressum Bart. Creeping St. Johns-wort. July–Aug. Wet ground, rare; Kankakee and Lawrence cos.

7. H. ellipticum Hook. St. Johns-wort. May–Aug. Roadsides, rare; Fulton and St. Clair cos.

8. H. spathulatum (Spach) Steud. Shrubby St. Johns-wort. July–Sept. Rocky stream banks and pastures; occasional in the s. half of the state. *H. prolificum* L.—G, J.

9. H. densiflorum Pursh. St. Johns-wort. July–Sept. Swampy situations, very rare; Alexander Co.

10. H. denticulatum Walt. St. Johns-wort. June–Aug. Moist woods, gravelly hills, rare; Jackson and Pope cos. (Including var. *recognitum* Fern. & Schub.)

11. H. sphaerocarpum Michx. Round-fruited St. Johns-wort. June–Sept. Leaves narrowly oblong or narrowly elliptic, flat, with lateral nerves. Rocky woods, hill prairies, roadsides; scattered throughout the entire state.

var. turgidum (Small) Svenson. June–Sept. Leaves linear, revolute, without any apparent lateral nerves. Dry fields, rare; Richland Co.

12. H. boreal (Britt.) Bickn. Northern St. Johns-wort. July–Sept. Sandy marshes, very rare; Iroquois Co.

13. H. mutilum L. Dwarf St. Johns-wort. July–Sept. Moist soil; occasional in all parts of Ill. (Including H. *mutilum* L. var. *parviflorum* [Willd.] Fern.)

14. H. gymnanthum Engelm. & Gray. Small St. Johns-wort. June–Sept. Moist soil; rare to occasional in the n. three-fourths of the state.

15. H. canadense L. Canadian St. Johns-wort. July–Sept. Moist sandy flats, occasional to rare; n. half of Ill.

16. H. majus (Gray) Britt. St. Johns-wort. July–Aug. Moist ground, not common; n. half of Ill.

17. H. gentianoides (L.) BSP. Pineweed. June–Oct. Sandy soil, often on dry bluffs, occasional; scattered in most parts of Ill.

18. H. drummondii (Grev. & Hook.) Torr. & Gray. Nits-and-lice. July–Sept. Bluffs, fields, and dry wooded slopes, occasional to common; s. half of Ill.

3. Triadenum Raf.

1. Leaves without punctations .... 1. T. tubulosum
1. Leaves punctate, at least on the lower surface 2
2. Leaves petiolate .... 2. T. walteri
2. Leaves sessile 3
3. Sepals obtuse, up to 5 mm long; styles up to 1.5 mm long   . 3. T. fraseri
3. Sepals acute, 5–8 mm long; styles 2–3 mm long .... 4. T. virginicum

1. T. tubulosum (Walt.) Gl. Marsh St. Johns-wort. Aug.–Sept. Swampy or marshy ground in woods, rare; Alexander, Johnson, Massac, Pope, and Pulaski cos. Hypericum tubulosum Walt.—F.

2. T. walteri (Gmel.) Gl. Marsh St. Johns-wort. July–Sept. Wooded swamps, not common; confined to the s. one-sixth of the state. Hypericum tubulosum (Walt.) Gl. var. walteri (Gmel.) Lott—F.

3. T. fraseri (Spach) Gl. Fraser's St. Johns-wort. July–Sept. Bogs and wooded swamps, rare; confined to the n. one-half of the state. Hypericum virginicum L. var. fraseri (Spach) Fern.—F.

4. T. virginicum (L.) Raf. Marsh St. Johns-wort. July–Sept. Bogs, very rare; Lake Co.

105. ELATINACEAE

1. Sepals 2–3, obtuse; petals 2–3; plants glabrous .... 1. Elatine
1. Sepals 5, acuminate; petals 5; plants pubescent .... 2. Bergia

1. Elatine L. Waterwort

1. E. brachysperma Gray. Waterwort. July–Oct. Shallow water, very rare; Menard and Sangamon cos. E. triandra Schk. var. brachysperma (Gray) Fassett—G.

2. Bergia L.

1. B. texana (Hook.) Seubert. June–Sept. Wet soil, very rare; St. Clair Co.

106. CISTACEAE

1. At least some of the flowers with 5 yellow petals; capsule 1-locular; style present 2

1. Flowers with 3 reddish petals; capsule partly 3-locular; style short or none .... 3. *Leachea*
2. Some of the flowers 1.5–2.5 cm broad; style very short; leaves lanceolate to oblanceolate .... 1. *Helianthemum*
2. None of the flowers as much as 1.5 cm broad; style slender, elongated; leaves scale-like .... 2. *Hudsonia*

1. Helianthemum Mill. Rockrose

1. Terminal petaliferous flower solitary; seeds papillate .. 1. *H. canadense*
1. Terminal petaliferous flowers two or more; seeds not papillate ....
.... 2. *H. bicknellii*

1. H. canadense (L.) Michx. Frostweed. June–July. Sandy woods, and prairies, occasional; in the n. half of the state; also St. Clair Co.
2. H. bicknellii Fern. Frostweed. June–July. Sandy woods, occasional; in the n. half of the state; also Madison and St. Clair cos.

2. Hudsonia L. Hudsonia. False Heather

1. H. tomentosa Nutt. Beach Heath. May–July. Flowers sessile or on very short bracteate pedicels. Sand dunes, hills, and blowouts, rare; Fulton, JoDaviess, and Lee cos.
var. intermedia Peck. Beach Heath. May–July. Flowers on leafless pedicels 1.5–6.0 mm long. Sandy soil, rare; JoDaviess Co.

3. Lechea Kalm ex L. Pinweed

1. Outer two sepals as long as or longer than the inner three 2
1. Outer two sepals conspicuously shorter than the inner three 4
2. Stems densely pubescent with spreading hairs; three inner sepals rough in appearance .... 1. *L. mucronata*
2. Stems sparsely to moderately pubescent with ascending hairs; three inner sepals not appearing rough 3
3. Cauline leaves lanceolate to oblong, basal leaves elliptic-ovate; capsule about as long as the calyx .... 2. *L. minor*
3. All leaves linear; capsule shorter than the calyx .... 3. *L. tenuifolia*
4. Leaves pubescent throughout on the lower surface .... 4. *L. stricta*
4. Leaves pubescent only on the veins of the lower surface 5
5. Seeds dark brown, smooth; pedicels mostly shorter than the calyx .....
.... 5. *L. pulchella*
5. Seeds pale brown, partly covered by a grayish membrane; pedicels as long as or longer than the calyx .... 6. *L. intermedia*

1. L. mucronata Raf. Hairy Pinweed. July–Sept. Sandy soil and gravelly slopes; occasional in the n. half of Ill., rare in the s. half. *L. villosa* Ell.—F, G, J.
2. L. minor L. Pinweed. July–Sept. Sandy soil and cliffs; occasional and sparsely distributed throughout Ill.
3. L. tenuifolia Michx. Narrow-leaved Pinweed. July–Sept. Leaves glabrous on the upper surface. Sandy soil and cliffs; occasional to common throughout the state.
var. occidentalis Hodgdon. Pinweed. July–Sept. Leaves pilose on the upper surface. Dry sandy woods, very rare; Peoria Co.

4. L. stricta Leggett. Pinweed. July–Sept. Sandy soil; sparsely scattered in the n. one-third of Ill.

5. L. pulchella Raf. Pinweed. July–Sept. Sandy soil; restricted to the n. half of Ill. *L. leggettii* Britt. & Holl.—F, G, J. (Including *L. leggettii* var. *moniliformis* [Bickn.] Hodgdon.)

6. L. intermedia Leggett. Pinweed. July–Oct. Sandy soil, rare; Cook, Lee, and Winnebago cos.

107. VIOLACEAE

1. Stamens united; sepals attached at the base, not auriculate ....
.... 1. *Hybanthus*
1. Stamens free; sepals attached above the base, auriculate ..... 2. *Viola*

1. Hybanthus Jacq. Green Violet

1. H. concolor (T. F. Forst.) Spreng. Green Violet. April–June. Rich woods; rather common in the s. cos., becoming rare and scattered northward. *Cubelium concolor* (Forst.) Raf.—G.

2. Viola L. Violet

1. Plants acaulescent, the leaves and peduncles seeming to arise from out of the ground 2
1. Plants with stems bearing alternate leaves and axillary flowers 25
2. Flowers basically blue 3
2. Flowers basically white 19
3. None of the leaves lobed 4
3. Some or all the leaves hastately or palmately lobed 14
4. Leaves less than half as long as broad .... 10. *V. sagittata*
4. Leaves more than half as long as broad 5
5. Style with downward turned beak .... 18. *V. odorata*
5. Style thickened at tip, not beaked nor turned downward 6
6. Spurred petals beardless within 7
6. Spurred petals bearded within 10
7. Petioles and leaves uniformly and densely villous .... 7. *V. sororia*
7. Petioles and leaves glabrous or only sporadically and sparsely pubescent 8
8. Lateral petals with hairs swollen at the tip; leaf blades somewhat pubescent on upper surface .... 2. *V. cucullata*
8. Lateral petals with hairs slender throughout; leaf blades entirely glabrous 9
9. Leaves cordate-ovate, rounded to short-pointed at the tip.............
.... 3. *V. pratincola*
9. Leaves elongate-triangular, tapering to an elongated tip ....
.... 5. *V. missouriensis*
10. Leaves and petioles glabrous (microscopically pubescent on upper surface of blade in *V. affinis*) 11
10. Leaves and petioles sparsely hirsute to pilose 12
11. Sepals obtuse; peduncles longer than petioles ..... 4. *V. nephrophylla*
11. Sepals acute; peduncles about equalling petioles .... 6. *V. affinis*

12. Leaves distinctly cordate; cleistogamous capsules on horizontal or arching peduncles 13
12. Leaves subcordate to nearly truncate; cleistogamous capsules on erect peduncles .... 9. *V. fimbriatula*
13. Petioles villous; leaves dark green, rather thick .... 7. *V. sororia*
13. Petioles hirsute, at least near the tip; leaves pale green, thin ....
.... 8. *V. septentrionalis*
14. All petals beardless .... 1. *V. pedata*
14. Lateral petals and sometimes the spurred petal bearded 15
15. Leaves hastately lobed 16
15. Leaves palmately lobed 17
16. Sepals ciliate .... 9. *V. fimbriatula*
16. Sepals glabrous .... 10. *V. sagittata*
17. Leaves cordate at base .... 11. *V. triloba*
17. Leaves cuneate or truncate at base 18
18. Leaves deeply divided into narrow segments; spurred petal villous within
.... 13. *V. pedatifida*
18. Leaves divided into broad segments; spurred petal glabrous within ....
.... 12. *V. viarum*
19. Leaves deeply palmately cleft .... 1. *V. pedata*
19. Leaves unlobed 20
20. Style with downward turned beak .... 18. *V. odorata*
20. Style thickened at tip, not beaked nor turned downward 21
21. Lateral petals bearded 22
21. Lateral petals beardless 23
22. Peduncles and leaves glabrous; seeds brown .... 3. *V. pratincola*
22. Peduncles and leaves pubescent; seeds buff .... 15. *V. incognita*
23. Leaves cordate at base .... 14. *V. macloskeyi* ssp. *pallens*
23. Leaves tapering to base 24
24. Leaves linear, lanceolate, or oblanceolate .... 16. *V. lanceolata*
24. Leaves oblong to ovate .... 17. *V. primulifolia*
25. Stipules not foliaceous; tip of style slender or only slightly swollen . 26
25. Stipules foliaceous; tip of style enlarged into a hollow, globose tip . 30
26. Flowers blue .... 22. *V. conspersa*
26. Flowers white or yellow 27
27. Flowers white 28
27. Flowers yellow 29
28. Stipules toothed .... 21. *V. striata*
28. Stipules entire .... 20. *V. canadensis* var. *rugulosa*
29. Leaves and at least the upper half of the stem densely pubescent ......
.... 19. *V. pubescens*
29. Leaves and stems glabrous or sparsely pubescent.....................
.... 19. *V. pubescens* var. *eriocarpa*
30. Base of stipules pinnately divided; upper leaves distinctly toothed .. 31
30. Base of stipules palmately divided; upper leaves entire or nearly so ....
.... 25. *V. rafinesquii*
31. Petals and sepals subequal, the petals pale yellow ..... 24. *V. arvensis*
31. Petals 2–3 times as long as the sepals, variously colored ....
.... 23. *V. X wittrockiana*

    1. V. pedata L. Birdfoot Violet. April–June. Petals lilac or violet. Cherty slopes, borders of woods; occasional throughout Ill. (Including var. *lineariloba* DC.)
    f. alba Thurb. April–June. Petals basically white. Dry woods; Hardin Co. *V. pedata* L. var. *lineariloba* DC. f *alba* (Thurb.) Britt.—F.

2. V. cucullata Ait. Marsh Blue Violet. April–June. Moist soil in woods and marshes; occasional in the n. half of Ill., apparently rare in the s. half.

3. V. pratincola Greene. Common Blue Violet. March–May. Flowers basically blue or purple. Woods, common; known from every Ill. co. *V. papilionacea* Pursh—F, G, J.

f. albiflora (Glover) Mohl. Confederate Violet. March–May. Flowers basically white, with purplish veins. Woods, occasional; scattered in Ill. *V. priceana* Pollard—J; *V. papilionacea* var. *priceana* (Pollard) Alexander—G; *V. papilionacea* f. *albiflora* Glover—F.

4. V. nephrophylla Greene. Northern Blue Violet. May–June. Usually moist soil; seemingly confined to ne. Ill.

5. V. missouriensis Greene. Missouri Violet. March–May. Low woods, occasional; scattered throughout the state.

6. V. affinis LeConte. Woodland Blue Violet. April–June. Low woods, occasional; scattered throughout Ill.

7. V. sororia Willd. Woolly Blue Violet. March–May. Woods, common; in every co.

8. V. septentrionalis Greene. Northern Blue Violet. May–June. Woods, rare; confined to the northernmost tier of cos.

9. V. fimbriatula Smith. Sand Violet. April–May. Dry woods, sandy soil, occasional; apparently restricted to the n. half of the state; also Richland Co.

10. V. sagittata Ait. Arrow-leaved Violet. March–June. Moist or dry woods, occasional; throughout the state.

11. V. triloba Schwein. Cleft Violet. April–June. Leaves with 3-5 usually shallow lobes. Woods and wooded slopes; occasional in the s. half of Ill., becoming rarer northward. *V. falcata* Greene—J.

var. dilatata (Ell.) Brainerd. Cleft Violet. April–June. Leaves with 5–7 very deep lobes. Woods; occasional in the s. two-thirds of Ill. *V. falcata* Greene—J.

12. V. viarum Pollard. May. Dry, gravelly hilltops, very rare; Kankakee and Peoria cos.

13. V. pedatifida G. Don. Prairie Violet. May–June. Prairies and dry woods; occasional in the n. half of Ill., becoming rare southward.

14. V. macloskeyi Lloyd ssp. pallens (Banks) M. S. Baker. Smooth White Violet. April–May. Moist soil, rare; confined to the n. one-eighth of Ill. (Including *V. blanda* Willd.) *V. pallens* (Banks) Brainerd—F, G, J.

15. V. incognita Brainerd. White Violet. May–June. Wooded slopes, rare; Cook, JoDaviess, and McHenry cos. (Including var. *forbesii* Brainerd.)

16. V. lanceolata L. Lance-leaved Violet. May–June. Leaf blades lanceolate, 3½ to 5 times as long as broad. Wet soil, occasional; confined to the n. half of the state.

var. vittata (Greene) Russell. Narrow-leaved Violet. May–June. Leaf blades linear, 6–15 times as long as broad. Moist sandy soil, rare; Kankakee Co. *V. vittata* Greene—G.

17. V. primulifolia L. Primrose Violet. May–June. Moist soil, rare; Kankakee Co.

18. V. odorata L. Sweet Violet. April–May. Introduced from Europe and occasionally escaped from cultivation; scattered throughout Ill.

19. V. pubescens Ait. Downy Yellow Violet. April–May. Leaves and at least the upper half of the stem densely pubescent; basal leaves 1–2. Rich woods, rare; confined to the n. half of the state. (Including var. *peckii* House, with glabrous capsules.)

var. eriocarpa (Schwein.) Russell. Smooth Yellow Violet. April–May. Leaves and stems glabrous or puberulent; basal leaves 5 or more. Woods, common; in every Ill. co. *V. eriocarpa* Schw.—G, J; *V. pennsylvanica* Michx.—F. (Including *V. pennsylvanica* var. *leiocarpa* [Fern. & Wieg.] Fern., with glabrous capsules.)

20. V. canadensis L. var. rugulosa (Greene) C. L. Hitchcock. White Violet. April–June. Rich woods, rare; Cook, JoDaviess, Lake, and Winnebago cos. *V. rugulosa* Greene—F, G, J.

21. V. striata Ait. Cream Violet. April–June. Moist soil in woods, meadows, and fields, common in the s. half of Ill., becoming rare in the n. half.

22. V. conspersa Reichenb. Dog Violet. April–June. Moist woods, rare; Cook, Lake, and Richland cos.

23. V. X wittrockiana Gams. Pansy. May–Aug. Native of Europe and occasionally escaped from cultivation in a few areas. *V. tricolor* L.—F, G, J.

24. V. arvensis Murr. Wild Pansy. April–Aug. Naturalized from Europe; rarely found as an adventive in Ill.

25. V. rafinesquii Greene. Johnny-jump-up. March–May. Naturalized from Europe and Asia; commonly escaped into waste areas in the s. three-fifths of Ill.; nearly absent in the n. cos. *V. kitaibeliana* Roem. & Schultes var. *rafinesquii* (Greene) Fern.—F.

108. PASSIFLORACEAE

1. *Passiflora* L. Passion-flower

1. Leaves shallowly 3-lobed, the lobes obtuse, entire; flowers greenish-yellow, up to 2.5 cm across; fruit purple .... 1. *P. lutea*
1. Leaves deeply 3- to 5-lobed, the lobes acute to acuminate, serrulate; flowers usually purple and white, 4 cm or more across; fruit yellow ....
.... 2. *P. incarnata*

1. P. lutea L. var. glabriflora Fern. Small Passion-flower. May–Sept. Woods and thickets; occasional to common in the s. half of the state, rare or absent elsewhere. *P. lutea* L.—J.

2. P. incarnata L. Large Passion-flower; Maypops. May–Sept. Dry soil, often in fields; occasional in the s. one-third of the state.

109. LOASACEAE

1. *Mentzelia* L.

1. Petals 5, 1.0–1.5 cm long; stamens about 20; capsules up to 1 cm long, containing about 9 seeds .... 1. *M. oligosperma*
1. Petals 10, 2–5 cm long; stamens numerous; capsule 2.5–4.0 cm long, containing several seeds 2
2. Petals 2–3 cm long; outer filaments dilated; capsule about 3 cm long ..
.... 2. *M. nuda*

2. Petals 4–5 cm long; all filaments filiform; capsule about 4 cm long . . . . .
. . . . 3. *M. decapetala*

1. M. oligosperma Nutt. Stickleaf. June–July. Limestone ledges, rare; confined to cos. bordering the Mississippi River.
2. M. nuda Torr. & Gray. July–Aug. Native of the w. U.S.; adventive along a railroad; Cook Co.
3. M. decapetala (Pursh) Urban & Gilg. July–Sept. Native of the w. U.S.; adventive along a railroad; Grundy Co.

110. CACTACEAE

1. Opuntia Mill. Prickly-pear Cactus

1. All roots fibrous . . . . 1. *O. compressa*
1. Some of the roots tuberous thickened . . . . 2. *O. macrorhiza*

1. O. compressa (Salisb.) Macbr. Prickly-pear. May–July. Sandy soil, bluff-tops; occasional throughout the state, except for the e. cent. cos. where it is rare or absent. *O. humifusa* Raf.—F; *O. rafinesquii* Engelm.—J.
2 O. macrorhiza Engelm. Prickly-pear. June–July. Sand prairie, very rare; Whiteside Co.

111. THYMELAEACEAE

1. Dirca L. Leatherwood

1. D. palustris L. Leatherwood. April–May. Rich, shaded woods; uncommon or rare throughout the state.

112. ELAEAGNACEAE

1. Leaves opposite; stamens 8; flowers unisexual . . . . 1. *Shepherdia*
1. Leaves alternate; stamens 4; flowers bisexual . . . . 2. *Elaeagnus*

1. Shepherdia Nutt.

1. S. canadensis (L.) Nutt. Canadian Buffalo-berry. May–July. Shores of Lake Michigan, rare; Cook and Lake cos.

2. Elaeagnus L. Oleaster

1. Fruit yellow; branchlets and leaves with silvery scales 1. *E. angustifolia*
1. Fruit pink or red; branchlets and leaves with brown and silvery scales . .
. . . . 2. *E. umbellata*

1. E. angustifolia L. Russian Olive. May–June. Native of Europe and Asia; rarely escaped from cultivation.

2. E. umbellata Thunb. May–June. Native of e. Asia; rarely escaped from cultivation; Williamson Co.

113. LYTHRACEAE

1. Plants more or less shrubby; leaves, or some of them, whorled ....
.... 1. *Decodon*
1. Plants herbaceous; leaves opposite, or the uppermost alternate (rarely but sometimes whorled in *Lythrum salicaria*) 2
2. Plants viscid-pubescent; petals 6; calyx tube spurred at base   2. *Cuphea*
2. Plants glabrous or variously pubescent, but not viscid 3
3. Calyx with 5, 6, or 7 lobes or teeth, the tube cylindric ...... 3. *Lythrum*
3. Calyx with 4 lobes or teeth, the tube hemispheric 4
4. Plants usually in water; petals absent; leaves up to 3 mm broad 4. *Peplis*
4. Plants usually not in water; petals 4; leaves over 3 mm broad 5
5. Flower usually one per leaf axil; leaves tapering to a short petiole or subsessile base; capsule dehiscing septicidally .... 5. *Rotala*
5. Flowers often 2 or more per leaf axil; leaves auriculate at sessile base; capsule dehiscing irregularly .... 6. *Ammannia*

1. Decodon J. F. Gmel.

1. D. verticillatus (L.) Ell. Swamp Loosestrife. July–Aug. Swamps, marshes, bogs; uncommon but scattered throughout the state. (Including var. *laevigatus* Torr. & Gray.)

2. Cuphea P. Br.

1. C. petiolata (L.) Koehne. Clammy Cuphea. July–Oct. Usually dry soil; occasional in the s. three-fourths of the state, rare or absent elsewhere.

3. Lythrum L.   Loosestrife

1. Flower solitary, axillary, up to 1.2 cm broad; stamens 4 ... 1. *L. alatum*
1. Flowers in terminal spikes, usually at least 1.2 cm broad; stamens 8 (–10)
.... 2. *L. salicaria*

1. L. alatum Pursh. Winged Loosestrife. June–Aug. Moist ground; occasional to common throughout the state.

2. L. salicaria L. Purple Loosestrife. June–Aug. Native of Europe; occasionally escaped into moist areas. (Including var. *tomentosum* [Mill.] DC.)

4. Peplis L.

1. P. diandra Nutt. Water Purslane. June–Aug. Shallow water; uncommon but scattered in the state.

5. Rotala L.

1. R. ramosior (L.) Koehne. Tooth-cup. July–Sept. Wet ground; occasional throughout the state.

6. Ammannia L.

1. Flowers sessile .... 1. *A. coccinea*
1. Flowers pedicellate .... 2. *A. auriculata*

1. A. coccinea Rottb. Long-leaved Ammannia. July–Aug. Wet soil; occasional throughout the state, except for the extreme n. cos.
2. A. auriculata Willd. Scarlet Loosestrife. May–June. Wet soil, rare; Cass and Jackson cos.

114. NYSSACEAE

1. Nyssa L. Tupelo

1. Most or all petioles 3 cm long or longer; pistillate flowers borne singly; fruit 2–3 cm long .... 1. *N. aquatica*
1. None of the petioles over 2.5 cm long; pistillate flowers 2 or more in a cluster; fruit up to 1.5 cm long .... 2. *N. sylvatica*

1. N. aquatica L. Tupelo Gum. April–May. Swamps, not common; in the s. one-sixth of the state; also Crawford, Richland, and Wabash cos.
2. N. sylvatica Marsh. Sour Gum. April–June. Leaves obtuse or acute; lower surface of leaves not papillose. Bogs and swamps, not common; in the extreme s. tip of the state; also Cook and Kankakee cos.
   var. caroliniana (Poir.) Fern. Sour Gum. April–June. Leaves acuminate; lower surface of leaves papillose. Mostly upland woods, occasional; confined to the s. one-third of the state.

115. MELASTOMACEAE

1. Rhexia L. Meadow Beauty

1. Stems 4-angled; plants with tubers; leaves rounded at the sessile base; neck of hypanthium shorter than the body at maturity ... 1. *R. virginica*
1. Stems cylindrical; plants without tubers; leaves narrowed to a short-petiolate base; neck of hypanthium longer than the body at maturity .....
.... 2. *R. mariana*

1. R. virginica L. Meadow Beauty. July–Sept. Sandy soils; not common but scattered throughout the state.
2. R. mariana L. Meadow Beauty. June–Sept. Sandy fields, rare; Mason, Massac, and Pope cos.

116. ONAGRACEAE

1. Calyx 2-lobed; petals 2; stamens 2 .... 1. *Circaea*
1. Calyx 4-lobed; petals 4 (or absent); stamens 4–12 2
2. Leaves opposite 3
2. Leaves alternate or basal 4
3. Leaves entire .... 2. *Ludwigia*
3. Leaves serrate .... 3. *Epilobium*
4. Petals yellow or greenish 5
4. Petals white or pink 7
5. Calyx divided to the tip of the ovary, persistent on the fruit 6
5. Calyx not divided to the tip of the ovary, deciduous from the fruit ......
.... 5. *Oenothera*
6. Stamens 8–12 .... 4. *Jussiaea*
6. Stamens 4 .... 2. *Ludwigia*
7. Petals at least 1 cm long 8
7. Petals less than 1 cm long 9
8. Calyx tube prolonged beyond the ovary; seeds without a coma ....
.... 5. *Oenothera*
8. Calyx tube not prolonged beyond the ovary; seeds with a coma ....
.... 3. *Epilobium*
9. Fruit indehiscent, with 1–4 seeds; seeds without a coma ..... 6. *Gaura*
9. Fruit dehiscent, with several seeds; seeds usually with a coma ....
.... 3. *Epilobium*

1. *Circaea* L.   Enchanter's Nightshade

1. Stems firm, often 40 cm tall or taller; calyx lobes (1.8–) 2.0–2.5 mm long;
bristles of fruit stiff; fruit 2-locular .... 1. *C. quadrisulcata*
1. Stems weak, up to 30 cm tall; calyx lobes up to 1.2 mm long; bristles of
fruit weak; fruit 1-locular .... 2. *C. alpina*

    1. C. quadrisulcata (Maxim.) Franch. & Sav. var. canadensis (L.)
Hara. Enchanter's Nightshade. June–Aug. Moist woods; common
throughout the state. *C. latifolia* Hill—J.
    2. C. alpina L. Small Enchanter's Nightshade. June–July. Cool ra-
vines, very rare; JoDaviess, Kane, and Lake cos.

2. *Ludwigia* L.   False Loosestrife

1. Leaves opposite; stems more or less trailing or creeping   1. *L. palustris*
1. Leaves alternate; stems erect 2
2. Flowers sessile, greenish, to 3 mm broad 3
2. Flowers short-pedicellate, yellow, over 1 cm across ... 4. *L. alternifolia*
3. Capsule up to 5 mm high, about twice the length of the calyx lobes;
bracts at base of calyx 3–4 mm long .... 2. *L. polycarpa*
3. Capsule 6–10 mm high, about four times the length of the calyx lobes;
bracts at base of calyx absent or up to 1 mm long .... 3. *L. glandulosa*

    1. L. palustris (L.) Ell. var. americana (DC.) Fern. & Grisc. Marsh
Purslane. July–Aug. Wet soil, particularly around ponds, along
streams, and in ditches; common throughout the state. *L. palustris*
(L.) Ell.–J.

2. L. polycarpa Short & Peter. False Loosestrife. June–Sept. Wet soil, particularly in ditches; occasional throughout the state.

3. L. glandulosa Walt. False Loosestrife. June–Sept. Swamps, low woods, not common; confined to the s. one-sixth of the state.

4. L. alternifolia L. Seedbox. June–Aug. Stems, leaves, and calyx glabrous. Wet ground; occasional throughout the state.

var. pubescens Palmer & Steyerm. Seedbox. June–Aug. Stems, leaves, and calyx pubescent. Wet ground; occasional in the southern-most cos.

3. *Epilobium L.   Willow Herb*

1. Calyx tube not prolonged beyond ovary; petals not notched . . . .
. . . . 1. *E. angustifolium*
1. Calyx tube prolonged beyond ovary; petals notched 2
2. Petals at least 1 cm long; stems softly hirsute; stigma deeply 4-lobed; seeds smooth . . . . 2. *E. hirsutum*
2. Petals up to 1 cm long; stems variously pubescent but not softly hirsute; stigma entire or slightly notched; seeds papillose 3
3. Leaves entire 4
3. Leaves serrulate or denticulate 5
4. Stems and leaves velvety-pubescent; most or all the leaves more than 4 mm wide . . . . 3. *E. strictum*
4. Stems and leaves merely canescent; most or all the leaves less than 4 mm wide . . . . 4. *E. leptophyllum*
5. Inflorescence and capsules glandular-pubescent; seeds short-beaked; coma white . . . . 5. *E. adenocaulon*
5. Inflorescence and capsules not glandular; seeds beakless; coma reddish-brown . . . . 6. *E. coloratum*

1. E. angustifolium L. Fireweed. June–Aug. In dunes and bogs, after a fire; confined to the n. one-fifth of the state.

2. E. hirsutum L. Hairy Willow Herb. Aug. Native of Europe; adventive in springy habitats and borders of ditches; Cook, DuPage, and Lake cos.

3. E. strictum Muhl. Downy Willow Herb. July–Sept. Boggy and springy habitats, not common; confined to the n. one-third of the state.

4. E. leptophyllum Raf. Bog Willow Herb. Aug.–Sept. Boggy and springy habitats, not common; confined to the n. one-half of the state.

5. E. adenocaulon Haussk. Northern Willow Herb. July–Sept. Bogs and calcareous springy areas; occasional in the n. one-fourth of the state. *E. glandulosum* Lehm. var. *adenocaulon* (Haussk.) Fern.—F.

6. E. coloratum Muhl. Cinnamon Willow Herb. Aug.–Sept. Bogs, marshes, wet ground; occasional throughout the state.

4. *Jussiaea L.   Primrose Willow*

1. Stems and leaves glabrous 2
1. Stems and leaves pubescent . . . . 3. *J. leptocarpa*
2. Plants creeping; petals 5; capsule cylindrical . . . . 1. *J. repens*
2. Plants more or less erect; petals 4; capsule 4-sided . . . . 2. *J. decurrens*

1. J. repens L. var. glabrescens Ktze. Creeping Primrose Willow. May–Oct. In shallow water or on muddy banks and shores; occasional to common in the s. half of the state. *J. repens* L.—G, J. (Including *J. diffusa* Forsk.)

2. J. decurrens (Walt.) DC. Erect Primrose Willow. July–Sept. Wet ground, rare; confined to the s. one-sixth of the state.

3. J. leptocarpa Nutt. Hairy Primrose Willow. July–Sept. Wet ground, particularly along the Mississippi River; confined to the s. one-fifth of the state.

5. *Oenothera* L. Evening Primrose

1. Petals white or rosy 2
1. Petals yellow (sometimes fading reddish) 3
2. Stems with white peeling epidermis; calyx tube 2–4 cm long; capsule 8-ribbed, 2.5–3.5 cm long .... 1. *O. nuttallii*
2. Stems without white peeling epidermis; calyx tube 1–2 cm long; capsule 4-winged and 8-ribbed, to 2 cm long .... 2. *O. speciosa*
3. Leaves linear or broader, never filiform, more than 1 mm broad 4
3. Leaves filiform, up to 1 mm broad .... 13. *O. linifolia*
4. Some or all the petals 4 cm long or longer; capsule at least 5 cm long .... 3. *O. missouriensis*
4. None of the petals 4 cm long; capsule up to 4 cm long 5
5. Some or all the leaves sinuate-pinnatifid .... 4. *O. laciniata*
5. None of the leaves sinuate-pinnatifid 6
6. Ovary and capsule sharply 4-angled 7
6. Ovary and capsule more or less terete 10
7. Petals up to 1 cm long; flower buds nodding; anthers to 2.5 mm long .. .. 5. *O. perennis*
7. Petals at least 1 cm long; flower buds erect; anthers 4 mm long or longer 8
8. Capsule glandular-pubescent .... 6. *O. tetragona*
8. Capsule eglandular 9
9. Stems with appressed hairs; calyx tube 1.5–2.5 cm long .... 7. *O. pilosella*
9. Stems with appressed hairs; calyx tube 0.5–1.5 cm long 8. *O. fruticosa*
10. Stigma deeply 4-lobed; main stem leaves mostly more than 5 mm broad 11
10. Stigma shallowly lobed or entire; main stem leaves up to 5 mm broad .. .. 12. *O. serrulata*
11. Capsules at least 4 mm thick at base; most of the main stem leaves 10 mm broad or broader 12
11. Capsules up to 3 mm thick at base; most of the main stem leaves up to 10 mm broad .... 11. *O. rhombipetala*
12. Petals 3.5 mm wide or wider .... 9. *O. biennis*
12. Petals up to 3 mm wide .... 10. *O. cruciata*

1. O. nuttallii Sweet. White Evening Primrose. June–July. Native of w. U. S.; rarely adventive in Illinois. *O. albicaulis* Nutt.—J.

2. O. speciosa Nutt. Showy Evening Primrose. June–July. Native of w. U.S.; occasionally adventive along roadsides.

3. O. missouriensis Sims. Missouri Primrose. May–Aug. Rocky glades, very rare; St. Clair Co.

4. O. laciniata Hill. Ragged Evening Primrose. May–July. Fields, prairies, roadsides; occasional throughout the state.

5. O. perennis L. Small Sundrops. June–Aug. Moist, open ground; Cook, Lake, Will, and Winnebago cos.

6. O. tetragona Roth. Four-angled Sundrops. June–Aug. Dry woods, rare; confined to the s. one-sixth of the state.

7. O. pilosella Raf. Prairie Sundrops. May–July. Prairies, fields; occasional throughout the state.

8. O. fruticosa L. var. linearis (Michx.) S. Wats. Shrubby Sundrops. May–Aug. Rocky woods, rare; confined to the s. one-sixth of the state.

9. O. biennis L. Evening Primrose. June–Oct. Plants variously pubescent, but not white-hairy. Fields, prairies, waste ground, common; in every co. (Including var. pycnocarpa [Atkinson & Bartlett] Wieg.)

var. canescens Torr. & Gray. Gray Evening Primrose. June–Oct. Plants white-hairy. Fields, prairies, waste ground; occasional throughout the state. O. strigosa (Rydb.) Mack. & Bush—F. (Including var. hirsutissima Gray.)

10. O. cruciata Nutt. Evening Primrose. July–Aug. Native in the N.E. states; adventive in Ill. in dry soil; Boone, Cook, and Kane cos. O. parviflora L.—F.

11. O. rhombipetala Nutt. Sand Primrose. June–Sept. Disturbed sandy soils; occasional throughout the state.

12. O. serrulata Nutt. Toothed Evening Primrose. May–Sept. Native of w. U.S.; rarely adventive in waste ground; Peoria and Randolph cos.

13. O. linifolia Nutt. Thread-leaved Sundrops. May–July. Sandstone cliffs, prairies; occasional in the s. one-fifth of the state.

6. Gaura L.

1. Ovary and capsule sessile 2
1. Ovary and capsule stipitate .... 4. *G. filipes*
2. Petals 5.5–15.0 mm long; calyx lobes 6–13 mm long. 3
2. Petals 1.5–3.0 mm long; calyx lobes 2.0–3.5 mm long .. 3. *G. parviflora*
3. Stems densely villous; about 50 percent of the pollen grains fertile ....
.... 1. *G. biennis*
3. Stems strigulose to short-villous; about 90 percent of the pollen grains
fertile .... 2. *G. longiflora*

1. G. biennis L. Butterfly-weed. July–Sept. Clearings in woodland and waste places; occasional in the n. two-thirds of the state, apparently absent elsewhere.

2. G. longiflora Spach. July–Oct. Waste places, often along roadsides and railway embankments; occasional throughout the state. (Including G. biennis L. var. pitcheri Torr. & Gray.)

3. G. parviflora Dougl. Small-flowered Gaura. June–July. Native of w. U.S.; adventive in fields, roadsides, edge of woods; occasional in the n. half of the state, less common elsewhere.

4. G. filipes Spach. Slender Gaura. July–Sept. Dry woods, rare; Hardin Co. (Including var. major Torr. & Gray.)

117. HALORAGIDACEAE

1. Leaves whorled or scattered, crowded; flowers 4-merous; fruit 4-angled
.... 1. *Myriophyllum*
1. Leaves alternate, rather remote; flowers 3-merous; fruit 3-angled ....
.... 2. *Proserpinaca*

1. Myriophyllum L.   Water Milfoil

1. Leaves verticillate and scattered; fruit with 4 dorsal tuberculate or undulate ridges 2
1. Leaves all verticillate; fruit without dorsal ridges or with smooth dorsal ridges 3
2. Bracts entire to pectinate; each section of the 4-angled fruit flattened, slightly papillose with a smooth dorsal ridge with undulate margins ....
.... 1. *M. hippuroides*
2. Bracts deeply toothed; each section of the 4-angled fruit rounded, smooth, with a tuberculate dorsal ridge .... 2. *M. pinnatum*
3. Bracts deeply pinnatifid; stamens 8 .... 3. *M. verticillatum*
3. Bracts entire to serrate to dentate; stamens 4 4
4. Bracts much exceeding the flower; fruit with smooth dorsal ridges .....
.... 4. *M. heterophyllum*
4. Bracts not exceeding the flower; fruit slightly tuberculate, without dorsal ridges .... 5. *M. exalbescens*

1. M. hippuroides Nutt. Mare's-tail Milfoil. June–Oct. Shallow water, very rare; Cook Co. *M. hippuroides* Nutt. var. *cheneyi* Fassett—G.

2. M. pinnatum (Walt.) BSP. Rough Water Milfoil. June–Oct. Rooted in muddy shores or in shallow waters of ponds and lakes; occasional throughout the state.

3. M. verticillatum L. var. pectinatum Wallr. Whorled Water Milfoil. June–Sept. Shallow water, not common; scattered throughout the state. *M. verticillatum* L.—J.

4. M. heterophyllum Michx. Water Milfoil. June–Sept. Quiet waters; occasional throughout the state.

5. M. exalbescens Fern. Spiked Water Milfoil. July–Sept. Quiet waters; occasional in the ne. cos., rare elsewhere.

2. Proserpinaca L.   Mermaid-weed

1. P. palustris L. Mermaid-weed. July–Oct. Shallow water and shores; occasional throughout the state. (Including var. *crebra* Fern. & Grisc.)

118. HIPPURIDACEAE

1. Hippuris L.   Mare's-tail

1. H. vulgaris L. Mare's-tail. June–Sept. Shores or shallow water, very rare; Kane, Lake, and McHenry cos.

119. ARALIACEAE

1. Leaves simple, palmately lobed, evergreen .... 1. *Hedera*
1. Leaves compound 2
2. Leaves alternate or basal, the ultimate divisions pinnately arranged ....
.... 2. *Aralia*
2. Leaves whorled, the leaflets palmately arranged .... 3. *Panax*

1. Hedera L.   Ivy

1. H. helix L. English Ivy. June–Sept. Native of Europe; commonly planted but rarely escaping from cultivation; Jackson Co.

2. Aralia L.   Spikenard

1. Shrub or small tree; parts of stem and leaves coarsely prickly ....
.... 1. *A. spinosa*
1. Herbs; leaves and stems without prickles, but sometimes bristly 2
2. Stems bristly .... 2. *A. hispida*
2. Stems without bristles 3
3. Leaflets 3–7 per leaf; peduncle appearing to arise directly from the ground .... 3. *A. nudicaulis*
3. Leaflets 9–21 per leaf; peduncle arising from the stem   . 4. *A. racemosa*

1. A. spinosa L. Hercules' Club; Devil's Walking-stick. July–Sept. Wooded slopes, occasional; confined to the s. one-third of the state.
2. A. hispida Vent. Bristly Sarsaparilla. June–Aug. Sandy soil, rare; Cook and Lake cos.
3. A. nudicaulis L. Wild Sarsaparilla. May–June. Rich wooded slopes, dune woods, rocky woods, bogs; occasional in the n. one-half of the state; also Macoupin Co.
4. A. racemosa L. American Spikenard. June–Aug. Rich woods, rocky woods; occasional throughout the state.

3. Panax L.   Ginseng

1. P. quinquefolius L. Ginseng. June–July. Rich woods, rocky woods; occasional throughout the state.

120. UMBELLIFERAE

1. All leaves simple 2
1. Some or all the leaves compound 5
2. Leaves spiny along the margins, mostly linear to lanceolate   1. *Eryngium*
2. Leaves not spiny along the margins, mostly ovate or oblong 3
3. Leaves sessile, perfoliate .... 2. *Bupleurum*
3. Leaves petiolate, not perfoliate 4
4. Plants prostrate; leaves entire or lobulate; flowers bluish, borne in heads
.... 1. *Eryngium*
4. Plants erect; leaves crenate; flowers yellow or purplish, borne in umbels
.... 3. *Thaspium*

5. Ovaries and fruits prickly 6
5. Ovaries and fruits glabrous, pubescent, or warty, not prickly 8
6. Stems glabrous; leaves palmately divided . . . . 4. *Sanicula*
6. Stems pubescent; leaves pinnately divided 7
7. Bracts of the primary involucre linear, entire . . . . 5. *Torilis*
7. Bracts of the primary involucre pinnately divided . . . . 6. *Daucus*
8. All leaves once-divided (basal ones sometimes undivided) 9
8. Leaves, or most of them, doubly compound 22
9. Leaflets entire 10
9. Leaflets toothed 13
10. Leaves palmately compound . . . . 7. *Cynosciadium*
10. Leaves pinnately compound 11
11. Leaflets capillary, up to 1 mm broad . . . . 8. *Ptilimnium*
11. Leaflets lanceolate to linear, more than 1 mm broad 12
12. Leaflets cross-septate; fruits up to 4 mm long . . . . . . 9. *Limnosciadium*
12. Leaflets not cross-septate; fruits 4.5–6.0 mm long . . . . 10. *Oxypolis*
13. Stems pubescent . . . . 11. *Heracleum*
13. Stems glabrous 14
14. All cauline leaves ternate, with 3 leaflets 15
14. Some of the cauline leaves with more than 3 leaflets 18
15. Flowers white 16
15. Flowers purple or yellow 17
16. Leaflets doubly toothed; fruits cylindrical . . . . 12. *Cryptotaenia*
16. Leaflets singly toothed; fruits flat . . . . 13. *Falcaria*
17. Fruit winged; central flower of the ultimate umbel pedicellate . . . .
. . . . 3. *Thaspium*
17. Fruit unwinged; central flower of the ultimate umbel sessile . . 14. *Zizia*
18. Leaflets palmately arranged . . . . 13. *Falcaria*
18. Leaflets pinnately arranged 19
19. Leaflets sparingly toothed, with no more than 8 teeth . . . . 10. *Oxypolis*
19. Leaflets many-toothed 20
20. Flowers yellow; plants of fields, roadsides, and waste ground . . . .
. . . . 15. *Pastinaca*
20. Flowers white; plants of swamps and wet woods 21
21. Fruit 2.5–3.0 mm long, with prominent ribs; leaflets serrate . . . 16. *Sium*
21. Fruit about 2 mm long, with inconspicuous ribs; leaflets sometimes laciniately toothed . . . . 17. *Berula*
22. Bulblets borne in the axils of the upper leaves . . . . 29. *Cicuta*
22. Bulblets absent 23
23. Stems pubescent only at the nodes 24
23. Stems glabrous or pubescent, but the hairs not confined to the nodes
. . . . 25
24. Flowers yellow . . . . 3. *Thaspium*
24. Flowers white . . . . 18. *Osmorhiza*
25. Ultimate leaf segments capillary, linear, or narrowly elliptic, usually less than 3.5 mm broad 26
25. Ultimate leaf segments broadly elliptic or broader, usually more than 3.5 mm broad 35
26. Flowers yellow 27
26. Flowers white 28
27. Sheaths of the petioles of the main cauline leaves at least 3 cm long . . .
. . . . 19. *Foeniculum*
27. Sheaths of the petioles of the main cauline leaves up to 3 cm long
. . . . 20. *Anethum*

28. Leaves all basal or nearly so .... 21. *Erigenia*
28. Leaves cauline 29
29. Bracts of the primary involucre numerous, often pinnately divided .. 30
29. Bracts of the primary involucre absent or few, not pinnately divided   31
30. Two finely divided stipules inserted beneath the sheathing petioles ....
.... 22. *Carum*
30. Stipules absent .... 8. *Ptilimnium*
31. Flowers, or some of them, zygomorphic .... 23. *Coriandrum*
31. Flowers actinomorphic 32
32. Ovaries and fruits warty .... 24. *Spermolepis*
32. Ovaries and fruits not warty 33
33. Umbel with 1–6 rays; involucral bracts elliptic to oblong ....
.... 25. *Chaerophyllum*
33. Umbel with at least 7 rays; involucral bracts linear or absent 34
34. Ultimate leaflets up to 1.5 mm broad; all leaves pinnately divided ......
.... 22. *Carum*
34. Ultimate leaflets 1.5–3.0 mm broad; at least the upper leaves ternately
divided .... 26. *Perideridia*
35. Flowers yellow or purple 36
35. Flowers white 39
36. Leaflets entire .... 27. *Taenidia*
36. Leaflets toothed or lobed 37
37. Leaflets of cauline leaves cuneate at base, rarely more than 10 mm
broad.. 28. *Polytaenia*
37. Leaflets of cauline leaves more or less rounded at base, usually more
than 10 mm broad 38
38. Fruit winged; central flower of the ultimate umbel pedicellate ....
.... 3. *Thaspium*
38. Fruit unwinged; central flower of the ultimate umbel sessile .. 14. *Zizia*
39. Leaves all or nearly all basal .... 21. *Erigenia*
39. Leaves borne along the stem 40
40. Fruits twice as long as broad or longer, at least 1 cm long.............
.... 18. *Osmorhiza*
40. Fruits never twice as long as broad, up to 1 cm long 41
41. Major leaflets toothed, but not divided into narrow segments 42
41. Major leaflets dissected into additional segments 44
42. Uppermost leaf sheaths longer than the blades .... 30. *Angelica*
42. Uppermost leaf sheaths shorter than the blades 43
43. Basal leaves with 9 leaflets; petals incurved at tip .... 31. *Aegopodium*
43. Basal leaves usually with more than 9 leaflets; petals not incurved at tip
.... 29. *Cicuta*
44. Fruits winged; central umbel barely or not overtopped by lateral branches... 32. *Conioselinum*
44. Fruits unwinged, but conspicuously ribbed; central umbel much over-
topped by lateral branches 45
45. Primary involucre absent; stems not spotted with purple .. 33. *Aethusa*
45. Primary involucre composed of linear bracts; stems usually spotted with
purple .... 34. *Conium*

1. Eryngium L.

1. Stems erect; leaves linear, with spinulose margins; corolla white ....
.... 1. *E. yuccifolium*

1. Stems prostrate; leaves lanceolate to ovate, not spinulose on the margins; corolla blue .... 2. *E. prostratum*

    1. E. yuccifolium Michx. Rattlesnake Master. July–Aug. Prairies, openings in woods; occasional throughout the state.
    2. E. prostratum Nutt. Eryngo. May–Nov. Edge of pond, very rare; Pope Co.

2. *Bupleurum* L.

    1. Bupleurum rotundifolium L. Thoroughwax. May–July. Native of Europe; rarely escaped from cultivation; St. Clair Co.

3. *Thaspium* Nutt. Meadow Parsnip

1. Stems pubescent at the nodes; leaflets ciliate along the margins; basal leaves 2- 3-ternate .... 1. *T. barbinode*
1. Stems glabrous; leaf margins not ciliate; basal leaves toothed or only once-ternate .... 2. *T. trifoliatum*

    1. T. barbinode (Michx.) Nutt. Hairy Meadow Parsnip. April–June. Moist woods, usually near streams; occasional in the n. half of the state, rare or absent elsewhere.
    2. T. trifoliatum (L.) Gray. Meadow Parsnip. April–June. Flowers purple. Prairies, rocky woods, thickets; not common, but scattered throughout the state.
    var. flavum Blake. Yellow Meadow Parsnip. April–June. Flowers yellow; occasional throughout the state.

4. *Sanicula* L. Snakeroot

1. Styles much longer than the bristles of the fruit; plants with rhizomes   2
1. Styles shorter than or merely equalling the bristles of the fruit; plants without rhizomes 3
2. Fruits sessile, at least 5 mm long; sepals of staminate flowers firm, 1–2 mm long .... 1. *S. marilandica*
2. Fruits stipitate, up to 4 mm long (excluding the stipe); sepals of staminate flowers soft, up to 1 mm long .... 2. *S. gregaria*
3. Pedicels of staminate flowers less than twice as long as the flowers; fruits stipitate, 3–5 mm long .... 3. *S. canadensis*
3. Pedicels more than twice as long as the flowers; fruits sessile, at least 5.5 mm long .... 4. *S. trifoliata*

    1. S. marilandica L. Black Snakeroot. May–July. Wooded slopes; occasional in the n. two-thirds of the state, rare or absent elsewhere.
    2. S. gregaria Bickn. Common Snakeroot. May–June. Woods; occasional to common throughout the state.
    3. S. canadensis L. Canadian Black Snakeroot. May–Aug. Woods; occasional throughout the state. (Including var. *grandis* Fern.)
    4. S. trifoliata Bickn. Large-fruited Black Snakeroot. June–July. Woods, not common; confined to the n. half of the state.

5. *Torilis Adans. Hedge Parsley*

1. T. japonica (Houtt.) DC. Hedge Parsley. June–Aug. Native of Europe and Asia; naturalized in waste ground; occasional throughout the state. (Including *T. arvensis* Link.)

6. *Daucus L. Carrot*

1. Primary involucral bracts bipinnatifid .... 1. *D. pusillus*
1. Primary involucral bracts once-pinnate .... 2. *D. carota*

1. D. pusillus Michx. Small Wild Carrot. April–June. Woods, rare; Jackson and Perry cos.
2. D. carota L. Wild Carrot; Queen-Anne's-Lace. May–Oct. All flowers white except for the central purple one or ones. Native of Europe; common throughout the state.
f. epurpuratus Farw. May–Oct. All flowers white or pinkish. Native of Europe; waste ground; occasional in the s. two-thirds of the state.
f. roseus Millsp. May–Oct. All flowers rose or purple. Native of Europe; waste ground, rare; Jackson Co.

7. *Cynosciadium DC.*

1. C. digitatum DC. May–June. Swampy woods, very rare; Jackson Co.

8. *Ptilimnium Raf. Mock Bishop's-weed*

1. Main leaves alternate or opposite; fruits about 1.5 mm long ....
.... 1. *P. costatum*
1. Main leaves verticillate; fruits 2–4 mm long .... 2. *P. nuttallii*

1. P. costatum (Ell.) Raf. Mock Bishop's-weed. July–Sept. Swampy ground, rare; confined to the s. one-fourth of the state.
2. P. nuttallii (DC.) Britt. Mock Bishop's-weed. June–Aug. Swampy ground, rare; Jackson, Pulaski, Randolph, St. Clair, and Union cos.

9. *Limnosciadium Math. & Const.*

1. L. pinnatum (DC.) Math. & Const. May–June. Native to the w. U.S.; rarely adventive in our w. cos. *Cynosciadium pinnatum* DC.—F.

10. *Oxypolis Raf. Cowbane*

1. O. rigidior (L.) Coulter & Rose. Cowbane. July–Sept. Leaflets lanceolate to oblong, at least 4 mm broad. Wet prairies; occasional throughout the state.
var. ambigua (Nutt.) Robins. Cowbane. July–Sept. Leaflets linear, up to 4 mm broad. Wet ground, rare; confined to the s. half of the state.

11. Heracleum L.   Cow Parsnip

1. H. maximum Bartr. Cow Parsnip. May–Aug. Low woods, rich woods; occasional in the n. two-thirds of the state, rare elsewhere. *H. lanatum* Michx.—G, J.

12. Cryptotaenia DC.   Honewort

1. C. canadensis (L.) DC. Honewort. May–Aug. Rocky woods, low ground; occasional throughout the state.

13. Falcaria Bernh.   Sickleweed

1. F. sioides (Wibel) Aschers. Sickleweed. July–Sept. Native of Europe; rarely adventive in Ill.; Schuyler Co.

14. Zizia Koch   Golden Alexanders

1. All of the basal leaves simple .... 1. *Z. aptera*
1. All or nearly all of the basal leaves ternately compound .... 2. *Z. aurea*

1. Z. aptera (Gray) Fern. Heart-leaved Meadow Parsnip. April–June. Prairies, rocky woods, not common; confined to the n. one-fifth of the state.
2. Z. aurea (L.) Koch. Golden Alexanders. April–June. Moist woods, prairies; occasional in the n. three-fourths of the state, less common in the s. one-fourth.

15. Pastinaca L.   Parsnip

1. P. sativa L. Parsnip. May–Oct. Native of Europe; naturalized in waste ground and along roads; common in the n. three-fourths of the state, occasional in the s. one-fourth, but in every co.

16. Sium L.   Water Parsnip

1. S. suave Walt. Water Parsnip. July–Sept. Marshes, ponds, low woods, swamps; occasional throughout the state.

17. Berula Hoffm.

1. B. pusilla (Nutt.) Fern. Water Parsnip. July–Sept. Marshes, rare; Kane, Kendall, McHenry, Mason, Peoria, Tazewell, and Woodford cos. *B. erecta* Cowles—G; *B. incisa* (Torr.) G. N. Jones—J.

18. Osmorhiza Raf.   Sweet Cicely

1. Styles 2 mm long or longer, exceeding the petals; roots strongly anise-scented .... 1. *O. longistylis*
1. Styles up to 1.5 mm long, shorter than the petals; roots weakly anise-scented .... 2. *O. claytonii*

1. O. longistylis (Torr.) DC. Anise-root. April–June. Stems glabrous or nearly so. Rich woods; occasional throughout the state.

var. villicaulis Fern. Sweet Cicely. April–June. Stems densely villous. Rich woods; less common than the preceding, but scattered throughout the state.

2. O. claytonii (Michx.) Clarke. Sweet Cicely. April–June. Rich woods; occasional throughout the state.

19. *Foeniculum Mill. Fennel*

1. F. vulgare Mill. Fennel. May–Sept. Native of Europe; occasionally cultivated but rarely escaped; Madison Co.

20. *Anethum L. Dill*

1. A. graveolens L. Dill. June–Aug. Native of Asia; occasionally cultivated and escaped; scattered throughout the state.

21. *Erigenia Nutt.*

1. E. bulbosa (Michx.) Nutt. Harbinger-of-Spring; Pepper-and-Salt. Feb.–April. Rich woods; occasional throughout the state.

22. *Carum L. Caraway*

1. C. carvi L. Caraway. May–July. Native of Europe; occasionally cultivated and escaped.

23. *Coriandrum L. Coriander*

1. C. sativum L. Coriander. May–June. Native of Europe; occasionally cultivated but rarely escaped; Jackson and Will cos.

24. *Spermolepis Raf.*

1. S. inermis (Nutt.) Math. & Constance. May–June. Sandy soil; occasional in the nw. one-fourth of the state, rare or absent elsewhere.

25. *Chaerophyllum L. Chervil*

1. Pedicel of fruit the same diameter throughout; fruits broadest near the middle .... 1. *C. procumbens*
1. Pedicel of fruit wider at the top than at the bottom; fruits broadest below the middle .... 2. *C. tainturieri*

1. C. procumbens (L.) Crantz. Wild Chervil. April–June. Rich woods, along streams, along railroads and highways; common throughout the state. (Including var. *shortii* Torr. & Gray.)

2. C. tainturieri Hook. Wild Chervil. March–May. Fields, roadsides, waste ground, not common; confined to the s. one-third of the state.

26. *Perideridia Reichenb.*

1. P. americana (Nutt.) Reichenb. April–July. Floodplains, thickets, rocky woods; occasional in the n. half of the state, becoming rare in the s. half.

338

27. Taenidia Drude   Yellow Pimpernel

1. T. integerrima (L.) Drude. Yellow Pimpernel. May–July. Prairies, dry, often rocky, woods; occasional throughout the state.

28. Polytaenia DC.

1. P. nuttallii DC. Prairie Parsley. April–June. Prairies, rocky woods; occasional throughout the state.

29. Cicuta L.   Water Hemlock

1. Upper leaves bearing bulblets in their axils; leaflets linear   1. *C. bulbifera*
1. None of the leaves bearing bulblets in their axils; leaflets narrowly lanceolate or broader .... 2. *C. maculata*

1. C. bulbifera L. Bulblet Water Hemlock. July–Sept. Marshes, swamps, not common; restricted to the n. half of the state; also Union Co.
2. C. maculata L. Water Hemlock. May–Sept. Marshes, wet prairies, moist woods; occasional to common throughout the state.

30. Angelica L.   Angelica

1. Uppermost leaf sheaths inflated to at least 2 cm broad .... 1. *A. atropurpurea*
1. Uppermost leaf sheaths slander, not more than 1 cm broad .... 2. *A. venenosa*

1. A. atropurpurea L. Angelica. May–Aug. Woodlands, thickets, marshes; occasional in the n. half of the state, rare elsewhere.
2. A. venenosa (Greenway) Fern. Wood Angelica. May–July. Prairies, rich woods, rocky woods, not common; confined to the s. one-fourth of the state.

31. Aegopodium L.   Goutweed

1. A. podagraria L. Goutweed. May–Aug. Native of Europe; infrequently cultivated and rarely escaped; DuPage and Lake cos.

32. Conioselinum Hoffm.   Hemlock Parsley

1. C. chinense (L.) BSP. Hemlock Parsley. July–Sept. Shaded, springy ground, very rare; Cook and Kane cos.

33. Aethusa L.

1. A. cynapium L. Fool's Parsley. June–Aug. Native of Europe; rarely escaped from cultivation; Cook and Kendall cos.

34. Conium L.   Poison Hemlock

1. C. maculatum L. Poison Hemlock. May–Aug. Native of Europe; naturalized in waste ground, fields, thickets, and along roads; occasional throughout the state.

121. CORNACEAE

1. *Cornus* L. Dogwood

1. Flowers surrounded by 4 conspicuous, petal-like bracts; fruits red ... 2
1. Flowers not subtended by 4 conspicuous, petal-like bracts; fruit blue or white 3
2. Herbs to 30 cm tall; flowers pedicellate; drupes globose, soft .... 1. *C. canadensis*
2. Trees to 12 m tall; flowers sessile; drupes ellipsoid, hard .. 2. *C. florida*
3. Leaves alternate .... 3. *C. alternifolia*
3. Leaves opposite 4
4. Leaves with spreading or curled pubescence on the lower surface .. 5
4. Leaves with appressed-pubescence or glabrous on the lower surface 7
5. Leaves broadly ovate to suborbicular, with 6–9 pairs of veins; drupes light blue .... 4. *C. rugosa*
5. Leaves elliptic to lance-ovate, with 3–4 pairs of veins; drupes white .. 6
6. Leaves finely pubescent above; branchlets red; pith white .... 5. *C. stolonifera*
6. Leaves scabrous above; branchlets gray; pith brown . 6. *C. drummondii*
7. Leaves conspicuously whitish or glaucous beneath; drupes white (bluish in *C. obliqua*) 8
7. Leaves green or rufescent beneath; drupes blue 10
8. Young twigs glabrous or nearly so; fruit white 9
8. Young twigs densely hairy; fruit blue .... 8. *C. obliqua*
9. Twigs red; cymes flat-topped, broader than high ..... 5. *C. stolonifera*
9. Twigs gray; cymes elongated, as high as broad .... 7. *C. racemosa*
10. Pith white; leaves glabrous beneath .... 9. *C. foemina*
10. Pith tawny; leaves with reddish pubescence beneath . 10. *C. amomum*

1. C. canadensis L. Bunchberry. May–June. Boggy woods, rare; Cook, Lake, LaSalle, and Ogle cos.

2. C. florida L. Flowering Dogwood. April–May. Rocky woods, wooded slopes, low woods; occasional to common in the s. half of the state. (Including f. *rubra* [Weston] Palmer & Steyerm.)

3. C. alternifolia L. f. Alternate-leaved Dogwood. May–June. Moist woods and slopes, not common; mostly in the n. half of the state.

4. C. rugosa Lam. Round-leaved Dogwood. May. Shaded, rocky slopes, not common; confined to the n. one-fourth of the state.

5. C. stolonifera Michx. Red Osier. June–July. Leaves glabrous or appressed-pubescent beneath. Open marshes, calcareous fens; occasional in the n. half of the state, rare elsewhere.

var. baileyi (Coult. & Evans) Drescher. Bailey's Dogwood. June–July. Leaves densely soft-pilose beneath. Dunes, rare; Cook and Lake cos. C. *baileyi* Coult. & Evans—J.

6. C. drummondii C. A. Mey. Rough-leaved Dogwood. May–June. Rocky woods, prairies, low ground; occasional to common in the s. three-fourths of the state, rare elsewhere.

7. C. racemosa Lam. Gray Dogwood. May–July. Moist woods, upland woods, prairies, roadsides; occasional to common throughout the state.

8. C. obliqua Raf. Pale Dogwood. May–July. Swamps, low woods, wet prairies; occasional to common in most of Ill.

9. C. foemina Mill. Stiff Dogwood. May–June. Swamps, low woods, occasional; restricted to the s. one-half of the state.

10. C. amomum Mill. Dogwood. June–July. Moist thickets, not common; confined to the extreme s. tip of the state.

122. ERICACEAE

1. Plants herbaceous 2
1. Plants woody (*Epigaea* only slightly so) 4
2. Plants devoid of chlorophyll; pollen borne singly .... 1. *Monotropa*
2. Plants with chlorophyll; pollen borne in tetrads 3
3. Leaves on stem; filaments hairy; flowers in corymbs .... 2. *Chimaphila*
3. Leaves basal or nearly so; filaments glabrous; flowers in racemes ......
.... 3. *Pyrola*
4. Ovary superior 5
4. Ovary inferior 11
5. Leaves entire 6
5. Leaves toothed 9
6. Plants upright 7
6. Plants trailing 8
7. Leaves deciduous; anthers awnless; flowers over 1.5 cm long, bright pink or rose; capsule elongated .... 4. *Rhododendron*
7. Leaves evergreen; anthers awned; flowers up to 1 cm long, pale pink or white; capsule depressed-globose .... 5. *Andromeda*
8. Fruit a capsule; flowers salverform; anthers awnless; leaves cordate; anthers splitting lengthwise .... 6. *Epigaea*
8. Fruit a drupe; flowers urn-shaped; anthers awned; leaves cuneate; anthers opening by a terminal pore .... 7. *Arctostaphylos*
9. Erect tree to 15 m tall; flowers in terminal racemes; leaves deciduous ..
.... 8. *Oxydendrum*
9. Trailing or ascending shrub to 35 cm tall; flowers in axillary leafy racemes or solitary; leaves evergreen 10
10. Fruit dry; leaves scurfy beneath .... 9. *Chamaedaphne*
10. Fruit fleshy; leaves not scurfy beneath .... 10. *Gaultheria*
11. Leaves with glandular dots; winter twigs predominantly black ....
.... 11. *Gaylussacia*
11. Leaves without glandular dots; winter twigs green, gray, or brown .....
.... 12. *Vaccinium*

1. Monotropa L. Indian Pipe

1. Flowers several per stem; plants somewhat pubescent 1.*M. hypopithys*
1. Flower solitary; plants glabrous .... 2. *M. uniflora*

1. M. hypopithys L. Pinesap. June–Oct. Rich woods, not common; scattered throughout the state. *M. lanuginosa* Michx.—J.
2. M. uniflora L. Indian Pipe. May–Sept. Rich woods; occasional throughout the state.

2. Chimaphila Pursh Pipsissewa

1. C. umbellata (L.) Bart. var. cisatlantica Blake. Pipsissewa. June–Aug. Dry woods, very rare; Lake, McHenry, and Winnebago cos. *C. corymbosa* Pursh—J.

3. Pyrola L. Wintergreen

1. Flowers borne along one side of raceme; style straight . 1. *P. secunda*
1. Flowers borne spirally in a raceme; style curved downward 2
2. Leaves thin, dull, the blades longer than the petiole; lobes of calyx up to 2 mm long . . . . 2. *P. elliptica*
2. Leaves coriaceous, shiny, the blades about the same length as the petiole; lobes of calyx over 2 mm long . . . . 3. *P. americana*

    1. P. secunda L. One-sided Pyrola. June–Aug. Moist woods, very rare; Cook and Winnebago cos.
    2. P. elliptica Nutt. Shinleaf. June–Aug. Moist woods, rare; confined to extreme n. Ill.
    3. P. americana Sweet. Wild Lily-of-the-Valley. June–Aug. Shaded, mossy, wooded slope, very rare; Ogle Co. *P. rotundifolia* L. var. *americana* (Sweet) Fern.—F, G.

4. Rhododendron L. Azalea

1. Pedicels, sepals, corolla, and capsules glandular; flowers fragrant . . . . .
. . . . 1. *R. prinophyllum*
1. Pedicels, sepals, corolla, and capsules not glandular; flowers essentially without an odor . . . . 2. *R. pericylymenoides*

    1. R. prinophyllum (Small) Millais. Pink Azalea. May. Cherty slopes, rare; Alexander and Union cos. *R. roseum* (Loisel.) Rehd.—F, G, J.
    2. R. periclymenoides (Michx.) Shinners. Pink Azalea. May. Cherty slopes, rare; Union Co. *R. nudiflorum* (L.) Torr.—F, G.

5. Andromeda L. Bog Rosemary

    1. A. glaucophylla Link. Bog Rosemary. May–June. Bogs, very rare; Lake and McHenry cos.

6. Epigaea L. Trailing Arbutus

    1. E. repens L. Trailing Arbutus. April–May. Woods, very rare; "northern Illinois," according to the collector. (Including var. *glabrifolia* Fern.)

7. Arctostaphylos Adans. Bearberry

    1. A. uva-ursi (L.) Spreng. var. coactilis Fern. & Macbr. Kinnikinnick. June–Aug. Sand dunes and black oak woods, rare; n. one-fourth of Ill.; also Peoria Co. *A. uva-ursi* (L.) Spreng.—J.

8. Oxydendrum DC. Sourwood

    1. O. arboreum (L.) DC. Sourwood. June–July. Naturalized from the se.; Cook Co.

9. Chamaedaphne Moench   Leatherleaf

1. C. calyculata (L.) Moench var. angustifolia (Ait.) Rehder. Leather-leaf. May. Bogs, rare; Cook, Kane, Lake, and McHenry cos. *C. calyculata* (L.) Moench—G, J.

10. Gaultheria L. Wintergreen

1. G. procumbens L. Checkerberry. June–Aug. Sandy soil of woods and bogs, rare; Cook, Lake, LaSalle, and Ogle cos.

11. Gaylussacia HBK. Huckleberry

1. G. baccata (Wang.) K. Koch. Black Huckleberry. May–June. Rocky woods and cliffs; occasional in the n. and s. cos., apparently absent from the cent. cos.

12. Vaccinium L. Blueberry

1. Corolla urceolate to campanulate, 5-lobed; stems upright; leaves eventually deciduous 2
1. Corolla deeply 4-lobed; stems prostrate; leaves evergreen .... 7. *V. macrocarpon*
2. Stamens long-exserted from the corolla 1. *V. stamineum* var. *neglectum*
2. Stamens included in the corolla 3
3. Flowers in leafy bracted racemes; anthers awned; pedicels jointed; berry dry .... 2. *V. arboreum*
3. Flowers solitary, in racemes, or in glomerules, but not conspicuously bracteate; anthers awnless; pedicels not jointed; berry juicy 4
4. Leaves with marginal spinulose teeth 3. *V. angustifolium* var. *laevifolium*
4. Leaves entire or serrulate, but without spinulose teeth 5
5. Corolla green, greenish-purple, or purple; leaves up to 5 cm long; low shrubs up to 1 m tall 6
5. Corolla white, sometimes tinged with pink; some or all the leaves over 5 cm long; tall shrubs over 1 m in height .... 6. *V. corymbosum*
6. Branchlets densely pubescent; berries strongly glaucous, sour......... .... 4. *V. myrtilloides*
6. Branchlets glabrous or somewhat pubescent; berries faintly glaucous (strongly so in a few specimens), sweet .... 5. *V. vacillans*

1. V. stamineum L. var. neglectum (Small) Deam. Deerberry. May–June. Edge of sandstone cliff, very rare; Pope Co. *V. neglectum* (Small) Fern.—G.

2. V. arboreum Marsh. Farkleberry. May–June. Bracteal leaves much smaller than foliage leaves. Sandstone cliffs, locally abundant; restricted to the s. three tiers of cos., plus Jersey and Randolph cos.

var. glaucescens (Greene) Sarg. Farkleberry. May–June. Bracteal leaves similar to foliage leaves. Sandstone cliffs, occasional; in the s. one-eighth of Ill.

3. V. angustifolium Ait. var. laevifolium House. Low-bush Blueberry. May–June. Sandy, open woods, slopes of dunes, bogs; confined to the n. one-fourth of the state. *V. angustifolium* Ait.—G, J.

4. V. myrtilloides Michx. Canada Blueberry. May–June. Tamarack

bogs, sandy or rocky slopes; Lake, LaSalle, Ogle, and Winnebago cos.

5. V. vacillans Torr. Low-bush Blueberry. April–May. Sandstone cliffs, open sandy woods; occasional in the s. one-sixth of the state, becoming rare or absent northward. (Including var. *crinitum* Fern.)

6. V. corymbosum L. High-bush Blueberry. May–June. Swamps and bogs; Cook, Lake, LaSalle, McHenry, Stephenson, and Winnebago cos.

7. V. macrocarpon Ait. American Cranberry. June–Aug. Bogs; Cook, Lake, McHenry, and Will cos. *Oxycoccus macrocarpus* (Ait.) Pursh—J.

123. PRIMULACEAE

1. Leaves entire or toothed 2
1. Leaves deeply dissected into threadlike divisions .... 9. *Hottonia*
2. Leaves all basal 3
2. Leaves cauline 5
3. Lobes of corolla reflexed; anthers exserted as a cone; calyx deeply divided; leaves entire, more than 2.5 cm long .... 1. *Dodecatheon*
3. Lobes of corolla spreading; anthers included; calyx tubular; leaves dentate or, if entire, less than 2.5 cm long 4
4. Leaves usually more than 2 cm long, dentate; corolla tube at least as long as the calyx, open at the throat; perennial .... 2. *Primula*
4. Leaves up to 2 cm long, entire; corolla tube shorter than the calyx, closed at the throat; annual .... 3. *Androsace*
5. Leaves, or most of them, alternate 6
5. Leaves, or most of them, opposite and whorled (occasional alternate leaves may be found in *Trientalis* [which also always has some whorled leaves], *Lysimachia quadrifolia,* and *L. terrestris*) 7
6. Ovary superior; flowers generally 4-merous; pedicels absent or less than 5 mm long; flower solitary .... 4. *Centunculus*
6. Ovary inferior; flowers 5-merous; pedicels 1–2 cm long; flowers in racemes .... 5. *Samolus*
7. Flowers scarlet (rarely blue; if white, the petals rounded at tip); capsule circumscissile .... 6. *Anagallis*
7. Flowers yellow or, if white, the petals usually pointed at the tip; capsule splitting lengthwise 8
8. Flowers yellow (white in *L. clethroides*); petals usually 5 or 6 .... 7. *Lysimachia*
8. Flowers white; petals usually 7 .... 8. *Trientalis*

1. Dodecatheon L. Shooting-star

1. Leaves tapering to the petiole, oblanceolate 2
1. Leaves abruptly contracted to a distinct petiole, the blades broadly ovate .... 3. *D. frenchii*
2. Capsule pale brown to yellow, with thin papery walls .... 1. *D. amethystinum*
2. Capsule dark reddish-brown, with rather thick, woody walls .... 2. *D. meadia*

1. D. amethystinum Fassett. Shooting-star. May–June. Bluffs and wooded slopes, rare; Carroll and JoDaviess cos. *D. radicatum*—G.

2. D. meadia L. Shooting-star. April–June. Woods, bluffs, meadows, and prairies; occasional to common throughout Ill. (Including f. *album* Macbr. and var. *brachycarpum* [Small] Fassett.)

3. D. frenchii (Vasey) Rydb. French's Shooting-star. April–May. Under overhanging sandstone cliffs, locally abundant; confined to the s. one-eighth of Ill. The type is from Jackson Co. *D. meadia* L. var. *frenchii* Vasey—F, G.

2. *Primula L. Primrose*

1. P. mistassinica Michx. Bird's-eye Primrose. May–June. Limestone cliffs, rare; JoDaviess Co.

3. *Androsace L.*

1. A. occidentalis Pursh. March–April. Cliffs, sandy soil, rocky woods; scattered in most parts of Ill.

4. *Centunculus L. Chaffweed*

1. C. minimum L. Chaffweed. May–Aug. Moist soil, rare; mostly in the cent. cost.

5. *Samolus L. Brookweed*

1. S. parviflorus Raf. Brookweed. May–Sept. Moist soil; occasional throughout Ill., except for the nw. cos.

6. *Anagallis L. Pimpernel*

1. A. arvensis L. Scarlet Pimpernel. June–Aug. Naturalized from Europe; waste ground, not common; scattered in Ill. (Including f. *caerulea* [Schreb.] Baumg.)

7. *Lysimachia* [Tourn.] L. Loosestrife

1. Corolla yellow 2
1. Corolla white .... 12. *L. clethroides*
2. Leaves epunctate; corolla lobes erose and apiculate; staminodia present
.... 3
2. Leaves punctate; corolla lobes entire; staminodia absent 6
3. Lateral nerves of blades obscure; blades firm .... 1. *L. quadriflora*
3. Lateral nerves of blades conspicuous; blades thin 4
4. Plants decumbent, rooting at the nodes .... 2. *L. radicans*
4. Plants erect, not rooting at the nodes 5
5. Blades ovate to ovate-lanceolate, rounded to subcordate at base ....
.... 3. *L. ciliata*
5. Blades narrow-lanceolate, elliptic, or linear, tapering to base (rarely somewhat rounded) .... 4. *L. lanceolata*
6. Flowers in dense axillary racemes; corolla lobes linear . 5. *L. thyrsiflora*

6. Flowers in terminal panicles or racemes or, if axillary, solitary or in panicles; corolla lobes lanceolate to orbicular (rarely linear in *L.* X *commixta*)
.... 7
7. Stems creeping; plants often evergreen .... 6. *L. nummularia*
7. Stems erect; plants not evergreen 8
8. Corolla without dark markings, crateriform 9
8. Corolla streaked with dark markings, rotate or saucer-shaped 10
9. Calyx dark glandular along the margin; corolla lobes not glandular-
ciliate; flowers in terminal or axillary panicles .... 7. *L. vulgaris*
9. Calyx not dark glandular along the margin; corolla lobes glandular-
ciliate; flowers in the axils of the upper leaves .... 8. *L. punctata*
10. Leaves in whorls of 3–7 .... 9. *L. quadrifolia*
10. Leaves opposite or alternate 11
11. Most or all the leaves (except the lowermost scale-like ones) opposite;
style 3–4 mm long; capsule 2.8–3.5 mm in diameter .... 10. *L. terrestris*
11. Most or all the leaves (except the lowermost scale-like ones) alternate;
style 5–6 mm long; capsule about 2 mm in diameter   11. *L.* X *commixta*

   1. L. quadriflora Sims. Loosestrife. June–Aug. Marshes, moist
prairies, bogs; occasional in the n. cos., becoming rare in the s. cos.
   2. L. radicans Hook. Creeping Loosestrife. June–Aug. Swampy
woods, rare; Pulaski and St. Clair cos.
   3. L. ciliata L. Fringed Loosestrife. June–Aug. Moist woods, bottomlands, thickets; common throughout the state.
   4. L. lanceolata Walt. Loosestrife. June–Aug. Basal rosettes from
slender rhizomes; leaves pale beneath. Moist woods, stream banks,
thickets; common throughout the state.
   var. hybrida (Michx.) Gray. July–Aug. Basal rosettes not from
slender rhizomes; leaves green beneath. Swampy woods, thickets,
meadows; occasional throughout the state. *L. hybrida* Michx.—F.
   5. L. thyrsiflora L. Tufted Loosestrife. May–July. Bottomland woods,
swamps, marshes, bogs; occasional in the n. half of Ill.; also Wabash
Co.
   6. L. nummularia L. Moneywort. June–Aug. Native of Europe; commonly naturalized in moist, shady areas; common throughout the
state.
   7. L. vulgaris L. June–Sept. Native of Europe; occasionally escaped
into moist fields.
   8. L. punctata L. Dotted Loosestrife. May–June. Native of Europe;
rarely escaped along wood's edge; Champaign and Cook cos.
   9. L. quadrifolia L. Whorled Loosestrife. May–Aug. Woods, roadsides, fields; confined to the n. one-fourth of the state.
   10. L. terrestris (L.) BSP. Swamp Candle. June–Aug. Swamps, bogs;
confined to the n. one-third of Ill.; Union Co.
   11. L. X commixta Fern. June–Aug. Wet areas, rare; scattered in Ill.
Reported to be a hybrid between *L. terrestris* (L.) BSP. and *L. thyrsiflora* L.
   12. L. clethroides Duby. July–Aug. Native of Japan; rarely escaped
from cultivation; Cook, Lake, and McHenry cos.

8. Trientalis L.   Star-flower

   1. T. borealis Raf. Star-flower. June–July. Moist woods, rare; Cook,
Lake, LaSalle, Ogle, and Winnebago cos.

9. Hottonia L.

1. H. inflata Ell. Featherfoil. June–Aug. Swamps, rare; Jackson, Johnson, and Union cos.

124. SAPOTACEAE

1. Bumelia Sw. Buckthorn

1. Branchlets and lower surface of leaves rusty-woolly; pedicels pubescent; fruit oblongoid to globose ..... 1. *B. lanuginosa* var. *oblongifolia*
1. Branchlets and lower surface of leaves silky at first, becoming smooth at maturity; pedicels glabrous; fruit ovoid .... 2. *B. lycioides*

1. B. lanuginosa (Michx.) Pers. var. oblongifolia (Nutt.) R. B. Clark. Woolly Buckthorn. June–July. Dry woods, rare; Hardin, Pulaski, and Monroe cos. *B. lanuginosa* (Michx.) Pers.—J.
2. B. lycioides (L.) Gaertn. f. Southern Buckthorn. June–Aug. Moist woods or rocky cliffs, rare; apparently confined to the one-fourth of the state.

125. EBENACEAE

1. Diospyros L. Persimmon

1. D. virginiana L. Common Persimmon. May–June. Dry woods, fields, roadsides, clearings; occasional to common in the s. two-thirds of the state, apparently absent in the n. one-third. (Including var. *pubescens* [Pursh] Dippel, with pubescent leaves and branches, var. *platycarpa* Sarg., with large, depressed-globose, early-ripening fruits, and f. *atra* Sarg., with dark purple fruits.)

126. STYRACACEAE

1. Calyx and corolla 4-parted; ovary inferior; fruit an elongated, winged nut .... 1. *Halesia*
1. Calyx and corolla 5-parted; ovary superior; fruit a globose drupe .... 2. *Styrax*

1. Halesia Ellis Silverbell

1. H. carolina L. Silverbell Tree. April–May. Moist woods, rare; Massac Co.

2. Styrax L. Storax

1. Leaves glabrous or puberulent beneath; flowers 3–4 in a cluster .... 1. *S. americana*

1. Leaves densely white-hairy beneath; flowers more than 4 in a raceme ..
.... 2. *S. grandifolia*

    1. S. americana Lam. Storax. April–May. Swampy woods, rare; confined to extreme s. Ill. and Lawrence Co.
    2. S. grandifolia Ait. Big Leaf Snowbell Bush. May. Along stream in woods, very rare; Alexander Co.

127. OLEACEAE

1. Leaves pinnately compound; fruit a samara; trees .... 1. *Fraxinus*
1. Leaves simple; fruit a capsule, drupe, or berry 2
2. Leaves somewhat serrulate; petals absent; fruit an ellipsoid drupe .....
.... 2. *Forestiera*
2. Leaves entire; petals 4; fruit a capsule or subglobose berry 3
3. Leaves truncate or subcordate; petioles 1–2 cm long; flowers lilac; fruit a capsule .... 3. *Syringa*
3. Leaves cuneate; petioles less than 1 cm long; flowers white; fruit a berry
.... 4. *Ligustrum*

1. *Fraxinus* [*Tourn.*] *L. Ash*

1. Leaflets sessile (except for the terminal one) .... 1. *F. nigra*
1. Leaflets petiolulate 2
2. Twigs 4-sided; body of samara flat .... 2. *F. quadrangulata*
2. Twigs more or less terete or angular, not 4-sided; body of samara terete
.... 3
3. Leaflets with petiolules winged nearly to base; leaves green on both sides .... 3. *F. pennsylvanica*
3. Leaflets with petiolules unwinged; leaves paler or brownish beneath 4
4. Leaves whitened beneath; body of fruit winged less than ⅓ its length ..
.... 4. *F. americana*
4. Leaves brownish beneath; body of fruit winged about ½ its length .....
.... 5. *F. tomentosa*

    1. F. nigra Marsh. Black Ash. May–June. Moist woods, springy slopes; occasional in the n. half of the state; also Jersey, Lawrence, and Wabash cos.
    2. F. quadrangulata Michx. Blue Ash. March–April. Moist woods, limestone cliffs; occasional in the n. three-fifths of the state, rare elsewhere.
    3. F. pennsylvanica Marsh. Red Ash. April–May. Branchlets and lower surface of leaves velvety-pubescent; leaflets entire; samaras 4–7 cm long. Moist woods; occasional to common throughout the state.
    var. austinii Fern. Green Ash. April–May. Branchlets and lower leaf surface velvety-pubescent; leaflets serrate; samaras less than 4 cm long. Moist woods, rare; confined to the s. half of the state.
    var. subintegerrima (Vahl) Fern. Green Ash. April–May. Branchlets and lower leaf surface essentially glabrous; leaflets serrate. Moist woods, common; throughout the state. *F. lanceolata* Borkh.—J.
    4. F. americana L. White Ash. April–May. Branchlets and petioles glabrous. Woods; common throughout the state.

var. biltmoreana (Beadle) J. Wright. Hairy White Ash. April–May. Branchlets and petioles velvety-pubescent. Rich woods, rare; confined to the s. one-third of the state.

5. F. tomentosa Michx. f. Pumpkin Ash. April–May. Swamps and low woods; occasional or rare in the s. two-fifths of the state.

2. Forestiera Poir. Swamp Privet

1. F. acuminata (Michx.) Poir. Swamp Privet. March–April. Swamps, low woods; occasional or rare in the s. three-fifths of the state.

3. Syringa L. Lilac

1. S. vulgaris L. Common Lilac. May–June. Native of Europe; rarely escaped in Ill.

4. Ligustrum L. Privet

1. L. vulgare L. Common Privet. June–July. Native of Europe; rarely escaped in Ill.

128. LOGANIACEAE

1. Leaves ovate-lanceolate to ovate; flowers large and showy, red and yellow; stamens 5 .... 1. *Spigelia*
1. Leaves linear or subulate; flowers small, white; stamens 4 ....
.... 2. *Polypremum*

1. Spigelia L.

1. S. marilandica L. Indian Pink. May–June. Woods; occasional in the s. one-fourth of the state.

2. Polypremum L.

1. P. procumbens L. June–Oct. Moist, sandy soil, very rare; Alexander Co.

129. GENTIANACEAE

1. Leaves whorled; flowering stem 1 m or more tall; each lobe of the corolla with a large nectariferous gland .... 1. *Swertia*
1. Leaves opposite, alternate, or reduced to scales 2
2. Leaves all scale-like .... 2. *Bartonia*
2. Some or all the leaves not scale-like 3
3. Lowermost leaves scale-like; calyx lobes 2 .... 3. *Obolaria*
3. None of the leaves scale-like; calyx lobes 4–5 4
4. Lobes of corolla longer than the tube; stem 4-angled .... 4. *Sabatia*

4. Lobes of corolla as long as or shorter than the tube; stem terete 5
5. Corolla red or pink, the tube up to 2 mm in diameter 5 . 5. *Centaurium*
5. Corolla blue, purple, or white, the tube at least 5 mm in diameter ......
.... 6. *Gentiana*

1. *Swertia* L.   Columbo

   1. S. caroliniensis (Walt.) Kuntze. American Columbo. May–June. Rich woods; occasional in the s. half of the state; also Cook Co. *Frasera caroliniensis* Walt.—J.

2. *Bartonia* Muhl.

1. Most or all the scale leaves opposite .... 1. *B. virginica*
1. Most or all the scale leaves alternate .... 2. *B. paniculata*

   1. B. virginica (L.) BSP. Yellow Bartonia. Aug.–Oct. Bogs, shaded sandstone cliffs, rare; in the n. one-fourth of the state; also Clark Co.
   2. B. paniculata (Michx.) Muhl. Screw-stem. Aug.–Sept. Shaded sandstone cliffs, very rare; Johnson Co.

3. *Obolaria* L.   Pennywort

   1. O. virginica L. Pennywort. April–May. Rich woods, rare; confined to the s. one-fourth of the state; also Crawford Co.

4. *Sabatia* Adans.   Rose Gentian

1. Tube of calyx 2–3 mm long; leaves at middle of stem clasping ....
.... 1. *S. angularis*
1. Tube of calyx 4–8 mm long; leaves at middle of stem not clasping .....
.... 2. *S. campestris*

   1. S. angularis (L.) Pursh. Marsh Pink. June–Sept. Moist soil; occasional in the s. two-thirds of the state, rare or absent elsewhere. (Including f. *albiflora* [Raf.] House.)
   2. S. campestris Nutt. Prairie Rose Gentian. July–Sept. Prairies, very rare; DuPage, Peoria, and Washington cos.

5. *Centaurium* Hill   Centaury

   1. C. pulchellum (Sw.) Druce. Centaury. June–Sept. Native of Europe; rarely adventive in Ill.; Cook Co.

6. *Gentiana* L.   Gentian

1. Corolla with petaloid outgrowths between the lobes 2
1. Corolla lacking petaloid outgrowths between the lobes 5
2. Corolla white or yellowish-white; margins of leaves glabrous   1. *G. alba*
2. Corolla blue or bluish-purple; margins of leaves ciliolate 3
3. Stems puberulent; anthers free; tube of corolla open at the top ....
.... 2. *G. puberulenta*
3. Stems glabrous; anthers coherent in a ring; tube of corolla nearly closed at the top 4

4. Lobes of corolla smaller than the appendages between them ....
.... 3. *G. andrewsii*
4. Lobes of corolla longer than the apendages between them ....
.... 4. *G. saponaria*
5. Corolla 2–6 cm long, the lobes fringed or dentate at the tip; seeds papillose 6
5. Corolla 1.0–2.0 (–2.5) cm long, the lobes neither fringed nor toothed;
seeds smooth .... 7. *G. quinquefolia*
6. Uppermost leaves narrowly ovate or ovate, rounded at the base; fringe
of corolla lobes uniformly at least 2 mm long .... 5. *G. crinita*
6. Uppermost leaves linear to linear-lanceolate; fringe of corolla lobes 2–4
mm long at the edges but reduced to short teeth at the summit ....
.... 6. *G. procera*

1. G. alba Gray. Pale Gentian. Aug.–Oct. Prairies, rich wooded slopes,
not common; scattered throughout the state. *G. flavida* Gray—F, G, J.

2. G. puberulenta Pringle. Downy Gentian. Sept.–Oct. Prairies; occasional in the n. two-thirds of the state, rare or absent elsewhere. *G.
puberula* Michx.—F, G, J.

3. G. andrewsii Griseb. Closed Gentian. Aug.–Oct. Moist woods,
prairies; occasional in the n. half of the state, rare in the s. half.

4. G. saponaria L. Soapwort Gentian. Aug.–Oct. Moist prairies,
sandy woods; occasional in the ne. cos.; also Gallatin and Pope cos.

5. G. crinita Froel. Fringed Gentian. Aug.–Nov. Marshes, not common; confined to the n. one-fourth of the state.

6. G. procera Holm. Small Fringed Gentian. Aug.–Oct. Marshes, not
common; confined to the n. one-fourth of the state.

7. G. quinquefolia L. var. occidentalis (Gray) Hitchc. Stiff Gentian.
Aug.–Oct. Meadows, prairies, calcareous woods; occasional in the n.
two-thirds of the state, absent elsewhere. *G. quinquefolia* L.—J.

130. MENYANTHACEAE

1. Leaves trifoliolate .... 1. *Menyanthes*
1. Leaves simple .... 2. *Nymphoides*

1. Menyanthes L. Buckbean

1. M. trifoliata L. var. minor Raf. Buckbean. May–June. Bogs,
marshes, not common; confined to the ne. cos.; also Peoria Co. *M.
trifoliata* L.—G, J.

2. Nymphoides Hill Floating Heart

1. N. peltata (Gmel.) Kuntze. Yellow Floating Heart. June–Sept. Native of Europe; rarely adventive in Ill. Montgomery Co.

131. APOCYNACEAE

1. Leaves alternate .... 1. *Amsonia*
1. Leaves opposite 2
2. Leaves evergreen; flowers solitary in the axils; corolla salverform ......
.... 2. *Vinca*
2. Leaves deciduous; flowers in cymes; corolla funnelform or campanulate
.... 3
3. Plants erect; corolla campanulate, white or pink .... 3. *Apocynum*
3. Plants twining; corolla funnelform, greenish-yellow .4. *Trachelospermum*

1. Amsonia Walt.

1. A. tabernaemontana Walt. Blue Star. April–June. Leaves elliptic to ovate, most of them at least 3 cm broad. Rocky woods, thickets; occasional in the s. three-fourths of the state.

var. salicifolia (Pursh) Woodson. Blue Star. May–June. Leaves lanceolate, to 3 cm broad, glaucous on the lower surface. Rocky woods; occasional in the s. half of the state.

var. gattingeri Woodson. Blue Star. May–July. Leaves lanceolate, to 3 cm broad, green on the lower surface. Rich woods, not common; confined to the s. and w. cos. of the state.

2. Vinca L. Periwinkle

1. Leaves to 3 cm long; calyx lobes glabrous; flower to 3 cm broad ....
.... 1. *V. minor*
1. Leaves over 3 cm long; calyx lobes ciliate; flower over 3 cm broad .....
.... 2. *V. major*

1. V. minor L. Common Periwinkle; Myrtle. April–May. Native of Europe; occasionally escaped from cultivation.

2. V. major L. Large Periwinkle. April–May. Native of Europe; rarely escaped from cultivation; Pope Co.

3. Apocynum L. Dogbane, Indian Hemp

1. Corolla 4–10 mm long, pink or pink-tinged, rarely white; seeds 2.5–4.0 mm long 2
1. Corolla 2–4 mm long, white or greenish-white; seeds 4–6 mm long .. 3
2. Flowers pendulous; corolla lobes recurved; seeds 2.5–4.0 mm long ....
.... 1. *A. androsaemifolium*
2. Flowers ascending; corolla lobes not recurved; seeds 3–4 mm long ....
.... 2. *A. X medium*
3. Leaves petiolate; corolla greenish-white; follicles over 10 cm long; coma of seed at least 2.5 cm long .... 3. *A. cannabinum*
3. Leaves sessile or nearly so; corolla white; follicles up to 10 cm long; coma of seed up to 2 cm long .... 4. *A. sibiricum*

1. A. androsaemifolium L. Spreading Dogbane. May–July. Woods, prairies; occasional to common throughout the state.

2. A. X medium Greene. Intermediate Dogbane. May–Aug. Woods,

rare; Peoria Co. Reputedly the hybrid between *A. androsaemifolium* L. and *A. cannabinum* L.

3. A. cannabinum L. Indian Hemp; Dogbane. May–Aug. Stems, leaves, and branches of the inflorescence glabrous or nearly so. Prairies, fields, rocky woods, common; in every co.

var. pubescens (Mitchell) A. DC. Hairy Dogbane. May–Aug. Stems, leaves, and branches of the inflorescence pubescent. Prairies, fields, rocky woods, common; in every co. *A. pubescens* R. Br.—J.

4. A. sibiricum Jacq. Indian Hemp. June–Aug. Leaves rounded at the sessile base. Prairies, fields, rocky woods, common; scattered throughout the state.

var. cordigerum (Greene) Fern. Indian Hemp. June–Aug. Leaves cordate at the clasping base. Moist fields, not common; scattered in Ill.

4. Trachelospermum Lemaire   Climbing Dogbane

1. T. difforme (Walt.) Gray. Climbing Dogbane. May–July. Swamps, wet woods; confined to the s. one-fourth of the state.

132. ASCLEPIADACEAE

1. Plants erect .... 1. *Asclepias*
1. Plants climbing or twining 2
2. Leaves cordate 3
2. Leaves rounded at the base .... 2. *Cynanchum*
3. Flowers whitish; plants glabrous .... 2. *Cynanchum*
3. Flowers greenish-yellow or reddish-purple; plants pubescent 3. *Matelea*

1. Asclepias L.   Milkweed

1. Most or all leaves alternate 2
1. Most or all leaves opposite or whorled 5
2. Leaves glabrous 3
2. Leaves pubescent 4
3. Corolla lobes spreading; flowers green and purple; leaves at least 1 cm
     broad .... 1. *A. viridis*
3. Corolla lobes reflexed; flowers green and white; leaves up to 5 mm
     broad .... 2. *A. stenophylla*
4. Latex present; flowers green and white; hood without a horn ....
     .... 3. *A. hirtella*
4. Latex absent; flowers orange or yellow; hood with an incurved horn ...
     .... 4. *A. tuberosa*
5. Leaves filiform to narrowly linear, up to 3 mm broad ... 5. *A. verticillata*
5. Leaves broader, more than 3 mm broad 6
6. Hood without a horn; corolla green, the hood green or purple 7
6. Hood with a horn; corolla not green or, if green, the hood white 8
7. Umbel solitary; stems and leaves hirsute .... 6. *A. lanuginosa*
7. Umbels several; stems and leaves glabrous or pilose .. 7. *A. viridiflora*
8. Leaves sessile 9
8. Leaves petiolate 10

9. Leaves obtuse, oblong or elliptic, broadest at or near the middle, clasping .... 8. *A. amplexicaulis*
9. Leaves subacute to acute, lanceolate to ovate, broadest at or near the base, not clasping .... 9. *A. meadii*
10. Leaves pubescent, at least on the lower surface 11
10. Leaves glabrous 13
11. Veins of leaf ascending; hoods yellow .... 10. *A. ovalifolia*
11. Veins of leaf horizontally spreading; hoods not yellow 12
12. Flowers dark purplish-red; follicles without soft spines .... 11. *A. purpurascens*
12. Flowers variously pale purple and green, or mixed with white; follicles usually with soft spines .... 12. *A. syriaca*
13. Never more than six pairs or whorls of leaves on the stem 14
13. At least seven pairs of leaves on the stem 15
14. Some of the leaves usually whorled; corolla pale pink, the lobes 4–6 mm long .... 13. *A. quadrifolia*
14. None of the leaves usually whorled; corolla white with a purple center, the lobes 8–12 mm long .... 14. *A. variegata*
15. Leaves broadly rounded or subcordate at base; lobes of corolla at least 1 cm long .... 15. *A. sullivantii*
15. Leaves tapering to base; lobes of corolla up to 1 cm long 16
16. Lobes of corolla greenish, 6–8 mm long, the hoods white or pink .... 16. *A. exaltata*
16. Lobes of corolla rose-purple to white, 3–5 mm long, the hoods rose-purple or white 17
17. Flowers white; follicles glabrous; seeds not comose ... 17. *A. perennis*
17. Flowers rose-purple (rarely white); follicles puberulent; seeds comose .. .. 18. *A. incarnata*

1. A. viridis Walt. Green-flowered Milkweed. May–July. Prairies; occasional in the s. one-third of the state. *Asclepiodora viridis* (Walt.) Gray—F, G, J.

2. A. stenophylla Gray. Narrow-leaved Green Milkweed. June–July. Dry upland woods, very rare; Adams Co. *Acerates angustifolia* (Nutt.) Decne.—G; *Asclepias angustifolia* Nutt.—J.

3. A. hirtella (Pennell) Woodson. Tall Green Milkweed. May–Aug. Prairies, fields; occasional throughout the state. *Acerates hirtella* Pennell—G.

4. A. tuberosa L. var. interior (Woodson) Shinners. Butterfly-weed. May–Sept. Woods, prairies, common; in every co. (Including f. *lutea* [Clute] Steyerm.) *A. tuberosa* L.—F, J.

5. A. verticillata L. Horsetail Milkweed. May–Sept. Dry rocky woods, prairies, fields, occasional to common; in every co.

6. A. lanuginosa Nutt. Woolly Milkweed. May–June. Rocky or gravelly prairies, rare; confined to the n. one-sixth of the state. *Acerates lanuginosa* (Nutt.) Dec.—G.

7. A. viridiflora Raf. Green Milkweed. May–Aug. Prairies, sands, gravels, and fields; occasional throughout the state. (Including var. *lanceolata* [Ives] Torr.) *Acerates viridiflora* (Raf.) Eat.—G.

8. A. amplexicaulis Sm. Sand Milkweed. May–July. Prairies, sandy areas; occasional in the n. three-fourths of the state, rare elsewhere.

9. A. meadii Torr. Mead's Milkweed. May–June. Mesic, virgin prairies, rare; Cook, Fulton, Hancock, Henderson, Peoria, Saline, and Tazewell cos.

10. A. ovalifolia Decne. Oval Milkweed. May–June. Prairies and dry woods, rare; Cook, Kankakee, Lake, and McHenry cos.

11. A. purpurascens L. Purple Milkweed. May–July. Woodland borders, prairies; occasional throughout the state.

12. A. syriaca L. Common Milkweed. May–Aug. Follicle with spiny processes up to 3 mm long. Fields, edge of prairies, waste ground; common in Ill.

var. kansana (Vail) Palmer & Steyerm. Common Milkweed. May–Aug. Follicle with spiny processes 3–10 mm long. Fields, waste ground; occasional and scattered in Ill.

13. A. quadrifolia Jacq. Whorled Milkweed. May–July. Rocky open woods; occasional in the s. two-thirds of the state, absent elsewhere.

14. A. variegata L. Variegated Milkweed. May–July. Rocky open woods, occasional; restricted to the s. one-fourth of the state.

15. A. sullivantii Engelm. Prairie Milkweed. June–July. Moist prairies; occasional in the n. two-thirds of the state, rare elsewhere.

16. A. exaltata L. Poke Milkweed. June–July. Woodland borders, uncommon; scattered in most parts of the state.

17. A. perennis Walt. White Milkweed. May–Sept. Wet woods, swamps; occasional in the s. one-fourth of the state; also along the Wabash River in se. Ill. and Richland Co.

18. A. incarnata L. Swamp Milkweed. June–Aug. Marshes, bogs, and ditches, common; in every co.

2. Matelea Aubl.   Climbing Milkweed

1. Flowers greenish-yellow; calyx and pedicels glabrous or nearly so; follicles smooth, angular; blooming usually in July and August . . . .
. . . . 1. *M. gonocarpa*
1. Flowers rose, maroon, or rarely cream; calyx and pedicels pubescent; follicles rounded, muricate; blooming usually in May and June 2
2. Petals 1.5–2.5 mm wide, rose or rarely cream . . . . 2. *M. obliqua*
2. Petals 3–6 mm wide, maroon . . . . 3. *M. decipiens*

1. M. gonocarpa (Walt.) Shinners. Climbing Milkweed. July–Aug. Rocky woods, rare; confined to the s. one-sixth of the state. *Gonolobus gonocarpos* (Walt.) Perry —F, G, J.

2. M. obliqua (Jacq.) Woodson. Climbing Milkweed. May–June. Rocky woods, very rare; Pope Co. *Gonolobus obliquus* (Jacq.) Schultes—F, G, J.

3. M. decipiens (Alex.) Woodson. May–June. Low floodplain woods, natural levee of Big Muddy River, very rare; Williamson Co. *Gonolobus decipiens* (Alex.) Perry—F, G, J.

3. Cynanchum L.

1. Leaves cordate; flowers white . . . . 1. *C. laeve*
1. Leaves rounded at base; flowers dark purple . . . . 2. *C. nigrum*

1. C. laeve (Michx.) Pers. Blue Vine. July–Sept. Moist woods, fields, thickets, roadsides; occasional to common in the s. two-thirds of the state, rare or absent elsewhere. *Ampelamus albidus* (Nutt.) Britt.—F, J.

2. C. nigrum (L.) Pers. Black Swallow-wort. June–Sept. Native of

Europe; rarely adventive in Ill.; Christian, Cook, DuPage, Kane, and Peoria cos.

133. CONVOLVULACEAE

1. Plants with chlorophyll 2
1. Plants without chlorophyll .... 6. *Cuscuta*
2. Leaves linear, up to 3 (–6) mm broad; style deeply 2-cleft .. 1. *Stylisma*
2. Leaves oblong to ovate, never linear, over 6 mm broad; style undivided or merely notched at the apex 3
3. Stigmas 2, non-capitate, without lobes 4
3. Stigma 1, capitate or with 2 or 3 lobes .... 5. *Ipomoea*
4. Stigmas elliptic, oblong, or flattened .... 2. *Jacquemontia*
4. Stigmas filiform or subulate 5
5. Calyx not concealed by 2 large bracts; fruit 2-locular .. 3. *Convolvulus*
5. Calyx concealed by 2 large bracts; fruit 1-locular .... 4. *Calystegia*

1. Stylisma Raf.

1. S. pickeringii (Torr.) Gray var. pattersonii (Fern. & Schub.) Myint. Aug. Sandy prairies, very rare; Cass, Henderson, and Mason cos. The original collection by *H. N. Patterson* in 1873 from Henderson Co., and a subsequent collection from the same area by V. H. Chase in 1934, are from the type locality. *Breweria pickeringii* (Torr.) Gray var. *pattersoni* Fern. & Schub.—F; *S. patersonii* (Fern. & Schub.) G. N. Jones—J.

2. Jacquemontia Choisy

1. J. tamnifolia (L.) Griseb. Aug.–Oct. Native of trop. Am.; adventive along a railroad; Grundy Co.

3. Convolvulus L.   Bindweed

1. C. arvensis L. Field Bindweed. May–Sept. Native of Europe; naturalized in waste areas; occasional to common throughout the state.

4. Calystegia R. Br.   Bindweed

1. Flowers double .... 1. *C. pubescens*
1. Flowers single 2
2. Plants upright, either erect or ascending .... 2. *C. spithamaea*
2. Plants trailing or climbing .... 3. *C. sepium*

1. C. pubescens Lindl. California Rose. May–Sept. Native of Asia; rarely adventive in Ill.; Champaign, DuPage, and Wabash cos. *Convolvulus japonicus* Thunb.—G, J; *Convolvulus pellitus* Ledeb. f. *anestius* Fern.—F.
2. C. spithamaea (L.) Pursh. Dwarf Bindweed. May–July. Dry soil; occasional in the n. four-fifths of the state. *Convolvulus spithamaeus* L.—F, G, J.

3. C. sepium (L.) R. Br. var. americana (Sims) Mohlenbr. American Bindweed. June–Aug. Moist soil, fields, waste areas; common throughout the state. *Convolvulus sepium* L.—F, G; *Convolvulus americanus* (Sims) Greene—J; *Convolvulus sepium* L. var. *americanus* Sims—G.

var. repens (L.) Mohlenbr. Trailing Bindweed. July. Apparently adventive in Ill. in a waste area; Lake Co. *Convolvulus repens* L.—J; *Convolvulus sepium* L. var. *repens* (L.) Gray—F, G.

var. fraterniflora (Mack. & Bush) Mohlenbr. Bindweed. June–Aug. Roadsides, fields; occasional in the s. three-fourths of the state; also DeKalb Co. *Convolvulus fraterniflorus* Mack. & Bush—J; *Convolvulus sepium* L. var. *fraterniflorus* Mack. & Bush—F, G.

5. *Ipomoea* L. *Morning-glory*

1. Flowers scarlet; stamens and style exserted .... 1. *I. coccinea*
1. Flowers white, pink, purple, or blue; stamens and style included 2
2. Perennials with glabrous or puberulent stems; sepals glabrous, obtuse to subacute; seeds pubescent .... 2. *I. pandurata*
2. Annuals with pubescent stems; sepals pubescent, acute to acuminate; seeds glabrous or nearly so 3
3. Calyx lobes linear-lanceolate, with long-tapering tips; corolla essentially 3.0–4.5 cm long, sky blue (when fresh) .... 3. *I. hederacea*
3. Calyx lobes oblong to lanceolate, acute to short-acuminate; corolla either less than 3 cm long or more than 4.5 cm long, not sky blue (when fresh) 4
4. Corolla less than 3 cm long, essentially white; ovary and capsule 2-locular .... 4. *I. lacunosa*
4. Corolla more than 4.5 cm long, usually not white; ovary and capsule 3-locular .... 5. *I. purpurea*

1. I. coccinea L. Red Morning-glory. July–Oct. Native of trop. Am.; occasionally naturalized in moist areas, particularly in the s. half of the state. *Quamoclit coccinea* (L.) Moench—G, J.

2. I. pandurata (L.) G. F. W. Mey. Wild Sweet Potato Vine. June–Oct. Generally in disturbed areas and thickets; occasional to common in the s. three-fourths of the state, less common elsewhere.

3. I. hederacea (L.) Jacq. Ivy-leaved Morning-glory. June–Oct. Native of trop. Am.; naturalized in waste areas; occasional to common throughout the state. (Including var. *integriuscula* Gray, with unlobed leaves.)

4. I. lacunosa L. Small White Morning-glory. July–Oct. Fields and near streams; occasional in the s. three-fifths of Ill., apparently absent elsewhere. (Including f. *purpurata* Fern.)

5. I. purpurea (L.) Roth. Common Morning-glory. July–Oct. Native of trop. Am.; naturalized in disturbed areas; occasional throughout the state.

6. *Cuscuta* L. *Dodder*

1. Sepals free to base ....•.................................... 2
1. Sepals united below into a tube 4
2. Flowers pedicellate, borne in rather loose cymes or panicles; seeds 1.4–1.5 mm long .... 1. *C. cuspidata*

2. Flowers sessile, borne in dense glomerules; seeds 1.7 mm long or longer 3
3. Bracts at base of sepals appressed; lobes of corolla obtuse; seeds 2.5–2.6 mm long .... 2. *C. compacta*
3. Bracts at base of sepals with recurved tips; lobes of corolla acute; seeds 1.7–1.8 mm long .... 3. *C. glomerata*
4. Most of the flowers with 4-lobed corollas 5
4. Most of the flowers with 5-lobed corollas 7
5. Corolla lobes erect; flowers sessile or on pedicels up to 0.5 mm long 6
5. Corolla lobes inflexed; flowers on pedicels usually at least 1 mm long .. .. 6. *C. coryli*
6. Scales absent, or reduced to minute teeth along the filaments; lobes of corolla acute, about as long as the corolla tube; seeds 1.3–1.4 mm long .... 4. *C. polygonorum*
6. Scales toothed from base to apex; lobes of corolla obtuse to subacute, shorter than the corolla tube; seeds 1.6–1.7 mm long 5. *C. cephalanthi*
7. Lobes of corolla obtuse, erect or spreading .... 7. *C. gronovii*
7. Lobes of corolla acute, the tips inflexed 8
8. Lobes of calyx obtuse; pedicels shorter than the flowers 9
8. Lobes of calyx acute; pedicels as long as or longer than the flowers ... . 10. *C. indecora*
9. Lobes of corolla acuminate; scales about half as long as corolla tube; seeds 1.0–1.2 mm long .... 8. *C. pentagona*
9. Lobes of corolla acute; scales about as long as corolla tube; seeds 1.5–1.6 mm long .... 9. *C. campestris*

    1. C. cuspidata Engelm. Dodder. July–Oct. Moist or dry areas; scattered throughout the state, but apparently not common.
    2. C. compacta Juss. Dodder. July–Oct. Moist areas; scattered throughout the state, but not common.
    3. C. glomerata Choisy. Dodder. July–Oct. Low, wet areas; occasional and scattered throughout the state.
    4. C. polygonorum Engelm. Dodder. July–Sept. Moist areas; occasional and scattered throughout the state.
    5. C. cephalanthi Engelm. Dodder. Aug.–Oct. Low, wet areas; occasional and scattered throughout the state.
    6. C. coryli Engelm. Dodder. July–Oct. Moist or dry areas; scattered throughout the state, but apparently not common.
    7. C. gronovii Willd. Dodder. July–Oct. Low, wet areas; common throughout the state.
    8. C. pentagona Engelm. Dodder. June–Oct. Dry, mostly disturbed, areas; occasional to common throughout the state.
    9. C. campestris Yuncker. Dodder. June–Oct. Mostly dry areas; scattered but not common.
    10. C. indecora Choisy. Dodder. July–Sept. Calyx lobes about half as long as corolla tube. Low areas; apparently only in the s. two-thirds of the state, but not common.
    var. neuropetala (Engelm.) Hitchc. Dodder. July–Sept. Calyx lobes about as long as corolla tube. Low areas; rare in the s. half of the state.

134. POLEMONIACEAE

1. Leaves pinnately compound or deeply pinnatifid 2
1. Leaves simple, entire 3
2. Leaves pinnately compound with lanceolate to oval leaflets; flowers blue, campanulate, to 1.5 cm long .... 1. *Polemonium*
2. Leaves deeply pinnatifid, the segments filiform; flowers red or pink, narrowly funnelform, at least 2.5 cm long .... 2. *Gilia*
3. All leaves alternate .... 3. *Collomia*
3. At least the lower cauline leaves opposite 4
4. Usually all leaves opposite; calyx actinomorphic; corolla 1.5 cm long or longer .... 4. *Phlox*
4. Uppermost leaves alternate; calyx slightly zygomorphic; corolla 8–12 mm long .... 5. *Microsteris*

1. *Polemonium L. Greek Valerian*

1. P. reptans L. Jacob's-ladder. April–June. Moist or sometimes dry woods; common throughout the state.

2. *Gilia R. & P.*

1. G. rubra (L.) Heller. Standing Cypress. June–Aug. Native of the s. U.S.; rarely escaped from cultivation; Clark and Woodford cos.

3. *Collomia Nutt.*

1. C. linearis Nutt. June–Aug. Native of the w. U.S.; adventive along railroads; occasional in the n. half of the state.

4. *Phlox L. Phlox*

1. Petals deeply notched or emarginate at the tip 2
1. Petals not notched or emarginate at the tip 3
2. Leaves narrowly linear; plants diffuse and much branched; petals deeply notched .... 1. *P. bifida*
2. Leaves lanceolate to narrowly ovate; plants erect or ascending, not diffusely branched; petals emarginate .... 2. *P. divaricata*
3. Calyx lobes longer than calyx tube; stamens and style about half as long as corolla tube 4
3. Calyx lobes equalling or shorter than calyx tube; stamens and style about as long as corolla tube 5
4. Leaves lanceolate to elliptic to narrowly ovate; plants often with sterile leafy shoots .... 2. *P. divaricata*
4. Leaves linear to narrowly lanceolate; plants without sterile leafy shoots .... 3. *P. pilosa*
5. Leaves with conspicuous lateral veins and reticulations; calyx teeth subulate .... 4. *P. paniculata*
5. Leaves without conspicuous lateral veins and reticulations; calyx teeth lanceolate 6
6. Flowers in panicles usually longer than broad; stems usually purple-spotted .... 5. *P. maculata*

6. Flowers in corymbs usually nearly as broad as long; stems green ......
.... 6. *P. glaberrima*

    1. P. **bifida** Beck. Cleft Phlox. March–June. Some of the pubescence glandular. Sandy soil, limestone cliffs, occasional; throughout the state, except for some of the cent. cos.
    var. **stellaria** (Gray) Wherry. Cleft Phlox. March–May. None of the pubescence glandular. Sandy areas, rare; Jackson Co.
    2. P. **divaricata** L. Blue Phlox. May–June. Corolla lobes emarginate. Woods; rare in the n. cos.
    ssp. **laphamii** (Wood) Wherry. Blue Phlox. April–June. Corolla lobes entire at tip. Woods, common; throughout the state. *P. divaricata* L.—J.
    3. P. **pilosa** L. Downy Phlox. May–Aug. Leaves pubescent, with some of the pubescence glandular. Dry woods and prairies; occasional throughout the state.
    ssp. **fulgida** (Wherry) Wherry. May–Aug. Leaves pubescent, but none of the pubescence glandular. Dry woods and prairies; occasional throughout the state. *P. pilosa* L.—J; *P. pilosa* var. *fulgida* Wherry—F.
    ssp. **sangamonensis** Levin & Smith. Sangamon Phlox. May–Aug. Leaves glabrous or nearly so. Roadsides, fields, woodlands; Champaign and Piatt cos.
    4. P. **paniculata** L. Garden Phlox. July–Sept. Rich woods; occasional in the s. four-fifths of the state.
    5. P. **maculata** L. Wild Sweet William. June–Aug. Moist woods; occasional in the n. half of the state.
    6. P. **glaberrima** L. ssp. **interior** (Wherry) Wherry. Smooth Phlox. May–Aug. Woods, prairies; occasional throughout the state. *P. glaberrima* L.—J; *P. glaberrima* var. *interior* Wherry—F.

5. *Microsteris* Greene

    1. M. **gracilis** (Dougl.) Greene. July–Aug. Native of the West Coast; rarely adventive in Ill.; Macon Co.

135. HYDROPHYLLACEAE

1. Leaves entire, with axillary spines; styles 2, free .... 1. *Hydrolea*
1. Leaves toothed, pinnately lobed, or pinnately compound; styles 2-cleft
.... 2
2. Some or all the cauline leaves opposite .... 2. *Ellisia*
2. All the cauline leaves alternate 3
3. Leaves palmately lobed .... 3. *Hydrophyllum*
3. Leaves pinnately lobed or compound 4
4. Lobes of corolla fimbriate .... 4. *Phacelia*
4. Lobes of corolla entire 5
5. Flowers up to 5 mm broad .... 4. *Phacelia*
5. Flowers 1 cm broad or broader 6
6. Branches of inflorescence glandular-pubescent .... 4. *Phacelia*
6. Branches of inflorescence lacking glandular-pubescence ....
.... 3. *Hydrophyllum*

360 HYDROPHYLLACEAE

1. Hydrolea L.

1. H. uniflora Raf. June–Sept. Swampy woods, ditches, rare; confined to the s. one-sixth of the state. *H. affinis* Gray—J.

2. Ellisia L.

1. E. nyctelea L. April–June. Wet ground; occasional in the n. three-fourths of the state, rare elsewhere.

3. Hydrophyllum L.   Waterleaf

1. Leaves palmately lobed 2
1. Leaves pinnately lobed or compound 3
2. Calyx with a reflexed appendage between two adjacent lobes; branches of inflorescence bristly hairy, with the hairs over 1 mm long ....
.... 1. *H. appendiculatum*
2. Calyx not appendaged between the lobes, or with minute teeth; branches of inflorescence pubescent with hairs up to 1 mm long ......
.... 2. *H. canadense*
3. Leaves 9- to 13-parted; stems hirsute .... 3. *H. macrophyllum*
3. Leaves 3- to 7-parted; stems glabrous or sparsely pubescent ....
.... 4. *H. virginianum*

1. H. appendiculatum Michx. Great Waterleaf. April–July. Rich woods; occasional throughout the state.
2. H. canadense L. Broad-leaf Waterleaf. May–July. Rich woods; occasional to rare throughout the state.
3. H. macrophyllum Nutt. Large-leaf Waterleaf. May–June. Rich woods; confined to the s. one-fourth of the state.
4. H. virginianum L. Virginia Waterleaf. April–July. Rich woods; occasional to common throughout the state.

4. Phacelia Juss.

1. Lobes of corolla fimbriate .... 1. *P. purshii*
1. Lobes of corolla not fimbriate 2
2. Flowers up to 5 mm broad .... 2. *P. ranunculacea*
2. Flowers at least 1 cm broad .... 3. *P. bipinnatifida*

1. P. purshii Buckley. Miami Mist. April–June. Woods, thickets; occasional in the s. one-half of the state.
2. P. ranunculacea (Nutt.) Constance. April–May. Rich woods; rare in the s. one-third of the state; also Adams Co.
3. P. bipinnatifida Michx. April–June. Rich, often rocky woods; occasional in the s. half of the state, rare elsewhere.

136. BORAGINACEAE

1. Stems and leaves glabrous 2
1. Stems and leaves pubescent 3

2. Leaves conspicuously veined, oblong to oval; flowers blue or rarely pinkish, 2–3 cm long .... 1. *Mertensia*
2. Leaves inconspicuously veined, linear to linear-oblong; flowers white, with a yellow eye, rarely blue, up to 6 mm long .... 2. *Heliotropium*
3. Ovary unlobed .... 2. *Heliotropium*
3. Ovary 4-lobed 4
4. Stamens exserted .... 3. *Echium*
4. Stamens included 5
5. Fruits echinate or prickly 6
5. Fruits smooth, wrinkled, rugose, reticulate, but not echinate or prickly 8
6. Corolla 8–12 mm broad; fruit 8–12 mm broad .... 4. *Cynoglossum*
6. Corolla 2–3 mm broad; fruit 3–5 mm broad 7
7. Racemes bracted throughout; pedicels not deflexed .... 5. *Lappula*
7. Racemes bracted only at the base; pedicels deflexed ...... 6. *Hackelia*
8. Corolla rotate .... 7. *Borago*
8. Corolla funnelform, salverform, or tubular 9
9. Leaves decurrent by wings along the stem .... 8. *Symphytum*
9. Leaves not decurrent 10
10. Corolla blue 11
10. Corolla greenish-white, white, orange, or yellow 13
11. All flowers bracted at the base; calyx tube hispid, the hairs neither glandular nor hooked .... 9. *Anchusa*
11. None of the flowers or only the lowermost flowers bracted at the base; calyx tube with appressed hairs or hooked or glandular hairs 12
12. All leaves petiolate .... 2. *Heliotropium*
12. None or only the basal leaves petiolate .... 10. *Myosotis*
13. All flowers subtended by bracts 14
13. None or only a few of the flowers subtended by bracts 17
14. Nutlets pubescent, neither white and shiny nor brown and pitted ....
.... 2. *Heliotropium*
14. Nutlets glabrous, white and shiny or brown and pitted 15
15. Leaves without lateral veins .... 11. *Lithospermum*
15. Leaves with distinct lateral veins 16
16. Stems spreading-hirsute; corolla 1–2 cm long .... 12. *Onosmodium*
16. Stems pubescent, but not spreading hirsute; corolla less than 1 cm long
.... 11. *Lithospermum*
17. Corolla yellow; leaves lanceolate to ovate-lanceolate .... 13. *Amsinckia*
17. Corolla white; leaves linear to oblong 18
18. Corolla 1–3 mm broad, with yellow appendages in the throat; nutlets obovoid; calyx tube with some glandular or hooked hairs   10. *Myosotis*
18. Corolla 5–10 mm broad, without yellow appendages in the throat; nutlets ovoid; calyx tube without glandular or hooked hairs . 14. *Plagiobothrys*

1. *Mertensia* Roth

1. M. *virginica* (L.) Pers. Bluebells. March–June. Rich woods; occasional throughout the state. (Including f. *rosea* Steyerm.)

2. *Heliotropium* L. Heliotrope

1. Plants glabrous throughout .... 1. *H. curassavicum*
1. Plants pubescent 2
2. Leaves linear, sessile or nearly so; flowers subtended by bracts ....
.... 2. *H. tenellum*

2. Leaves ovate or oval, petiolate; flowers bractless 3
3. Flowers blue; plants hirsute or hispid .... 3. *H. indicum*
3. Flowers white; plants hoary-pubescent .... 4. *H. europaeum*

1. H. curassavicum L. Seaside Heliotrope. May–Sept. Native of the s. U.S. and trop. Am.; probably adventive in Ill.; bottomlands, very rare; St. Clair Co.

2. H. tenellum (Nutt.) Torr. Slender Heliotrope. June–Aug. Limestone ledges, very rare; Monroe Co.

3. H. indicum L. Indian Heliotrope. July–Nov. Native of Asia; moist, disturbed areas; occasional in the s. half of the state, apparently absent elsewhere.

4. H. europaeum L. European Heliotrope. June–Sept. Native of Europe; rarely escaped from cultivation; Cook Co.

3. Echium L. Viper's Bugloss

1. E. vulgare L. Blueweed. May–Sept. Native of Europe; adventive in disturbed areas, particularly along railroads; occasional in the n. half of the state, absent elsewhere.

4. Cynoglossum L. Hound's-tongue

1. Corolla pale blue to white; stem rough-hairy; leaves only on the lower half of the stem .... 1. *C. virginianum*
1. Corolla reddish-purple; stem softly hairy; leaves borne all along the stem .... 2. *C. officinale*

1. C. virginianum L. Wild Comfrey. April–June. Rich woods, wooded slopes; occasional in the s. one-third of the state, rare or absent elsewhere.

2. C. officinale L. Common Hound's-tongue. May–July. Native of Europe; adventive in pastures, fields, roadsides; occasional to common in the n. two-thirds of the state, rare elsewhere.

5. Lappula Moench Stickseed

1. Each margin of the nutlet with two rows of prickles; fruits mostly 3–4 mm long .... 1. *L. echinata*
1. Each margin of the nutlet with one row of prickles; fruits mostly 2–3 mm long .... 2. *L. redowskii*

1. L. echinata Gilib. Beggar's Lice. May–Sept. Native of Europe; adventive in waste areas; occasional in the n. half of the state; also Coles and St. Clair cos. *L. myosotis* Moench—J.

2. L. redowskii (Hornem.) Greene var. occidentalis (Wats.) Rydb. Stickseed. May–Sept. Native of the w. U.S.; rarely adventive in waste ground; McHenry and Rock Island cos. *L. occidentalis* (S. Wats.) Greene—J.

6. Hackelia Opiz Stickseed

1. Fruits globose; basal leaves cordate at base; cauline leaves ovate-oblong .... 1. *H. virginiana*

1. Fruits pyramidal; basal leaves cuneate at base; cauline leaves linear to oblong .... 2. *H. americana*

    1. H. virginiana (L.) I. M. Johnston. Stickseed. June–Sept. Moist or dry woods; occasional throughout the state.
    2. H. americana (Gray) Fern. Stickseed. June–July. Rocky slopes, rare; JoDaviess Co.

7. Borago L.   Borage

    1. B. officinalis L. Borage. July–Sept. Native of Europe; occasionally cultivated but rarely escaped; Champaign, Cook, and Hancock cos.

8. Symphytum L.   Comfrey

    1. S. officinale L. Common Comfrey. June–Aug. Native of Europe; occasionally escaped from cultivation into waste ground.

9. Anchusa L.   Alkanet

    1. A. officinalis L. Common Alkanet. May–Oct. Native of Europe; rarely adventive in waste ground; Kane Co.

10. Myosotis L.   Forget-me-not

1. Corolla white, 1–2 mm across; lobes of calyx 2-lipped .. 1. *M. virginica*
1. Corolla blue, sometimes with a yellow center, 2–9 mm across; lobes of calyx equal or nearly so 2
2. Corolla 5–9 mm broad, blue with a yellow center 3
2. Corolla 2–4 mm broad, blue but usually without a yellow center 4
3. Calyx with appressed hairs, the lobes much shorter than the tube ......
.... 2. *M. scorpioides*
3. Calyx with hooked hairs, the lobes about as long as the tube ....
.... 3. *M. sylvatica*
4. All flowers bractless; pedicels longer than the calyx .... 4. *M. arvensis*
4. At least the lower flowers bracted; pedicels shorter than the calyx .....
.... 5. *M. stricta*

    1. M. virginica (L.) BSP. Scorpion Grass. April–May. Fruiting pedicels more or less erect, the lowest to 20 mm apart. Dry woods, moist woods, fields; occasional to common throughout the state. *M. verna* Nutt.—F, G, J.
    var. macrosperma (Engelm.) Fern. Scorpion Grass. April–May. Fruiting pedicels spreading, the lowest 20 mm or more apart. Dry woods, moist woods, fields; occasional in the s. one-third of the state. *M. macrosperma* Engelm.—F, G, J.
    2. M. scorpioides L. Forget-me-not. April–Oct. Native of Europe; frequently cultivated, rarely escaped into waste ground.
    3. M. sylvatica Hoffm. Garden Forget-me-not. June–July. Native of Europe; commonly cultivated but rarely escaped; Jackson Co.
    4. M. arvensis (L.) Hill. Field Scorpion Grass. June–Aug. Native of Europe; rarely adventive in Ill.; Cook Co.

5. M. stricta Link. Small-flowered Forget-me-not. May–Aug. Native of Europe; rarely adventive in Ill.; St. Clair Co.

11. *Lithospermum* L.   Gromwell

1. Corolla white, greenish-white, or pale yellow, up to 8 mm long 2
1. Corolla bright yellow or orange, 15 mm long or longer 4
2. Leaves linear, the lateral veins not evident; flowers white; nutlets brown or gray, pitted .... 1. *L. arvense*
2. Leaves lanceolate to ovate, the lateral veins prominent; flowers greenish-white to pale yellow; nutlets white, shiny 3
3. Most or all the cauline leaves 2 cm broad or broader, ovate ....
.... 2. *L. latifolium*
3. Most or all the cauline leaves up to 2 cm broad, lanceolate ....
.... 3. *L. officinale*
4. Leaves linear, acute, most of them not exceeding 5 mm in width; corolla tube at least twice as long as the calyx, the lobes toothed 4. *L. incisum*
4. Leaves linear-lanceolate to oblong, more or less obtuse, most of them more than 5 mm in width; corolla tube less than twice as long as the calyx, the lobes entire 5
5. Lobes of calyx during flowering up to 6 mm long, lengthening to 8 mm during fruiting; stems softly pubescent; corolla tube glabrous within at the base .... 5. *L. canescens*
5. Lobes of calyx during flowering up to 11 mm long, lengthening to 15 mm long during fruiting; stems somewhat rough-pubescent; corolla tube pubescent within at the base .... 6. *L. caroliniense*

1. L. arvense L. Corn Gromwell. April–June. Native of Europe; adventive in fields and waste ground; in every co.

2. L. latifolium Michx. American Gromwell. April–June. Rich woods; occasional throughout the state.

3. L. officinale L. European Gromwell. June–July. Native of Europe; rarely adventive in Ill.; Lake Co.

4. L. incisum Lehm. Yellow Puccoon. April–June. Prairies; occasional in the n. half of the state, becoming rare in the s. half.

5. L. canescens (Michx.) Lehm. Hoary Puccoon. April–June. Prairies; occasional to common throughout the state.

6. L. caroliniense (Walt.) MacM. Hairy Puccoon. April–June. Sandy prairies, sandy woods; occasional to common in the n. half of the state, less common in the s. half. (Including *L. croceum* Fern.)

12. *Onosmodium* Michx.   Marbleseed

1. Nutlets constricted at the base .... 1. *O. hispidissimum*
1. Nutlets not constricted at the base 2
2. Nutlets not strongly pitted, usually 3.5–4.5 mm long . 2. *O. occidentale*
2. Nutlets strongly pitted, usually not more than 3 mm long .. 3. *O. molle*

1. O. hispidissimum Mack. Marbleseed. May–July. Rocky prairies, rocky woods, often in calcareous areas, not common; scattered throughout the state. (Including var. *macrospermum* Mack. & Bush.)

2. O. occidentale Mack. Marbleseed. May–July. Rocky prairies, rocky woods, not common; confined to the w. half of the state.

3. O. molle Michx. Marbleseed. May–July. Limestone ridges, rare; confined to the s. one-sixth of the state.

13. Amsinckia Lehm. Fiddle-neck

1. Calyx lobes with long, white hairs .... 1. *A. barbata*
1. Calyx lobes without long, white hairs .... 2. *A. spectabilis*

1. A. barbata Greene. Tarweed. June–July. Native of w. N.Am.; rarely adventive in waste ground.
2. A. spectabilis Fisch. & Mey. Fiddle-neck. June–July. Native of the w. U.S.; rarely adventive in Ill. along railroads; Champaign and De-Kalb cos.

14. Plagiobothrys Fisch. & Mey.

1. P. hirtus (Greene) I. M. Johnston var. figuratus (Piper) I. M. Johnston. June–Aug. Native of the w. U.S.; rarely adventive in Ill.; St. Clair Co. *Allocarya figurata* Piper—J.

137. VERBENACEAE

1. Corolla 2-lipped; calyx teeth 4; flowers in a dense, rounded head ......
.... 1. *Lippia*
1. Corolla nearly regular; calyx lobes 5; flowers various, but not in a dense, rounded head .... 2. *Verbena*

1. Lippia L.

1. Leaves acute to acuminate at the apex .... 1. *L. lanceolata*
1. Leaves broadly rounded at the apex .... 2. *L. cuneifolia*

1. L. lanceolata Michx. Fog-fruit. May–Sept. Wet soil; common throughout the state; in every co. (Including var. *recognita* Fern. & Grisc.) *Phyla lanceolata* (Michx.) Greene—G, J.
2. L. cuneifolia (Torr.) Steud. Hoary Fog-Fruit. May–Sept. Native to the w. U.S.; rarely adventive in Ill.; Menard Co. *Phyla cuneifolia* Torr.) Greene—G.

2. Verbena L. Vervain

1. Corolla at least 1 cm long; calyx at least 8 mm long 2
1. Corolla up to 8 mm long; calyx up to 5 mm long 3
2. Corolla tube about twice as long as the calyx; calyx lobes deltoid, short
.... 1. *V. peruviana*
2. Corolla tube slightly longer than the calyx; calyx lobes setaceous, elongated .... 2. *V. canadensis*
3. Stems often spreading or sprawling; bracts slightly to much longer than the flowers 4
3. Stems erect; bracts not longer than the flowers 6
4. Stems spreading or sprawling; bracts 5 mm long or longer 5

4. Stems more or less erect; bracts 3–4 mm long .... 5. *V.* X *perriana*
5. Bracts 6 mm long or longer; inflorescence 1 cm broad or broader .....
.... 3. *V. bracteata*
5. Bracts about 5 mm long; inflorescence up to 1 cm broad 4. *V.* X *deamii*
6. Leaves up to 2 cm broad 7
6. Leaves, or most of them, more than 2 cm broad 9
7. Flowers up to 6 mm broad; stem glabrous or sparsely strigose ....
.... 6. *V. simplex*
7. Flowers up to 10 mm broad; stem densely hairy 8
8. Leaves lanceolate .... 7. *V.* X *blanchardii*
8. Leaves elliptic to elliptic-ovate .... 8. *V.* X *moechina*
9. Leaves sessile or on petioles less than 5 mm long; spikes 1–3 per stem,
7–10 mm thick .... 9. *V. stricta*
9. Leaves petiolate, the petioles at least 5 mm long; spikes usually more
than 3 per stem, mostly less than 7 mm thick 10
10. Some or all the flowers overlapping, the calyx of one reaching beyond
the base of the calyx of the flower above 11
10. None or few of the flowers overlapping, the calyx of one at most reaching only to the base of the calyx of the flower above 12
11. Calyx 3–4 mm long; leaves sparsely pubescent .... 10. *V. hastata*
11. Calyx 4–5 mm long; leaves softly pubescent ....11. *V.* X *rydbergii*
12. Flowers white 13
12. Flowers blue .... 14. *V. engelmannii*
13. Stems hirtellous or nearly glabrous; each flower not reaching the
base of the one immediately above ....12. *V. urticifolia*
13. Stems long-hairy; each flower reaching just to the base of the calyx
of the flower immediately above .... 13. *V.* X *illicita*

1. V. peruviana Britt. Peruvian Vervain. June–Aug. Native of S.Am.; frequently cultivated but rarely escaped; Kane and Kankakee cos.

2. V. canadensis Britt. Rose Verbena. March–Oct. Rocky woods, edge of fields and prairies; occasional in the s. two-thirds of the state, rare or absent elsewhere.

3. V. bracteata Lag. & Rodr. Creeping Vervain. April–Oct. Waste ground; occasional to common throughout the state.

4. V. X deamii Moldenke. July–Aug. Known only from Henderson, Peoria, and Wash cos. Reputed to be a hybrid between *V. bracteata* and *V. stricta.* The type is from Peoria Co.

5. V. X perriana Moldenke. June–Aug. Dry soil; Adams, Cass, Monroe, Wabash, and Woodford cos. Reputed to be a hybrid between *V. bracteata* and *V. urticifolia.*

6. V. simplex Lehm. Narrow-leaved Vervain. May–Sept. Prairies, fields, waste ground; occasional and scattered throughout the state.

7. V. X blanchardii Moldenke. July–Aug. Dry fields; Cook Co. Re-puted to be a hybrid between *V. hastata* and *V. simplex.* The type is from Cook Co.

8. V. X moechina Moldenke. June–July. Dry ground, particularly in pastures; Adams, Hardin, Peoria, and Winnebago cos. Reputed to be a hybrid between *V. simplex* and *V. stricta.*

9. V. stricta Vent. Hoary Vervain. May–Sept. Prairies, pastures, fields, common; in every co. (Including f. *albiflora* Wadmond and f. *roseiflora* Benke.)

10. V. hastata L. Blue Vervain. June–Oct. Wet woods, wet prairies, wet waste ground, common; in every co. (Including f. *rosea* Cheney.)

11. V. X rydbergii Moldenke. June–Sept. Wet ground; scattered

throughout the state. Reputed to be a hybrid between *V. hastata* and *V. stricta*. The type is from St. Clair Co.

12. V. urticifolia L. White Vervain. June–Oct. Leaves hirsute on the lower surface; nutlets about 2 mm long, corrugated on back. Fields, thickets, disturbed woods, common in every co.

var. leiocarpa Perry & Fern. June–Oct. Leaves velutinous on the lower surface; nutlets about l.5 mm long, not corrugated on the back. Low ground; Cook and Kane cos.

13. V. X illicita Moldenke. June–Sept. Low ground; scattered throughout the state. Reputed to be a hybrid between *V. stricta* and *V. urticifolia*.

14. V. X engelmannii Moldenke. July–Sept. Woods, fields; scattered throughout the state. Reputed to be a hybrid between *V. hastata* and *V. urticifolia*.

138. PHRYMACEAE

1. Phryma L. Lopseed

1. P. leptostachya L. Lopseed. June–Sept. Rich woods, occasional; scattered throughout the state.

139. LABIATAE

1. Lobes of corolla almost equal, not bilabiate 2
1. Lobes of corolla bilabiate 7
2. All flowers borne in the axils of the leaves 3
2. Some or all the flowers in terminal inflorescences 4
3. Flowers white; fertile stamens 2 .... 4. *Lycopus*
3. Flowers pink or purple; fertile stamens 4 .... 3. *Mentha*
4. Leaves entire or nearly so 5
4. Leaves toothed 6
5. Calyx more or less equally 5-cleft; stamens included; ovary 5-lobed ....
.... 1. *Isanthus*
5. Calyx more or less bilabiate; stamens excluded; ovary 4-lobed ....
.... 2. *Trichostema*
6. Corolla 4-lobed .... 3. *Mentha*
6. Corolla 5-lobed .... 28. *Perilla*
7. Ovary 4-lobed, the style not basal 8
7. Ovary deeply 4-parted, the style basal 9
8. Upper lip of corolla seemingly absent; leaves petiolate; stems at least 2.5 dm tall; creeping stolons absent .... 5. *Teucrium*
8. Upper lip of corolla very short; leaves, or the upper ones, sessile; stems less than 2.5 dm tall; creeping stolons present .... 6. *Ajuga*
9. Calyx crested on the upper side 10
9. Calyx not crested on the upper side 11
10. Calyx bilobed, the tips rounded .... 7. *Scutellaria*
10. Calyx 5-lobed, the tips pointed .... 16. *Ocimum*
11. Stamens included within the corolla tube .... 8. *Marrubium*
11. Stamens exserted or ascending under the lip, not confined to the tube of the corolla 12

12. Anther-bearing stamens 2 13
12. Anther-bearing stamens 4 19
13. Stamens exserted beyond the corolla 14
13. Stamens included 17
14. Flowers in axillary verticils or clusters or cymes; flowers not yellow (if yellow, then with purple spots) 15
14. Flowers in terminal panicles; corolla light yellow ...... 12. *Collinsonia*
15. Calyx very unequally lobed .... 11. *Blephilia*
15. Calyx equally 5-lobed 16
16. Corolla up to 1 cm long; flowers in loose cymes .... 9. *Cunila*
16. Corolla more than 1 cm long; flowers in dense capitate clusters ....
.... 10. *Monarda*
17. Bracts longer than the clusters of flowers they subtend .. 13. *Hedeoma*
17. Bracts shorter than the clusters of flowers they subtend 18
18. Flowers mostly up to 10 in a rather loose whorl .... 14. *Salvia*
18. Flowers mostly more than 10 in dense whorls .... 11. *Blephilia*
19. Leaves entire 20
19. Leaves serrate, crenate, dentate, or lobed 23
20. Plants forming dense mats; leaves obtuse at the apex. .... 15. *Thymus*
20. Plants erect; leaves acute to acuminate at the apex 21
21. Leaves ovate; upper lip of corolla 4-lobed .... 16. *Ocimum*
21. Leaves linear to lanceolate; upper lip of corolla 2-lobed, emarginate, or entire 22
22. Calyx bilabiate (nearly equally 5-lobed in *S. hortensis*); stamens strongly curved .... 17. *Satureja*
22. Calyx more or less equally 5-lobed; stamens straight or nearly so ......
.... 18. *Pycnanthemum*
23. Some or all the cauline leaves cordate 24
23. None of the cauline leaves cordate 31
24. Flowers yellow, greenish-yellow, or white dotted with purple 25
24. Flowers purple, rose-purple, or blue 27
25. Stem glabrous or nearly so .... 19. *Agastache*
25. Stem pubescent 26
26. Stem densely canescent; calyx 5-toothed, the tube puberulent; corolla white, dotted with purple .... 20. *Nepeta*
26. Stem more or less hirsute; calyx 4-parted, the tube hirsute; corolla yellow-white .... 21. *Synandra*
27. Stems creeping or trailing .... 22. *Glechoma*
27. Stems erect 28
28. All flowers borne in axillary clusters .... 23. *Lamium*
28. Some or all the flowers in terminal inflorescences 29
29. Lower leaf surface densely short-white-hairy .... 19. *Agastache*
29. Lower leaf surface glabrous or pubescent, but not densely short-white-hairy 30
30. Calyx 15-nerved; inflorescence crowded, the flower clusters touching
.... 19. *Agastache*
30. Calyx 5- to 10-nerved; inflorescence interrupted, most or all the flower clusters separated .... 24. *Stachys*
31. All flowers borne in axillary clusters 32
31. Some or all the flowers borne in terminal inflorescences 34
32. Flowers and fruits on pedicels 1–2 mm long .... 25. *Melissa*
32. Flowers and fruits sessile 33
33. Cauline leaves often lobed (except *L. marrubiastrum*); upper lip of corolla woolly .... 26. *Leonurus*

33. Cauline leaves merely toothed or entire; upper lip of corolla not woolly
.... 27. *Galeopsis*
34. Calyx teeth spinescent .... 26. *Leonurus*
34. Calyx teeth not spinescent 35
35. Flowers and fruits on pedicels 36
35. Flowers and fruits sessile 37
36. Calyx teeth more or less equal; stamens 4 .... 29. *Physostegia*
36. Calyx 2-lipped; stamens 2 .... 28. *Perilla*
37. Calyx with lobes more or less equal, not bilabiate 38
37. Calyx bilabiate 40
38. Flowers in dense or loose inflorescences longer than broad 39
38. Flowers in dense heads as broad as or broader than long ....
.... 18. *Pycnanthemum*
39. Flowers much longer than the subtending bracts; stems glabrous;
leaves sessile .... 29. *Physostegia*
39. Flowers not longer than or only a little longer than the subtending
bracts; stems pubescent or, if glabrous, the leaves petiolate 24. *Stachys*
40. Bracts spinescent .... 30. *Dracocephalum*
40. Bracts not spinescent 41
41. Flowers in heads as broad as or broader than long; corolla less than 1
cm long .... 18. *Pycnanthemum*
41. Flowers in heads longer than broad; corolla 1 cm long or longer.......
.... 31. *Prunella*

1. Isanthus Michx. False Pennyroyal

1. I brachiatus (L.) BSP. False Pennyroyal. July–Oct. Rocky woods, prairies; occasional throughout the state.

2. Trichostema L. Blue Curls

1. T. dichotomum L. Blue Curls. Aug.–Oct. Woods, edge of fields, not common; scattered in the state.

3. Mentha L. Mint

1. All flowers in axils of the leaves 2
1. At least some of the flowers in terminal inflorescences 4
2. Uppermost leaves not more than three times longer than the cluster of
flowers .... 1. *M.* X *cardiaca*
2. Uppermost leaves at least five times longer than the cluster of flowers 3
3. Stems glabrous, or at least not retrorsely hairy .... 2. *M.* X. *gentilis*
3. Stems retrorsely hairy .... 3. *M. arvensis*
4. Leaves sessile or nearly so; spikes not interrupted 5
4. Leaves petiolate; spikes more or less interrupted 7
5. Plants glabrous or nearly so .... 4. *M. spicata*
5. Plants downy-pubescent 6
6. Spikes dense, mostly 1.2 cm thick or thicker ... 5. *M.* X *alopecuroides*
6. Spikes very slender, mostly less than 1 cm thick ..... 6. *M. rotundifolia*
7. Calyx lobes pubescent 8
7. Calyx lobes glabrous .... 8. *M. citrata*
8. Leaves sharply serrate, not crisped .... 7. *M.* X *piperita*
8. Leaves lacerate, crisped .... 9. *M. crispa*

1. M. X cardiaca Gerarde. Little-leaved Mint. July–Oct. Native o
Europe; occasionally adventive in Ill.; Champaign, Christian, Coles
Lake, Whiteside, and Woodford cos. Reputed to be a hybrid betweer
M. spicata and *M. arvensis.*

2. M. X gentilis L. Red Mint. July–Sept. Native of Europe; rarely ad
ventive in Ill.; Henry Co. Reputed to be a hybrid between *M. spicata*
and *M. arvensis.*

3. M. arvensis L. Field Mint. July–Sept. Leaves cuneate at base, the
petioles longer than the cluster of flowers. Waste ground, very rare.

var. villosa (Benth.) S. R. Stewart. Field Mint. July–Sept. Marshes
low ground; occasional to common in most of the state. *M. canadensis* L.—J.

4. M. spicata L. Spearmint. June–Oct. Native of Europe; frequently
planted and occasionally escaped; throughout the state.

5. M. X alopecuroides Hull. Foxtail Mint. July–Oct. Native o
Europe; rarely adventive in Ill.; Cook, Effingham, Fayette, Kane, Piatt
and Schuyler cos. Reputed to be a hybrid between *M. longifolia* (L.
Huds. and *M. rotundifolia.*

6. M. rotundifolia (L.) Huds. Apple Mint. June–Sept. Native o
Europe; rarely adventive in Ill.; Hancock Co.

7. M. X piperita L. Peppermint. June–Oct. Native of Europe; oc
casionally adventive in moist waste ground. Reputed to be a hybrid
between *M. spicata* and *M. aquatica* L.

8. M. citrata Ehrh. Bergamot Mint. July–Oct. Native of Europe; no
commonly adventive in moist ground.

9. M. crispa L. Curly Mint. July–Sept. Native of Europe; adventive ir
Cook and Lake cos.

4. Lycopus L.   Water Horehound

1. Some or all the leaves pinnatifid to pinnate . . . . 1. *L. americanus*
1. Leaves merely toothed, never pinnatifid 2
2. Calyx teeth broadly triangular, obtuse to subacute, up to 1 mm long . 3
2. Calyx teeth narrowly triangular to subulate, acute, 1–2 mm long 4
3. Stem puberulent; base of plant not tuberous; stamens more or less in
   cluded . . . . 2. *L. virginicus*
3. Stem glabrous or nearly so; base of plant tuberous; stamens exserted
   . . . . 3. *L. uniflorus*
4. Middle and lower leaves petiolate; base of plant not tuberous . . . .
   . . . . 4. *L. rubellus*
4. Middle and lower leaves sessile; base of plant tuberous . . . . 5. *L. aspe*

1. L. americanus Muhl. Common Water Horehound. June–Oct. We
ground, common; in every co.

2. L. virginicus L. Bugle Weed. July–Oct. Wet ground, occasional
scattered throughout the state.

3. L. uniflorus Michx. Northern Bugle Weed. Aug.–Sept. Marshes
calcareous fens, around lakes; occasional in the n. half of the state
also Pope Co.

4. L. rubellus Moench. Stalked Water Horehound. July–Oct. Sten
and lower leaf surface glabrous or nearly so. Low woods, we
meadows, not common; scattered throughout the state.

var. arkansanus (Fresn.) Benner. July–Oct. Stem and lower leaf sur
face pubescent. Swampy woods, rare; Jackson and Union cos.

5. L. asper Greene. Rough Water Horehound. July–Aug. Usually in stagnant water, rare; confined to the n. one-fourth of the state.

5. *Teucrium* L. Germander

1. T. canadense L. var. virginicum (L.) Eat. American Germander. June–Sept. Calyx and bracts without glandular hairs; lower leaf surface appressed-pubescent. Low woods, wet prairies, wet ditches and fields, common; in every co.
var. occidentale (Gray) McClintock & Epling. Gray Germander. July–Sept. Calyx and bracts glandular-pubescent; lower leaf surface spreading-pubescent. Low woods, wet prairies, wet fields; occasional in the n. three-fourths of the state. (Including *T. occidentale* Gray var. *boreale* [Bickn.] Fern.) *T. occidentale* Gray—F, J.

6. *Ajuga* L. Bugleweed

1. Leaves glabrous or slightly pubescent; plants stoloniferous 1. *A. reptans*
1. Leaves densely soft-pubescent; plants tufted, without stolons . . . .
. . . . 2. *A. genevensis*

1. A. reptans L. Carpet Bugle. May–July. Native of Europe; frequently planted, but rarely escaped; Jackson, Kendall, and Lake cos.
2. A. genevensis L. Geneva Bugleweed. May–July. Native of Europe; occasionally cultivated, seldom escaped; Cook, DuPage, Jackson, Lake, and McHenry cos.

7. *Scutellaria* L. Skullcap

1. Flowers solitary in the axils of the leaves or of the reduced leaf-like bracts 2
1. Flowers 2-several in racemes 4
2. Corolla at least 1.5 cm long . . . . 1. *S. epilobiifolia*
2. Corolla up to 1.2 cm long 3
3. Most of the leaves 2 cm long or longer; stolons without moniliform tubers; stems (at maturity) glabrous . . . . 2. *S. nervosa*
3. Most of the leaves up to 2 cm long; stolons with moniliform tubers; stems (at maturity) pubescent . . . . 3. *S. parvula*
4. Corolla up to 1 cm long; racemes produced from the axils of the leaves, lateral . . . . 4. *S. lateriflora*
4. Corolla 1 cm long or longer; most of the racemes terminal 5
5. Leaves cordate at the base . . . . 5. *S. ovata*
5. Leaves rounded to cuneate at the base 6
6. Calyx with glandular hairs . . . . 6. *S. elliptica*
6. Calyx with eglandular hairs . . . . 7. *S. incana*

1. S. epilobiifolia Muhl. Marsh Skullcap. June–Sept. Marshes; confined to the n. half of the state. *S. galericulata* L.—G, J.
2. S. nervosa Pursh. Veiny Skullcap. April–July. Low woods; occasional in the s. three-fourths of the state; also Carroll and JoDaviess cos. (Including var. *calvifolia* Fern. and f. *alba* Steyerm.)
3. S. parvula Michx. Small Skullcap. May–July. Pubescence glandular; lower leaf surface with sessile glands and long hairs. Rocky

woods, prairies, fields, limestone barrens; occasional throughout the state.

var. australis Fassett. Small Skullcap. May–July. Pubescence glandular; lower leaf surface with long hairs only. Prairies, rocky woods; occasional in the s. cos. *S. australis* (Fassett) Epling—G.

var. leonardii (Epling) Fern. Small Skullcap. May–July. Pubescence eglandular. Prairies, rocky woods; scattered throughout the state. *S. leonardii* Epling—G, J.

4. S. lateriflora L. Mad-dog Skullcap. June–Oct. Marshes, swampy woods, borders of rivers and streams; occasional to common throughout the state.

5. S. ovata Hill. Heart-leaved Skullcap. May–Oct. Some or all the leaves over 4 cm long; plants generally taller than 25 cm; uppermost bracts longer than the calyx. Rocky woods; occasional in the s. one-third of the state.

var. versicolor (Nutt.) Fern. Heart-leaved Skullcap. May–Oct. Some or all the leaves over 4 cm long; plants generally taller than 25 cm; uppermost bracts shorter than the calyx. Rocky woods, rich woods; occasional throughout the state. *S. ovata* Hill—J.

var. rugosa (Wood) Fern. Heart-leaved Skullcap. May–Oct. None of the leaves 4 cm long; plants generally less than 25 cm tall. Limestone woods, rare; Monroe and Randolph cos.

6. S. elliptica Muhl. Hairy Skullcap. May–July. Dry, rocky woods; restricted to the s. one-sixth of the state; also Sangamon Co. (Including var. *hirsuta* [Short] Fern.)

7. S. incana Biehler. Downy Skullcap. June–Sept. Dry, rocky woods; occasional in the s. three-fourths of the state.

8. Marrubium L.   Horehound

1. M. vulgare L. Common Horehound. May–Sept. Native of Europe and Asia; naturalized in fields, pastures, and along roads; occasional to common throughout the state.

9. Cunila L.   Dittany

1. C. origanoides (L.) Britt. Dittany. July–Nov. Dry woods, sandstone cliffs; occasional to common in the s. one-third of the state.

10. Monarda L.   Wild Bergamot

1. Only one head of flowers on each stem or stem-branch 2
1. Heads of flowers two or more on each stem or stem-branch 5
2. Leaves sessile or nearly so . . . . 1. *M. bradburiana*
2. Leaves on petioles 5 mm long or longer 3
3. Bracts bright red or whitish; tip of upper lip of corolla beardless 4
3. Bracts green or pink-tinged; tip of upper lip of corolla bearded . . . .
. . . . 4. *M. fistulosa*
4. Bracts bright red; corolla red; throat of calyx without a beard . . . .
. . . . 2. *M. didyma*
4. Bracts whitish; corolla white or pink; throat of calyx bearded . . . .
. . . . 3. *M. clinopodia*
5. Corolla yellow, with purple spots; calyx teeth triangular . 5. *M. punctata*
5. Corolla white or pink; calyx teeth subulate . . . . 6. *M. citriodora*

1. M. bradburiana Beck. Monarda. April–June. Dry woods, bluffs, roadsides; occasional in the s. half of the state, rare in the n. half. *M. russeliana* Nutt.—F.

2. M. didyma L. Oswego Tea. June–Sept. Native of the e.U.S.; escaped from cultivation into woodlands; Cook, Hancock, Lake, and Wabash cos.

3. M. clinopodia L. Bee Balm. June–July. Woods; scattered in the s. three-fourths of the state.

4. M. fistulosa L. Wild Bergamot. May–Aug. Dry woods, fields, prairies, along roads; common throughout the state. (Including var. *mollis* [L.] Benth.)

5. M. punctata L. var. villicaulis Pennell. Horsemint. June–Oct. Stems villous or pilose. Sandy fields and woods, dunes, prairies; occasional in the n. half of the state, rare southward.

var. lasiodonta Gray. Horsemint. June–Oct. Stems canescent-pubescent. Sandy soil; Madison and St. Clair cos. *M. punctata* L. ssp. *occidentalis* Epling—G.

6. M. citriodora Cerv. Lemon Mint. May–Aug. Native of the w. U.S.; known as an escape from cultivation in Cook Co.

11. Blephilia Raf.   Pagoda Plant

1. At least the upper leaves cuneate to the nearly sessile base   1. *B. ciliata*
1. Upper leaves petiolate, rounded at the base .... 2. *B. hirsuta*

1. B. ciliata (L.) Benth. Pagoda Plant. May–Aug. Open woods, fields, prairies; occasional throughout the state.

2. B. hirsuta (Pursh) Benth. Pagoda Plant. May–Sept. Rich woods; occasional throughout the state.

12. Collinsonia L.

1. C. canadensis L. Richweed. July–Sept. Rocky woods; occasional in the s. tip of the state; also Champaign and Clark cos.

13. Hedeoma Pers.

1. Leaves linear, entire, sessile; plants faintly aromatic ..... 1. *H. hispida*
1. Leaves oblong-ovate to elliptic, more or less toothed, petiolate; plants strongly aromatic .... 2. *H. pulegioides*

1. H. hispida Pursh. Rough Pennyroyal. May–July. Rocky woods, prairies; occasional in the n. half of the state, rare in the s. half of the state.

2. H. pulegioides (L.) Pers. American Pennyroyal. July–Sept. Rocky woods, fields, roadsides; occasional to common throughout the state.

14. Salvia L.   Sage

1. Most leaves in a basal rosette 2
1. Most leaves cauline 3
2. Whorls of flowers more or less separated; leaves lyrate-pinnatifid; calyx not viscid ....1. *S. lyrata*

2. Whorls of flowers very remote; leaves crenate; calyx viscid ....
.... 2. *S. pratensis*
3. Leaves not more than 2 cm broad; upper lip of calyx without teeth .. 4
3. Leaves more than 2 cm broad; upper lip of calyx 3-toothed 5
4. Leaves denticulate or serrate; corolla 15 mm long or longer; flowers 6 or
more per whorl .... 3. *S. azurea* var. *grandiflora*
4. Leaves entire or sparsely serrate; corolla 8–12 mm long; flowers 1–3 per
whorl .... 4. *S. reflexa*
5. Upper surface of leaves glabrous; calyx much longer than pedicel in
fruit .... 5. *S. sylvestris*
5. Upper surface of leaves pubescent; calyx about as long as the pedicel
in fruit .... 6. *S. verticillata*

1. S. lyrata L. Cancer-weed. April–June. Rich, open woods; occasional in the s. one-sixth of the state.

2. S. pratensis L. Meadow Sage. June–July. Native of Europe; escaped to an open woods in Lake Co.

3. S. azurea Lam. var. grandiflora Benth. Blue Sage. Native of the w. U.S. Prairies, edge of woods; Champaign, DuPage, Johnson, Piatt, Pope, and Tazewell cos. *S. pitcheri* Torr.—G, J.

4. S. reflexa Hornem. Rocky Mountain Sage. May–Oct. Native of the w. U.S. Dry woods, pastures, fields; occasional throughout the state except for the southernmost cos.

5. S. sylvestris L. Wild Sage. June–Sept. Native of Europe; adventive in pastures and along roads; Cook and Winnebago cos.

6. S. verticillata L. June–Sept. Native of Europe; adventive along a railroad; Scott Co.

15. Thymus L. Thyme

1. T. serpyllum L. Creeping Thyme. July–Aug. Native of Europe; occasionally cultivated but rarely escaped in Ill.; Hancock and Henderson cos.

16. Ocimum L. Basil

1. O. basilicum L. Basil. Aug.–Sept. Native of Africa and Asia; occasionally cultivated but rarely escaped in Ill.; Hancock Co.

17. Satureja L. Savory

1. Annual from fibrous roots; all leaves lance-linear; style branches equal
.... 1. *S. hortensis*
1. Perennial from stolons; leaves from stolons elliptic to ovate; style
branches very unequal .... 2. *S. arkansana*

1. S. hortensis L. Summer Savory. Aug.–Oct. Native of Europe and Asia; rarely escaped from cultivation; DuPage and Peoria cos.

2. S. arkansana (Nutt.) Briq. Low Calamint. May–Oct. Calcareous fens, rocky soil, sand flats; occasional in the ne. one-fourth of the state. *Clinopodium arkansanum* (Nutt.) G. N. Jones—J; *Satureja glabella* (Michx.) Briq. var. *angustifolia* (Torr.) Svenson—G.

18. *Pycnanthemum Michx. Mountain Mint*

1. Bracts whitened, at least on the upper surface 2
1. Bracts green 5
2. Calyx regular, the teeth up to 1 mm long .... 1. *P. muticum*
2. Calyx bilabiate, the lower teeth 1–2 mm long 3
3. Lower and middle cauline leaves green on the upper surface; calyx teeth acute to acuminate 4
3. Lower and middle cauline leaves whitened on the upper surface; calyx teeth obtuse .... 4. *P. albescens*
4. Corolla rose to purple; at least some part of the lower leaf surface hirsute .... 2. *P. pycnanthemoides*
4. Corolla white with purple spots; lower leaf surface not hirsute ....
.... 3. *P. incanum*
5. Stems glabrous; largest leaves up to 5 (–5.5) mm broad 5.*P. tenuifolium*
5. Stems pubescent, at least on the angles; largest leaves (5–) 6–40 mm broad 6
6. Outermost bracts glabrous or nearly so on the upper surface 7
6. Outermost bracts pubescent on the upper surface .... 8. *P. pilosum*
7. Calyx teeth 1–2 mm long, pilose throughout; leaves short-petiolate .....
.... 6. *P. torrei*
7. Calyx teeth up to 1 mm long, pubescent only at the tip; leaves sessile ..
.... 7. *P. virginianum*

1. P. muticum (Michx.) Pers. Mountain Mint. July–Sept. Low woods, very rare; Henderson and Wabash cos.

2. P. pycnanthemoides (Leavenw.) Fern. Mountain Mint. July–Aug. Open woods; confined to the s. one-fifth of the state.

3. P. incanum (L.) Michx. Gray Mountain Mint. July–Sept. Dry woods; confined to the s. one-fourth of the state. (Including var. *loomisii* [Nutt.] Fern.)

4. P. albescens Torr. & Gray. White Mountain Mnt. July–Sept. Cherty slopes, very rare; Union Co.

5. P. tenuifolium Schrad. Slender Mountain Mint. June–Sept. Woods, fields, prairies; occasional to common throughout the state. *P. flexuosum* [Walt.] BSP.—G, J.

6. P. torrei Benth. Mountain Mint. June–Oct. Dry woods, very rare; Alexander Co.

7. P. virginianum (L.) Dur. & Jacks. Common Mountain Mint. July–Sept. Marshes, calcareous fens, prairies; occasional to common in the n. half of the state, becoming less common southward.

8. P. pilosum Nutt. Hairy Mountain Mint. July–Sept. Dry woods, prairies; occasional in the cent. cos., less common to rare in the n. and s. cos.

19. *Agastache Clayt. Giant Hyssop*

1. Corolla yellow; calyx teeth obtuse to subacute .... 1. *A. nepetoides*
1. Corolla purple or blue; calyx teeth acute to acuminate 2
2. Stems pubescent; corolla purple .... 2. *A. scrophulariaefolia*
2. Stems glabrous; corolla blue .... 3. *A. foeniculum*

1. A. nepetoides (L.) Ktze. Yellow Giant Hyssop. July–Sept. Open woods; occasional throughout the state.

2. A. scrophulariaefolia (Willd.) Ktze. Purple Giant Hyssop. July-Sept. Open woods; occasional in the n. half of the state; also Wabas̄ Co. (Including var. *mollis* [Fern.] Heller.)

3. A. foeniculum (Pursh) Ktze. Blue Giant Hyssop. June–Sept. Dr soil, very rare; Menard Co.

20. Nepeta L.   Catnip

1. N. cataria L. Catnip. June–Sept. Native of Europe; naturalized iᴎ fields, open woods, along roads and railroads; occasional to commoᴎ throughout the state.

21. Synandra Nutt.

1. S. hispidula (Michx.) Baill. Synandra. May–June. Rich woods̄ rare; Jackson and Williamson cos.

22. Glechoma L.   Ground Ivy

1. G. hederacea L. Ground Ivy. April–July. Corolla more than 1.5 cᴍ long. Native of Europe; naturalized in moist ground; not commoᴎ DuPage, Lawrence, and Peoria cos.

var. micrantha Moricand. Ground Ivy. April–July. Corolla up to 1.! cm long. Native of Europe; naturalized in moist ground and lawns̄ occasional throughout the state, becoming abundant northward. *G heterophylla* Waldst. & Kit.—J.

23. Lamium L.   Dead Nettle

1. At least the upper leaves sessile .... 1. *L. amplexicaul*
1. All leaves petiolate. ....
2. Flowers up to 15 mm long .... 2. *L. purpureuᴍ*
2. Flowers more than 15 mm long .... 3. *L. maculatuᴍ*

1. L. amplexicaule L. Henbit. Feb.–Nov. Native of Europe, Asia, anᴅ Africa; naturalized in waste ground; occasional to common throughᴏ out the state. (Including f. *albiflorum* D. M. Moore.)

2. L. purpurem L. Purple Dead Nettle. April–Oct. Native of Europᴇ and Asia; naturalized in waste ground; occasional in the s. half of thᴇ state, uncommon in the n. half.

3. L. maculatum L. Spotted Dead Nettle. May–Aug. Native oᴦ Europe; adventive in waste ground; Cook, DuPage, and Kane cos.

24. Stachys L.   Hedge Nettle

1. Stems pubescent on the sides ....
1. Stems glabrous on the sides, although sometimes pubescent on thᴇ angles ....
2. Calyx teeth triangular, shorter than the tube; leaves deeply cordate a the base .... 1. *S. riddell*
2. Calyx teeth lance-subulate, as long as the tube; leaves rounded or sha� lowly cordate at the base .... 2. *S. palustris̄*
3. Leaves conspicuously nerved; margins of leaves sharply toothed .... ᐟ

3. Leaves obscurely nerved; margins of leaves low-toothed ....
.... 5. *S. hyssopifolia* var. *ambigua*
4. Calyx uniformly pubescent throughout .... 3. *S. clingmanii*
4. Calyx glabrous or merely with setae on the nerves ..... 4. *S. tenuifolia*

1. S. riddellii House. Heart-leaved Hedge Nettle. June–July. Rich woods, very rare; Hardin Co.

2. S. palustris L. var. pilosa (Nutt.) Fern. Woundwort. June–Sept. Leaves obtuse to subacute, oblong to oval, sessile or on petioles less than 1 cm long. Moist soil, rare; Hancock Co.

var. homotricha Fern. Woundwort. June–Sept. Leaves acuminate, lanceolate, sessile or on petioles less than 1 cm long. Wet prairies, swampy or marshy soil; occasional to common in the n. half of the state, rare elsewhere. Including var. *nipigonesis* Jennings.) *S. arenicola* Britt.—J.

var. phaneropoda Weath. Woundwort. July–Sept. Leaves acuminate, on petioles 1 cm long or longer. Moist soil, rare; confined to extreme w. Ill.

3. S. clingmanii Small. Hedge Nettle. July–Sept. Rich, rocky woods, rare; Alexander, Hardin, and Massac cos.

4. S. tenuifolia Willd. Smooth Hedge Nettle. June–Sept. Leaves glabrous on both surfaces; calyx glabrous. Moist soil; occasional throughout the state.

var. hispida (Pursh) Fern. Hairy Hedge Nettle. June–Sept. Leaves hispid, at least on the veins beneath; calyx setose. Low woods, swamps, marshes; occasional in the n. half of the state, rare elsewhere. (Including var. *platyphylla* Fern.) *S. hispida* Pursh—G, J.

5. S. hyssopifolia Michx. var. ambigua Gray. Hyssop Hedge Nettle. June–Aug. Moist soil; occasional in the cent. cos., rare in the n. and s. cos. *S. aspera* Michx.—J.

25. Melissa L. Balm

1. M. officinalis L. Balm. June–Aug. Native of Europe; rarely adventive in Ill.; Jackson and Wabash cos.

26. Leonurus L. Motherwort

1. Some of the leaves palmately lobed; corolla bearded within ....
.... 1. *L. cardiaca*
1. All leaves unlobed; corolla beardless .... 2. *L. marrubiastrum*

1. L. cardiaca L. Motherwort. May–Aug. Native of Europe and Asia; naturalized in disturbed, shaded areas; occasional in the n. half of the state, less common in the s. half.

2. L. marrubiastrum L. Lion's-tail. June–Aug. Native of Europe and Asia; naturalized in disturbed soil; occasional in the n. two-thirds of the state.

27. Galeopsis L. Hemp Nettle

1. Stems bristly-pubescent at the nodes and swollen beneath the nodes ..
.... 1. *G. tetrahit*
1. Stems appressed-pubescent at the nodes and not swollen beneath the nodes .... 2. *G. ladanum*

1. G. tetrahit L. Common Hemp Nettle. June–Sept. Native of Europe and Asia; naturalized in waste places; Boone, Cook, and Henderson cos.

2. G. ladanum L. Red Hemp Nettle. June–Sept. Native of Europe; rarely adventive in waste ground; Cook Co.

28. Perilla L.

1. P. frutescens L. Beefsteak Plant. Aug.–Oct. Native of Asia; disturbed soil, particularly in damp areas; common in the s. one-fourth of the state, less common northward. (Including var. *crispa* [Benth.] Deane.)

29. Physostegia Benth.   False Dragonhead

1. Leaves more or less undulate .... 1. *P. intermedia*
1. Leaves definitely serrate 2
2. Leaves (at least the upper) broadly rounded at base, the teeth rarely more than 1 mm long; corolla rarely longer than 1.5 cm 2. *P. parviflora*
2. Leaves cuneate or subcuneate at base, the teeth regularly more than 1 mm long; corolla 1.5–3.0 cm long 3
3. Upper leaves abruptly reduced in size; spike appearing pedunculate . 4
3. Upper leaves gradually reduced in size; spike appearing sessile ....
.... 5. *P. speciosa*
4. Broadest leaves never exceeding a width of 1 cm; flowers remote ......
.... 3. *P. angustifolia*
4. At least some of the leaves over 1 cm broad; some of the flowers overlapping .... 4. *P. virginiana*

1. P. intermedia (Nutt.) Engelm. & Gray. False Dragonhead. May–July. Edge of slough, rare; Adams Co. *Dragocephalum intermedium* Nutt.—G.

2. P. parviflora Nutt. Small-flowered False Dragonhead. July–Oct. Moist prairies, rare; Kane, Lake, and Lee cos. *Draocephalum nuttallii Britt*—G.

3. P. angustifolia Fern. Narrow-leaved False Dragonhead. June–Sept. Low prairies; occasional in the n. half of the state.

4. P. virginiana (L.) Benth. False Dragonhead. May–Sept. Moist soil, particularly prairies; occasional throughout the state. (Including f. *candida* Benke.)

5. *P. speciosa (Sweet) Sweet. False Dragonhead. May–*Sept. Low prairies; occasional throughout the state. *P. virginiana* var. *speciosa* (Sweet) Gray—F; *Dracocephalum formosius* (Lunell) Rydb.—G.

30. Dracocephalum L.   Dragonhead

1. D. parviflorum Nutt. American Dragonhead. May–Aug. Native of the w. U.S. Dry soil, along railroads; not common, confined to the n. half of the state. *Moldavica parviflora* (Nutt.) Adans.—G.

31. Prunella L.   Self-heal

1. P. vulgaris L. Self-heal. May–Sept. At least the upper leaves rounded at the base. Native of Europe; naturalized in lawns, fields, and waste ground; occasional throughout the state.

var. lanceolata (Bart.) Fern. Self-heal. May–Sept. At least the upper leaves cuneate at the base. Disturbed woods, pastures, meadows, common; in every co.

140. SOLANACEAE

1. Plants woody, at least at base, often climbing or trailing 2
1. Plants herbaceous throughout, not climbing or trailing 3
2. At least some of the leaves with lobes near the base; flowers in terminal or axillary cymes .... 1. *Solanum*
2. None of the leaves lobed near the base; flowers solitary in the axils .... 2. *Lycium*
3. Leaves compound (the deeply pinnatifid, yellow-flowered, densely prickly *Solanum rostratum* might be sought here) 4
3. Leaves simple, although often deeply lobed 5
4. Flowers yellow; stems pilose .... 3. *Lycopersicum*
4. Flowers white, purplish, or blue; stems glabrous or appressed pubescent .... 1. *Solanum*
5. Stems and sometimes the leaves prickly .... 1. *Solanum*
5. Stems and leaves without prickles 6
6. None of the leaves petiolate, the uppermost clasping .. 4. *Hyoscyamus*
6. Some or all the leaves petiolate, the uppermost not clasping (although broadly rounded and sessile in some species of *Petunia*) 7
7. Leaves pinnatifid at least halfway to midvein .... 1. *Solanum*
7. Leaves entire or toothed, but never divided halfway to midvein 8
8. Corolla pale blue, campanulate .... 5. *Nicandra*
8. Corolla white or yellow (sometimes purplish in *Solanum americanum*), rotate or funnelform 9
9. Stamens exserted .... 1. *Solanum*
9. Stamens included 10
10. Corolla 7 cm long or longer; fruit prickly .... 6. *Datura*
10. Corolla up to 3 cm long; fruit not prickly 11
11. Flowers campanulate or rotate; fruit a berry enclosed in an inflated calyx .... 7. *Physalis*
11. Flowers funnelform; fruit a capsule 12
12. Flowers greenish-yellow; stamens all alike .... 8. *Nicotiana*
12. Flowers variously colored, but usually not greenish-yellow; one of the stamens considerably smaller than the others .... 9. *Petunia*

1. Solanum L.   Nightshade

1. Plants prickly 2
1. Plants not prickly 5
2. Corolla yellow; calyx prickly .... 1. *S. rostratum*
2. Corolla purple or white; calyx not prickly 3
3. Stems hirsute, the stellate hairs sessile .... 2. *S. carolinense*
3. Stems with short tomentum, the stellate hairs stipitate 4
4. Leaves linear-lanceolate to narrowly oblong, silvery-gray; calyx lobes linear .... 3. *S. elaeagnifolium*
4. Leaves ovate, green; calyx lobes ovate .... 4. *S. torreyi*
5. Plants woody, climbing; berry red .... 5. *S. dulcamara*
5. Plants herbaceous, not climbing; berry green or black 6

6. Leaves compound .... 6. *S. tuberosum*
6. Leaves simple, although sometimes deeply pinnatifid 7
7. Leaves deeply pinnatifid; berry green .... 7. *S. triflorum*
7. Leaves entire; berry black .... 8. *S. americanum*

1. S. rostratum Dunal. Buffalo-bur. July–Oct. Native of the w. U.S.; naturalized in waste ground; occasional throughout the state.

2. S. carolinense L. Horse-nettle. June–Oct. Open woods and waste places; in every co. (Including f. *albiflorum* [Kuntze] Benke.)

3. S. elaeagnifolium Cav. Silvery Horse-nettle. July–Oct. Native of the w. U.S.: adventive in waste areas; Adams, Cook, and Crawford cos.

4. S. torreyi Gray. July–Sept. Native of the w. U.S.; adventive in waste areas; Henry Co.

5. S. dulcamara L. Bittersweet Nightshade. June–Oct. Native of Europe; naturalized in waste areas, thickets, and boggy areas; common in the n. half of the state, occasional in the s. half. (Including var. *villosissimum* Desv. and f. *albiflorum* House).

6. S. tuberosum L. Potato. June–Aug. Native of S. Am.; rarely escaped from cultivation.

7. S triflorum Nutt. June–Aug. Cut-leaved Nightshade. Native of the w. U.S.; adventive in waste ground; occasional in the n. cos.

8. S. americanum Mill. Black Nightshade. May–Nov. Woods, waste ground; in every co. *S. nigrum* L.—G, J.

2. Lyceum L.   Matrimony Vine

1. Leaves gray-green; lobes of calyx obtuse .... 1. *L. halimifolium*
1. Leaves dark green; lobes of calyx acute .... 2. *L. chinense*

1. L. halimifolium Mill. Common Matrimony Vine. June–Sept. Native of Europe; occasionally naturalized in waste ground; scattered throughout the state.

2. L. chinense Mill. Chinese Matrimony Vine. June–Sept. Native of Asia; rarely adventive; Champaign Co.

3. Lycopersicum Mill.

1. L. esculentum Mill. Tomato. May–Sept. Native of trop. Am.; rarely escaped from cultivation.

4. Hyoscyamus L.   Henbane

1. H. niger L. Black Henbane. June–Aug. Native of Europe; rarely escaped in Ill.; McHenry and Peoria cos.

5. Nicandra Adans.   Apple-of-Peru

1. N. physalodes (L.) Pers. Apple-of-Peru. July–Sept. Native of Peru; naturalized in waste ground; scattered throughout the state.

6. Datura L.   Jimsonweed

1. Plants glabrous; leaves coarsely toothed; corolla to 10 cm long ....
.... 1. *D. stramonium*

1. Plants pubescent; leaves entire or undulate; corolla 15–20 cm long ....
.... 2. *D. innoxia*

   1. D. stramonium L. Jimsonweed. July–Oct. Stem green; corolla
white. Native of Asia; naturalized in waste areas; in every co.
   var. tatula (L.) Torr. Purple Jimsonweed. July–Oct. Stem purple; corolla purplish. Native of Asia; naturalized in waste areas; occasional
throughout the state.
   2. D. innoxia Mill. Aug.–Oct. Native of trop. Am.; naturalized in
waste areas; occasional throughout the state. *D. meteloides* DC.—G.

7. *Physalis* L. Ground Cherry

1. Flowers yellow; calyx during fruiting green or brownish 2
1. Flowers white; calyx during fruiting red or scarlet .... 14. *P. alkekengi*
2. Stems glabrous or nearly so, never uniformly villous, hispid, hirsute, or
   glandular-hairy 3
2. Stems uniformly villous, hispid, hirsute, or glandular-hairy 8
3. Annuals without rhizomes; leaves mostly conspicuously dentate 4
3. Perennials with rhizomes; leaves entire or merely repand-dentate ... 6
4. Corolla with black spots near base; peduncles up to 6 mm long; calyx
   often puberulent, with deltoid teeth; berry purple .... 1. *P. ixocarpa*
4. Corolla without a darkened center; peduncles 10 mm long or longer;
   calyx usually glabrous, with acute to attenuate teeth; berry yellow ... 5
5. Calyx in fruit strongly 10-angled; peduncles 1–2 cm long 2. *P. angulata*
5. Calyx in fruit obscurely 10-angled; peduncles nearly all over 2 cm long
   .... 3. *P. pendula*
6. Leaves rather thick, lanceolate to linear, on petioles up to 3 cm long;
   berry yellow .... 4. *P. longifolia*
6. Leaves membranaceous, ovate to ovate-oblong, at least some of them
   on petioles up to 6 cm long; berry red or purple 7
7. Leaves opaque; fruiting calyx ovoid, to 3 cm long, sunken at the base ..
   .... 5. *P. subglabrata*
7. Leaves translucent; fruiting calyx pyramidal, to 6 cm long, deeply
   sunken at the base .... 6. *P. macrophysa*
8. Pubescence of stem glandular-viscid 9
8. Pubescence of stem not glandular-viscid 12
9. Annuals without rhizomes; fruiting calyx abruptly acuminate 10
9. Perennials with rhizomes; fruiting calyx merely acute 11
10. Fruiting calyx long-tapering, 3–4 cm long; reticulations between the lateral nerves obscure .... 7. *P. barbadensis*
10. Fruiting calyx short-tapering, 2–3 cm long; reticulations between the lateral nerves prominent .... 8. *P. pruinosa*
11. Leaves more or less tapering to the base, most of them less than 6 cm
   long; berry red .... 9. *P. virginiana*
11. Leaves more or less rounded or cordate at the base, many of them over
   6 cm long; berry yellow .... 10. *P. heterophylla*
12. Peduncles during flowering up to 5 mm long; annuals without rhizomes
   .... 11. *P. pubescens*
12. Peduncles during flowering 1 cm long or longer; perennials with rhizomes 13
13. Stems villous; calyx in fruit shallowly or deeply sunken at the base;
   leaves mostly oblong to ovate-lanceolate to ovate. 14
13. Stems hispid or hirsute; calyx in fruit not sunken at the base; leaves lanceolate to oblanceolate .... 13. *P. lanceolata*

14. Leaves more or less tapering to the base 15
14. Leaves more or less rounded or cordate at the base   10. *P. heterophylla*
15. Some or all the hairs branched; calyx in fruit mostly 4 cm long or longer, shallowly sunken at the base .... 12. *P. pumila*
15. All the hairs unbranched; calyx in fruit mostly less than 4 cm long, deeply sunken at the base .... 9. P. virginiana

1. P. ixocarpa Brotero. Tomatillo. July–Aug. Native of the sw. U.S., Mex., and S. Am.; rarely adventive in Ill.; DuPage, Lake, and Wabash cos.

2. P. angulata L. Ground Cherry. July–Sept. Native of the s. U.S.; infrequently escaped to waste areas in the cent. cos.

3. P. pendula Rydb. Ground Cherry. July–Sept. Native of the sw. U.S.; rarely adventive in Ill.; Alexander, Cook, DuPage, Jackson, and Union cos.

4. P. longifolia Nutt. Ground Cherry. June–Aug. Native of the w. U.S.: rarely adventive in Ill.; Jackson Co.

5. P. subglabrata Mack. & Bush. Smooth Ground Cherry. June–Sept. Disturbed areas; common throughout the state. *P. longifolia* Nutt.—G.

6. P. macrophysa Rydb. Large-fruited Ground Cherry. June–July. Native of the sw. U.S.; adventive in waste ground; Champaign and Peoria cos.

7. P. barbadensis Jacq. Ground Cherry. June–Sept. Native of trop. Am.; rarely escaped in Ill. Jackson Co.

8. P. pruinosa L. Ground Cherry. June–Oct. Disturbed areas, occasional; scattered throughout the state.

9. P. virginiana Mill. Ground Cherry. May–Aug. Woods, disturbed areas; occasional throughout the state.

10. P. heterophylla Nees. Ground Cherry. May–Sept. At least some of the pubescence on the stem glandular. Woods, disturbed areas; occasional throughout the state.

var. ambigua (Gray) Rydb. Ground Cherry. May–Sept. Stems without glandular hairs; leaves thick, dentate. Waste areas; occasional to common throughout the state.

var. nyctaginea (Dunal) Rydb. Ground Cherry. May–Sept. Stems without glandular hairs; leaves thin, usually entire. Waste areas, not common; apparently confined to the s. half of the state.

11. P. pubescens L. Annual Ground Cherry. June–Oct. Disturbed areas, occasional; throughout the state.

12. P. pumila Nutt. Dwarf Ground Cherry. June–Aug. Dry hillside, very rare; Peoria Co.

13. P. lanceolata Michx. Narrow-leaved Ground Cherry. May–Sept. Native of the w. U.S.; rarely adventive in Ill. Champaign, Mason, and Union cos.

14. P. alkekengi L. Chinese Lantern. July–Sept. Native of Asia; rarely escaped from cultivation.

8. Nicotiana L. Tobacco

1. N. rustica L. Wild Tobacco. Sept.–Oct. Native of Peru; adventive in waste areas; Cook, Menard, and Peoria cos.

9. *Petunia* Juss. *Petunia*

1. Corolla white, salverform .... 1. *P. axillaris*
1. Corolla red, violet, or purple, funnelform 2
2. Corolla 3–4 cm long .... 2. *P. violacea*
2. Corolla 5–9 cm long .... 3. *P.* X *hybrida*

    1. P. axillaris (Lam.) BSP. White Petunia. July–Sept. Native of S. Am.; rarely escaped from cultivation; Jackson and Lake cos.
    2. P. violacea Lindl. Violet Petunia. June–Aug. Native of S. Am.; occasionally escaped from cultivation.
    3. P. X hybrida Vilm. Garden Petunia. June–Aug. Native of S. Am.; rarely escaped from cultivation.

141. SCROPHULARIACEAE

1. Tree with ovate, cordate leaves up to 30 cm long and nearly as broad or broader; corolla to 6 cm long, violet and yellow .... 1. *Paulownia*
1. Herbs with various leaves but not as above; corolla various, but not as above 2
2. Leaves whorled .... 2. *Veronicastrum*
2. Leaves opposite or alternate 3
3. All leaves opposite (excluding bracteal leaves) 4
3. At least some of the leaves alternate (excluding bracteal leaves) 22
4. Anther-bearing stamens 2 5
4. Anther-bearing stamens 4 7
5. Calyx and corolla lobes each 4; fruit heart-shaped .... 3. *Veronica*
5. Calyx and corolla lobes each 5; fruit subacute or rounded at tip, not heart-shaped 6
6. Sterile stamens 2; bractlets absent .... 5. *Lindernia*
6. Sterile stamens absent, or minute; each flower subtended by 2 bractlets .... 4. *Gratiola*
7. One elongated sterile stamen present 8
7. Sterile stamen absent, or represented only by a gland as broad as long 9
8. Flowers sessile .... 6. *Chelone*
8. Flowers pedicellate .... 7. *Penstemon*
9. Some or all of the leaves pinnatifid or pinnately lobed (excluding *Gerardia auriculata* which has 2 small basal lobes) 10
9. Leaves toothed or entire, not pinnatifid 13
10. Corolla lavender, to 5 mm long; plants never exceeding 20 cm in height .... 8. *Conobea*
10. Corolla yellow, at least 15 mm long; plants more than 20 cm tall ... 11
11. Stamens included 12
11. Stamens exserted .... 11. *Seymeria*
12. Corolla bilabiate; calyx asymmetrical; capsule flattened . 9. *Pedicularis*
12. Corolla not bilabiate; calyx usually symmetrical; capsule globose to ovoid .... 10. *Gerardia*
13. Flowers 1–2 in the axils of the cauline leaves 14
13. Flowers 3 or more in spikes, racemes, panicles, or corymbs 18
14. Leaves entire 15
14. Leaves toothed 17
15. Leaves orbicular .... 12. *Bacopa*
15. Leaves filiform to linear to elliptic 16

16. Calyx lobes 5; corolla slightly 2-lipped; seeds usually more than 4 per capsule .... 10. *Gerardia*
16. Calyx lobes 4; corolla strongly 2-lipped; seeds 2–4 per capsule ....
.... 13. *Melampyrum*
17. Calyx 5-parted nearly to base; flowers up to 1 cm long, white with purple lines .... 12. *Bacopa*
17. Calyx tubular, with 5 lobes; flowers over 1 cm long, violet, or, if less than 1 cm long, yellow .... 14. *Mimulus*
18. Flowers sessile 19
18. Flowers pedicellate 20
19. Corolla campanulate; leaves usually with a pair of lobes at the base, otherwise entire .... 10. *Gerardia*
19. Corolla salverform; leaves toothed .... 15. *Buchnera*
20. Stems creeping; sterile stamens absent .... 16. *Mazus*
20. Stems erect; sterile stamen reduced to a gland or scale 21
21. All leaves petiolate; corolla purple or greenish-purple; coarse perennials
.... 17. *Scrophularia*
21. Middle and upper leaves sessile; corolla blue and white or violet-purple
.... 18. *Collinsia*
22. Corolla strongly spurred at base 23
22. Corolla not spurred at base (occasionally saccate at base in *Antirrhinum*) 26
23. Stems glabrous 24
23. Stems pubescent 25
24. Plants trailing or twining; leaves suborbicular, lobed or crenulate ......
.... 19. *Cymbalaria*
24. Plants erect; leaves linear to lanceolate, entire .... 20. *Linaria*
25. Stems spreading or prostrate; fruit glabrous; leaves ovate to oblong ...
.... 21. *Kickxia*
25. Stems erect; fruit puberulent; leaves linear to linear-lanceolate ....
.... 22. *Chaenorrhinum*
26. Stamens 5; corolla more or less actinomorphic .... 23. *Verbascum*
26. Stamens 2 or 4; corolla zygomorphic 27
27. Leaves and lower part of the stem glabrous; leaves entire ....
.... 24. *Antirrhinum*
27. Leaves and stems pubescent; leaves toothed or lobed 28
28. Leaves merely toothed; calyx about as long as the greenish-yellow corolla .... 25. *Wulfenia*
28. Some or all the leaves pinnatifid or deeply lobed; calyx shorter than the scarlet, red, or yellow corolla 29
29. Margins of the lobed leaves toothed; leaves divided into 12 or more lobes .... 9. *Pedicularis*
29. Margins of the lobed leaves entire; leaves divided into 3–7 lobes ....
.... 26. *Castilleja*

1. *Paulownia Sieb. & Zucc.*

1. P. tomentosa (Thunb.) Steud. Princess Tree. April–May. Native of China; adventive in the s. one-third of the state.

2. *Veronicastrum Fab.*

1. V. virginicum (L.) Farw. Culver's-root. June–Sept. Woods, prairies; occasional throughout the state. (Including f. *villosum* [Raf.] Pennell.)

3. Veronica L.   Speedwell

1. Flowers solitary in the axils of the cauline leaves 2
1. Flowers in racemes or spikelike inflorescences (rarely solitary in *V. ser-pyllifolia*) 6
2. Corolla white; leaves glabrous .... 1. *V. peregrina*
2. Corolla blue or lilac; leaves pubescent 3
3. Corolla 8–12 mm across; capsule 6–10 mm long .... 2. *V. persica*
3. Corolla up to 5 mm across; capsule less than 5 mm long 4
4. Leaves reniform-suborbicular, 3- to 7-lobed, broader than long; capsule not compressed .... 3. *V. hederaefolia*
4. Leaves ovate to oval, or the upper ones linear to lanceolate, longer than broad; capsule compressed 5
5. Pedicels 1–2 mm long; seeds smooth .... 4. *V. arvensis*
5. Pedicels 4–12 mm long; seeds rugulose .... 5. *V. polita*
6. Stems glabrous (or with upcurved hairs less than 0.5 mm long in *V. ser-pyllifolia*). 7
6. Stems pubescent 10
7. Racemes terminal; leaves up to 1.5 cm long .... 6. *V. serpyllifolia*
7. Racemes axillary; leaves, or most of them, over 1.5 cm long 8
8. Leaves linear to linear-lanceolate; axis of raceme zigzag 7. *V. scutellata*
8. Leaves lanceolate to oblong to ovate; axis of raceme straight 9
9. Leaves sessile, sometimes clasping; pedicels glandular-pubescent .... 8. *V. comosa*
9. Leaves petiolate; pedicels eglandular .... 9. *V. americana*
10. Racemes terminal; leaves acuminate .... 10. *V. longifolia*
10. Racemes axillary; leaves obtuse to subacute 11
11. Leaves petiolate; raceme densely flowered; pedicels shorter than the calyx .... 11. *V. officinalis*
11. Leaves sessile or very short-petiolate; raceme loosely flowered; pedicels about as long as the calyx or longer .... 12. *V. chamaedrys*

1. V. peregrina L. White Speedwell. April–Aug. Stem and capsules glabrous. Fields, disturbed soil, common; in every co.

var. xalapensis (HBK.) St. John. April–Aug. Stem and capsules glandular-pubescent. Fields, disturbed soil, occasional but not as common as the preceding.

2. V. persica Poir. Bird's-eye Speedwell. April–Aug. Native of Europe and Asia; occasionally naturalized in Ill.; Coles, DuPage, Jackson, Lake, Peoria, and Wabash cos.

3. V. hederaefolia L. Ivy-leaved Speedwell. March–April. Native of Europe; adventive in a grassy meadow; Jackson and Union cos.

4. V. arvensis L. Corn Speedwell. March–Aug. Native of Europe; fields, woods, waste ground, common; in every co.

5. V. polita Fries. Speedwell. March–May. Native of Europe; adventive in lawns, occasional; Jackson, Lawrence, Piatt, Randolph, and Saline cos. *V. didyma* Tenore—J; *V. agrestis* L.—G.

6. V. serpyllifolia L. Thyme-leaved Speedwell. April–July. Native of Europe; adventive in lawns and meadows; scattered throughout the state.

7. V. scutellata L. Marsh Speedwell. April–July. Moist soil, often around ponds, rare; confined to the n. half of the state.

8. V. comosa Richter. Water Speedwell. June–Sept. Wet ditches, shores; occasional in the n. half of the state, rare or absent elsewhere. *V. catenata* Pennell—J; *V. salina* Schur.—G.

9. V. americana (Raf.) Schwein. American Brooklime. June–Aug. Swampy ground, rare; confined to the n. half of the state.

10. V. longifolia L. Garden Speedwell. May–Sept. Native of Europe; rarely escaped from cultivation; Cook Co.

11. V. officinalis L. Common Speedwell. May–Sept. Native of Europe; adventive in grassy areas and open woods; Cook, DuPage, and Richland cos.

12. V. chamaedrys L. Germander Speedwell. May–July. Native of Europe; rarely adventive in Ill.; DuPage Co.

4. Gratiola L. Hedge Hyssop

1. Leaves toothed; sterile stamens absent or minute; corolla pale yellow or white; annuals 2
1. Leaves entire or remotely toothed; sterile stamens 2; corolla golden-yellow .... 3. *G. aurea*
2. Pedicels to 10 mm long (in fruit); stems glabrous, more or less fleshy .. .. 1. *G. virginiana*
2. Pedicels over 10 mm long (in fruit); stems more or less pubescent, not fleshy .... 2. *G. neglecta*

1. G. virginiana L. Round-fruited Hedge Hyssop. April–Oct. Moist soil; occasional to uncommon throughout the state.

2. G. neglecta Torr. Clammy Hedge Hyssop. May–Oct. Moist soil; occasional throughout the state.

3. G. aurea Muhl. Goldenpert. June–Sept. Wet ground, very rare; Cook Co.

5. Lindernia All. False Pimpernel

1. Pedicels of all flowers longer than the subtending leaves .... 1. *L. anagallidea*
1. Pedicels of the lowermost flowers shorter than the subtending leaves .. .. 2. *L. dubia*

1. L. anagallidea (Michx). Pennell. Slender False Pimpernel. June–Oct. Moist soil; occasional throughout the state, except for the nw. cos. where it is rare or absent.

2. L. dubia (L.) Pennell. False Pimpernel. June–Oct. All pedicels shorter than the subtending leaves. Moist ground; common throughout the state.

var. riparia (Raf.) Fern. False Pimpernel. June–Oct. Uppermost pedicels longer than the subtending leaves. Moist ground; common throughout the state. *L. dubia* (L.) Pennell—J.

6. Chelone L. Turtlehead

1. Corolla greenish-yellow to creamy-white; leaves sessile or on petioles up to 5 mm long .... 1. *C. glabra*
1. Corolla rose-purple; leaves on petioles 5 mm long or longer .... 2. *C. obliqua*

1. C. glabra L. White Turtlehead. July–Oct. Wet soil; occasional in the n. half of the state, rare in the s. half. (Including var. *linifolia* Coleman, and f. *tomentosa* [Raf.] Pennell.)

2. C. obliqua L. var. speciosa Pennell & Wherry. Pink Turtlehead. Aug.–Oct. Low woods, swampy meadows, not common; known from the s. one-third of the state; also Adams, Fulton, Hancock, Henderson, and Tazewell cos.; adventive in DuPage Co. *C. obliqua* L.—G, J.

7. *Penstemon* Mitchell   Beard-tongue

1. Corolla 3.5–5.0 cm long 2
1. Corolla up to 3 cm long 3
2. Leaves glaucous, broadly ovate, entire .... 1. *P. grandiflorus*
2. Leaves green, narrowly ovate, sharply toothed .... 2. *P. cobaea*
3. Corolla glandular-pubescent within .... 3. *P. tubaeflorus*
3. Corolla eglandular within 4
4. The two lips of the corolla nearly equal in size and shape; sterile filament sparsely to moderately pubescent 5
4. The two lips of the corolla unequal in size and shape; sterile filament densely bearded 8
5. Anthers pubescent with short stiff hairs 6
5. Anthers glabrous .... 7. *P. calycosus*
6. Sepals during anthesis 5–8 mm long; capsules 9–12 mm long .... 4. *P. digitalis*
6. Sepals during anthesis 2–5 mm long; capsules 6–9 mm long 7
7. Leaves acuminate, sharply serrate; sepals at fruiting time 5 mm long or longer .... 5. *P. alluviorum*
7. Leaves obtuse to subacute, entire or sparsely toothed; sepals at fruiting time 3–4 mm long .... 6. *P. deamii*
8. Corolla up to 2.3 cm long; stems without glandular pubescence 9
8. Corolla more than 2.3 cm long; stems glandular-pubescent .... 10. *P. hirsutus*
9. Leaves pubescent; throat of corolla longer than the tube   8. *P. pallidus*
9. Leaves glabrous or nearly so; throat of corolla about as long as the tube .... 9. *P. arkansanus*

1. P. grandiflorus Nutt. Large-flowered Bear-tongue. May–June. Sandy soil, very rare; Henderson Co.; also adventive in McHenry Co.

2. P. cobaea Nutt. June. Native s. and w. of Ill.; in a flat, dry meadow in Ill.; Kane Co.

3. P. tubaeflorus Nutt. Beard-tongue. May–June. Prairies, dry woods; not common in the s. half of the state, rare in the n. half.

4. P. digitalis Nutt. Foxglove Beard-tongue. May–July. Woods, thickets; occasional throughout the state.

5. P. alluviorum Pennell. Lowland Beard-tongue. May–June. Swampy woods, rare; Union Co.

6. P. deamii Pennell. Deam's Beard-tongue. May–June. Dry woods, rare; Jackson, Monroe, Pope, Union, and Wabash cos.

7. P. calycosus Small. Smooth Beard-tongue. May–July. Wooded slopes, edge of woods; occasional throughout the state.

8. P. pallidus Small. Pale Beard-tongue. April–July. Dry woods, prairies; common in the s. half of the state, occasional in the n. half.

9. P. arkansanus Pennell. Beard-tongue. April–June. Wooded slopes, very rare; Jackson and Randolph cos. *P. pallidus* Small—G.

10. P. hirsutus (L.) Willd. Hairy Beard-tongue. June–July. Gravelly prairies, wooded slopes, not common; Cook, Grundy, Hardin, Union, and Will cos.

8. Conobea Aubl.

1. C. multifida (Michx.) Benth. May–Oct. Moist soil, fields, prairies; occasional in the s. four-fifths of the state, rare or absent elsewhere. *Leucospora multifida* (Michx.) Nutt.—G, J.

9. Pedicularis L. Lousewort

1. Most of the cauline leaves opposite, sessile; corolla creamy or yellowish-white; flowers appearing from August to October ...... 1. *P. lanceolata*
1. Most of the cauline leaves alternate, petiolate; corolla yellow; flowers appearing from April to June .... 2. *P. canadensis*

1. P. lanceolata Michx. Swamp Wood Betony. Aug.–Oct. Calcareous fens, wet meadows; occasional in the n. half of the state, uncommon in the s. half.
2. P. canadensis L. Lousewort. April–June. Woods, prairies; occasional throughout the state. (Including var. *dobbsii* Fern.)

10. Gerardia L. Gerardia

1. Corolla yellow; some or all the leaves pinnately divided 2
1. Corolla pink or purple; all leaves entire, or merely 3-lobed 4
2. Stems glabrous and glaucous .... 1. *G. flava*
2. Stems pubescent and usually not glaucous 3
3. Corolla 4.5 cm long or longer; pubescence not glandular; capsule rusty-pubescent 1.2–2.5 cm long .... 2. *G. grandiflora*
3. Corolla 3–4 cm long; pubescence glandular; capsule glandular-pubescent, up to 1.2 cm long .... 3. *G. pedicularia*
4. Leaves with two basal lobes; stems retrorsely hispid ... 4. *G. auriculata*
4. Leaves without lobes; stems glabrous or scabrous 5
5. Plants yellow-green, usually not turning black upon drying; calyx strongly reticulate-nerved; seeds yellow 6
5. Plants deep green or purplish, usually turning black upon drying; calyx weakly reticulate-nerved, or nerveless; seeds dark brown to black ... 7
6. Branches with solitary flowers; stems terete .... 5. *G. gattingeri*
6. Branches with 2 or more flowers; stems more or less angled ....
.... 6. *G. skinneriana*
7. Pedicels more than twice as long as the calyx .... 7. *G. tenuifolia*
7. Pedicels less than twice as long as the calyx 8
8. Capsule ellipsoid, longer than broad; pedicels 5–8 mm long 8. *G. aspera*
8. Capsule globose or subglobose, about as broad as long; pedicels to 5 mm long 9
9. Axillary fascicles of leaves as long as the main leaves; stem scabrous ..
.... 9. *G. fasciculata*
9. Axillary fascicles of leaves much shorter than the main leaves or lacking; stem smooth or nearly so 10
10. Calyx up to half as long as the capsule; style to 1 cm long; flowers usually up to 2 cm long .... 10. *G. paupercula*
10. Calyx half as long as the capsule or longer; style 1.5–2.0 cm long; flowers usually over 2 cm long .... 11. *G. purpurea*

1. G. flava L. Smooth False Foxglove. June–Sept. Rocky woods; occasional in the s. half of the state, less common in the n. half. (Including var. *macrantha* [Pennell] Fern.) *Aureolaria flava* (L.) Farw.—G, J.

2. G. grandiflora Benth. var. pulchra (Pennell) Fern. Yellow False Foxglove. July–Sept. Woods; occasional in the n. two-thirds of the state, absent elsewhere. *Aureolaria grandiflora* (Benth.) Pennell—G, J.

3. G. pedicularia L. var. ambigens Fern. Clammy False Foxglove. Aug.–Sept. Sandy woods; occasional in the ne. cos., absent elsewhere. *Aureolaria pedicularia* (L.) Raf.—G, J.

4. G. auriculata Michx. Auriculate False Foxglove. Aug.–Sept. Dry prairies; occasional in the n. three-fourths of the state, absent elsewhere. *Tomanthera auriculata* (Michx.) Raf.—G.

5. G. gattingeri Small. Round-stemmed False Foxglove. Aug.–Oct. Dry, often rocky, woods; uncommon to occasional throughout the state.

6. G. skinneriana Wood. Pale False Foxglove. July–Sept. Dry prairies, rocky woods; not common but scattered throughout the state.

7. G. tenuifolia Vahl. Slender False Foxglove. Aug.–Oct. Calyx up to 1 mm long; axillary fascicles of leaves rarely present. Moist soil; occasional throughout the state.

var. macrophylla Benth. Aug.–Oct. Calyx 1–2 mm long; auxillary fascicles of leaves commonly present. Moist soil; occasional throughout the state. (Including var. *parviflora* Nutt.)

8. G. aspera Dougl. Rough False Foxglove. Aug.–Sept. Prairies, not common but scattered throughout the state.

9. G. fasciculata Ell. False Foxglove. Aug.–Sept. Edge of woods, sandy open ground, not common; confined to the s. one-sixth of the state.

10. G. paupercula (Gray) Britt. False Foxglove. Aug.–Sept. Moist soil, rare; scattered throughout the state. *G. purpurea* L. var. *paupercula* Gray—G. (Including var. *borealis* [Pennell] Deam.)

11. G. purpurea L. False Foxglove. July–Sept. Moist soil; occasional in the n. three-fourths of the state, less common in the s. one-fourth.

11. Seymeria Pursh

1. S. macrophylla Nutt. Mullein Foxglove. June–Sept. Rich woods, rocky slopes, dry woods, thickets; occasional throughout the state. *Dasistoma macrophylla* (Nutt.) Raf.—G, J.

12. Bacopa Aubl. Water Hyssop

1. Leaves entire, more or less orbicular; pedicels bractless . . . .
. . . . 1. *B. rotundifolia*
1. Leaves serrate, oblanceolate; pedicels with 2 bracts . . 2. *B. acuminata*

1. B. rotundifolia (Michx.) Wettst. Water Hyssop. May–Sept. Wet ground or in shallow water; occasional in the s. half of the state, rare or absent elsewhere.

2. B. acuminata (Walt.) Robins. July–Aug. Wet roadside ditch, very rare; Wabash Co. *Mecardonia acuminata* (Walt.) Small—G, J.

13. Melampyrum L. Cow-wheat

1. M. lineare Desr. var. latifolium Bart. Cow-wheat. June–Aug. Bogs, marshes, dunes, very rare; Cook Co. *M. lineare* Desr.—J.

14. *Mimulus* L.   Monkey-flower

1. Corolla yellow, up to 1.2 cm long; leaves oval to suborbicular, entire or denticulate .... 1. *M. glabratus*
1. Corolla purple, at least 2 cm long; leaves lanceolate to elliptic to ovate, serrate 2
2. Leaves petiolate; stem winged .... 2. *M. alatus*
2. Leaves sessile or even clasping; stem wingless .... 3. *M. ringens*

    1. M. glabratus HBK. var. fremontii (Benth.) Grant. Yellow Monkey-flower. June–Oct. Around springs, rare; confined to the n. half of the state. *M. geyeri* Torr.—J.
    2. M. alatus Ait. Winged Monkey-flower. June–Sept. Wet ground; common in the s. half of the state, becoming rare in the n. half. A hybrid between this species and *M. ringens* L. has been collected in Cass Co.
    3. M. ringens L. Monkey-flower. June–Sept. Most or all the cauline leaves rounded or clasping at the base. Wet ground; occasional in the n. three-fourths of the state, rare elsewhere.
    var. minthodes (Greene) Grant. Monkey-flower. June–Sept. Most or all the cauline leaves tapering to the base. Wet ground; occasional in the westernmost cos. of the state.

15. *Buchnera* L.

    1. B. americana L. Blue Hearts. July–Sept. Prairies, fields, not common; scattered in Ill.

16. *Mazus* Lour.

    1. M. japonicus (Thunb.) Kuntze. July–Oct. Native of Asia; adventive in lawns; Cook Co.

17. *Scrophularia* L.   Figwort

1. Petioles unwinged; sterile stamen purple or brown; flowering from July to October .... 1. *S. marilandica*
1. Petioles winged; sterile stamen greenish-yellow; flowering from May to June .... 2. *S. lanceolata*

    1. S. marilandica L. Late Figwort. July–Oct. Usually rich woods; occasional throughout the state. (Including f. *neglecta* [Rydb.] Pennell.)
    2. S. lanceolate Pursh. Early Figwort. May–June. Woods, fence rows; occasional in the n. half of the state, otherwise very rare.

18. *Collinsia* Nutt.

1. Lower lip of corolla blue; at least some of the upper leaves clasping; capsule with 2–4 seeds .... 1. *C. verna*
1. Lower lip of corolla violet or purple; none of the leaves clasping; capsule with 6 or more seeds .... 2. *C. violacea*

    1. C. verna Nutt. Blue-eyed Mary. April–June. Rich woods; occasional throughout the state, but often abundant where found.

2. C. violacea Nutt. Violet Collinsia. May–June. Sandy hillside, rare; Shelby Co.

19. Cymbalaria Hill

1. C. muralis Gaertn. Kenilworth Ivy. June–Oct. Native of Europe and Asia; occasionally escaped from cultivation; Champaign and Kane cos.

20. Linaria Mill. Toadflax

1. Corolla yellow or yellow and orange, 2–4 cm long 2
1. Corolla blue or violet, up to 1.2 cm long .... 3. *L. canadensis*
2. Leaves linear to linear-lanceolate; corolla 2–3 cm long, yellow and orange .... 1. *L. vulgaris*
2. Leaves ovate to oblong; corolla 3.5–4.0 cm long, yellow 2. *L. dalmatica*

1. L. vulgaris Hill. Butter-and-Eggs. May–Nov. Native of Europe; commonly escaped from cultivation throughout the state.
2. L. dalmatica (L.) Mill. June–Sept. Native of Europe; rarely escaped from cultivation; Winnebago Co.
3. L. canadensis (L.) Dumort. Blue Toadflax. April–Sept. Flowers 5–10 mm long; seeds smooth. Sandy soil; occasional in the n. half of the state, uncommon in the s. half.
var. texana (Scheele) Pennell. South Blue Toadflax. April–Sept. Flowers 10–17 mm long; seeds rugulose. Sandy soil, not common; confined to the s. tip of the state. *L. texana* Scheele—J.

21. Kickxia Dumort. Cancerwort

1. K. elatine (L.) Dumort. Canker-root. May–Oct. Native of Europe; adventive in waste ground; not common but scattered in Ill.

22. Chaenorrhinum Reichenb.

1. C. minus (L.) Lange. Dwarf Snapdragon. June–Sept. Native of Europe; adventive in Ill., particularly along railroads.

23. Verbascum L. Mullein

1. Plants pubescent, not glandular; capsule densely tomentose; flowers in spikes 2
1. Plants glabrous or glandular-pubescent; capsule puberulent or glabrous; flowers in racemes 3
2. Flowers up to 2 (–2.5) cm across; spike crowded; leaves usually not clasping .... 1. *V. thapsus*
2. Flowers over 2.5 cm across; spike interrupted, at least below; some of the leaves usually clasping .... 2. *V. phlomoides*
3. Pedicels 1 cm long or longer, longer than the capsules .. 3. *V. blattaria*
3. Pedicels up to 5 mm long, shorter than the capsules ... 4. *V. virgatum*

1. V. thapsus L. Woolly Mullein. May–Sept. Native of Europe; commonly adventive in Ill.; in every co.

2. V. phlomoides L. Clasping Mullein. July–Aug. Native of Europe; rarely adventive in Ill.; Cook and DuPage cos.

3. V. blattaria L. Moth Mullein. May–Sept. Native of Europe; often adventive in Ill.; scattered throughout the state. (Including f. *erubescens* Brugger, and f. *albiflora* [Don] House.)

4. V. virgatum Stokes. Mullein. Native of Europe; rarely adventive in Ill.; Coles and Pulaski cos.

24. Antirrhinum L.   Snapdragon

1. Perennial; leaves oblong to ovate; corolla at least 2 cm long   1. *A. majus*
1. Annual; leaves lanceolate; corolla up to 1.2 cm long .... 2. *A. orontium*

1. A. majus L. Common Snapdragon. June–Oct. Native of Europe; rarely escaped from cultivation; Jackson Co.

2. A. orontium L. Lesser Snapdragon. Sept. Native of Europe; escaped into a cultivated oat field; DuPage Co.

25. Wulfenia Jacq.

1. W. bullii (Eat.) Barnh. Kitten Tails. May–June. Sandy soil, prairies, not common; primarily along the Illinois and Mississippi rivers in the nw. one-fourth of the state. *Synthyris bullii* (Eat.) Heller—J; *Besseya bullii* (Eat.) Rydb.—G.

26. Castilleja Mutis   Paintbrush

1. Bracts scarlet, orange, or yellow; corolla up to 3.5 cm long ....
.... 1. *C. coccinea*
1. Bracts green; corolla 4 cm long or longer .... 2. *C. sessiliflora*

1. C. coccinea L. Indian Paintbrush. April–July. Bracts scarlet or orange. Prairies, sandy woods; occasional in the n. three-fourths of the state, absent elsewhere.

f. lutescens Farw. Bracts yellow. Prairies, sandy woods; scattered in the n. three-fourths of the state.

2. C. sessiliflora Pursh. Downy Yellow Painted Cup. May–July. Prairies, rare; DuPage, Lake, McHenry, and Winnebago cos.

142. BIGNONIACEAE

1. Vines; leaves compound 2
1. Trees; leaves simple .... 3. *Catalpa*
2. Leaflets 7–11; tendrils absent .... 1. *Campsis*
2. Leaflets 2; tendrils present .... 2. *Bignonia*

1. Campsis Lour.   Trumpet Creeper

1. C. radicans (L.) Seem. Trumpet Creeper. June–Aug. Roadsides, fields, edge of woods; native and common in the s. half of the state, adventive and rarer in the n. half.

2. *Bignonia* L. *Cross-vine*

1. B. capreolata L. Cross-vine. April–June. Low woods, wooded slopes; confined to the s. one-fourth of the state.

3. *Catalpa Scop.*

1. Lowest corolla lobe not notched; fruit 5–10 mm in diameter; crushed leaves with an unpleasant odor .... 1. *C. bignonioides*
1. Lowest corolla lobe notched; fruit 12 mm in diameter or broader; crushed leaves without an unpleasant odor .... 2. *C. speciosa*

    1. C. bignonioides Walt. Catalpa. May–June. Native to the se. U.S.; escaped along roads, railroads, and streams; occasional throughout the state.
    2. C. speciosa Warder. Catalpa. May–June. Low woods; native in the s. one-fourth of the state, escaped from cultivation in most parts of Ill.

143. MARTYNIACEAE

1. *Proboscidea Schmidel Unicorn-plant*

    1. P. louisianica (Mill.) Thell. Proboscis-flower. June–Sept. Low ground, particularly along rivers; occasional throughout the state, where it probably is naturalized.

144. OROBANCHACEAE

1. Calyx deeply split on the lower side, 3- to 4-toothed on the upper side; stamens exserted .... 1. *Conopholis*
1. Calyx more or less equally 4- or 5-toothed; stamens included 2
2. Corolla 5-lobed; flower solitary .... 3. *Orobanche*
2. Corolla 4-lobed; flowers spicate 3
3. Calyx 4-lobed .... 3. *Orobanche*
3. Calyx 5-lobed 4
4. Corolla up to 1 cm long .... 2. *Epifagus*
4. Corolla 1.5–2.0 cm long .... 3. *Orobanche*

1. *Conopholis Wallr. Squaw-root*

    1. C. americana (L. f.) Wallr. Cancer-root. May–July. Wooded ravines; parasitic on oak roots; rare; scattered in the n. two-thirds of the state.

2. *Epifagus Nutt. Beech-drops*

    1. E. virginiana (L.) Bart. Beech-drops. Aug.–Oct. Parasitic on beech roots; restricted to the s. one-fourth of the state; also Clark, Crawford, and Lawrence cos.

3. Orobanche L. Broomrape

1. Flowers in spikes; each flower subtended by 1–3 bracts; corolla 4-lobed
.... 2
1. Flower solitary, not subtended by bracts; corolla 5-lobed 3
2. Calyx 4-lobed; each flower subtended by 3 bracts .... 1. *O. ramosa*
2. Calyx 5-lobed; each flower subtended by 1–2 bracts . 2. *O. ludoviciana*
3. Corolla creamy-white to lilac; calyx lobes lance-subulate, longer than
the tube .... 3. *O. uniflora*
3. Corolla purple; calyx lobes deltoid, equalling or shorter than the tube ..
.... 4. *O. fasciculata*

1. O. ramosa L. Broomrape. Aug. Native of Europe; parasitic on
hemp roots, very rare; Champaign Co.
2. O. ludoviciana Nutt. Broomrape. Aug.–Sept. Parasitic on roots of
various members of the Compositae, rare; Lee, Menard, White, and
Wabash cos.
3. O. uniflora L. One-flowered Broomrape. May–June. Parasitic on
roots of various plants; scattered throughout the state.
4. O. fasciculata Nutt. Clustered Broomrape. April–Aug. Parasitic on
roots of various members of the Compositae, not common; confined
to the n. half of the state.

145. LENTIBULARIACEAE

1. Utricularia L. Bladderwort

1. Leaves entire, rarely seen .... 1. *U. cornuta*
1. Leaves pinnately divided, conspicuous 2
2. Fruiting pedicels ascending or erect 3
2. Fruiting pedicels arched-recurving 4
3. Corolla up to 1.2 cm long; leaf segments terete .... 2. *U. gibba*
3. Corolla 1.2 cm long or longer; leaf segments flat ..... 3. *U. intermedia*
4. Corolla 1.5–2.5 cm long; leaf segments terete .... 4. *U. vulgaris*
4. Corolla 5–8 mm long; leaf segments flat .... 5. *U. minor*

1. U. cornuta Michx. Horned Bladderwort. July–Sept. Bogs and wet
sands, rare; Cook and Lake cos.
2. U. gibba L. Humped Bladderwort. Aug.–Oct. Swamps, ponds,
lakes, and ditches, not common; in s. cos. along the Mississippi
River; also Cook, Lake, and McHenry cos.
3. U. intermedia Hayne. Flat-leaved Bladderwort. July–Aug. Shallow
water, rare; Lake, McHenry, Ogle, and Tazewell cos.
4. U. vulgaris L. Common Bladderwort. July–Aug. Shallow water,
not common; scattered throughout the state.
5. U. minor L. Small Bladderwort. June–Aug. Shallow water, very
rare; Lake and McHenry cos.

146. ACANTHACEAE

1. Corolla 2-lipped, less than 2 cm long; stamens 2 2
1. Corolla nearly actinomorphic, not 2-lipped, at least 2.5 cm long; stamens 4 .... 3. *Ruellia*
2. Petioles at least 1 cm long; flowers bracteate .... 1. *Dicliptera*
2. Petioles mostly less than 1 cm long; flowers ebracteate ..... 2. *Justicia*

1. *Dicliptera Juss.*

1. D. brachiata (Pursh) Spreng. Aug.–Oct. Bottomland woods, rare; Massac Co.

2. *Justicia L.   Water Willow*

1. Leaves linear to lanceolate .... 1. *J. americana*
1. Leaves elliptic to oblong .... 2. *J. ovata*

1. J. americana (L.) Vahl. Water Willow. May–Oct. In and along streams, occasional; throughout the state. *Dianthera americana* L.—J.
2. J. ovata (Walt.) Lindau. Water Willow. May–June. Swamps, bottomland woods, rare; Alexander and Pulaski cos. *J. humilis* Michx. var. *lanceolata* (Chapm.) Gl.—G.

3. *Ruellia L.   Wild Petunia*

1. Calyx lobes not more than 1.5 mm broad; stems and leaves pubescent throughout 2
1. Calyx lobes at least 2 mm broad; stems and leaves glabrous or sparsely pubescent .... 4. *R. strepens*
2. Leaves sessile, or with petioles at most 3 mm long ...... 1. *R. humilis*
2. Leaves with petioles longer than 3 mm 3
3. Flowers pedunculate, solitary or in cymes; capsules puberulent .... 2. *R. pedunculata*
3. Flowers sessile or subsessile in glomerules; capsules glabrous .... 3. *R. caroliniensis*

1. R. humilis Nutt. Wild Petunia. May–Oct. Corolla up to 4.5 cm long, the tube up to 2.5 cm long. Prairies, dry woods, bluffs; occasional throughout the state. (Including f. *alba* [Steyerm.] Fern., f. *grisea* Fern., and var. *frondosa* Fern.)
var. longiflora (Gray) Fern. Wild Petunia. May–Oct. Corolla 5 cm long or longer, the tube 3 cm long or longer. Prairies, dry woods, bluffs; less common than the preceding but scattered throughout the state. (Including var. *expansa* Fern.)
2. R. pedunculata Torr. Wild Petunia. May–Sept. Rocky woods; confined to the s. one-fourth of the state.
3. R. caroliniensis (Walt.) Steud. var. dentata (Nees) Fern. Wild Petunia. May–Oct. Dry woods, not common; confined to the s. one-fourth of the state. (Including var. *salicina* Fern.)
4. R. strepens L. Smooth Ruellia. May–Oct. Corolla 3 cm long or longer, the tube open. Rich or low woods; occasional in the s. four-fifths of the state. (Including f. *alba* Steyerm.)

f. cleistantha (Gray) McCoy. Smooth Ruellia. May–Oct. Corolla less than 3 cm long, the tube more or less closed. Rich or low woods; scattered in the s. one-third of the state.

147. PLANTAGINACEAE

1. Plantago L. Plantain

1. Leaves all basal 2
1. Leaves opposite .... 12. *P. indica*
2. Leaves linear, never broader than 1 cm 3
2. Leaves lanceolate, elliptic, oval, or ovate, some or all usually over 1 cm broad 6
3. Bracts much exceeding the flowers, very conspicuous ... 1. *P. aristata*
3. Bracts shorter than to barely exceeding the flowers, not conspicuous 4
4. Bracts, sepals, and inflorescence tomentose .... 2. *P. purshii*
4. Bracts, sepals, and inflorescence glabrous or nearly so 5
5. Capsule with 4 seeds; sepals obovate .... 3. *P. pusilla*
5. Capsule with 10 or more seeds; sepals ovate .... 4. *P. heterophylla*
6. Leaves cordate at the base, with lateral veins arising from the mid-vein; flowering stalks hollow .... 5. *P. cordata*
6. Leaves rounded or cuneate at the base, not cordate; flowering stalks solid 7
7. Leaves lanceolate, oblanceolate, or elliptic; seeds 1–3 (–4) per capsule 8
7. Leaves ovate or oval; seeds usually 4 or more per capsule 11
8. Spikes ellipsoid; leaves lanceolate .... 6. *P. lanceolata*
8. Spikes cylindrical; leaves obovate to elliptic 9
9. Flowers fragrant; corolla spreading, leaving the capsule exposed .... 7. *P. media*
9. Flowers not fragrant; corolla ascending, closed over the concealed capsule 10
10. Leaves mostly entire; sepals obtuse; seeds pale brown, up to 2 mm long .... 8. *P. virginica*
10. Leaves with a few coarse teeth; sepals attenuate at tip; seeds red, 2.5–3.0 mm long. .... 9. *P. rhodosperma*
11. Seeds flat; leaves densely canescent; seeds never more than 4 per capsule .... 7. *P. media*
11. Seeds turgid; leaves glabrous or nearly so on at least one surface, at least not densely canescent; seeds usually more than 4 per capsule 12
12. Base of petioles purplish; sepals acute or subacute, 2.5–3.0 mm long .. .. 10. *P. rugelii*
12. Base of petioles green; sepals obtuse, up to 2 mm long ... 11. *P. major*

1. P. aristata Michx. Bracted Plantain. May–Nov. Pastures, fields, upland woods, cliffs, waste ground; common throughout the state.
2. P. purshii Roem. & Schultes. Salt-and-pepper Plant. May–Aug. Sandy soil, not common; known from the n. one-half of the state; also Christian Co.
3. P. pusilla Nutt. Small Plantain. April–June. Leaves more or less entire; spikes up to 6 cm long; seeds 1.2 mm long. Fields, pastures, sandstone cliffs; occasional in the s. half of the state.

var. major Engelm. Small Plantain. April–June. Leaves usually
toothed; spikes more than 6 cm long; seeds 1.7–1.8 mm long. Sandstone cliffs, rare; confined to the s. one-fourth of the state.

4. P. heterophylla Nutt. Small Plantain. April–May. Sandy soil, very
rare; Pulaski and Union cos.

5. P. cordata Lam. Heartleaf Plantain. April–July. Along streams in
woods; at one time scattered throughout most of the state, but apparently rare now.

6. P. lanceolata L. Buckhorn. April–Oct. Native of Europe; naturalized on waste ground; in every co.

7. P. media L. Hoary Plantain. June–Sept. Native of Europe; rarely
adventive in Ill.; roadside; Cook Co.

8. P. virginica L. Dwarf Plantain. April–June. Fields, roadsides,
sandstone cliffs; occasional to common throughout the state.

9. P. rhodosperma Dcne. Red-seeded Plantain. May–June. Native to
the sw. U.S.; rarely adventive in Ill.; field; Williamson Co.

10. P. rugelii Dcne. Rugel's Plantain. May–Oct. Fields, woods, waste
ground, common; in every co.

11. P. major L. Common Plantain. May–Oct. Native of Europe; naturalized in waste ground; common in the n. one-third of the state, less
common elsewhere.

12. P. indica L. Whorled Plantain. July–Oct. Native of Europe; adventive in waste ground; Champaign, Cook, Kane, Kankakee, Lake,
McHenry, and Winnebago cos.

148. RUBIACEAE

1. Shrubs; flowers in dense, globose, pedunculate heads 1. *Cephalanthus*
1. Herbs; flowers solitary, or 1–3 together, or in cymes or, if in heads, the
head not pedunculate 2
2. Leaves in whorls of 4–8 3
2. Leaves opposite 4
3. Flowers pink or blue, subtended by an involucre .... 2. *Sherardia*
3. Flowers white, yellow, greenish, or maroon, not subtended by an involucre .... 3. *Galium*
4. Flowers sessile 5
4. Flowers pedicellate 6
5. Flowers 1–3 per axil .... 4. *Diodia*
5. Flowers several in an axillary glomerule .... 5. *Spermacoce*
6. Leaves evergreen, about as broad as long; fruit a red berry; flowers in
pairs, with their ovaries united .... 6. *Mitchella*
6. Leaves deciduous, longer than broad; fruit a capsule; flowers solitary or
in cymes, their ovaries not united .... 7. *Houstonia*

1. Cephalanthus L. Buttonbush

1. C. occidentalis L. Buttonbush. June–Aug. Branchlets and leaves
glabrous. Wet ground; occasional throughout the state.

var. pubescens Raf. Buttonbush. June–Aug. Branchlets and leaves
soft-pubescent. Wet ground; less common than the preceding.

2. *Sherardia* L.

1. S. arvensis L. Field Madder. May–Sept. Native of Europe; rarely adventive in waste ground; Jackson Co.

3. *Galium* L.   Bedstraw

1. Fruits and ovaries bristly or pubescent 2
1. Fruits and ovaries smooth 9
2. Leaves in whorls of 4 3
2. Leaves in whorls of 6–8 8
3. Leaves up to 1 cm long; flowers solitary in the axils .... 1. *G. virgatum*
3. Leaves, or some of them, over 1 cm long; flowers in cymes or panicles 4
4. Stems and leaves glabrous, or the leaf margins slightly scabrous 5
4. Stems and leaves pilose 7
5. Leaves linear-lanceolate; flowers bright white, in panicles 2. *G. boreale*
5. Leaves ovate-lanceolate to lanceolate; flowers greenish-white, yellowish, or becoming purplish, in cymes 6
6. Leaves ovate-lanceolate; flowers greenish-white, the lobes of the corolla usually more or less hairy .... 3. *G. circaezans*
6. Leaves lanceolate; flowers yellow, becoming purple, the lobes of the corolla glabrous .... 4. *G. lanceolatum*
7. All flowers sessile; leaves 3-nerved .... 3. *G. circaezans*
7. Some or all flowers pedicellate; leaves 1-nerved .... 5. *G. pilosum*
8. Stems smooth or retrorsely scabrous; margins of leaves with ascending cilia .... 6. *G. triflorum*
8. Stems retrorsely bristly; margins of leaves retrorsely scabrous ....
.... 7. *G. aparine*
9. Plants erect 10
9. Plants matted or loosely ascending 12
10. Leaves in whorls of 4 .... 2. *G. boreale*
10. Leaves in whorls of 6–8 11
11. Corolla yellow; leaves roughish .... 8. *G. verum*
11. Corolla white; leaves smooth .... 9. *G. mollugo*
12. Corolla 3-lobed; upper part of stems scabrous, but not retrorsely bristly
.... 13
12. Corolla 4-lobed; stems glabrous or retrorsely bristly 14
13. Pedicels smooth; leaves in whorls of 4–6 on the same plant ....
.... 10. *G. tinctorium*
13. Pedicels scabrous; leaves mostly in whorls of 4 .... 11. *G. trifidum*
14. Stems retrorsely bristly 15
14. Stems smooth 16
15. Leaves mostly in whorls of 6; flowers white .... 12. *G. asprellum*
15. Leaves mostly in whorls of 4; flowers yellow ... 13. *G. pedemontanum*
16. Leaves in whorls of 6, mostly linear .... 14. *G. concinnum*
16. Leaves in whorls of 4, mostly elliptic-oblong to oblanceolate 17
17. Leaves conspicuously reflexed; inflorescence over-topped by lateral branches; fruits up to 1.5 mm in diameter .... 15. *G. labradoricum*
17. Leaves ascending; inflorescence not over-topped by lateral branches; fruits more than 2 mm in diameter .... 16. *G. obtusum*

1. G. virgatum Nutt. Dwarf Bedstraw. April–June. Limestone ledges, rare; Monroe Co.
2. G. boreale L. Northern Bedstraw. May–July. Moist meadows, prairies, and fields; occasional in the n. one-fourth of the state, ab-

sent elsewhere. (Including var. *intermedium* DC. and var. *hyssopifolium* [Hoffm.] DC.)

3. G. circaezans Michx. Wild Licorice. May–July. Rich, often rocky, woods; occasional to common throughout the state. (Including var. *hypomalacum* Fern.)

4. G. lanceolatum Torr. Wild Licorice. June–July. Rich woods, rare; Cook and Kane cos.

5. G. pilosum Ait. Hairy Bedstraw. June–Aug. Sandy woods; occasional to common throughout the state.

6. G. triflorum Michx. Sweet-scented Bedstraw. May–Sept. Moist woods; occasional to common throughout the state.

7. G. aparine L. Goosegrass. May–July. Woods; common throughout the state. (Including var. *vaillantii* [DC.] W. D. J. Koch.)

8. G. verum L. Yellow Bedstraw. June–Aug. Native of Europe; rarely adventive in waste ground; DuPage and LaSalle cos.

9. G. mollugo L. White Bedstraw. June–Sept. Native of Europe; rarely adventive in waste ground; Champaign, Cook, DuPage, Henry, and Kane cos.

10. G. tinctorium L. Stiff Bedstraw. May–Sept. Swamps, marshes; occasional in the ne. corner of the state, rare in the extreme s. cos., absent elsewhere.

11. G. trifidum L. Small Bedstraw. July–Aug. Moist soil, not common; apparently confined to the ne. cos. and a few cos. in cent. Ill. along the Illinois River.

12. G. asprellum Michx. Rough Bedstraw. July–Aug. Moist thickets; occasional in the n. one-fourth of the state.

13. G. pedemontanum All. Bedstraw. May–June. Native of Europe; adventive in waste ground; Champaign Co.

14. G. concinnum Torr. & Gray. Shining Bedstraw. June–July. Dry woods; common throughout the state.

15. G. labradoricum Wieg. Bog Bedstraw. Bogs, marshes, rare; Boone, Lake, and McHenry cos.

16. G. obtusum Bigel. Wild Madder. May–July. Moist ground; common throughout the state.

4. *Diodia* L. Buttonweed

1. Corolla 7–10 mm long; fruits 7–10 mm long .... 1. *D. virginiana*
1. Corolla 4–6 mm long; fruits 2.5–4.0 mm long .... 2. *D. teres*

1. D. virginiana L. Large Buttonweed. June–Aug. Swamps, wet woods, not common; confined to the s. one-fourth of the state.

2. D. teres Walt. Rough Buttonweed. July–Aug. Fields, roadsides, woods; occasional to common throughout the state. (Including var. *setifera* Fern. & Grisc.)

5. *Spermacoce* L. Buttonweed

1. S. glabra Michx. Smooth Buttonweed. June–Oct. Wet ground; occasional in the s. half of the state, rare or absent elsewhere.

6. *Mitchella* L. Partridge-berry

1. M. repens L. Partridge-berry. Rocky woods, swampy woods; occasional in the s. and n. cos., rare in the cent. cos.

7. Houstonia L. Bluets

1. Flower one per peduncle; stamens included; corolla salverform 2
1. Flowers in corymbs, cymes, or panicles; stamens exserted; corolla funnelform 4
2. Corolla lobes 2.5–5.0 mm broad, very pale blue .... 1. *H. caerulea*
2. Corolla lobes to 2.5 mm broad, usually dark purple 3
3. Calyx lobes 2–4 mm long, about as long as the corolla tube 2. *H. minima*
3. Calyx lobes 1–2 mm long, up to ⅔ as long as the corolla tube ....
.... 3. *H. pusilla*
4. Calyx lobes shorter than calyx tube; middle and upper stipules setiform
.... 4. *H. nigricans*
4. Calyx lobes longer than calyx tube; middle and upper stipules deltoid or rounded 5
5. Lowermost leaves 3- to 5-nerved .... 5. *H. purpurea*
5. Lowermost leaves 1-nerved .... 6. *H. longifolia*

1. H. caerulea L. Bluets. April–June. Sandy soil of fields, prairies, and open woods; not common but scattered in all parts of the state except the nw. cos; very rare in the s. cos.

2. H. minima Beck. Tiny Bluets. April–May. Rocky woods, prairies, fields; occasional in the s. and w. cos., apparently absent elsewhere.

3. H. pusilla Schoepf. Star Violet, Small Bluets. March–April. Rocky woods, fields; occasional in the s. one-fourth of the state. *H. patens* Ell.—F. (Including f. *albiflora* Standl.)

4. H. nigricans (Lam.) Fern. Narrow-leaved Bluets. May–Oct. Rocky ledges, hill prairies; confined to the s. cos. along the Mississippi River.

5. H. purpurea L. Broad-leaved Bluets. May–June. Leaves ovate-lanceolate to broadly ovate, mostly 10 mm broad or broader. Rocky woods, prairies; occasional in the s. half of the state.

var. calycosa Gray. Bluets. May–June. Leaves broadly lanceolate, mostly less than 12 mm broad. Rocky woods, occasional; confined to the s. one-fourth of the state. *H. lanceolata* (Poir.) Britt.—F, J.

6. H. longifolia Gaertn. Long-leaved Bluets. April–July. Cauline leaves oblong to oblanceolate, mostly 2–5 mm broad; basal leaves eciliate; capsules as high as or higher than broad. Rocky woods, prairies, fields; occasional in the southernmost and northernmost cos., absent elsewhere.

var. ciliolata (Torr.) Torr. Bluets. April–July. Cauline leaves oblong to oblanceolate, mostly 2–5 mm broad; basal leaves ciliate; capsules as high as or higher than broad. Rocky woods, prairies, fields; occasional in the southernmost and northernmost cos., absent elsewhere. *H. canadensis* Willd.—F, G.

var. tenuifolia (Nutt.) Wood. Bluets. April–Oct. Cauline leaves linear, 1.0–2.5 mm broad; capsules broader than high. Rocky woods; occasional in the s. one-fourth of the state. *H. tenuifolia* Nutt.—F, G, J.

149. CAPRIFOLIACEAE

1. Small trees or erect or climbing shrubs 2
1. Creeping or ascending herbs 6

2. Leaves pinnately compound .... 1. *Sambucus*
2. Leaves simple 3
3. Mature leaves entire (leaves of vigorous shoots rarely pinnatifid) 4
3. Leaves toothed or lobed 5
4. Corolla tubular or funnelform, mostly zygomorphic; berries several-seeded .... 2. *Lonicera*
4. Corolla campanulate, actinomorphic; drupes 2-seeded ....
.... 3. *Symphoricarpos*
5. Corolla tubular; inflorescence 3-flowered; fruit a several-seeded capsule
.... 4. *Diervilla*
5. Corolla spreading; inflorescence almost always more than 3-flowered; fruit a 1-seeded drupe .... 5. *Viburnum*
6. Stems creeping; leaves evergreen; flowers pedicellate; fruit a 1-seeded capsule .... 6. *Linnaea*
6. Stems ascending; leaves deciduous; flowers sessile; fruit a 2- to 5-seeded drupe .... 7. *Triosteum*

1. Sambucus L. Elderberry

1. Cymes umbel-like, flat; twigs with white pith; fruit dark purple ....
.... 1. *S. canadensis*
1. Cymes panicle-like, ovoid; twigs with brown pith; fruit usually bright red
.... 2. *S. pubens*

1. S. canadensis L. Elderberry. June–July. Woodlands, along roads, common; in every co. (Including var. *submollis* Rehder.)
2. S. pubens Michx. Red-berried Elder. April–May. Moist woods, not common; Boone, Cook, Lake, LaSalle, McHenry, and Winnebago cos.

2. Lonicera L. Honeysuckle

1. Some of the leaves connate 2
1. None of the leaves connate 5
2. Corolla at least 3.5 cm long, red, the lobes more or less equal ....
.... 1. *L. sempervirens*
2. Corolla up to 3 cm long, white, yellow, orange, or purple, the lobes 2-lipped 3
3. Uppermost connate leaves broader than long, glaucous on the upper surface .... 2. *L. prolifera*
3. Uppermost connate leaves longer than broad, not glaucous 4
4. Corolla orange-yellow or yellow; leaves gray on the lower surface ......
.... 3. *L. flava*
4. Corolla yellow tinged with purple or brick-red; leaves whitened on the lower surface .... 4. *L. dioica*
5. Plants twining or climbing; corolla mostly 3 cm long or longer ....
.... 5. *L. japonica*
5. Plants erect or ascending; corolla up to 2 cm long 6
6. Peduncles much longer than the petioles 7
6. Peduncles shorter than the petioles .... 10. *L. maackii*
7. Corolla pubescent on the outside 8
7. Corolla glabrous on the outside 9
8. Bractlets about half as long as the ovaries; filaments pubescent ....
.... 6. *L. xylosteum*
8. Bractlets about as long as the ovaries; filaments glabrous 7. *L. morrowi*

9. Peduncles longer than the flowers .... 8. *L. tatarica*
9. Peduncles mostly shorter than the flowers ....9. *L. X bella*

1. L. sempervirens L. Trumpet Honeysuckle. April–July. Escaped from cultivation, mostly along roads; occasional and scattered throughout the state.
2. L. prolifera (Kirchn.) Rehd. Grape Honeysuckle. May–June. Wooded slopes, rocky banks; occasional to common in the n. three-fifths of the state.
3. L. flava Sims. Yellow Honeysuckle. April–May. Rocky woods, sandstone cliffs, rare; Jackson and Pope cos.
4. L. dioica L. Red Honeysuckle. May–June. Woods, not common; Cook, Kane, and Lake cos.
5. L. japonica Thunb. Japanese Honeysuckle. May–June. Native of Asia; woods, thickets; very common in the s. one-third of the state, less common elsewhere.
6. L. xylosteum L. European Fly Honeysuckle. May. Native of Europe and Asia; rarely escaped from cultivation; Coles and Cook cos.
7. L. morrowi Gray. May–June. Native of Japan; rarely escaped from cultivation; Carroll, Cook, and Lake cos.
8. L. tatarica L. Tartarian Honeysuckle. May–June. Native of Europe and Asia; occasionally escaped from cultivation in the n. half of the state, much rarer in the s. half.
9. *L.* X bella Zabel. May–June. Native of Asia; occasionally escaped from cultivation in the n. one-fourth of the state; also Macon and Peoria cos. Reputed to be a hybrid between *L. morrowi* and *L. tatarica*.
10. L. maackii Maxim. Amur Honeysuckle. May–June. Native of Asia; escaped into woodlands; DuPage and Kankakee cos.

3. *Symphoricarpos Duham. Snowberry*

1. Corolla 3–4 mm long, greenish and purplish; drupes coral-pink to purple
.... 1. *S. orbiculatus*
1. Corolla 5–9 mm long, pink; drupes white or greenish-white 2
2. Flowers pedicellate; style included .... 2. *S. albus*
2. Flowers sessile; style exserted .... 3. *S. occidentalis*

1. S. orbiculatus Moench. Coralberry. July–Aug. Disturbed woods, pastures; occasional throughout the state.
2. S. albus (L.) Blake var. laevigatus (Fern.) Blake. Garden Snowberry. Native of the w. U.S.; rarely escaped from cultivation. *S. rivularis* Suksd.—J.
3. S. occidentalis Hook. Wolfberry. June–Aug. Dry open ground; occasional in the n. one-fifth of the state.

4. *Diervilla Mill. Bush Honeysuckle*

1. D. lonicera Mill. Bush Honeysuckle. May–June. Sandy woods; occasional in the n. one-fourth of the state; also Coles and Piatt cos.

5. *Viburnum* L. *Viburnum*

1. Leaves more or less 3-lobed, palmately nerved 2
1. Leaves serrulate to dentate, pinnately nerved 4
2. Leaves dotted beneath; young branchlets more or less pilose; drupes purple-black at maturity .... 1. *V. acerifolium*
2. Leaves not dotted beneath; young branchlets glabrous or nearly so; drupes red or orange 3
3. Petiole with convex glands; stipules broadened or thickened at the tip .... 2. *V. trilobum*
3. Petiole with concave glands; stipules slender-tipped ...... 3. *V. opulus*
4. Pubescence of stellate hairs; winter buds without scales . 4. *V. lantana*
4. Pubescence of simple hairs or absent; winter buds with 1–2 pairs of outer scales 5
5. Leaves serrulate; petioles flat; stone of drupe flat; buds with 1 pair of outer scales 6
5. Leaves dentate; petioles terete; stone of drupe grooved or ridged; buds with 2 pairs of outer scales 8
6. Petiole and midvein of leaf on the lower surface rufous-pubescent .... 5. *V. rufidulum*
6. Petioles and midvein of leaf on the lower surface glabrous or without rufous pubescence 7
7. Leaves obtuse to acute at the apex; margins of the petioles neither undulate nor revolute .... 6. *V. prunifolium*
7. Leaves acuminate at the apex; margins of the petioles undulate and revolute .... 7. *V. lentago*
8. Leaves deeply cordate; leaves with 40 or more teeth; bark peeling off .. .. 8. *V. molle*
8. Leaves subcordate, rounded, or subcuneate at the base; leaves with up to 40 (–44) teeth; bark not peeling off 9
9. Petioles on leaves subtending the inflorescence up to 7 mm long .... 9. *V. rafinesquianum*
9. Petioles on leaves subtending the inflorescence 1.5 cm long or longer 10
10. Stipules present, subulate; branches of inflorescence and lower surface of leaves somewhat pubescent .... 10. *V. dentatum*
10. Stipules absent; branches of inflorescence and lower surface of leaves glabrous except on the nerves .... 11. *V. recognitum*

1. V. acerifolium L. Maple-leaved Arrowwood. May–June. Moist woods, occasional; confined to the ne. one-fourth of the state.

2. V. trilobum Marsh. High-bush Cranberry. June–July. Moist woods, rare; restricted to the n. one-third of the state.

3. V. opulus L. European High-bush Cranberry. June–July. Native of Europe; frequently cultivated and sometimes escaped, particularly in the ne. cos.

4. V. lantana L. Wayfaring Tree. May–June. Native of Europe; rarely escaped from cultivation; DuPage, Kane, and Kendall cos.

5. V. rufidulum Raf. Southern Black Haw; Rusty Nannyberry. April–May. Rocky woods; occasional in the s. one-fourth of the state, extending northward in cos. along the Mississippi River to Pike Co.

6. V. prunifolium L. Black Haw; Nannyberry. April–May. Low woods, wooded slopes; occasional throughout the state.

7. V. lentago L. Nannyberry. May–June. Moist woods; occasional in the n. half of the state; apparently adventive in Jackson Co.

8. V. molle Michx. Arrowwood. May–June. Rocky banks of streams, dry hillsides, rare; Adams and Peoria cos.

9. V. rafinesquianum Schultes. Downy Arrowwood. May–June. Wooded slopes, rocky woods; occasional in the n. half of the state, extending southward to Clark and Coles cos.

10. V. dentatum L. var. deamii (Rehd.) Fern. Southern Arrowwood. May–June. Low woods, rare; restricted to the s. one-fourth of the state.

11. V. recognitum Fern. Smooth Arrowwood. May–June. Woodlands, along streams; occasional in the s. half of the state; also Cook, DuPage, and Lake cos. *V. dentatum* L. var. *indianense* (Rehd.) Gl.—G.

6. *Linnaea* Gron. ex. L. Twinflower

1. L. americana Forbes. Twinflower. June–Aug. Bogs, very rare; Cook Co.

7. *Triosteum* L. Horse Gentian

1. Stems setose-hispid; calyx lobes glabrous except for the bristly margins .... 1. *T. angustifolium*
1. Stems glabrous or villous or glandular-pubescent; calyx lobes pubescent throughout 2
2. At least 3 or more pairs of leaves connate at the base; style exserted; corolla yellow, green, or purplish .... 2. *T. perfoliatum*
2. None or only up to 2 (–3) pairs of leaves connate at the base; style included; corolla orange or red 3
3. Stem with some glandular pubescence .... 3. *T. aurantiacum*
3. Stem without glandular pubescence .... 4. *T. illinoense*

1. T. angustifolium L. Yellow-flowered Horse Gentian. April–May. Low ground, rocky woods, not common; confined to the s. one-fourth of the state.

2. T. perfoliatum L. Late Horse Gentian. May–July. Dry woods, thickets; occasional in the n. two-thirds of the state, less common in the s. one-third.

3. T. aurantiacum Bickn. Early Horse Gentian. May–June. Dry woods, rich woods; occasional in the n. two-thirds of the state. *T. perfoliatum* L. var. *aurantiacum* (Bickn.) Wieg.—G.

4. T. illinoense (Wieg.) Rydb. Illinois Horse Gentian. May–June. Dry woods, rich woods; occasional throughout the state. *T. perfoliatum* L. var. *illinoense* Wieg.—G; *T. aurantiacum* Bickn. var. *illinoense* (Wieg.) Palmer & Steyerm.—F. The type was collected in Henderson Co.

150. ADOXACEAE

1. *Adoxa* L.

1. A. moschatellina L. Moschatel. May–July. Moist cliffs, rare; JoDaviess Co.

151. VALERIANACEAE

1. Cauline leaves, or some of them, pinnately compound; calyx divided into 5–15 setiform lobes .... 1. *Valeriana*
1. Cauline leaves entire or toothed, not pinnately compound; calyx lobes reduced to short teeth, or absent .... 2. *Valerianella*

1. Valeriana L. Valerian

1. Flowers unisexual, white; leaves parallel-veined; root tuberous ....
.... 1. *V. ciliata*
1. Flowers perfect, pink or rose; leaves pinnately veined; roots fibrous . 2
2. Corolla pale pink, 1 cm long or longer; most of the basal leaves simple
.... 2. *V. pauciflora*
2. Corolla rose, 4–5 mm long; all of the basal leaves pinnately compound
.... 3. *V. officinalis*

    1. V. ciliata Torr. & Gray. Valerian. May–June. Wet prairies, calcareous fens; occasional in the n. one-fifth of the state. *V. edulis* Nutt. ssp. *ciliata* (Torr. & Gray) F. G. Mey.—G.
    2. V. pauciflora Michx. Pink Valerian. May–June. Low woods; occasional in the s. half of the state; also Vermilion Co.
    3. V. officinalis L. Garden Heliotrope. June–July. Native of Europe; occasionally cultivated but infrequently escaped; Cook, Kane, and Lake cos.

2. Valerianella Mill.

1. Corolla 1.5–2.0 mm long; bracts and bractlets spinulose-ciliate 2
1. Corolla 3–5 mm long; bracts and bractlets eciliate, except sometimes at the tip 3
2. Corolla lobes bluish; stamens included or shortly exserted   1. *V. olitoria*
2. Corolla lobes white; stamens long-exserted .... 2. *V. radiata*
3. Fruits ovoid-ellipsoid, about 1 mm broad .... 3. *V. intermedia*
3. Fruits orbicular or globose, 2–3 mm broad 4
4. Fruits flattened, 3.0–3.5 mm long, about 3 mm broad ... 4. *V. patellaria*
4. Fruits not flattened, 2.0–2.5 mm long, about 2 mm broad 5. *V. umbilicata*

    1. V. olitoria (L.) Poll. Corn Salad. May. Native of Europe; rarely adventive in waste ground; LaSalle Co.
    2. V. radiata (L.) Dufr. Corn Salad. April–May. Seed-bearing cavity of fruit broader than the combined width of the two empty cells; fruits 1.4–1.5 mm long, 0.7–0.8 mm broad. Wet fields, prairies, low woods; occasional to common in the s. half of the state.
    var. missouriensis Dyal. Corn Salad. April–May. Seed-bearing cavity of fruit not as broad as the combined width of the two empty cells; fruits 1.7–1.8 mm long, about 1.2 mm broad. Wet ground, rare; Jackson Co.
    3. V. intermedia Dyal. Corn Salad. May–June. Wet ground, rare; Kankakee, LaSalle, and Will cos. *V. radiata* (L.) Dufr. var. *intermedia* (Dyal) Gl.—G.
    4. V. patellaria (Sulliv.) Wood. Corn Salad. May–June. Wet ground, rare; LaSalle Co.

5. V. umbilicata (Sulliv.) Wood. Corn Salad. May–June. Wet ground, rare; Kankakee and LaSalle cos.

152. DIPSACACEAE

1. Dipsacus L.   Teasel

1. Leaves pinnatifid or bipinnatifid, ciliate on the margins   1. *D. laciniatus*
1. Leaves toothed, prickly on the margins .... 2. *D. sylvestris*

1. D. laciniatus L. Cut-leaved Teasel. July–Sept. Native of Europe; adventive in waste ground; Cook, DuPage, Jackson, Johnson, Kane, Lake, and Will cos.
2. D. sylvestris Huds. Common Teasel. June–Oct. Native of Europe; occasionally adventive in waste ground throughout the state.

153. CUCURBITACEAE

1. Leaves pinnately cleft, often more than halfway to the midvein ....
.... 1. *Citrullus*
1. Leaves unlobed or shallowly lobed 2
2. Plants covered with soft, sticky hairs .... 2. *Lagenaria*
2. Plants glabrous or pubescent, but without soft, sticky hairs 3
3. Ovaries and fruits not spinulose 4
3. Ovaries and fruits spinulose 6
4. Corolla more than 5 cm long and more than 5 cm broad . 3. *Cucurbita*
4. Corolla up to 3.5 cm long and up to 3.5 cm broad 5
5. Corolla yellow, 2.5–3.5 cm long and broad .... 4. *Cucumis*
5. Corolla white or greenish, at most only 1 cm long and broad ....
.... 5. *Melothria*
6. Plants pubescent; corolla 5-lobed; ovary 1-locular; fruit 1-seeded ......
.... 6. *Sicyos*
6. Plants glabrous or nearly so; corolla 6-lobed; ovary 2-locular; fruit 4-seeded .... 7. *Echinocystis*

1. Citrullus Neck   Melon

1. C. vulgaris Schrad. Watermelon. May–Sept. Native of Africa; occasionally escaped from cultivation, but seldom persisting.

2. Lagenaria Seringe   Gourd

1. L. vulgaris Seringe. Gourd. May–Sept. Native of the E. Hemisphere; rarely escaped from cultivation; Hancock Co.

3. Cucurbita L.   Gourd

1. Leaves stiff, thick, often nearly twice as long as broad   1. *C. foetidissima*
1. Leaves flexible, thin, never twice as long as broad .... 2. *C. pepo*

1. C. foetidissima HBK. Missouri Gourd. May–Sept. Native of the w. U.S.; occasional as an adventive along railroads.

2. C. pepo L. var. ovifera (L.) Alef. Pear Gourd. May–Sept. Native of trop. Am.; moist soil, very rare; Union Co.

4. Cucumis L.   Cucumber

1. C. sativus L. Cucumber. May–Sept. Native of Asia; rarely escaped from cultivation.

5. Melothria L.

1. M. pendula L. Creeping Cucumber. June–Sept. Rocky soil, rare; Alexander and Pope cos.

6. Sicyos L.   Bur Cucumber

1. S. angulatus L. Bur Cucumber. July–Sept. Moist soil of fields and woods, occasional; scattered throughout the state.

7. Echinocystis Torr. & Gray

1. E. lobata (Michx.) Torr. & Gray. Wild Balsam-apple. July–Sept. Moist soil; occasional in the n. three-fourths of Ill., apparently absent in the s. one-fourth.

154. CAMPANULACEAE

1. Flowers actinomorphic; corolla tube not split down one side 2
1. Flowers zygomorphic; corolla tube split down one side ..... 3. *Lobelia*
2. Leaves sessile, often clasping; flowers solitary and sessile in the axils ..
.... 1. *Specularia*
2. Leaves petiolate, or at least not clasping at the base; flowers in spikes or racemes or, if solitary, then pedicellate .... 2. *Campanula*

1. Specularia Fab.   Venus' Looking-glass

1. Bracts suborbicular to ovate; seeds up to 0.7 mm long 2
1. Bracts linear to lanceolate; seeds 0.7–1.0 mm long ... 3. *S. leptocarpa*
2. Capsule with pores near apex .... 1. *S. biflora*
2. Capsule with pores midway between base and apex ... 2. *S. perfoliata*

1. S. biflora (R. & P.) Fisch. & Mey. April–June. Venus' Looking-glass. Dry, often disturbed, soil; confined to the s. one-sixth of the state.

2. S. perfoliata (L.) A. DC. April–Aug. Venus' Looking-glass. Dry, often disturbed, soil, common; probably in every co.

3. S. leptocarpa (Nutt.) Gray. May–July. Fields, rare; Logan and Will cos.

2. *Campanula* L. Bellflower

1. Cauline leaves linear to narrowly lanceolate (ovate, cordate basal leaves may be present at base of plant in *C. rotundifolia*) 2
1. Cauline leaves oblong-lanceolate to ovate 4
2. Stems glabrous or rarely closely puberulent; corolla 1.5 cm long or longer .... 1. *C. rotundifolia*
2. Stems scabrous with retrorse hairs on the angles; corolla up to 1.2 cm long 3
3. Corolla white, 5–8 mm long; capsule up to 2 mm long 2. *C. aparinoides*
3. Corolla bluish, 10–12 mm long; capsule 3–5 mm long ... 3. *C. uliginosa*
4. Corolla campanulate; capsule with pores nearly basal ....
.... 4. *C. rapunculoides*
4. Corolla rotate; capsule with pores nearly apical ...... 5. *C. americana*

    1. C. rotundifolia L. Bellflower. June–Sept. Stems and cauline leaves smooth. Woods, hill prairies, sandstone cliffs; occasional in the n. half of the state; also Jackson Co. *C. intercedens* Witasek—J.
    var. velutina A. DC. Bellflower. July–Sept. Stems and cauline leaves closely puberulent; crevices of cliffs; JoDaviess Co.
    2. C. aparinoides Pursh. Marsh Bellflower. June–Aug. Marshes and bogs; occasional in the n. half of the state.
    3. C. uliginosa Rydb. Marsh Bellflower. June–Aug. Marshes and bogs; occasional in the n. half of the state. *C. aparinoides* Pursh var. *grandiflora* Holz—G.
    4. C. rapunculoides L. European Bellflower. July–Sept. Native of Europe; adventive in waste areas; apparently confined to the n. half of the state.
    5. C. americana L. American Bellflower. June–Oct. Woods, common; in every co. (Including var. *illinoensis* [Fresn.] Farw.)

3. *Lobelia* L. Lobelia

1. Flowers 1.5 cm. long or longer 2
1. Flowers up to 1.5 cm long 5
2. Flowers red or deep rose 3
2. Flowers blue (rarely white) 4
3. Flowers red; calyx glabrous or puberulent ....1. *L. cardinalis*
3. Flowers deep rose; calyx hirsute .... 2. *L. cardinalis* X *siphilitica*
4. Auricles at base of calyx 2–5 mm long; stems glabrous or sparsely hirsute; flowers 2.0–3.3 cm long .... 3. *L. siphilitica*
4. Auricles at base of calyx less than 2 mm long; stems densely puberulent throughout; flowers 1.5–2.5 cm long .... 4. *L. puberula*
5. Cauline leaves linear to narrowly lanceolate; lower lip of corolla glabrous .... 5. *L. kalmii*
5. Cauline leaves oblong, lanceolate, or obovate 6
6. Lower part of stems villous or hirsute; capsules inflated, completely enclosed by the calyx .... 6. *L. inflata*
6. Stems glabrous or short-pubescent; capsules not inflated, partly exserted from the calyx .... 7. *L. spicata*

    1. L. cardinalis L. Cardinal-flower. July–Sept. Wet ground; occasional to common throughout the state.
    2. L. cardinalis X siphilitica Schneck. Sept. Wabash Co.

3. L. siphilitica L. Blue Cardinal-flower. Aug.–Oct. Wet ground; occasional to common throughout the state. (Including var. *ludoviciana* A. DC.)

4. L. puberula Michx. Downy Lobelia. Aug.–Oct. Wet ground; occasional in the s. one-third of the state. (Including var. *simulans* Fern.)

5. L. kalmii L. Kalm's Lobelia. July–Sept. Springy areas and dunes; occasional in the n. one-third of the state.

6. L. inflata L. Indian Tobacco. June–Oct. Woods, fields, disturbed areas; in every co.

7. L. spicata Lam. Spiked Lobelia. July–Aug. Appendages between calyx lobes up to 1 mm long or absent; leaves mostly spreading. Dry woods, prairies; occasional throughout the state. (Including var. *hirtella* Gray and var. *campanulata* McVaugh.)

var. leptostachys (A. DC.) Mack. & Bush. Spiked Lobelia. June–Aug. Appendages between calyx lobes 1–5 mm long; leaves mostly ascending. Dry soil; occasional throughout the state. *L. leptostachys* A. DC.—J.

155. COMPOSITAE

1. Heads with both ray- and disc-flowers or with disc-flowers only; sap not milky 2
1. Heads with ray-flowers only; sap milky 75
2. Heads with both ray- and disc-flowers 3
2. Heads with disc-flowers only (some plants with small and inconspicuous rays are also keyed out here) 40
3. Pappus of capillary bristles 4
3. Pappus of rigid awns, small chaffy scales, reduced to a mere crown, or absent 11
4. Rays yellow (white in one species of *Solidago*) 5
4. Rays not yellow 10
5. Involucral bracts in one series 6
5. Involucral bracts in more than one series 7
6. Leaves opposite; involucral bracts with 3–7 conspicuous glands; rays small, inconspicuous .... 23. *Dyssodia*
6. Leaves alternate; involucral bracts glandless; rays conspicuous .... 45. *Senecio*
7. Pappus double .... 26. *Heterotheca*
7. Pappus single 8
8. Leaves white woolly beneath .... 42. *Inula*
8. Leaves not white woolly beneath 9
9. Heads numerous; involucre 2.5–9.0 mm high .... 27. *Solidago*
9. Heads few; involucre 1–2 cm high .... 28. *Haplopappus*
10. Involucral bracts in 3–5 series, either foliaceous or with chartaceous base and herbaceous tip .... 31. *Aster*
10. Involucral bracts in 1–2 series, neither foliaceous nor with chartaceous base and herbaceous tip .... 32. *Erigeron*
11. Rays yellow or orange 12
11. Rays not yellow or orange 29
12. Pappus of 2-several awns or bristles (sometimes deciduous) 13
12. Pappus of a few short teeth, scales, or absent 18

13. Involucral bracts with recurved tips, often resinous ...... 24. *Grindelia*
13. Involucral bracts without recurved tips, never resinous 14
14. Pappus of 2 (–3) smooth awns .... 14. *Verbesina*
14. Pappus of 2-several awns (if two, then barbed) 15
15. Awns of pappus barbed or chaffy 16
15. Awns of pappus not barbed nor chaffy .... 22. *Gaillardia*
16. Leaves alternate .... 25. *Xanthocephalum*
16. Leaves opposite or whorled 17
17. Achenes beaked; ray-flowers orange .... 17. *Cosmos*
17. Achenes not beaked; ray-flowers yellow .... 16. *Bidens*
18. Leaves, at least the lower, opposite or whorled, not all basal 19
18. Leaves all alternate or basal 23
19. Achenes flattened 20
19. Achenes not at all or scarcely flattened 21
20. Rays usually 8; involucral bracts in two series; disc-flowers fertile ......
.... 15. *Coreopsis*
20. Rays numerous; involucral bracts in more than two series; disc-flowers
sterile .... 2. *Silphium*
21. Leaves deeply angulate lobed or lyrate-pinnatifid .... 1. *Polymnia*
21. Leaves entire or serrate 22
22. Involucral bracts obtuse; ray-flowers fertile, persistent .... 7. *Heliopsis*
22. Involucral bracts acute to acuminate; ray-flowers sterile, deciduous ....
.... 12. *Helianthus*
23. Stems scapose 24
23. Stems leafy 25
24. Leaves large, ovate, toothed; scapes 2–3 m tall; heads several ....
.... 2. *Silphium*
24. Leaves small, spatulate, entire; scapes 5–15 cm tall; heads solitary .....
.... 20. *Hymenoxys*
25. Receptacle chaffy 26
25. Receptacle naked .... 21. *Helenium*
26. Stem usually winged .... 14. *Verbesina*
26. Stem never winged 27
27. Receptacle flat or convex .... 12. *Helianthus*
27. Receptacle conic or columnar 28
28. Achenes flattened, sharp margined or winged .... 11. *Ratibida*
28. Achenes 4-sided, not sharp margined nor winged .... 9. *Rudbeckia*
29. Leaves opposite 30
29. Leaves alternate and/or basal 32
30. Leaves deeply angulate lobed or lyrate-pinnatifid .... 1. *Polymnia*
30. Leaves serrate 31
31. Leaves ovate, petiolate .... 18. *Galinsoga*
31. Leaves lanceolate, sessile .... 8. *Eclipta*
32. Leaves all basal .... 29. *Bellis*
32. Leaves cauline and basal 33
33. Leaves dissected, incised, or coarsely and irregularly toothed 34
33. Leaves entire, regularly serrate, or dentate 36
34. Rays 4–6; heads numerous .... 34. *Achillea*
34. Rays 10–30; heads few 35
35. Leaves cut into filiform divisions .... 33. *Anthemis*
35. Leaves incised or coarsely and irregularly toothed .36. *Chrysanthemum*
36. Rays purple or purple with yellow 37
36. Rays white or pink 38
37. Rays reflexed, entirely purple .... 10. *Echinacea*

37. Rays not reflexed, purple, often with yellow toward tip .. 22. *Gaillardia*
38. Rays numerous .... 30. *Boltonia*
38. Rays 1–5 39
39. Pappus of two slender awns; leaves entire or serrate ... 14. *Verbesina*
39. Pappus of 2–3 inconspicuous scales; leaves dentate .... 3. *Parthenium*
40. Pappus of capillary bristles 41
40. Pappus of rigid awns, small chaffy scales, reduced to a mere crown, or absent 64
41. Flowers pink, purple, blue, or yellow (rarely white) 42
41. Flowers white or cream 55
42. Twining herbaceous vine .... 47. *Mikania*
42. Erect herb 43
43. Leaves opposite or whorled, cauline 44
43. Leaves alternate or basal 45
44. Involucral bracts with 3–7 conspicuous glands; flowers yellow .... 23. *Dyssodia*
44. Involucral bracts glandless; flowers not yellow .... 46. *Eupatorium*
45. Leaves prickly 46
45. Leaves not prickly 49
46. Heads 1-flowered, in capitate clusters .... 52. *Echinops*
46. Heads many-flowered, distinct 47
47. Pappus bristles plumose .... 55. *Cirsium*
47. Pappus bristles not plumose 48
48. Receptacle with numerous bristles .... 54. *Carduus*
48. Receptacle without bristles .... 56. *Onopordum*
49. Involucral bracts pectinate or tipped with a rigid spine .. 57. *Centaurea*
49. Involucral bracts not pectinate nor tipped with a rigid spine 50
50. Involucral bracts with hooked tips .... 53. *Arctium*
50. Involucral bracts without hooked tips 51
51. Flowers yellow; involucral bracts in one series .... 45. *Senecio*
51. Flowers not yellow; involucral bracts in two series 52
52. Pappus bristles plumose or barbellate .... 49. *Liatris*
52. Pappus bristles not plumose nor barbellate 53
53. Pappus double, the outer series shorter .... 50. *Vernonia*
53. Pappus single, subequal 54
54. Heads 2- to 5-flowered, aggregated in dense clusters subtended by several foliaceous bracts .... 51. *Elephantopus*
54. Heads many-flowered, not aggregated in dense clusters subtended by foliaceous bracts .... 39. *Pluchea*
55. Leaves prickly 56
55. Leaves not prickly 57
56. Heads 1-flowered, in capitate clusters .... 52. *Echinops*
56. Heads many-flowered, distinct .... 55. *Cirsium*
57. Involucral bracts scarious 58
57. Involucral bracts not scarious 59
58. Leaves mostly basal, cauline leaves small .... 40. *Antennaria*
58. Leaves all or mostly cauline .... 41. *Gnaphalium*
59. Involucral bracts in one series (often with a few small bracts at base of head) 60
59. Involucral bracts in more than one series 61
60. Pappus smooth .... 43. *Erechtites*
60. Pappus scabrous .... 44. *Cacalia*
61. Leaves, at least lower, opposite or whorled .... 46. *Eupatorium*
61. Leaves all alternate 62

62. Involucral bracts striate; achenes nearly round; leaves resinous-dotted
.... 48. *Brickellia*
62. Involucral bracts not striate; achenes flattened; leaves not resinous-dotted 63
63. Involucral bracts in 3–5 series, either foliaceous or with chartaceous base and herbaceous tip; rays broad, few .... 31. *Aster*
63. Involucral bracts in 1–2 series, neither foliaceous nor with chartaceous base and herbaceous tip; rays narrow, numerous .... 32. *Erigeron*
64. Flowers green or greenish 65
64. Flowers yellow or white 69
65. Staminate and pistillate flowers in separate heads 66
65. Staminate and pistillate flowers in same head, or all flowers perfect   67
66. Involucre of pistillate heads tuberculate or with straight spines ....
.... 5. *Ambrosia*
66. Involucre of pistillate heads with hooked spines .... 6. *Xanthium*
67. Heads numerous, in spikes, racemes, or panicles 68
67. Heads few to solitary .... 35. *Matricaria*
68. Heads in long terminal bracted spikes; leaves opposite, at least the lower entire or serrate .... 4. *Iva*
68. Heads in panicles or racemes; leaves alternate, lobed or incised, or occasionally entire or toothed .... 38. *Artemisia*
69. Crushed plant smelling like pineapple .... 35. *Matricaria*
69. Crushed plant not smelling like pineapple 70
70. Flowers white 71
70. Flowers yellow or yellowish 72
71. Leaves alternate, pinnatifid .... 19. *Hymenopappus*
71. Leaves opposite, entire or lobed .... 13. *Melanthera*
72. Leaves opposite 73
72. Leaves alternate or basal 74
73. Pappus of 2–4 awns; achenes flattened; disc-flowers fertile   16. *Bidens*
73. Pappus absent; achenes scarcely flattened; disc-flowers sterile ....
.... 1. *Polymnia*
74. Leaves pinnately dissected .... 37. *Tanacetum*
74. Leaves shallowly lobed, spiny margined .... 58. *Cnicus*
75. Pappus of capillary bristles only (an outer series of small scales often overlooked) 76
75. Pappus not entirely of capillary bristles 87
76. Pappus plumose 77
76. Pappus not plumose, of simple capillary bristles 80
77. Leaves grass-like; plume-branches of pappus interwebbed.............
.... 66. *Tragopogon*
77. Leaves not grass-like; plume branches of pappus not interwebbed .. 78
78. Stems leafy .... 65. *Picris*
78. Stems scapose 79
79. Receptacle chaffy .... 63. *Hypochaeris*
79. Receptacle naked .... 64. *Leontodon*
80. Achenes round or angled, scarcely flattened 81
80. Achenes strongly flattened 86
81. Heads usually several on each stem; stems leafy 82
81. Heads solitary; stems scapose 85
82. Flowers whitish or purplish; heads pendent .... 72. *Prenanthes*
82. Flowers yellow to orange-red; heads not pendent 83
83. Achenes filiform-beaked .... 70. *Pyrrhopappus*
83. Achenes more or less narrowed toward apex, but scarcely beaked .. 84

84. Pappus white .... 71. *Crepis*
84. Pappus sordid to tawny .... 73. *Hieracium*
85. Achenes smooth or nearly so, 10-nerved; leaves entire .. 61. *Microseris*
85. Achenes muricate toward apex, 4- to 5-nerved; leaves lobed to toothed
(rarely subentire) .... 67. *Taraxacum*
86. Achenes filiform-beaked or narrowed toward apex, only rarely beakless;
flowers blue or purple .... 69. *Lactuca*
86. Achenes beakless; flowers yellow .... 68. *Sonchus*
87. Flowers blue (rarely white or pink); pappus of minute narrow scales only
.... 59. *Cichorium*
87. Flowers yellow or orange; pappus various, but not composed of minute
narrow scales only, or absent 88
88. Involucre with short bracts at base imitating an outer series 60. *Lapsana*
88. Involucre without short bracts at base .... 62. *Krigia*

1. *Polymnia* L. Leafcup

1. Leaves pinnately lobed; achenes 3-angled, not striate . 1. *P. canadensis*
1. Leaves subpalmately lobed; achenes slightly flattened laterally, striate ..
.... 2. *P. uvedalia*

1. P. canadensis L. Leafcup. June–Nov. Moist or dry woods; common throughout Ill. (Including f. *radiata* [Gray] Fassett.)
2. P. uvedalia (L.) L. Bear's-foot; Yellow Leafcup. July–Sept. Rich woods; occasional in s. Ill., extending n. to St. Clair Co; also Vermilion Co. (Including var. *densipilis* Blake.)

2. *Silphium* L. Rosinweed

1. Leaves connate-perfoliate; stem conspicuously 4-angled ....
.... 1. *S. perfoliatum*
1. Leaves not connate-perfoliate; stem nearly round or only obtusely angled 2
2. Leaves deeply pinnatifid or bipinnatifid .... 2. *S. laciniatum*
2. Leaves entire or toothed 3
3. Leaves chiefly basal, large, cordate, long-petiolate; stems 1–3 m tall ....
.... 3. *S. terebinthinaceum*
3. Leaves cauline, relatively small, lanceolate to ovate, sessile; stems
0.5–1.5 m tall .... 4. *S. integrifolium*

1. S. perfoliatum L. Cup-plant. July–Aug. Open woods and low ground; common throughout Ill.
2. S. laciniatum L. Compass-plant; Rosinweed. July–Aug. Prairies; common throughout Ill.
3. S. terebinthinaceum Jacq. Prairie-dock. July–Sept. Prairies; common throughout Ill. (Including var. *lucy-brauniae* Steyerm.)
4. S. integrifolium Michx. Rosinweed. July–Aug. Prairies; common throughout Ill. (Including var. *deamii* Perry.)

3. *Parthenium* L.

1. Leaves pinnately dissected .... 1. *P. hysterophorus*
1. Leaves toothed .... 2. *P. integrifolium*

1. P. hysterophorus L. Santa Maria. Aug.–Oct. Adventive from trop. Am. Waste ground, rare; Cook Co.

2. P. integrifolium L. American Feverfew; Wild Quinine. July–Sept. Prairies and dry woods; common throughout Ill.

4. Iva L.   Marsh-elder

1. Heads in racemose spikes, bracteate .... 1. *I. annua*
1. Heads in paniculate spikes, ebracteate. .... 2. *I. xanthifolia*

1. I. annua L. Marsh-elder; Sumpweed. Aug.–Oct. Moist ground; occasional throughout Ill., chiefly in the s. half. *I. ciliata* Willd—F, G. J.

2. I. xanthifolia Nutt. Burweed Marsh-elder. July–Sept. Moist waste ground; occasional in the n. half of Ill.

5. Ambrosia L.   Ragweed

1. Pistillate involucre with only 1 row of tubercles or erect spines near apex
.... 2
1. Pistillate involucre with several rows of spines on surface ....
.... 5. *A. tomentosa*
2. Staminate heads sessile in solitary terminal spikes ..... 1. *A. bidentata*
2. Staminate heads pedicellate in 1-several slender racemes 3
3. Leaves palmately 3- to 5-lobed or entire .... 2. *A. trifida*
3. Leaves pinnatifid or bipinnatifid 4
4. Leaves petiolate, bipinnatifid, glabrous or nearly so above; involucre with 5–7 spines .... 4. *A. artemisiifolia*
4. Leaves sessile or subsessile, pinnatifid, harsh above; involucre unarmed or with blunt tubercles .... 3. *A. psilostachya*

1. A. bidentata Michx. Ragweed. July–Oct. Fields and waste ground; common in s. half of Ill.

2. A. trifida L. Giant Ragweed. July–Oct. Petioles of at least the upper leaves more or less wing margined; ribs or fruits ending in short spines. Fields and waste ground; in every co. (Including f. *integrifolia* [Muhl.] Fern.)

var. texana Scheele. Giant Ragweed. July–Oct. Petioles wingless; ribs of fruits ending in blunt to almost obsolete tubercles. Adventive from sw. Waste ground; occasional in s. Ill.

3. A. psilostachya DC. Western Ragweed. July–Oct. Dry sandy soil in fields and waste ground; occasional in n. three-fifths of Ill. (Including var. *coronopifolia* [Torr. & Gray] Farw.) *A. coronopifolia* Torr. & Gray—J.

4. A. artemisiifolia L. Common Ragweed. Aug.–Oct. Fields and waste ground; in every co. (Including var. *elatior* [L.] Desc. and var. *elatior* f. *villosa* Fern. & Grisc.)

5. A. tomentosa Nutt. False Ragweed. July–Oct. Adventive from w. Waste ground, rare; La Salle and McHenry cos. *Franseria discolor* Nutt.—F, G, J.

Reported hybrids: *A. artemisiifolia X A. trifida,* Champaign Co.; *A. bidentata X A. trifida,* Clay, Jersey, Sangamon, St. Clair, and Vermilion cos.

6. *Xanthium L.   Cocklebur*

1. Leaves lanceolate, tapering to base, with 3-parted axillary spine ....
.... 1. *X. spinosum*
1. Leaves ovate, base cordate or deltoid, without axillary spine ....
.... 2. *X. strumarium*

1. X. spinosum L. Spiny Cocklebur. Aug.–Oct. Adventive from trop. Am. Waste ground, rare; Alexander, Cook, and Pulaski cos.

2. X. strumarium L. Common Cocklebur. Our plants belong to the following varieties:

var. canadensis (Mill.) Torr. & Gray. Aug.–Oct. Body of fruit and prickles glandular-hispidulous. Waste ground and along rivers; common throughout Ill. (Including *X. commune* Britt., *X. italicum* Moretti, *X. oviforme* Wallr., and *X. variens* Greene.)

var. glabratum (DC.) Cronq. Aug.–Oct. Body of fruit glabrous or merely glandular. Waste ground and along rivers; common throughout Ill. (Including *X. chasei* Fern., *X. chinense* Mill., *X. globosum* Shull, *X. inflexum* Mack. & Bush, and *X. pennsylvanicum* Mill.)

7. *Heliopsis Pers.   Ox-eye*

1. H. helianthoides (L.) Sweet. False Sunflower. July–Aug. Open woods and prairies; common throughout Ill. (Including var. *scabra* [Dunal] Fern.)

8. *Eclipta L.*

1. E. alba (L.) Hassk. Yerba de Tajo. July–Oct. Wet ground and waste ground; common in the s. three-fourths of the state, rare or absent elsewhere.

9. *Rudbeckia L.   Coneflower*

1. Leaves clasping .... 1. *R. amplexicaulis*
1. Leaves not clasping 2
2. Disc greenish-yellow .... 2. *R. laciniata*
2. Disc brown or purple 3
3. Lower stem leaves 3-lobed or 3-parted 4
3. Lower stem leaves merely toothed or entire 5
4. Chaff of receptacle glabrous .... 3. *R. triloba*
4. Chaff of receptacle pubescent toward tip .... 4. *R. subtomentosa*
5. Pappus absent; style-branches elongate, acute .... 5. *R. hirta*
5. Pappus present, but consisting of a minute short crown, commonly toothed on angles; style-branches short, blunt .... 6. *R. fulgida*

1. R. amplexicaulis Vahl. Coneflower. June–Aug. Waste ground along railroads, rare; Cook and Jackson cos.

2. R. laciniata L. Goldenglow. July–Sept. Moist ground; occasional throughout Ill.

3. R. triloba L. Brown-eyed Susan. Aug.–Oct. Woods; common throughout Ill.

4. R. subtomentosa Pursh. Fragrant Coneflower. Aug.–Sept. Prairies and low open woods; occasional throughout Ill.

5. R. hirta L. Black-eyed Susan. June–Sept. Open woods and fields; common throughout Ill. Occasional specimens with pale yellow rays having white to cream tips (f. *flavescens* Clute) are observed. (Including *R. bicolor* Nutt., *R. serotina* Nutt., and *R. serotina* var. *lanceolata* [Bisch.] Fern. & Schub.)

6. R. fulgida Ait. Orange Coneflower. July–Oct. Cauline leaves elliptic, sessile or narrowed to winged petiole. Woods, rare; Pope Co. (Including *R. spathulata* Michx. and *R. tenax* Boynt. & Beadle.)

var. sullivantii (Boynt. & Beadle) Cronq. Sullivant's Orange Coneflower. July–Oct. Cauline leaves ovate, sharply toothed, petiolate. Moist ground; occasional in e. Ill. (Including *R. deamii* Blake, *R. palustris* Eggert, and *R. speciosa* Boynt. & Beadle.) *R. sullivantii* Boynt. & Beadle—J.

var. missouriensis (Engelm.) Cronq. Missouri Orange Coneflower. July–Oct. Cauline leaves linear-spatulate, entire. Dry open ground, rare; Monroe and Randolph cos. *R. missouriensis* Engelm.—F, G.

10. Echinacea Moench. Coneflower

1. Leaves ovate, rounded at base, serrate or dentate, upper frequently entire .... 1. *E. purpurea*
1. Leaves lanceolate to linear, attenuate to base, all entire ... 2. *E. pallida*

1. E. purpurea (L.) Moench. Purple Coneflower. July–Aug. Open woods and prairies; occasional throughout Ill.

2. E. pallida (Nutt.) Nutt. Pale Coneflower. July–Aug. Open woods and prairies; occasional throughout Ill.

E. simulata McGregor, a species similar to *E. pallida*, except for being a diploid and having yellow pollen while *E. pallida* is a tetraploid and has white pollen, has been reported from Ill. Its status as a distinct species is dubious.

11. Ratibida Raf. Prairie Coneflower

1. Disc columnar, equalling or exceeding rays, 2–4 times as long as wide .... 1. *R. columnifera*
1. Disc ellipsoid, shorter than rays, 1.0–1.5 times as long as wide .... 2. *R. pinnata*

1. R. columnifera (Nutt.) Woot. & Standl. Long-headed Coneflower. July–Aug. Adventive from w. U.S. Along railroads; scattered in the n. two-thirds of Ill.

2. R. pinnata (Vent.) Barnh. Drooping Coneflower; Yellow Coneflower. July–Aug. Prairies; common in the n. three-fourths of the state, occasional elsewhere.

12. Helianthus L. Sunflower

1. Leaves with conspicuously bristly ciliate margins; rays less than 1 cm long .... 1. *H. ciliaris*
1. Leaves without conspicuously bristly ciliate margins; rays normally over 1 cm long 2
2. Disc red or purple 3
2. Disc yellow 8

3. Leaves linear 4
3. Leaves lanceolate to cordate 5
4. Stem glabrous, usually glaucous .... 2. *H. salicifolius*
4. Stem strigose to pilose .... 3. *H. angustifolius*
5. Receptacle flat or nearly so 6
5. Receptacle convex or low-conic 7
6. Involucral bracts ovate, abruptly contracted above; receptacular bracts not bearded at apex .... 4. *H. annuus*
6. Involucral bracts lanceolate, tapering; receptacular bracts conspicuously bearded at apex .... 5. *H. petiolaris*
7. Leaves tapering to short thick petiole, or sessile .... 6. *H. rigidus*
7. Leaves abruptly slender petiolate .... 7. *H. silphioides*
8. Stems scapose, or with 3–5 pairs of cauline leaves smaller than the basal leaves .... 8. *H. occidentalis*
8. Stems leafy to inflorescence 9
9. Stems glabrous or nearly so below inflorescence, often glaucous .. 10
9. Stems pubescent 14
10. Heads 1.5–3.0 cm wide, disc 0.4–1.0 cm wide, rays 1.0–1.5 cm long .....
.... 9. *H. microcephalus*
10. Heads 4–9 cm wide, disc 1.0–2.5 cm wide, rays 2–4 cm long 11
11. Leaves thin, membranaceous .... 10. *H. decapetalus*
11. Leaves thick, firm 12
12. All leaves opposite except uppermost 13
12. Middle and upper leaves alternate, lower opposite or whorled ....
.... 13. *H. grosseserratus*
13. Leaves sessile or, if petiolate, the petioles not over 0.5 cm long ....
.... 11. *H. divaricatus*
13. Leaves evidently petiolate, the petioles 0.5–1.0 cm long 12. *H. strumosus*
14. Leaves densely grayish-pubescent on both surfaces. ..... 14. *H. mollis*
14. Leaves not densely grayish-pubescent on both surfaces 15
15. Leaves broadly lanceolate to ovate; petioles 2–8 cm long ....
.... 15. *H. tuberosus*
15. Leaves lanceolate; petioles, when present, not over 1.5 cm long 16
16. All leaves opposite except uppermost .... 16. *H. hirsutus*
16. Upper and middle leaves alternate, lower opposite 17
17. Stem with spreading hirsute hairs .... 17. *H. giganteus*
17. Stem with short appressed or subappressed hairs .. 18. *H. maximilianii*

1. H. ciliaris DC. Ciliate Sunflower. Aug.–Sept. Adventive from w. Waste ground, rare; St. Clair Co.

2. H. salicifolius A. Dietr. Willow-leaved Sunflower. Aug.–Oct. Adventive from w. Waste ground, rare; Cook Co.

3. H. angustifolius L. Narrow-leaved Sunflower. Aug.–Oct. Moist ground, rare; Massac and Pope cos.

4. H. annuus L. Common Sunflower; Garden Sunflower. Aug–Sept. Adventive and introduced from w., often escaped from cultivation. Fields and waste ground; occasional throughout Ill.

5. H. petiolaris Nutt. Petioled Sunflower. June–Sept. Adventive from w. Sandy soil in waste ground and fields; occasional throughout Ill.

6. H. rigidus (Cass.) Desf. Prairie Sunflower. July–Sept. Prairies; occasional throughout Ill. (Including var. *subrhomboideus* [Rydb.] Fern.) *H. laetiflorus* Pers.—F, G.

7. H. silphioides Nutt. Silphium Sunflower. July–Oct. Prairies, rare; Alexander and St. Clair cos.

8. H. occidentalis Riddell. Western Sunflower. July–Sept. Sandy soil; common in n. half of Ill., extending s. to St. Clair Co.

9. H. microcephalus Torr. & Gray. Small Wood Sunflower. Aug–Oct. Dry open woods; occasional in s. Ill.

10. H. decapetalus L. Pale Sunflower; Thin-leaved Sunflower. Aug.–Oct. Dry, open woods; occasional throughout Ill.

11. H. divaricatus L. Woodland Sunflower. July–Sept. Open woods; common throughout Ill.

12. H. strumosus L. Pale-leaved Sunflower. July–Sept. Open woods; common throughout Ill.

13. H. grosseserratus Martens. Sawtooth Sunflower. July–Oct. Prairies, edges of woods, fencerows; common throughout Ill.

14. H. mollis Lam. Downy Sunflower. Aug.–Sept. Prairies; common nearly throughout Ill. (Including var. *cordatus* S. Wats.)

15. H. tuberosus L. Jerusalem Artichoke. Aug.–Oct. Upper and middle leaves alternate, moderately and inconspicuously pubescent below with mostly appressed hairs. Moist ground; common throughout Ill.

var. subcanescens Gray. Aug.–Oct. All leaves opposite except uppermost, densely pubescent below with loose or spreading hairs. Moist ground; occasional throughout Ill. *H. tomentosus* Michx.—J.

16. H. hirsutus Raf. Bristly Sunflower. Aug.–Sept. Open woods and fields; occasional throughout Ill. (Including var. *trachyphyllus* Torr. & Gray and var. *stenophyllus* Torr. & Gray.)

17. H. giganteus L. Tall Sunflower. Aug.–Oct. Moist ground, rare; Kankakee and Tazewell cos. *H. tomentosus* Michx.—F.

18. H. maximilianii Schrader. Maximillian's Sunflower. July–Aug. Adventive from w. Waste ground; occasional in Ill. (Including *H. dalyi* Britt.)

Reported hybrids: *H. X doronicoides* Lam. (*H. mollis X H. giganteus*). Sangamon Co.; *H. X luxurians* E. E. Wats. (*H. giganteus X H. grosseserratus*), Cook, Effingham, Lake, and Will cos.

13. Melanthera Rohr.

1. M. nivea (L.) Small. June–Oct. Floodplain woods, rare; Massac and Pulaski cos. *M. hastata* Michx.—J.

14. Verbesina L. Crownbeard. Wingstem

1. Stem usually winged, at least in the upper part; leaves sessile 2
1. Stem wingless; leaves petiolate .... 4. *V. encelioides*
2. Rays yellow, 2–3 cm long 3
2. Rays white, 4–7 mm long .... 3. *V. virginica*
3. Stem simple; heads few (2–8); involucral bracts erect 1. *V. helianthoides*
3. Stem usually branched toward top; heads many (10–100); involucral bracts deflexed .... 2. *V. alternifolia*

1. V. helianthoides Michx. Yellow Crownbeard. May–July. Open woods and prairies; common in s. three-fifths of Ill., absent elsewhere.

2. V. alternifolia (L.) Britt. Yellow Ironweed. Aug.–Sept. Moist soil in open woods; occasional throughout Ill. *Actinomeris alternifolia* (L.) DC.—F.

3. V. virginica L. Tickweed; Frostweed. July–Aug. Dry, open woods, rare; confined to the extreme se. cos.

4. V. encelioides (Cav.) Benth. & Hook. Golden Crownbeard. June–Aug. Adventive from w. Waste ground, rare; St. Clair Co. (Including var. *exauriculata* Robins. & Greenm.)

15. *Coreopsis L. Tickseed*

1. Leaves undivided or rarely with 1–2 short lateral lobes 2
1. Leaves 3- to 5-lobed or undivided 3
2. Leaves mostly basal, linear to oblanceolate .... 1. *C. lanceolata*
2. Leaves produced to middle of stem or higher, ovate to elliptic-lanceolate
.... 2. *C. pubescens*
3. Leaves sessile, deeply 3-lobed to or below middle .... 3. *C. palmata*
3. Leaves usually petiolate, divided into 3–5 segments 4
4. Ligules of ray-flowers reddish-brown at base or throughout; disc-flowers reddish-brown 5
4. Ligules of ray-flowers yellow; disc flowers yellow or reddish-brown .. 6
5. Achenes linear, wingless; leaf segments linear to linear-lanceolate .....
.... 4. *C. tinctoria*
5. Achenes obovate, cartilaginous-margined; leaf segments lanceolate to orbicular .... 5. *C. basalis*
6. Leaf segments elliptic-lanceolate; achenes 5–7 mm long; disc-flowers yellow or reddish-brown .... 6. *C. tripteris*
6. Leaf segments linear to linear-lanceolate; achenes 1–4 mm long; disc-flowers yellow .... 7. *C. grandiflora*

1. C. lanceolata L. Tickseed Coreopsis; Sand Coreopsis. June–July. Dry sandy or rocky soils; scattered throughout Ill., but not common. (Including var. *villosa* Michx.)

2. C. pubescens Ell. Star Tickseed. June–Sept. Open woods; rare in s. Ill.

3. C. palmata Nutt. Prairie Coreopsis. June–July. Prairies and open woods; common in the n. three-fourths of Ill., occasional elsewhere.

4. C. tinctoria Nutt. Golden Coreopsis. July–Sept. Adventive from w. and escaped from cultivation. Waste ground; scattered throughout Ill., but not common. (Including f. *atropurpurea* [Hook.] Fern.)

5. C. basalis (Otto & Dietr.) Blake. June–Sept. Introduced from sw. and escaped from cultivation. Waste ground, rare; Lake Co.

6. C. tripteris L. Tall Tickseed. Aug.–Sept. Open woods; common throughout Ill. (Including var. *deamii* Standl.)

7. C. grandiflora Hogg. Large-flowered Coreopsis. June–Aug. Adventive from west and escaped from cultivation. Waste ground; occasional throughout Ill.

16. *Bidens L. Beggar-ticks. Sticktight*

1. Plants aquatic, with submerged leaves finely dissected .... 1. *B. beckii*
1. Plants subaquatic or terrestrial, without submerged, finely dissected leaves 2
2. Rays 1 cm or more long 3
2. Rays less than 0.5 cm long or absent 5
3. Leaves simple .... 2. *B. cernua*
3. Leaves pinnately dissected 4

4. Achenes cuneate, 2 mm or less wide, margins ciliate ... 3. *B. coronata*
4. Achenes obovate, more than 2 mm wide, margins scarious or less commonly only ciliate .... 4. *B. aristosa*
5. Leaves simple or sometimes 3-lobed 6
5. Leaves pinnately dissected. 7
6. Outer involucral bracts rarely much exceeding disc; achenes (3-) 4-angled, tuberculate, sparsely antrorsely appressed-pubescent; corolla 5-lobed; stamens exserted .... 5. *B. connata*
6. Outer involucral bracts 2–5 times as long as disc; achenes flat, smooth, glabrous or rarely sparsely pubescent; corolla 4-lobed; stamens included .... 6. *B. comosa*
7. Achenes linear .... 7. *B. bipinnata*
7. Achenes flat 8
8. Achenes downwardly barbed (upwardly barbed in a variety of *B. frondosa*); outer involucral bracts with ciliate margins 9
8. Achenes upwardly barbed; outer involucral bracts with smooth margins .... 10. *B. discoidea*
9. Outer involucral bracts 5–8; achenes sparsely pubescent 8. *B. frondosa*
9. Outer involucral bracts 10–20; achenes glabrous or nearly so ....
.... 9. *B. vulgata*

1. B. beckii Torr. Water Marigold. Aug.–Sept. Ponds and slow streams, rare; Cook, Lake, and St. Clair cos. *Megalodonta beckii* (Torr.) Greene—F.

2. B. cernua L. Sticktight; Nodding Bur Marigold. July–Oct. Wet ground; occasional throughout Ill. (Including var. *integra* Wieg. and var. *elliptica* Wieg.)

3. B. coronata (L.) Britt. Tall Swamp Marigold; Beggar-ticks; Sticktight. July–Oct. Wet ground; occasional throughout Ill. (Including var. *tenuiloba* [Gray] Sherff and var. *trichosperma* [Michx.] Fern.)

4. B. aristosa L. Swamp Marigold; Tickseed Sunflower. Aug.–Oct. Outer involucral bracts 8–10, entire, merely finely ciliate. Wet ground; common in the s. two-thirds of the state, occasional elsewhere. Occasionally specimens with awns downwardly barbed (f. *fritcheyi* [Fern.] Wunderlin) or with awns absent or minute (f. *mutica* [Gray] Wunderlin) are observed.

var. retrorsa (Sherff) Wunderlin. Swamp Marigold; Tickseed Sunflower. Aug.–Oct. Outer involucral bracts 12–20, coarsely ciliate, appearing almost serrulate. Wet ground; occasional throughout Ill. (Our plants belong to f. *involucrata* [Nutt.] Wunderlin.) *B. polylepis* Blake—F, G, J.

5. B. connata Muhl. Swamp Beggar-ticks. Sept.–Oct. Wet ground; occasional throughout Ill. (Including var. *petiolata* [Nutt.] Farw.) *B. tripartita* L. (in part)—G.

6. B. comosa (Gray) Wieg. Beggar-ticks. Sept.–Oct. Wet ground; common throughout Ill. *B. tripartita* L. (in part)—G.

7. B. bipinnata L. Spanish Needles. Aug.–Sept. Waste ground and open woods; occasional in the s. three-fourths of the state, rare or absent elsewhere.

8. B. frondosa L. Common Beggar-ticks; Sticktight. Aug.–Oct. Moist waste ground; common throughout Ill.

9. B. vulgata Greene. Tall Beggar-ticks; Sticktight. Aug.–Oct. Moist waste ground; occasional throughout Ill. (Including f. *puberula* [Wieg.] Fern.)

10. B. discoidea (Torr. & Gray) Britt. Swamp Beggar-ticks; Sticktight. Aug.–Oct. Wet ground; occasional in the s. two-thirds of Ill.

17. Cosmos Cav. Cosmos

1. C. sulphureus Cav. Cosmos. Aug.–Oct. Adventive from sw. Waste ground, rare; Will Co.

18. Galinsoga R. & P. Peruvian Daisy

1. Pappus of disc-flowers slightly fringed, tapering to awn-tips; pappus of ray-flowers well developed .... 1. *G. ciliata*
1. Pappus of disc-flowers conspicuously fringed, not tapering to awn-tips; pappus of ray-flowers absent or minute .... 2. *G. parviflora*

1. G. ciliata (Raf.) Blake. Peruvian Daisy; Quickweed. June–Oct. Naturalized from trop. Am. Gardens and waste ground; occasional throughout Ill.
2. G. parviflora Cav. Peruvian Daisy. June–Oct. Naturalized from trop. Am. Gardens and waste ground; rare in Ill.

19. Hymenopappus L'Hér.

1. H. scabiosaeus L'Hér. May–June. Open sandy woods and prairies, rare; Cass, Iroquois, Kankakee, and Mason cos.

20. Hymenoxys Cass.

1. H. acaulis (Pursh) Parker. Lakeside Daisy; Four-nerved Starflower. May–July. Dry gravelly banks and fields, rare; Mason and Will cos. (Our plants belong to var. *glabra* [Nutt.] Parker.) *Actinea herbacea* (Greene) B. L. Robins.—F, J.

21. Helenium L. Sneezeweed

1. Leaves linear to linear-filiform, not decurrent .... 1. *H. amarum*
1. Leaves linear-lanceolate to ovate, decurrent 2
2. Disc depressed-globose; disc-flowers yellow .... 2. *H. autumnale*
2. Disc globose; disc-flowers purplish. .... 3. *H. flexuosum*

1. H. amarum (Raf.) Rock. Bitterweed. Aug.–Oct. Fields and waste ground; occasional in s. Ill., extending n. to Pike and Champaign cos. *H. tenuifolium* Nutt.—F, G, J.
2. H. autumnale L. Autumn Sneezeweed. Aug.–Oct. Wet ground; common throughout Ill. (Including var. *canaliculatum* [Nutt.] Torr. & Gray and var. *parviflorum* [Nutt.] Fern.)
3. H. flexuosum Raf. Purple-headed Sneezeweed. June–Sept. Fields; common in s. half of Ill., rare in n. Ill. *H. nudiflorum* Nutt.—F, j.

22. Gaillardia Foug. Gaillardia. Blanket-flower

1. Ray- and disc-flowers yellow; chaff-like development on the receptacle among the disc-flowers reduced to soft short scales, or obsolete ....
.... 1. *G. aestivalis*

1. Ray-flowers yellow and purple, disc-flowers purple, reddish-brown, or purplish-brown; chaff-like bristles on the receptacle among the disc-flowers firm and well developed, equalling or longer than achenes . . . . .
. . . . 2. *G. pulchella*

    1. G. aestivalis (Walt.) Rock. Gaillardia. July–Oct. Prairies, very rare; a single collection from Alexander Co., made in 1874. *G. lutea* Greene—F; *G. lanceolata* Michx. var. *flavovirens* C. Mohr.—G.
    2. G. pulchella Foug. Blanket-flower. June–July. Introduced from w. and escaped from cultivation. Waste ground; rarely escaped.

23. *Dyssodia* Cav. Fetid Marigold

    1. D. papposa (Vent.) Hitchc. Fetid Marigold. Sept.–Oct. Waste ground and fields; occasional throughout Ill.

24. *Grindelia* Willd. Gumweed. Tarweed

    1. G. squarrosa (Pursh) Dunal. Gumweed; Tarweed. July–Sept. Upper and middle leaves 2–4 times as long as wide, ovate to oblong. Fields and waste ground; occasional in the n. half of the state, rare elsewhere.
    var. serrulata (Rydb.) Steyerm. Gumweed; Tarweed. July–Sept. Upper and middle leaves 5–8 times as long as wide, linear-oblong to oblanceolate. Fields and waste ground; occasional in the n. half of the state, rare elsewhere.

25. *Xanthocephalum* Willd. Broomweed

1. Pappus of disc-flowers composed of 10–12 minute scales   1. *X. texanum*
1. Pappus of disc-flowers composed of 5–8 linear strap-shaped projections equalling corolla . . . . 2. *X. dracunculoides*

    1. X. texanum (DC.) Shinners. Broomweed. July–Oct. Adventive from w. Waste ground, rare; St. Clair Co.
    2. X. dracunculoides (DC.) Shinners. Broomweed. July–Oct. Adventive from w. Waste ground, rare; Tazewell Co. *Gutierrezia dracunculoides* (DC.) Blake—F, G, J.

26. *Heterotheca* Cass. Golden Aster

1. Achenes of ray-flowers with capillary pappus . . . . 1. *H. villosa*
1. Achenes of ray-flowers without pappus . . . . 2. *H. latifolia*

    1. H. villosa (Pursh) Shinners. Golden Aster. June–Sept. Sandy soil; occasional throughout Ill. (Our plants belong to var. *camporum* [Greene] Wunderlin.) *Chrysopsis camporum* Greene—F, J; *Chrysopsis villosa* (Pursh) Nutt. var. *camporum* (Greene) Cronq.—G.
    2. H. latifolia Buckl. Camphorweed; Golden Aster. July–Oct. Sandy soil, rare; Alexander, Henry, and Union cos. (Our plants belong to var. *macgregoris* Wagenkn.) *H. subaxillaris* (Lam.) Britt. & Rusby—F, G, J.

27. Solidago L. Goldenrod

20. Basal leaves usually not present at flowering .... 20. *S. ulmifolia*
21. Leaves 3-nerved 22
21. Leaves 1-nerved 24
22. Leaves ovate to elliptic 23
22. Leaves lanceolate or oblanceolate .... 23. *S. canadensis*
23. Middle and upper leaves sessile or nearly so, not rough to touch ......
.... 21. *S. drummondii*
23. All except the uppermost leaves evidently petiolate, rough to touch ....
.... 22. *S. radula*
24. Stem grayish puberulent; leaves oblanceolate .... 24. *S. nemoralis*
24. Stem hirsute, occasionally subglabrous below; leaves lanceolate to elliptic .... 25. *S. rugosa*

1. S. graminifolia (L.) Salisb. Grass-leaved Goldenrod. Aug.–Oct. Leaves 3- to 5-nerved, glabrous or nearly so to hirtellous, not resinous or conspicuously glandular-punctate, thin; heads glomerate, rarely pedunculate; involucre 3–5 mm high. Moist ground; common throughout Ill. (Including var. *nuttallii* [Greene] Fern. [= *S. hirtella* (Greene) Bush] and var. *media* [Greene] Harris [= *S. media* (Greene) Bush.]

var. gymnospermoides (Greene) Croat. Grass-leaved Goldenrod. Aug.–Oct. Leaves 1-nerved, rarely faintly 3-nerved, glabrous, usually resinous and conspicuously glandular-punctate, firm; heads pedunculate, rarely glomerate; involucre 4.5–6.0 mm high. Open sandy ground; occasional in w. and n. Ill. *S. gymnospermoides* (Greene) Fern.—F, G.

var. remota (Greene) Harris. Grass-leaved Goldenrod. Aug.–Sept. Leaves 1-nerved, rarely faintly 3-nerved, glabrous or nearly so, more or less resinous, not conspicuously glandular-punctate; involucre 3.0–4.5 mm high. Open sandy soil near Lake Michigan, rare; Cook, Kankakee, and Lake cos. *S. remota* (Greene) Friesner—F, G.

2. S. rigida L. Rigid Goldenrod. Aug.–Sept. Prairies; common throughout Ill., except for the extreme s. tip.

3. S. riddellii Frank. Riddell Goldenrod. Aug.–Sept. Moist ground; occasional in the n. half of the state, rare in the s. half.

4. S. ohioensis Riddell. Ohio Goldenrod. Aug.–Oct. Moist ground; occasional in ne. Ill., extending s. to Peoria and Woodford cos.

5. S. sciaphila Steele. Cliff Goldenrod. Aug.–Oct. Calcareous cliffs, rare; Carroll, JoDaviess, LaSalle, and Ogle cos.

6. S. caesia L. Woodland Goldenrod; Blue-stem Goldenrod. Aug.–Oct. Woods; occasional throughout Ill. (Including f. *axillaris* [Pursh] House.)

7. S. flexicaulis L. Broadleaf Goldenrod. Aug.–Oct. Woods; occasional nearly throughout Ill. *S. latifolia* L.—F, G.

8. S. bicolor L. White Goldenrod. Aug.–Oct. Flowers white to cream. Dry open woods; scattered throughout Ill., but not common.

var. concolor Torr. & Gray. Hispid Goldenrod. Aug.–Oct. Flowers yellow-orange. Dry open woods; Alexander and Jackson cos. *S. hispida* Muhl.—F, G, J.

9. S. buckleyi Torr. & Gray. Buckley's Goldenrod. Sept.–Oct. Woods; rare in s. Ill.

10. S. petiolaris Ait. Goldenrod. Sept.–Oct. Open rocky woods; occasional in sw. Ill. (Including var. *wardii* [Britt.] Fern.)

11. S. sphacelata Raf. Goldenrod. Aug.–Oct. River bluffs, rare; Pope Co.

12. S. missouriensis Nutt. Missouri Goldenrod. Aug.–Sept. Prairies; common throughout Ill. (Including var. *fasciculata* Holz and var. *glaberrima* [Martens] Rosend. & Cronq.) S. *glaberrima* Martens—J.

13. S. juncea Ait. Early Goldenrod. July–Aug. Open woods, fields, and waste ground; common throughout Ill. (Including f. *scabrella* [Torr. & Gray] Fern.)

14. S. uliginosa Nutt. Swamp Goldenrod. Aug.–Sept. Swamps and bogs; occasional in n. one-half of Ill. (Including var. *linoides* [Torr. & Gray] Fern.)

15. S. speciosa Nutt. Showy Goldenrod. Aug.–Oct. Open woods and prairies; common throughout Ill. (Including var. *rigidiuscula* Torr. & Gray and var. *jejunifolia* [Steele] Cronq.)

16. S. patula Muhl. Spreading Goldenrod. Aug.–Oct. Wet ground; occasional throughout Ill.

17. S. gigantea Ait. Late Goldenrod. Aug.–Oct. Moist ground; common throughout Ill. (Including var. *serotina* [Kuntze] Cronq.)

18. S. sempervirens L. Seaside Goldenrod. Aug.–Oct. Adventive from e. coast. Waste ground, rare; Cook Co.

19. S. arguta Ait. Goldenrod. July–Oct. Dry open woods, rare; Union Co. S. *strigosa* Small (in part)—F, G, J.

20. S. ulmifolia Muhl. Elm-leaved Goldenrod. Aug.–Oct. Woods; common throughout Ill.

21. S. drummondii Torr. & Gray. Drummond's Goldenrod. Sept.–Oct. River bluffs; occasional in sw. Ill.

22. S. radula Nutt. Rough Goldenrod. Aug.–Oct. Dry open woods; occasional in w. and s. Ill. (Including var. *laeta* [Greene] Fern. and var. *stenolepis* Fern.)

23. S. canadensis L. Tall Goldenrod. Aug.–Oct. Open woods and fields; common throughout Ill. (Including var. *scabra* [Muhl.] Torr. & Gray, S. *altissima* L., and var. *hargeri* Fern.)

24. S. nemoralis Ait. Field Goldenrod. Aug.–Oct. Open woods and fields; common throughout Ill. (Including var. *longipetiolata* [Mack. & Bush] Palmer & Steyerm, and var. *decemflora* [DC.] Fern.)

25. S. rugosa Mill. Rough-leaved Goldenrod. Aug.–Oct. Moist ground, rare; Jackson, Johnson, Lawrence, Pope, and Randolph cos.

28. *Haplopappus* Cass.

1. H. ciliatus (Nutt.) DC. Aug.–Sept. Adventive from west. Waste ground; occasional in sw. Ill.

29. *Bellis* L.   English Daisy

1. B. perennis L. English Daisy. May–June. Introduced from Europe and escaped from cultivation. Waste ground, rare; Champaign and McLean cos.

30. *Boltonia* L'Hér.   Boltonia

1. Leaves lanceolate to oblanceolate, 5–20 mm wide; disc 6–10 mm wide
.... 1. *B. asteroides*
1. Leaves linear, 1–5 mm wide; disc 3–6 mm wide .... 2. *B. diffusa*

1. B. asteroides (L.) L'Hér. False Aster. July–Oct. Leaves not decurrent. Moist ground; common throughout Ill. (Including var. *recognita*

[Fern. & Grisc.] Cronq., *B. recognita* [Fern. & Grisc.] G. N. Jones, var. *latisquama* [Gray] Cronq., *B. latisquama* Gray, and *B. latisquama* var. *microcephala* Fern. & Grisc.)

var. decurrens (Torr. & Gray) Engelm. False Aster. July–Oct. Leaves decurrent. Moist ground; occasional along Illinois River. *B. latisquama* var. *decurrens* (Torr. & Gray) Fern. & Grisc.—F; *B. decurrens* (Torr. & Gray) Wood—G, J.

2. B. diffusa Ell. False Aster. July–Sept. Moist to dry open woods and fields; occasional in s. half of Ill. (Including var. *interior* Fern. & Grisc., *B. interior* [Fern. & Grisc.] G. N. Jones.)

31. Aster L.   Aster

1. A. macrophyllus L. Big-leaved Aster. Aug.–Sept. Dry open woods, rare; Boone, Cook, and Lake cos. (Including var. *pinguifolius* Burgess, var. *velutinus* Burgess, var. *sejunctus* Burgess, and var. *ianthinus* [Burgess] Fern.)

2. A. furcatus Burgess. Forked Aster. Aug.–Oct. Dry woods; occasional in n. half of Ill. (Including f. *erythractis* Benke.)

3. A. schreberi Nees. Schreber's Aster. Aug.–Oct. Woods, rare; Marshall, Peoria, and Tazewell cos. *A. chasei* G. N. Jones—J.

4. A. anomalus Engelm. Blue Aster. Sept.–Oct. Dry woods; occasional in sw. Ill., n. to Peoria and Woodford cos.

5. A. shortii Lindl. Short's Aster. Aug.–Oct. Dry woods; common throughout Ill. Rarely rose–flowered specimens (f. *gronemannii* Benke) rather than blue-flowered ones are observed.

6. A. azureus Lindl. Sky-blue Aster. Sept.–Oct. Prairies and dry open woods; occasional throughout Ill. (Including f. *laevicaulis* Fern. and var. *poaceus* [Burgess] Fern.)

7. A. cordifolius L. Blue Wood Aster; Heart-leaved Aster. Aug.–Oct. Dry woods; occasional throughout Ill. (Including var. *moratus* Shinners.)

8. A. sagittifolius Wedem. ex Willd. Arrow-leaved Aster. Aug.–Oct. Stem glabrous or sparsely pubescent along decurrent lines on upper part. Dry woods; common throughout Ill.

var. drummondii (Lindl.) Shinners. Drummond's Aster. Aug.–Oct. Stem evenly and usually densely short-pubescent throughout. Dry open woods; common throughout Ill. (Including *A. texanus* Burgess, *A. sagittifolius* f. *hirtellus* [Lindl.] Shinners, and *A. drummondii* var. *rhodactis* Benke.) *A. drummondii* Lindl.—G, J.

9. A. novae-angliae L. New England Aster. Aug.–Oct. Moist ground; common throughout Ill. Rarely rose-flowered (f. *rosarius* House) or white-flowered (f. *geneseensis* House) specimens rather than the normal violet-purple ones are found.

10. A. oblongifolius Nutt. Aromatic Aster. Aug.–Oct. Dry open woods; occasional throughout Ill., except for the e. cent. cos. (Including var. *angustatus* Shinners and f. *roseoligulatus* [Benke] Shinners.)

11. A. patens Ait. Spreading Aster. Aug.–Oct. Open woods; occasional in s. third of Ill. (Including var. *phlogifolius* Muhl.)

12. A. puniceus L. Swamp Aster. Aug.–Oct. Stem evenly hispid above. Moist ground; occasional in the n. two-thirds of the state, rare elsewhere.

var. lucidulus Gray. Shiny Swamp Aster. Aug.–Oct. Stems puberulent in lines above. Moist ground; occasional in n. half of Ill. *A. lucidulus* (Gray) Wieg.—F, G. J.

13. A. laevis L. Smooth Aster. Aug.–Oct. Moist sandy soil in woods; moist prairies; common throughout Ill. Rarely white-flowered specimens (f. *beckwithiae* House) are observed.

14. A. prenanthoides Muhl. Aster. Sept.–Oct. Moist ground; rare in n. two-fifths of Ill. absent elsewhere.

15. A. novi-belgii L. New Belgium Aster. Aug.–Sept. Moist ground, rare; Cook, Lake, and McHenry cos. *A. longifolius* Lam.—J.

16. A. sericeus Vent. Silky Aster. Sept.–Oct. Prairies; sand barrens; occasional in the n. half of the state, rare or absent elsewhere.

17. A. umbellatus Mill. Flat-top Aster. Aug.–Oct. Moist ground; occasional in n. half of Ill., absent elsewhere.

18. A. ptarmicoides (Nees) Torr. & Gray. Aster. Aug.–Sept. Ray-flowers white. Dry sandy soil; occasional in n. half of Ill., absent elsewhere.

var. lutescens (Lindl.) Gray. Aster. Aug.–Sept. Ray-flowers yellow. Dry prairies, rare; Cook Co.

19. A. brachyactis Blake. Rayless Aster. July–Sept. Native of the w. U.S. Waste ground, rare; Cook Co.

20. A. parviceps (Burgess) Mack. & Bush. Aug.–Oct. Prairies and open woods; occasional in n. three-fifths of Ill., absent elsewhere.

21. A. pilosus Willd. Hairy Aster. Aug.–Oct. Stems and often leaves pilose. Waste ground; common throughout Ill. (Including var. *platyphyllus* [Torr. & Gray] Blake.)

var. pringlei (Gray) Blake. Pringle's Aster. Aug.–Oct. Stems and leaves glabrous or nearly so. Dry sandy soil, rare; Lake Co. (Including var. *demotus* Blake.) *A. pringlei* (Gray) Britt.—J.

22. A. ericoides L. Heath Aster. July–Oct. Stem with appressed or ascending short hairs. Prairies and dry open ground; occasional throughout Ill. Rarely blue-flowered (f. *caeruleus* [Benke] Blake) or rose-flowered (f. *gramsii* Benke) specimens are observed.

var. prostratus (Ktze.) Blake. Heath Aster. Sept.–Oct. Stems with spreading or slightly reflexed hairs. Prairies and dry open ground; common throughout Ill. (Including f. *prostratus* [Ktze.] Fern. and f. *exiguus* Fern.) *A. exiguus* (Fern.) Rydb.—J.

23. A. vimineus Lam. Sept.–Oct. Moist open ground; occasional throughout Ill. (Including var. *subdumosus* Wieg.)

24. A. dumosus L. Rice Button Aster; Bushy Aster. Aug.–Sept. Moist sandy soil, rare; Champaign, Cook, Kankakee, McDonough, Monroe and Will cos. (Including var. *cordifolius* [Michx.] Torr. & Gray and var. *strictior* Torr. & Gray.)

25. A. praealtus Poir. Willow Aster. Sept.–Oct. Moist ground; common throughout Ill. (Including var. *angustior* Wieg. and var. *subasper* [Lindl.] Wieg.)

26. A. linariifolius L. Flax-leaved Aster. Sept.–Oct. Dry open soil; occasional in n. half of Ill., extending s. to St. Clair Co.

27. A. junciformis Rydb. Rush Aster. Aug.–Sept. Bogs and calcareous fens; occasional in ne. Ill., extending s. to Peoria Co.

28. A. turbinellus Lindl. Aster. Sept.–Oct. Prairies, dry open woods; occasional in s. three-fifths of Ill.

29. A. tataricus L. f. Tartarian Aster. Sept.–Oct. Introduced from e. Asia and escaped from cultivation. Waste ground, rare; Jersey, St. Clair, and Sangamon cos.

30. A. ontarionis Wieg. Ontario Aster. Sept.–Oct. Moist ground; common throughout Ill.

31. A. lateriflorus (L.) Britt. Side-flowered Aster. Sept.–Oct. Woods; common throughout Ill.

32. A. simplex Willd. Paniced Aster. Aug.–Oct. Moist ground; common throughout Ill. (Including var. *interior* [Wieg.] Cronq. and var. *ramosissimus* [Torr. & Gray] Cronq.)

Reported hybrids: *A.* X *amythystinus* Nutt. (*A. ericoides* X *A. novae-angliae*). Champaign and Peoria cos.; *A. ericoides* var. *prostratus* X *A. pilosus,* Champaign, Cook, and Du Page cos.; *A. laevis* X *A. simplex,* McHenry Co.

32. *Erigeron L. Fleabane*

1. Rays conspicuous, more than 5 mm long 2
1. Rays, inconspicuous, less than 3 mm long 5
2. Leaves clasping 3
2. Leaves not clasping 4
3. Heads few, 2.5–3.5 cm in diameter; rays 50–100, about 1 mm wide . . . . .
. . . . 1. *E. pulchellus*
3. Heads several, 1.5–2.0 cm in diameter; rays 150–200, about 0.5 mm wide
. . . . 2. *E. philadelphicus*

4. Cauline leaves many; basal leaves ovate, coarsely dentate; middle part
of stem with long, spreading hairs .... 3. *E. annuus*
4. Cauline leaves few; basal leaves spatulate, entire or nearly so; middle
part of stem usually with short, appressed hairs .... 4. *E. strigosus*
5. Stem diffusely branched from near the base; rays purplish ....
.... 5. *E. divaricatus*
5. Stem simple or nearly so up to inflorescence; rays white ....
.... 6. *E. canadensis*

1. E. pulchellus Michx. Robin's Plantain. April–June. Open woods;
occasional throughout Ill.

2. E. philadelphicus L. Marsh Fleabane. May–June. Open woods,
fields, and waste ground; common throughout Ill. (Including f. *sca-
turicola* [Fern.] Cronq. and f. *angustatus* Vict. & Rosseau.)

3. E. annuus (L.) Pers. Daisy Fleabane; White Top. June–Oct. Open
woods, fields , and waste ground; common throughout Ill.

4. E. strigosus Muhl. Daisy Fleabane; White Top. May–July. Open
woods, fields, waste ground; common throughout Ill. (Including var.
septentrionalis [Fern. & Wieg.] Fern. and var. *beyrichii* [Fisch. & Mey.]
Gray.)

5. E. divaricatus Michx. Dwarf Fleabane. June–Sept. Waste ground;
occasional throughout Ill. *Conyza ramosissima* Cronq.—G.

6. E. canadensis L. Horseweed; Muleweed. May–Oct. Waste
ground; in every co. *Conyza canadensis* (L.) Cronq.—G.

33. Anthemis L.   Dogfennel. Chamomile

1. Ray-flowers yellow .... 1. *A. tinctoria*
1. Ray-flowers white 2
2. Achenes glandular-tuberculate .... 2. *A. cotula*
2. Achenes not glandular-tuberculate 3
3. Achenes obtusely 3-angled .... 3. *A. nobilis*
3. Achenes 10-ribbed .... 4. *A. arvensis*

1. A. tinctoria L. Yellow Chamomile. June–Sept. Introduced from
Europe and escaped from cultivation. Waste ground, rare; Cook, Du-
Page, and Winnebago cos.

2. A. cotula L. Dogfennel; Mayweed. May–Sept. Naturalized from
Europe. Waste ground; common throughout Ill.

3. A. nobilis L. Garden Chamomile. June–Sept. Introduced from
Europe and escaped from cultivation, rare; Cook and DuPage cos.

4. A. arvensis L. Field Chamomile; Corn Chamomile. May–Aug. Naturalized from Europe; occasional throughout Ill. (Including var. *agrestis* [Wallr.] DC.)

34. Achillea L.   Yarrow. Milfoil

1. A. millefolium L. Common Yarrow; Milfoil. May–Aug. Leaves and
stems arachnoid to glabrescent; corymb flat-topped. Naturalized from
Europe. Fields and waste ground; common throughout Ill. Rarely
rose-purple flowered specimens (f. *rosea* Rand & Redf.) are observed.

subsp. lanulosa (Nutt.) Piper. Western Yarrow. June–Aug. Leaves
and stems densely woolly; corymb round-topped. Adventive from w.
Fields and waste ground; occasional throughout Ill. *A. lanulosa*
Nutt.—F. J.

35. *Matricaria L. Wild Chamomile*

1. Heads with evident white rays 2
1. Heads rayless .... 3. *M. matricarioides*
2. Achenes with 2 marginal and 1 ventral, strongly callous-thickened, almost winglike ribs .... 1. *M. maritima*
2. Achenes with 2 marginal and 3 ventral, raised, but not winglike ribs ....
.... 2. *M. chamomilla*

    1. M. maritima L. Scentless Chamomile. June–Sept. Naturalized from Europe. Waste ground, rare; DuPage Co. (Our plants belong to var. *agrestis* [Knaf] Wilmott.)
    2. M. chamomilla L. Chamomile. May–Oct. Naturalized from Europe. Roadsides and waste ground; occasional in s. half of Ill. extending n. to Woodford Co.
    3. M. matricarioides (Less.) Porter. Pineapple-weed. May–Aug. Naturalized from Pacific coast. Roadsides and waste ground; common throughout Ill.

36. *Chrysanthemum L. Chrysanthemum*

1. Heads solitary or few, 3–5 cm broad .... 1. *C. leucanthemum*
1. Heads several or numerous, 0.5–1.5 cm broad 2
2. Leaves pinnatifid or bipinnatifid; rays 3–8 mm long .. 2. *C. parthenium*
2. Leaves crenate or with few reduced basal pinnae; rays less than 1 mm long, or absent .... 3. *C. balsamita*

    1. C. leucanthemum L. Ox-eye Daisy. May–Aug. Naturalized from Europe and Asia. Fields and waste ground; common throughout Ill. (Including var. *pinnatifidum* Lecog. & Lamotte.)
    2. C. parthenium (L.) Bernh. Feverfew. June–Sept. Escaped from cultivation or naturalized from Europe. Roadsides and waste ground; occasional in n. half of Ill.
    3. C. balsamita L. Costmary. Sept–Oct. Escaped from cultivation or naturalized from Europe. Waste ground, not common. (Including var. *tanacetoides* Boiss.)

37. *Tanacetum L. Tansy. Golden-buttons*

    1. T. vulgare L. Tansy; Golden-buttons. July–Sept. Naturalized from Europe. Fields and waste ground; occasional throughout Ill.

38. *Artemisia L. Wormwood*

1. Leaves white-tomentose, at least below 2
1. Leaves glabrous or nearly so 7
2. Leaves lanceolate to linear 3
2. Leaves pinnatifid 4
3. Leaves regularly serrate .... 1. *A. serrata*
3. Leaves entire or irregularly few-toothed .... 2. *A. ludoviciana*
4. Leaves green and glabrous above, densely white-tomentose below .....
.... 3. *A. vulgaris*
4. Leaves silvery-sericeous on both sides, or eventually subglabrous above in age, or the basal leaves grayish-hairy 5
5. Basal leaves not grayish-hairy 6

5. Basal leaves grayish-hairy .... 7. *A. campestris*
6. Divisions of leaves oblong to lanceolate, 1.5–5.0 mm wide ....
.... 4. *A. absinthium*
6. Divisions of leaves linear to filiform, 1 mm wide or less ... 5. *A. frigida*
7. Leaves or their divisions linear to filiform, entire margined 8
7. Leaves or their divisions lanceolate, sharply toothed 10
8. Involucral bracts glabrous 9
8. Involucral bracts tomentose .... 8. *A. abrotanum*
9. Leaves mostly entire, occasionally irregularly lobed or lower 3- to 5-lobed .... 6. *A. dracunculus*
9. Leaves mostly pinnatifid .... 7. *A. campestris*
10. Inflorescence dense, spicate .... 9. *A. biennis*
10. Inflorescence loose, paniculate .... 10. *A. annua*

1. A. serrata Nutt. Aug.–Sept. Moist ground; occasional in n. half of Ill.

2. A. ludoviciana Nutt. Western Mugwort; White Sage. July–Sept. Leaves glabrate above, tomentose below. Waste ground; occasional in the n. three-fifths of Ill.

var. gnaphalodes (Nutt.) T. & G. White Sage. July–Sept. Leaves white-tomentose on both surfaces. Waste ground; occasional throughout Ill. (Including var. *latifolia* [Besser] Torr.) *A. gnaphalodes* Nutt.—J

3. A. vulgaris L. Common Mugwort. July–Oct. Introduced from Europe and escaped from cultivation. Waste ground, rare; Champaign, Cook, and Henry cos. (Including var. *latiloba* Ledeb.)

4. A. absinthium L. Common Wormwood; Absinth. July–Sept. Introduced from Europe and escaped from cultivation. Waste ground, rare; Cook, DeKalb, Kane, Lake, McHenry, Peoria, and Wabash cos.

5. A. frigida Willd. Wormwood. July–Sept. Adventive from w. Waste ground, rare; Cook Co.

6. A. dracunculus L. False Tarragon. July–Sept. Prairies; occasional in n. half of Ill. *A. dracunculoides* Pursh—J; *A. glauca* Pall.—F.

7. A. campestris L. Beach Wormwood. July–Sept. Sandy soil; occasional in n. half of Ill. (Our plants belong to ssp. *caudata* [Michx.] Hall & Clem.) (Including *A. caudata* var. *calvens* Lunell.) *A. caudata* Michx.—F. G. J.

8. A. abrotanum L. Southern Wormwood. Aug.–Sept. Introduced from Europe and rarely persisting from cultivation. Waste ground, rare; Cook Co.

9. A. biennis Willd. Biennial Wormwood. Aug.–Oct. Adventive from nw. U.S. Waste ground; occasional throughout Ill.

10. A. annua L. Annual Wormwood. Aug.–Oct. Introduced from Europe and naturalized. Waste ground; occasional throughout Ill.

39. Pluchea Cass.   Marsh Fleabane. Stinkweed

1. P. camphorata (L.) DC. Camphorweed; Stinkweed. July–Oct. Swamps and sloughs; occasional in s. Ill., extending n. to DeWitt Co.

40. Antennaria Gaertn.   Pussytoes. Ladies'-tobacco

1. Basal leaves prominently 1-nerved or obscurely 3-nerved, usually less than 1.5 cm wide .... 1. *A. neglecta*

1. Basal leaves prominently 3- to 5-nerved, usually 1.5 cm or more wide ..
.... 2. *A. plantaginifolia*

1. A. neglecta Greene. Pussytoes; Ladies'-tobacco. April–May. Basal leaves cuneate-spatulate, gradually tapering to the sessile base. Fields and open woods; common throughout Ill. (Including f. *simplex* [Peck] Fern., *A. campestris* Rydb., *A. petaloides* var. *scariosa* Fern., and *A. petaloides* var. *subcorymbosa* Fern.)

var. attenuata (Fern.) Cronq. Pussytoes; Ladies'-tobacco. May–June. Basal leaves obovate, abruptly contracted below the middle into a petiole-like base. Fields and open woods, rare; DeKalb, Henry, and Kendall cos. (Including *A. brainerdii* Fern., *A. rupicola* Fern., *A. neodioica* Greene, *A. neodioica* var. *interjecta* Fern., *A. neodioica* var. *chlorophylla* Fern., and *A. neodioica* var. *grandis* Fern.) *A. neodioica* Greene—J; *A. neodioica* var. *attenuata* Fern.—F.

2. A. plantaginifolia (L.) Richards. Pussytoes; Ladies'-tobacco. April–May. Upper surface of leaves arachnoid at first, tardily glabrate in age; involucre of pistillate plants 5–7 mm high; upper part of stem without purple glands. Open woods; common throughout Ill. (Including var. *petiolata* [Fern.] Heller.)

var. arnoglossa (Greene) Cronq. Pussytoes; Ladies'-tobacco. April–June. Upper surface of leaves glabrous or nearly so at first; involucre of pistillate plants 7–10 mm high; upper part of stem usually with purple glands. Open woods; occasional throughout Ill. *A. parlinii* Fern.—F, J; *A. parlinii* var. *arnoglossa* (Greene) Fern.—F.

var. ambigens (Greene) Cronq. Pussytoes; Ladies'-tobacco. May–June. Upper surface of leaves arachnoid at first, tardily glabrate in age; involucre of pistillate plants 7–10 mm high; upper part of stem without purple glands. Fields and open woods; common throughout Ill. (Including *A. fallax* Greene, *A. fallax* var. *calophylla* [Greene] Fern., *A. farwellii* Greene, and *A. munda* Fern.)

41. *Gnaphalium* L. Cudweed. Everlasting

1. Inflorescence a narrow spike-like panicle; pappus bristles united in a ring at base and deciduous as a whole .... 1. *G. purpureum*
1. Inflorescence a cymose or paniculate cluster; pappus bristles not united in a ring at base and deciduous as a whole 2
2. Involucral bracts white or whitish; achenes smooth 3
2. Involucral bracts brown; achenes scabrous .... 4. *G. uliginosum*
3. Leaves decurrent .... 2. *G. macounii*
3. Leaves not decurrent .... 3. *G. obtusifolium*

1. G. purpureum L. Early Cudweed. May–July. Fields and open woods; occasional throughout Ill., except ne. counties, more common in s. third of state.

2. G. macounii Greene. Western Cudweed; Clammy Cudweed. July–Sept. Sandy soil in open woods and fields, rare; Clark Co.

3. G. obtusifolium L. Sweet Everlasting; Catsfoot. Aug.–Oct. Meadows and open woods; common throughout Ill.

4. G. uliginosum L. Low Cudweed. June–Sept. Naturalized from Europe. Roadsides and pastures, rare; Cook and Lake cos.

42. Inula L. Elecampane

1. I. helenium L. Elecampane. July–Aug. Introduced from Europe and escaped from cultivation. Fields, waste ground, and open woods; not common in Ill.

43. Erechtites Raf. Fireweed

1. E. hieracifolia (L.) Raf. Fireweed. Aug.–Oct. Moist woods, bogs, waste ground, and recently burned areas; common throughout Ill. (Including var. *praealta* [Raf.] Fern.)

44. Cacalia L. Indian-plantain

1. Heads 20- to 40-flowered; involucral bracts 12–15 .... 1. *C. suaveolens*
1. Heads 5-flowered; involucral bracts 5 2
2. Lower leaves reniform, lobed or coarsely angulate-dentate 3
2. Lower leaves lance-ovate, entire, shallowly dentate, or crenate ....
.... 4. *C. tuberosa*
3. Leaves glaucous below; stem round or nearly so .... 2. *C. atriplicifolia*
3. Leaves not glaucous below; stem conspicuously grooved. ....
.... 3. *C. muhlenbergii*

   1. C. suaveolens L. Sweet Indian-plantain. July–Aug. Wet ground; not common in n. Ill., rare in s. Ill.
   2. C. atriplicifolia L. Pale Indian-plantain. July–Oct. Moist or dry open woods and prairies; occasional throughout Ill.
   3. C. muhlenbergii (Sch.-Bip.) Fern. Great Indian-plantain. July–Sept. Woods; not common, but scattered throughout Ill.
   4. C. tuberosa Nutt. Prairie Indian-plantain. June–July. Wet prairies, marshes, and bogs; not common, but scattered throughout the state.

45. Senecio L. Butterweed. Ragwort. Groundsel. Squaw-weed

1. Leaves chiefly basal, basal leaves crenate, dentate, or entire, median stem leaves often pinnatifid 2
1. Stems nearly equally leafy throughout; leaves all pinnatifid or coarsely sinuate-dentate 5
2. Basal leaves with winged petioles .... 1. *S. obovatus*
2. Basal leaves without winged petioles 3
3. Basal leaves cordate to subcordate .... 2. *S. aureus*
3. Basal leaves oblanceolate to oval, not cordate 4
4. Leaves and stems, especially nodes, floccose-tomentose at maturity; peduncles tomentose .... 3. *S. plattensis*
4. Leaves and stems glabrous or nearly so at maturity; peduncles glabrous or nearly so .... 4. *S. pauperculus*
5. Involucral bracts black-tipped; rays absent .... 5. *S. vulgaris*
5. Involucral bracts not black-tipped; rays present, 3–10 mm long 6
6. Stem glabrous or obscurely floccose-tomentose at nodes; rays 6–10 mm long .... 6. *S. glabellus*
6. Stem glandular hairy; rays about 3 mm long .... 7. *S. viscosus*

   1. S. obovatus Muhl. Round-leaved Groundsel. April–June. Open rich woods and rocky outcroppings, rare; Champaign, Clark, and Ver-

milion cos. (Including var. *rotundus* Britt. and f. *elongatus* [Pursh] Fern.)

2. S. aureus L. Golden Ragwort; Squaw-weed. April–May. Wet ground; occasional throughout Ill. (Including var. *intercursus* Fern., var. *gracilis* [Pursh] Hook., and var. *aquilonius* Fern.)

3. S. plattensis Nutt. Prairie Groundsel; Prairie Ragwort. May–June. Dry woods and prairies; common in n. and w. Ill., occasional or rare elsewhere.

4. S. pauperculus Michx. Northern Ragwort. May–June. Open woods, and meadows; common in n. half of Ill., extending s. to Macoupin Co. (Including var. *balsamitae* [Muhl.] Fern.)

5. S. vulgaris L. Common Groundsel. June–July. Naturalized from Europe. Waste ground; occasional in n. half of Ill.

6. S. glabellus Poir. Butterweed. April–June. Moist shady ground along streams; common in s. half of Ill., extending n. to Peoria Co.

7. S. viscosus L. Sticky Groundsel. July–Sept. Naturalized from Europe. Waste ground, rare; Cook and Jackson cos.

46. *Eupatorium* L.   Thoroughwort

1. Leaves whorled 2
1. Leaves opposite, or the upper ones alternate 4
2. Heads 9- to 22-flowered; inflorescence flat-topped .... 1. *E. maculatum*
2. Heads 5- to 7-flowered; inflorescence convex 3
3. Stems solid, purple only at nodes .... 2. *E. purpureum*
3. Stems hollow, purple throughout .... 3. *E. fistulosum*
4. Leaves distinctly petiolate 5
4. Leaves sessile, subsessile, or connate-perfoliate 8
5. Involucral bracts acuminate, filiform-tipped .... 4. *E. coelestinum*
5. Involucral bracts rounded to acute, not filiform-tipped 6
6. Flowers pink to pale purple .... 5. *E. incarnatum*
6. Flowers white 7
7. Involucral bracts broadly rounded, densely pubescent . 6. *E. serotinum*
7. Involucral bracts acute, glabrous or sparsely pubescent . 7. *E. rugosum*
8. Leaves tapering to base .... 8. *E. altissimum*
8. Leaves cuneate, subcordate, truncate, or connate-perfoliate 9
9. Leaves connate-perfoliate (truncate in one form) .... 9. *E. perfoliatum*
9. Leaves cuneate to subcordate .... 10 *E. sessilifolium*

1. E. maculatum L. Spotted Joe-Pye-weed. July–Sept. Moist ground; common in n. Ill., rare in s. Ill.

2. E. purpureum L. Purple Joe-Pye-weed; Green-stemmed Joe-Pye-weed; Sweet Joe-Pye-weed. July–Aug. Open woods; common throughout Ill.

3. E. fistulosum Barratt. Hollow Joe-Pye-weed; Trumpet-weed. July–Sept. Low, wet ground; Alexander, Coles, Jackson, and Pope cos.

4. E. coelestinum L. Mistflower; Blue Boneset; Wild Ageratum. July–Oct. Moist ground; common in s. two-thirds of Ill., absent elsewhere, except for Cook and DuPage cos. Rarely white-flowered specimens (f. *album* E. J. Alex.) or reddish-purple-flowered specimens (f. *illinoense* Benke) rather than the normal blue- to violet-flowered specimens are observed.

5. E. incarnatum Walt. Aug.–Oct. Wet woods and swamps, rare; Alexander, Pulaski, and Union cos.

6. E. serotinum Michx. Late Boneset. Aug–Oct. Moist open woods and clearings; common throughout Ill.

7. E. rugosum Houtt. White Snakeroot. July–Sept. Woods; common throughout Ill. (Including var. *tomentellum* [Robinson] Blake.)

8. E. altissimum L. Tall Boneset; Tall Thoroughwort. Aug.–Oct. Open woods and prairies; common throughout Ill.

9. E. perfoliatum L. Common Boneset; Thoroughwort. Aug.–Oct. Wet ground; common throughout Ill. Rarely plants with flowers purple and involucral bracts purplish-tinged (f. *purpureum* Britton), leaves three at a node rather than two (f. *trifolium* Fassett), and leaf bases truncate (f. *truncatum* [Muhl.] Fassett), are observed.

10. E. sessilifolium L. Upland Boneset. Aug.–Oct. Woods; occasional in the s. half of Ill., uncommon elsewhere. (Our plants belong to var. *brittonianum* Porter.)

Reported hybrids: *E.* X *polyneuron* (F. J. Herm.) Wunderlin (*E. perfoliatum* X *E. serotinum*). *E. perfoliatum* L. var. *cuneatum* (Engelm. ex Torr. & Gray) Engelm.—F, G (in part); *E. serotinum* Michx. var. *polyneuron* F. J. Herm.—G (in part).

47. Mikania Willd. Climbing Hempweed

1. M. scandens (L.) Willd. Climbing Hempweed. Aug.–Oct. Low woods, swamps, and banks of streams; not common and confined to s. Ill.

48. Brickellia Ell.

1. B. eupatorioides (L.) Shinners. False Boneset. Aug.–Oct. Prairies and open woods; occasional to common throughout Ill. (Including var. *corymbulosa* [Torr. & Gray] Shinners and var. *ozarkana* [Shinners] Shinners.) *Kuhnia eupatorioides* L.—F, G, J.

49. Liatris Schreb. Blazing-star

1. Pappus evidently plumose 2
1. Pappus merely barbellate 4
2. Heads 10- to 60-flowered; corolla lobes pubescent within 3
2. Heads 4- to 6-flowered; corolla lobes glabrous within ... 3. *L. punctata*
3. Involucral bracts rounded to mucronate, appressed ... 1. *L. cylindracea*
3. Involucral bracts acuminate, squarrose .... 2. *L. squarrosa*
4. Heads cylindrical, 3- to 18-flowered 5
4. Heads campanulate to hemispherical, 18- to 100-flowered (rarely 12–15) 6
5. Stem pubescent in inflorescence .... 4. *L. pycnostachya*
5. Stem glabrous in inflorescence .... 5. *L. spicata*
6. Involucral bracts herbaceous, green, ciliate margined, puberulent .... 6. *L. scabra*
6. Involucral bracts scarious margined, usually colored, glabrous or nearly so 7
7. Heads uniform in size; corolla pubescent toward base within .... 7. *L. aspera*
7. Terminal head distinctly larger than others; corolla glabrous within .... 8. *L. ligulistylis*

1. L. cylindracea Michx. Blazing-star. Aug.–Sept. Prairies; occasional throughout Ill. except for the southernmost cos. where it is rare or absent.

2. L. squarrosa (L.) Michx. Blazing-star. June–Sept. Dry open woods and prairies; occasional in s. Ill. Rarely white-flowered specimens (f. *alba* Evers & Thieret) are observed. (Including var. *hirsuta* [Rydb.] Gaiser.)

3. L. punctata Hook. Blazing-star. Aug.–Oct. Adventive from w. Along railroads, rare; DuPage Co. (Including var. *nebraskana* Gaiser.)

4. L. pycnostachya Michx. Prairie Blazing-star. July–Sept. Prairies; common throughout Ill.

5. L. spicata (L.) Willd. Marsh Blazing-star. July–Sept. Prairies and wet meadows; not common, but scattered throughout Ill. Rarely white-flowered specimens (f. *albiflora* Britt.) are observed.

6. L. scabra (Greene) K. Schum. Blazing-star. Sept.–Oct. Prairies and open woods; occasional in the s. one-third of Ill.; also Coles Co.

7. L. aspera Michx. Rough Blazing-star. Sept.–Oct. Prairies; common throughout Ill. Rarely white-flowered specimens (f. *benkei* [Macb.] Fern.) are observed. (Including var. *intermedia* [Lunell] Gaiser and *L.* X *sphaeroidea* Michx.)

8. L. ligulistylis (A. Nels.) K. Schum. Blazing-star. Aug.–Sept. Mesic to moist prairies, rare; Cook and Montgomery cos. (Including *L.* X *nieuwlandii* [Lunell] Gaiser.)

Reported hybrids: *L.* X *ridgewayi* Standl. (*L. pycnostachya* X *L. squarrosa*), Richland Co.; *L.* X *steelei* Gaiser (*L. aspera* X *L. spicata*), Lake Co.; *L.* X *gladewitzii* (Farwell) Shinners (*L. aspera* X *L. cylindracea*), Lake Co.; *L. cylindracea* X *L. spicata*, Lake Co.

50. Vernonia Schreb.   Ironweed

1. Tips of involucral bracts long-filiform; heads 12–20 mm wide . . . .
. . . . 1. *V. arkansana*
1. Tips of involucral bracts obtuse to abruptly acuminate; heads 4–12 mm wide 2
2. Leaves tomentose to tomentellous beneath 3
2. Leaves glabrous or scabrous-hirtellous beneath (sometimes tomentellous on veins) 4
3. Tips of involucral bracts obtuse to mucronate, erect or slightly spreading . . . . 2. *V. missurica*
3. Tips of involucral bracts abruptly acuminate, squarrose . 3. *V. baldwinii*
4. Leaves glabrous, conspicuously punctate beneath; outer pappus of short capillary bristles . . . . 4. *V. fasciculata*
4. Leaves scabrous-hirtellous, scarcely punctate beneath; outer pappus of scale-like bristles . . . . 5. *V. gigantea*

1. V. arkansana DC. Aug.–Sept. Low open woods and prairies, very rare; Champaign Co. (original site since destroyed). *V. crinita* Raf.— F, G.

2. V. missurica Raf. Missouri Ironweed. July–Sept. Low open woods and prairies; common throughout Ill. Rarely rose-flowered (f. *carnea* Standl.) and white-flowered (f. *swinkii* Steyerm.) specimens are observed.

3. V. baldwinii Torr. Baldwin's Ironweed. July–Sept. Prairies and open ground; occasional in s. one-half of Ill. (Including var. *interior* (Small) Schub.).

4. V. fasciculata Michx. Common Ironweed. July–Aug. Low ground and prairies; common in n. half of Ill. occasional in s. half.

5. V. gigantea (Walt.) Trel. Tall Ironweed. July–Sept. Low woods and open ground; common throughout most of Ill. (Including var. *taeniotricha* Blake). *V. altissima* Nutt.—F, G, J.

51. *Elephantopus* L.   *Elephant's-foot*

1. E. carolinianus Willd. Elephant's-foot. Aug.–Sept. Dry woods; occasional in s. third of Ill.

52. *Echinops* L.   *Globe Thistle*

1. E. sphaerocephalus L. Globe Thistle. July–Sept. Naturalized from Europe. Waste ground, rare; Coles, Cook, DuPage, Kankakee, and Will cos.

53. *Arctium* L.   *Burdock*

1. Heads more or less corymbose, long-peduncled; petioles strongly angled 2
1. Heads more or less racemose, sessile or short-peduncled; petioles slightly angled .... 3. *A. minus*
2. Petioles solid; involucre glabrous or nearly so .... 1. *A. lappa*
2. Petioles hollow; involucre tomentose .... 2. *A. tomentosum*

1. A. lappa L. Great Burdock. July–Oct. Naturalized from Eurasia. Waste ground; occasional in n. Ill.; also Coles Co.

2. A. tomentosum Mill. Cotton Burdock. June–Sept. Naturalized from Europe. Fields and waste ground; Cook and Morgan cos.

3. A. minus (Hill) Bernh. Common Burdock. July–Sept. Naturalized from Eurasia. Waste ground; occasional to common throughout Ill. Occasional white-flowered (f. *pallidum* Farw.) or laciniate-leaved (f. *laciniatum* Clute) specimens are observed. (Including *A. nemorosum* Lej. & Court.)

54. *Carduus* L.   *Plumeless Thistle. Musk Thistle*

1. Heads solitary, nodding, 3–5 cm wide; stem unwinged immediately below head .... 1. *C. nutans*
1. Heads clustered, not nodding, 1.5–2.5 cm wide; stem winged immediately below head .... 2. *C. acanthoides*

1. C. nutans L. Musk Thistle; Nodding Thistle. June–Oct. Naturalized from Europe. Fields and waste ground; occasional throughout Ill. (Our plants belong to var. *leiophyllus* [Petronic] Arenes.)

2. C. acanthoides L. Plumeless Thistle. July–Sept. Naturalized from Europe. Fields and waste ground; rare in n. Ill., absent elsewhere.

55. *Cirsium* Mill.   *Thistle*

1. Outer and middle involucral bracts conspicuously spine-tipped 2
1. Outer and middle involucral bracts without or with at most a short spine less than 1 mm long 7

2. Stem conspicuously winged with decurrent leaf bases ... 1. *C. vulgare*
2. Stem scarcely or not at all winged with decurrent leaf bases 3
3. Leaves white-tomentose below 4
3. Leaves not white-tomentose below .... 6. *C. pumilum*
4. Leaves pinnately divided into linear lobes; stems persistently white-tomentose .... 2. *C. pitcheri*
4. Leaves pinnately lobed or merely toothed; stems not persistently white-tomentose 5
5. Involucre 2.5–3.5 cm high; peduncles leafy 6
5. Involucre 1.5–2.0 cm high; peduncles naked or nearly so ....
.... 5. *C. carolinianum*
6. Leaves pinnately lobed, margins revolute .... 3. *C. discolor*
6. Leaves merely toothed or shallowly lobed .... 4. *C. altissimum*
7. Heads 1.0–1.5 cm wide; flowers unisexual .... 7. *C. arvense*
7. Heads 2–3 cm wide; flowers perfect .... 8. *C. muticum*

1. C. vulgare (Savi) Tenore. Bull Thistle. July–Aug. Naturalized from Europe. Fields and waste ground; common throughout Ill.

2. C. pitcheri (Torr.) Torr. & Gray. Dune Thistle. June–July. Sand dunes near Lake Michigan, rare; Cook and Lake cos.

3. C. discolor (Mulh.) Spreng. Field Thistle. Aug.–Sept. Fields, open woods, and waste ground; common throughout Ill. Rarely white-flowered specimens (f. *albiflorum* [Britt.] House) are observed.

4. C. altissimum (L.) Spreng. Tall Thistle. Aug.–Sept. Woods; occasional throughout Ill.

5. C. carolinianum (Walt.) Fern. & Schub. Carolina Thistle. June–July. Dry open woods, very rare; Pope Co.

6. C. pumilum (Nutt.) Spreng. Small Prairie Thistle; Hill's Thistle. June–July. Dry prairies; occasional to rare in n. three-fifths of Ill. *C. hillii* (Canby) Fern.—F, G, J.

7. C. arvense (L.) Scop. Canada Thistle. June–Oct. Naturalized from Europe. Leaves merely toothed or shallowly lobed, weakly prickly. Fields and waste ground; abundant throughout Ill. (Including var. *mite* Wimm. & Grab., var. *intergrifolium* Wimm. & Grab., and var. *vestitum* Wimm. & Grab.) *C. setosum* (Willd.) Bieb.—J.

var. horridum Wimm. & Grab. Canada Thistle. June–Oct. Naturalized from Europe. Leaves deeply pinnatifid, strongly prickly. Fields and waste ground; occasional throughout Ill., more common than var. *arvense*.

8. C. muticum Michx. Swamp Thistle. Aug.–Sept. Wet ground; calcareous fens; occasional in n. Ill., extending s. to Wabash Co.

56. *Onopordum* L.   Scotch Thistle

1. O. acanthium L. Scotch Thistle. July–Aug. Naturalized from Europe. Waste ground, rare; Champaign and Cook cos.

57. *Centaurea* L.   Star Thistle. Knapweed

1. Involucral bracts spine tipped 2
1. Involucral bracts not spine tipped 4
2. Central spine of involucral bract stout, 1–3 cm long 3
2. Center spine of involucral bract slender, 1.5–4.0 mm long . 3. *C. diffusa*
3. Stems winged; flowers yellow .... 1. *C. solstitalis*

3. Stems not winged; flowers purple .... 2. *C. calcitrapa*
4. Leaves pinnatifid with linear segments .... 4. *C. maculosa*
4. Leaves entire, toothed, or larger ones few lobed 5
5. Involucral bracts entire margined 6
5. Involucral bracts with pectinate or erose margin 7
6. Outer involucral bracts with scarious tip, inner with plumose hairy tip
.... 5. *C. repens*
6. All involucral bracts similar, coriaceous, and without scarious margin ..
.... 6. *C. moschata*
7. Involucre 2–4 cm high; pappus 6–12 mm long .... 7. *C. americana*
7. Involucre 1.0–1.5 cm high; pappus 3 mm long or less 8
8. Outer involucral bracts with pectinate or erose margins to middle or
below 9
8. Outer involucral bracts with pectinate margins only near tip 11. *C. dubia*
9. Involucral bracts lanceolate, without enlarged tip .... 8. *C. cyanus*
9. Involucral bracts with enlarged tip 10
10. Outer involucral bracts with pectinate margin, inner with erose margin
.... 9. *C. nigra*
10. All involucral bracts with erose margins .... 10. *C. lacea*

1. C. solstitalis L. Yellow Star Thistle; Barnaby's Thistle. July–Sept. Naturalized from Mediterranean region. Waste ground; Jackson Co.

2. C. calcitrapa L. Star Thistle. June–Oct. Naturalized from Mediterranean region. Waste ground, rare; Grundy Co.

3. C. diffusa Lam. July–Sept. Naturalized from se. Europe. Waste ground, rare; Marshall and Will cos.

4. C. maculosa Lam. Spotted Knapweed. July–Sept. Naturalized from Europe. Waste ground; occasional in the n. half of the state, absent elsewhere.

5. C. repens L. Russian Knapweed. June–Aug. Naturalized from Caucasus region. Waste ground, rare; Winnebago Co.

6. C. moschata L. Sweet Sultan. July–Sept. Introduced from sw. Asia and occasionally escaped from cultivation. Waste ground, rare; Champaign Co.

7. C. americana Nutt. American Basket Flower. July–Sept. Adventive from sw. U.S. and occasionally escaped from cultivation. Waste ground, rare; Lawrence and Wabash cos.

8. C. cyanus L. Bachelor's Button; Corn Flower. July–Sept. Introduced from Europe and escaped from cultivation. Waste ground; occasional throughout Ill.

9. C. nigra L. Black Knapweed. July–Sept. Naturalized from Europe. Waste ground, rare; Boone and Cook cos.

10. C. jacea L. Brown Knapweed. July–Sept. Naturalized from Europe. Waste ground, rare; Champaign, DuPage, and Winnebago cos.

11. C. dubia Suter. Tyrol Knapweed. July–Sept. Naturalized from Europe. Waste ground, rare; DuPage, Gallatin, Lake, and Stephenson cos. (Our plants belong to subsp. *vochinensis* [Bernh.] Heyek.) *C. vochinensis* Bernh.—F, J.

58. Cnicus L.   Blessed Thistle

1. C. benedictus L. Blessed Thistle. May–Sept. Introduced from Europe and escaped from cultivation. Waste ground, rare; Champaign Co.

59. Cichorium L.   Chicory

1. C. intybus L. Chicory. June–Oct. Naturalized from Europe. Waste ground; common throughout Ill. (Including f. *roseum* Neum. and f. *album* Neum.)

60. Lapsana L.   Nipplewort

1. L. communis L. Nipplewort. June–Sept. Naturalized from Europe. Waste ground; not common.

61. Microseris D. Don   Prairie Dandelion

1. M. cuspidata (Pursh) Sch.-Bip. Prairie Dandelion. April–June. Dry prairies, rare; n. half of Ill. *Agoseris cuspidata* (Pursh) D. Dietr.—F, J.

62. Krigia Schreb.   Dwarf Dandelion

1. Pappus with capillary bristles 2
1. Pappus minute or absent .... 4. *K. oppositifolia*
2. Pappus with 20–40 bristles 3
2. Pappus of 5–7 ovate scales and 5 (rarely 10) alternating bristles ....
.... 3. *K. virginica*
3. Plants scapose; heads solitary .... 1. *K. dandelion*
3. Plants with a few cauline leaves; heads several .... 2. *K. biflora*

1. K. dandelion (L.) Nutt. Potato Dandelion; Dwarf Dandelion. April–June. Open woods; occasional in s. third of Ill.
2. K. biflora (Walt.) Blake. False Dandelion. May–Sept. Open woods and prairies; common throughout Ill. (Including f. *glandulifera* Fern.)
3. K. virginica (L.) Willd. Dwarf Dandelion. Apr.–Aug. Sandy soil; occasional in n. and w. cent. Ill., rare elsewhere.
4. K. oppositifolia Raf. Dwarf Dandelion. March–April Sandy soil; occasional in s. third of Ill. *Serinia oppositifolia* (Raf.) Kuntze—F, J.

63. Hypochaeris L.   Cat's-ear

1. Leaves hispid; all achenes beaked .... 1. *H. radicata*
1. Leaves glabrous or pubescent only on midrib; outer achenes beakless, inner beaked .... 2. *H. glabra*

1. H. radicata L. Cat's-ear. May–Aug. Naturalized from Europe. Waste ground, very rare; Champaign, Cook, and McHenry cos.
2. H. glabra L. Smooth Cat's-ear. May–Aug. Naturalized from Europe. Waste ground, very rare; Jackson Co.

64. Leontodon L.   Hawkbit

1. Pappus of all flowers with a single row of plumose bristles ....
.... 1. *L. autumnalis*
1. Pappus of two types, inner double, with plumose and setiform bristles, outer reduced to a short irregular crown .... 2. *L. leysseri*

1. L. autumnalis L. Fall Dandelion. June–Oct. Naturalized from Europe. Waste ground, rare; Champaign Co., not collected since 1891.

2. L. leysseri (Wallr.) G. Beck. Hawkbit. June–Sept. Naturalized from Europe. Waste ground, rare; Cook Co.

65. *Picris L.   Bitterweed*

1. Outer involucral bracts ovate, 3.5–8.0 mm wide; achenes with distinct slender beak .... 1. *P. echioides*
1. Outer involucral bracts linear-lanceolate, less than 3 mm wide; achenes with little or no beak .... 2. *P. hieracioides*

1. P. echioides L. Bristly Ox-tongue. July–Sept. Naturalized from Europe. Waste ground, rare; Hancock Co.

2. P. hieracioides L. Cat's-ear. July–Sept. Naturalized from Europe. Waste ground, rare; Menard Co.

66. *Tragopogon L.   Goat's-beard*

1. Flowers purple .... 1. *T. porrifolius*
1. Flowers yellow 2
2. Involucral bracts longer than flowers; peduncles enlarged below head .... 2. *T. dubius*
2. Involucral bracts equalling or shorter than flowers; peduncles not or scarcely enlarged below head .... 3. *T. pratensis*

1. T. porrifolius L. Salsify; Vegetable-oyster. June–Aug. Naturalized from Europe. Fields and waste ground; occasional throughout Ill.

2. T. dubius Scop. Sand Goat's-beard. May–June. Naturalized from Europe. Fields and waste ground; common throughout Ill. *T. major* Jacq.—F.

3. T. pratensis L. Common Goat's-beard. May–Sept. Naturalized from Europe. Fields and waste ground; common throughout Ill.
Reported hybrid: (*T. pratensis* X *T. porrifolius*), Kane Co.

67. *Taraxacum Zinn   Dandelion*

1. Achenes greenish-brown .... 1. *T. officinale*
1. Achenes red to reddish-brown .... 2. *T. laevigatum*

1. T. officinale Weber. Common Dandelion. March–Nov. Naturalized from Europe. Fields, lawns, and waste ground; very common throughout Ill.

2. T. laevigatum (Willd.) DC. Red-seeded Dandelion. April–July. Naturalized from Europe. Waste ground; common throughout Ill. *T. erythrospermum* Andrz.—F, J.

68. *Sonchus L.   Sow Thistle*

1. Heads 3.5 cm in diameter; flowers bright yellow .... 1. *S. arvensis*
1. Heads 1.0–2.5 cm in diameter; flowers pale yellow 2
2. Basal auricles of leaves acute; achenes longitudinally ribbed and papillate .... 2. *S. oleraceus*

2. Basal auricles of leaves rounded; achenes longitudinally ribbed, otherwise smooth .... 3. *S. asper*

      1. S. arvensis L. Field Sow Thistle. July–Sept. Involucres and peduncles with coarse gland-tipped hairs. Naturalized from Europe. Fields and waste ground; occasional in n. half of Ill.
      var. glabrescens Guenth., Gram., & Winn. Smooth Sow Thistle. July–Sept. Involucres and peduncles glabrous or nearly so. Naturalized from Europe. Fields and waste ground; abundant in n. half of Ill., extending s. to Clay Co. *S. uliginosus* Bieb.—F, G, J.
      2. S. oleraceus L. Common Sow Thistle. July–Oct. Naturalized from Europe. Fields and waste ground; common throughout Ill. (Including f. *integrifolium* [Wallr.] Beck and f. *lacerus* [Willd.] Beck.)
      3. S. asper (L.) Hill. Spiny Sow Thistle. July–Oct. Naturalized from Europe. Fields and waste ground; occasional throughout Ill. (Including f. *inermis* [Bisch.] Beck and f. *glandulosus* Beck.)

69. *Lactuca* L.   Lettuce

1. Leaves broadly ovate .... 1. *L. sativa*
1. Leaves elliptic to linear-lanceolate 2
2. Achenes with 1 (–3) prominent median nerves on each face 3
2. Achenes with several prominent nerves on each face 4
3. Involucre 10–15 mm high in fruit; achenes 4.5–6.0 mm long (beak included) .... 2. *L. canadensis*
3. Involucre 15–22 mm high in fruit; achenes 7–10 mm long (beak included) .... 3. *L. ludoviciana*
4. Flowers yellow; achenes with a filiform beak nearly as long as or longer than the body 5
4. Flowers blue to white; achenes with a short stout beak less than half to nearly as long as body or beak absent 6
5. Leaves prickly margined; achenes spinulose above. ..... 4. *L. serriola*
5. Leaves not prickly margined; achenes not spinulose above 5. *L. saligna*
6. Pappus brown .... 6. *L. biennis*
6. Pappus white 7
7. Leaves toothed, commonly pinnately lobed; involucre 10–13 mm high; heads 0.6–1.0 cm in diameter .... 7. *L. floridana*
7. Leaves entire margined, lower commonly pinnately lobed; involucre 16–20 mm high; heads 1–2 cm in diameter .... 8. *L. tatarica*

      1. L. sativa L. Garden Lettuce. July–Sept. Introduced from Eurasia and escaped from cultivation, not naturalized. Waste ground; occasional throughout Ill.
      2. L. canadensis L. Wild Lettuce. June–Aug. Waste ground; common throughout Ill. (Including f. *angustata* Wieg.; var. *obovata* Wieg.; var. *obovata* f. *steelei* [Britt.] Fern.; var. *obovata* f. *stenopoda* Wieg.; var. *longifolia* [Michx.] Farw.; var. *longifolia* f. *angustipes* Wieg.; var. *latifolia* Ktze.; var. *latifolia* f. *villicaulis* Fern.; and var. *latifolia* f. *exauriculata* Wieg.)
      3. L. ludoviciana (Nutt.) DC. Prairie Lettuce. July–Sept. Dry prairies; rare in n. half of Ill., extending s. to Monroe Co. Blue-flowered specimens (f. *campestris* [Greene] Fern.) are generally found more frequently than the typical yellow-flowered specimens.
      4. L. serriola L. Prickly Lettuce. July–Sept. Naturalized from Europe.

Waste ground; common throughout Ill. (Including var. *integrata* [Gren. & Godr.] Farw.) *L. scariola* L.—F, J.

5. L. saligna L. Willow Lettuce. July–Aug. Naturalized from Europe. Waste ground; common in s. two-thirds of Ill., absent elsewhere. (Including f. *ruppiana* [Wallr.] Beck.)

6. L. biennis (Moench) Fern. Tall Blue Lettuce. Aug.–Sept. Moist open woods; occasional throughout Ill. (Including f. *integrifolia* [Torr. & Gray] Fern.)

7. L. floridana (L.) Gaertn. Woodland Lettuce. July–Sept. Woods; common throughout Ill. (Including f. *leucantha* Fern. and var. *villosa* [Jacq.] Cronq.)

8. L. tatarica (L.) C. A. Mey. Blue Lettuce. July–Aug. Adventive from w. U.S. Waste ground; Cook, DuPage, Kane, and Lake cos. (Our plants belong to ssp. *pulchella* [Pursh] Steb.). *L. pulchella* [Pursh] DC.—F, G, J.

Reported hybrids: *L. canadensis* X *L. ludoviciana*, Cook Co.; *L.* X *morssii* Robins. (*L. canadensis* X *L. biennis*), Ill., fide Fernald.

70. Pyrrhopappus DC.   False Dandelion

1. P. carolinianus (Walt.) DC. False Dandelion. May–June. Dry woods and prairies; occasional in s. half of Ill.

71. Crepis L.   Hawksbeard

1. C. capillaris (L.) Wallr. Hawksbeard. June–July. Naturalized from Europe. Waste ground, rare; Cook, DuPage, and Lake cos.

72. Prenanthes L.   Rattlesnake Root

1. Involucral bracts pubescent 2
1. Involucral bracts glabrous 4
2. Inflorescence narrow and elongate 3
2. Inflorescence corymbose-paniculate .... 3. *P. crepidinea*
3. Flowers purplish; stems and leaves glabrous and usually glaucous .....
.... 1. *P. racemosa*
3. Flowers cream-colored; stems and leaves pubescent ......2. *P. aspera*
4. Involucral bracts 7–10, mostly 8; flowers 8–15, mostly 13 per head; pappus reddish-borwn .... 4. *P. alba*
4. Involucral bracts 4–6, mostly 5; flowers 5–6 per head; pappus straw-colored .... 5. *P. altissima*

1. P. racemosa Michx. Glaucous White Lettuce. Aug.–Sept. Involucral bracts 7–10, mostly 8; flowers 9–16 per head, mostly 13. Moist ground; occasional throughout Ill.

subsp. multiflora Cronq. White Lettuce. Aug.–Sept. Involucral bracts 10–14, mostly 13; flowers 17–26 per head, mostly 21. Moist ground; occasional throughout Ill.

2. P. aspera Michx. Rough White Lettuce. Aug.–Sept. Prairies; occasional throughout Ill.

3. P. crepidinea Michx. Great White Lettuce. Aug.–Oct. Moist open woods; occasional throughout Ill.

4. P. alba L. White Lettuce; Rattlesnake Root. Aug.–Sept. Woods; occasional in n. half of Ill., absent elsewhere.

5. P. altissima L. Tall White Lettuce. Aug.–Sept. Woods; occasional in e. and s. Ill.; also Cook and Lake cos. (Our plants belong to var. *cinnamomea* Fern.)

73. Hieracium L. Hawkweed

1. Plants with well-developed cluster of basal leaves at flowering time . 2
1. Plants without well-developed cluster of basal leaves at flowering time 4
2. Basal leaves rounded to cordate at base; pappus white. 1. *H. murorum*
2. Basal leaves tapering to base; pappus sordid 3
3. Flowers yellow .... 2. *H. pratense*
3. Flowers red-orange .... 3. *H. aurantiacum*
4. Stem leafy nearly to summit; inflorescence leafy bracted 5
4. Stem leafless or nearly so 6
5. Lower leaves largest, others progressively reduced upwards, becoming sessile .... 4. *H. scabrum*
5. Middle and lower leaves similar in size and shape .... 5. *H. canadense*
6. Leaves and lower part of stem with hairs 1–2 cm long   6. *H. longipilum*
6. Leaves and lower part of stem with hairs less than 1 cm long ....
.... 7. *H. gronovii*

1. H. murorum L. Golden Lungwort. June–Aug. Naturalized from Europe. Waste ground; rare; Sangamon Co.

2. H. pratense Tausch. King Devil. June–Aug. Naturalized from Europe. Waste ground, rare; Cook, Lake, and McHenry cos.

3. H. aurantiacum L. Orange Hawkweed; Devil's Paint Brush. June–July. Naturalized from Europe. Waste ground, rare; Lake and Ogle cos.

4. H. scabrum Michx. Hairy Hawkweed. Aug.–Sept. Open ground and dry woods; occasional throughout Ill. (Including var. *intonsum* Fern. & St. John.)

5. H. canadense Michx. Canada Hawkweed. Aug.–Sept. Dry woods, sand barrens; occasional in n. one-fourth of Ill., absent elsewhere. (Including var. *fasciculatum* [Prush] Fern.)

6. H. longipilum Torr. Hairy Hawkweed. July–Sept. Field and open woods; occasional throughout Ill.

7. H. gronovii L. Hairy Hawkweed. July–Sept. Dry open woods; occasional in the s. three-fourths of Ill., rare elsewhere.

Sequence and Summary of the Families

	Family	Genera	Species	Lesser Taxa	Hybrids
1.	Equisetaceae	1	8		4
2.	Lycopodiaceae	1	6	1	1
3.	Selaginellaceae	1	2		
4.	Isoetaceae	1	2		
5.	Ophioglossaceae	2	8	2	
6.	Osmundaceae	1	3		
7.	Hymenophyllaceae	1	1		
8.	Polypodiaceae	18	40	4	10
9.	Marsileaceae	1	1		
10.	Salviniaceae	1	2		
11.	Taxaceae	1	1		
12.	Pinaceae	2	10		
13.	Taxodiaceae	1	1		
14.	Cupressaceae	2	4	1	
15.	Typhaceae	1	2		
16.	Sparganiaceae	1	5		
17.	Ruppiaceae	1	1		
18.	Zannichelliaceae	1	1		
19.	Najadaceae	1	5		
20.	Potamogetonaceae	1	20		2
21.	Juncaginaceae	2	3		
22.	Alismaceae	3	12		
23.	Butomaceae	1	1		
24.	Hydrocharitaceae	3	5		
25.	Poaceae	87	287	42	5
26.	Cyperaceae	13	225	18	2
27.	Araceae	5	6	1	
28.	Lemnaceae	4	14		
29.	Xyridaceae	1	2		

Family	Genera	Species	Lesser Taxa	Hybrids
30. Commelinaceae	2	8		
31. Pontederiaceae	3	4		
32. Juncaceae	2	24	4	
33. Liliaceae	31	63	3	
34. Smilacaceae	1	9	2	
35. Dioscoreaceae	1	2		
36. Iridaceae	4	14	1	1
37. Burmanniaceae	1	1		
38. Orchidaceae	6	43	2	2
39. Saururaceae	1	1		
40. Salicaceae	2	31	5	4
41. Myricaceae	1	1		
42. Juglandaceae	2	12	5	
43. Betulaceae	5	11		2
44. Fagaceae	3	22	1	15
45. Ulmaceae	3	10	5	
46. Moraceae	5	7	1	
47. Urticaceae	5	9	1	
48. Santalaceae	1	1		
49. Loranthaceae	1	1		
50. Aristolochiaceae	2	3	2	
51. Polygonaceae	3	43	5	
52. Chenopodiaceae	9	29	2	
53. Amaranthaceae	4	17		
54. Nyctaginaceae	1	4		
55. Phytolaccaceae	1	1		
56. Aizoaceae	1	1		
57. Portulacaceae	3	6		
58. Caryophyllaceae	17	48		
59. Ceratophyllaceae	1	2		
60. Nymphaeaceae	2	3	1	
61. Nelumbonaceae	1	1		
62. Cabombaceae	2	2		
63. Ranunculaceae	17	57	9	
64. Berberidaceae	4	6		
65. Menispermaceae	3	3		
66. Magnoliaceae	2	2		
67. Annonaceae	1	1		
68. Lauraceae	2	2	2	
69. Papaveraceae	11	21		
70. Capparidaceae	2	4	1	

Family	Genera	Species	Lesser Taxa	Hybrids
71. Cruciferae	36	76	11	
72. Resedaceae	1	1		
73. Sarraceniaceae	1	1		
74. Droseraceae	1	2		
75. Crassulaceae	1	6		
76. Saxifragaceae	11	23		
77. Hamamelidaceae	2	2		
78. Platanaceae	1	1		
79. Rosaceae	23	132	9	1
80. Leguminosae	46	128	9	2
81. Linaceae	1	6		
82. Oxalidaceae	1	5		
83. Geraniaceae	2	8		
84. Zygophyllaceae	2	2		
85. Rutaceae	2	2	1	
86. Simaroubaceae	1	1		
87. Polygalaceae	1	7	2	
88. Euphorbiaceae	9	34	4	
89. Callitrichaceae	1	3		
90. Limnanthaceae	1	1		
91. Anacardiaceae	2	6	2	
92. Aquifoliaceae	1	3		
93. Celastraceae	2	8		
94. Staphyleaceae	1	1		
95. Aceraceae	1	7	3	
96. Hippocastanaceae	1	4	1	
97. Sapindaceae	2	2		
98. Balsaminaceae	1	2		
99. Rhamnaceae	3	8	2	
100. Vitaceae	3	12	3	
101. Tiliaceae	1	2	1	
102. Sterculiaceae	1	1		
103. Malvaceae	11	24	1	
104. Hypericaceae	3	23	3	
105. Elatinaceae	2	2		
106. Cistaceae	3	9	2	
107. Violaceae	2	25	5	1
108. Passifloraceae	1	2		
109. Loasaceae	1	3		
110. Cactaceae	1	2		
111. Thymelaeaceae	1	1		

Family	Genera	Species	Lesser Taxa	Hybrids
112. Elaeagnaceae	2	3		
113. Lythraceae	6	8		
114. Nyssaceae	1	2	1	
115. Melastomaceae	1	2		
116. Onagraceae	6	32	2	
117. Haloragidaceae	2	6		
118. Hippuridaceae	1	1		
119. Araliaceae	3	6		
120. Umbelliferae	34	46	5	
121. Cornaceae	1	10	1	
122. Ericaceae	12	22	1	
123. Primulaceae	9	21	1	1
124. Sapotaceae	1	2		
125. Ebenaceae	1	1		
126. Styracaceae	2	3		
127. Oleaceae	4	8	3	
128. Loganiaceae	2	2		
129. Gentianaceae	6	14		
130. Menyanthaceae	2	2		
131. Apocynaceae	4	7	4	1
132. Asclepiadaceae	3	23	1	
133. Convolvulaceae	6	21	3	
134. Polemoniaceae	5	10	4	
135. Hydrophyllaceae	4	9		
136. Boraginaceae	14	32	1	
137. Verbenaceae	2	9	1	7
138. Phrymaceae	1	1		
139. Labiatae	31	77	13	4
140. Solanaceae	9	32	3	1
141. Scrophulariaceae	26	74	6	
142. Bignoniaceae	3	4		
143. Martyniaceae	1	1		
144. Orobanchaceae	3	6		
145. Lentibulariaceae	1	5		
146. Acanthaceae	3	7	2	
147. Plantaginaceae	1	12	1	
148. Rubiaceae	7	28	4	
149. Caprifoliaceae	7	31		1
150. Adoxaceae	1	1		
151. Valerianaceae	2	8	1	
152. Dipsacaceae	1	2		

Family	Genera	Species	Lesser Taxa	Hybrids
153. Cucurbitaceae	7	8		
154. Campanulaceae	3	14	2	1
155. Compositae	73	290	25	15
Totals	830	2699	265	83

Additional Taxa

The following taxa new to the Illinois flora have been discovered too late to be incorporated in the text.

BORAGINACEAE—BORAGE FAMILY

Asperugo procumbens L. Madwort. June–July. Native of Europe; waste ground; Kane Co.

CALYCANTHACEAE—CALYCANTHUS FAMILY

Calycanthus floridus L. Strawberry Bush. May. Native of se. U.S.; escaped from cultivation into a woods; Jackson Co.

CARYOPHYLLACEAE—PINK FAMILY

Cerastium pumilum Curtis. Dwarf Chickweed. April. Native of Europe; adventive in waste ground; Jackson Co.

CELASTRACEAE—BITTERSWEET FAMILY

Euonymus alatus (Thunb.) Sieb. Burning-bush. May–June. Native of Asia; commonly cultivated; adventive in woodland; Coles Co.

COMMELINACEAE—SPIDERWORT FAMILY

Tradescantia subaspera Ker var. montana (Shuttlew.) Anders. & Woodson. June–July. Moist woods; Jackson Co.

COMPOSITAE—SUNFLOWER FAMILY

Aster undulatus L. Aug.–Oct. Dry woodlands; Saline Co.
Grindelia lanceolata Nutt. Tarweed. July–Sept. Rocky prairies; Grundy Co.

CYPERACEAE—SEDGE FAMILY

Eleocharis parvula (Roem. & Schultes) Link. Dwarf Spikerush. July–Sept. Sandy soil along margin of lake; Coles Co.

DIOSCOREACEAE—WILD YAM FAMILY

Dioscorea batatas Dcne. Chinese Yam. June–Sept. Native of Asia; escaped into waste ground; Jackson Co.

ERICACEAE—HEATH FAMILY

Chimaphila maculata (L.) Pursh. Spotted Wintergreen. June–July. Dry woods; Pope Co.

GENTIANACEAE—GENTIAN FAMILY

Gentiana septemfida Pall. Gentian. July–Sept. Native of Asia; escaped from cultivation into waste ground; Cook Co.

LILIACEAE—LILY FAMILY

Lycoris radiata Herb. Surprise Lily. July–Sept. Native of Japan; escaped from cultivation to roadsides; Jackson Co.

LINACEAE—FLAX FAMILY

Linum perenne L. ssp. lewisii (Pursh) Hult. Flax. June–Sept. Native of w. U.S.; escaped from cultivation into waste ground; Grundy Co.

MARSILEACEAE—WATER CLOVER FAMILY

Pilularia americana A. Br. June–Oct. Shallow water and shoreline of lake; Pope Co.

OLEACEAE—OLIVE FAMILY

Ligustrum obtusifolium Sieb. & Zucc. Blunt-leaved Privet. June–July. Native of Asia; escaped from cultivation into waste ground; Jackson Co.

ORCHIDACEAE—ORCHID FAMILY

Isotria medeoloides (Pursh) Raf. Small Whorled Pogonia. May. Dry, rocky woods; Randolph Co.

Spiranthes romanzoffiana Cham. Hooded Ladies'-tresses. Aug.–
Sept. Meadows and thickets; Coles, Cook, and Peoria cos.

POACEAE—GRASS FAMILY

Pennisetum alopecuroides (L.) Spreng. Aug–Oct. Native of Asia;
adventive along a stream; Crawford Co.

RANUNCULACEAE—BUTTERCUP FAMILY

Eranthis hyemalis Salisb. Winter Aconite. April–May. Native of
Asia; escaped from cultivation; St. Clair Co.
Ranunculus arvensis L. Spiny-fruited Buttercup. May–June. Na-
tive of Europe; escaped along a railroad; Jackson Co.

RHAMNACEAE—BUCKTHORN FAMILY

Rhamnus davurica Pall. Buckthorn. May–June. Native of Asia; escaped from cultivation; DuPage Co.

ROSACEAE—ROSE FAMILY

Filipendula ulmaria (L.) Maxim. Queen-of-the-Meadow. June–Aug.
Native of Europe; escaped from cultivation into waste ground; Kane
Co.
Potentilla canadensis L. Cinquefoil. May–June. Dry soil; DeKalb
Co.
Rosa wichuriana Crepin. Memorial Rose. June–July. Native of
Asia; escaped from cultivation into waste ground; Jackson Co.

SCROPHULARIACEAE—FIGWORT FAMILY

Verbascum pulverulenta Schrad. July–Oct. Native of Europe; ad-
ventive in waste ground; Lake Co.
Veronica agrestis L. Field Speedwell. June–Sept. Native of
Europe; adventive in waste ground; Kane Co.

UMBELLIFERAE—CARROT FAMILY

Spermolepis echinata (Nutt.) Heller. May–June. Native of s.w.
U.S.; adventive along a railroad; Jackson Co.

Glossary

Index

Glossary

Acaulescent. Seemingly without aerial stems.

Achene. A type of one-seeded, dry, indehiscent fruit with the seed coat not attached to the mature ovary wall.

Actinomorphic. Having radial symmetry; regular, in reference to a flower.

Acuminate. Gradually tapering to a point.

Acute. Sharply tapering to a point.

Adaxial. Toward the axis; when referring to a leaf, the upper surface.

Ament. A spike of unisexual, apetalous flowers; a catkin.

Anther. The terminal part of a stamen which bears pollen.

Anthesis. Flowering time.

Antrorse. Projecting forward.

Apical. At the apex.

Apiculate. Abruptly short-pointed at the tip.

Appressed. Lying flat against the surface.

Areole. A small area between leaf veins.

Aril. An appendage of the seed, usually enclosing the seed.

Arillate. Bearing an aril.

Aristate. Bearing an awn.

Articulated. Jointed.

Attenuate. Gradually becoming narrowed.

Auricle. An ear-like lobe.

Auriculate. Bearing an ear-like process.

Awn. A bristle usually terminating a structure.

Axillary. Borne from an axil.

Beard. A tuft of stiff hairs.

Berry. A type of fruit where the seeds are surrounded only by fleshy material.

Bidentate. Having two teeth.

Bifid. Two-cleft.

Bifoliolate. Bearing two leaflets.

Bifurcate. Forked.

Biglandular. Bearing two glands.

Bilabiate. Two-lipped.

Bipinnate. Divided once into distinct segments, with each segment in turn divided into distinct segments.

Bipinnatifid. Divided part way to the center, with each lobe again divided partway to the center.

Bisexual. Referring to a flower which contains both stamens and pistils.

Biternate. Divided into three segments two times.

Bract. An accessory structure at the base of many flowers, usually appearing leaflike.

Bracteate. Bearing one or more bracts.

Bracteole. A secondary bract.

Bractlet. A small bract.

Bristle. A stiff hair or hair-like growth; a seta.

Bulblet. A small bulb.

Callosity. Any hardened thickening.

Callus. A hard swollen area at the outside base of a lemma or palea.

Calyx. The outermost segments of the perianth of a flower, composed of sepals.

Campanulate. Bell-shaped.

Canescent. Grayish-hairy.

Capillary. Thread-like.

Capitate. Forming a head.

Capsule. A dry, dehiscent fruit composed of more than one carpel.

Carpel. A simple pistil, or one member of a compound pistil.

Cartilaginous. Firm but flexible.

Caruncle. A fleshy outgrowth near the point of attachment of a seed.

Catkin. A spike of unisexual, apetalous flowers; an ament.

Caudate. With a tail-like appendage.

Caudex. The woody base of a perennial plant.

Cauline. Belonging to a stem.

Cespitose. Growing in tufts.

Chaffy. Covered with scales.

Chartaceous. Papery.

Cilia. Marginal hairs.

Ciliate. Bearing cilia.

Ciliolate. Bearing small cilia.

Circumscissile. Usually referring to a fruit which dehisces by a horizontal, circular line.

Claw. A narrow, basal stalk, particularly of a petal.

Coherent. The growing together of like parts.

Coma. A tuft of hairs at the end of a seed.

Compressed. Flattened.

Concave. Curved on the inner surface; opposed to convex.

Connate. Union of like parts.

Connective. That portion of a stamen between the two anther halves.

Convex. Curved on the outer surface; opposed to concave.

Convolute. Rolled lengthwise.

Coralline. Having the texture of coral.

Cordate. Heart-shaped.

Coriaceous. Leathery.

Corm. An underground, vertical stem with scaly leaves, differing from a bulb by lacking fleshy leaves.

Corolla. The segments of a flower just within the calyx, composed of petals.

Corona. A crown of petal-like structures.

Corrugated. Folded or wrinkled.

Corymb. A type of inflorescence where the pedicellate flowers are arranged along an elongated axis but with the flowers all attaining about the same height.

Corymbiform. Shaped like a corymb.

Cotyledon. A seed leaf.

Crateriform. Cone-shaped but sunken in the center at the top.

Crenate. With round teeth.

Crenulate. With small, round teeth.

Crest. A ridge.

Crisped. Curled.

Cucullate. Hood-shaped.

Culm. The stem which terminates in an inflorescence.

Cuneate. Wedge-shaped; tapering to the base.

Cupular. Shaped like a small cup.

Cuspidate. Terminating in a very short point.

Cyathium. A cup-like structure enclosing flowers.

Cyme. A type of broad and flattened inflorescence in which the central flowers bloom first.

Cymose. Bearing a cyme.

Cystolith. A stone-like concretion in the epidermis of some plants.

Deciduous. Falling away.

Decumbent. Lying flat, but with the tip ascending.

Decurrent. Adnate to the petiole or stem and then extending beyond the point of attachment.

Deflexed. Turned downward.

Dehiscent. Splitting at maturity.

Deltoid. Triangular.

Dentate. With sharp teeth, the tips of which project outward.

Denticulate. With small, sharp teeth, the tips of which project outward.

Diffuse. Loosely spreading.

Digitate. Radiating from a common point, like the fingers from a hand.

Dilated. Swollen; expanded.

Dimorphic. Having two forms.

Dioecious. With staminate flowers on one plant, pistillate flowers on another.

Disarticulate. To come apart; to become disjointed.

Disk. An enlarged outgrowth of the receptacle.

Divergent. Spreading apart.

Drupe. A type of fruit in which the seed is surrounded by a hard, dry covering which, in turn, is surrounded by fleshy material.

Ebracteate. Without bracts.

Echinate. Spiny.

Eciliate. Without cilia.

Eglandular. Without glands.

Ellipsoid. Referring to a solid object which is broadest at the middle, gradually tapering to both ends.

Elliptic. Broadest at middle, gradually tapering to both ends.

Emarginate. Having a shallow notch at the extremity.

Epidermis. The outermost layer of cells.

Epunctate. Without dots.

Erose. With an irregularly notched margin.

Exudate. Secreted material.

Falcate. Sickle-shaped.

Fascicle. Cluster.

Ferruginous. Rust-colored.

Fetid. Ill-smelling.

Fibrous. Referring to roots borne in tufts.

Filament. That part of the stamen supporting the anther.

Filiform. Thread-like.

Fimbriate. Fringed.

Flaccid. Weak; flabby.

Flexible. Able to be bent readily.

Flexuous. Zigzag.

Floret. A small flower.

Follicle. A type of dry, dehiscent fruit which splits along one side at maturity.

Friable. Breaking easily into small particles.

Frond. The leaf of a fern; the vegetative structure in the Lemnaceae.

Funnelform. Shaped like a funnel.

Galea. A hooded portion of a perianth.

Gemmae. Bulblets or buds capable of developing into new plants.

Geniculate. Bent.

Gibbous. Swollen on one side.

Glabrate. Becoming smooth.

Glabrous. Without pubescence or hairs.

Gland. An enlarged, spherical body functioning as a secretory organ.

Glandular. Bearing glands.

Glaucous. With a whitish covering which can be rubbed off.

Globose. Round; globular.

Glochidium. A process bearing barbs.

Glomerule. A small compact cluster.

Glume. A sterile scale subtending a spikelet.

Glutinous. Covered with a sticky secretion.

Grain. The fruit of most grasses.

Hastate. Spear-shaped; said of a leaf which is triangular with spreading basal lobes.

Hirsute. With stiff hairs.

Hirsutulous. With minute stiff hairs.

Hirtellous. Finely hirsute.

Hispid. With rigid hairs.

Hispidulous. With minute rigid hairs.

Hoary. Grayish-white, usually referring to pubescence.

Hood. That part of an organ, usually of a flower, which is strongly concave and arching.

Horn. An accessory structure found in certain flowers.

Hyaline. Transparent.

Hypanthium. A development of the receptacle beneath the calyx.

Hypogynium. A structure subtending the ovary in *Scleria*.

Imbricate. Overlapping.

Indehiscent. Not splitting open at maturity.

Indurate. Hard.

Indusium. An outgrowth of the blade which covers or partially covers the sori.

Inferior. Referring to the position of the ovary when it is surrounded by the adnate portion of the floral tube or is embedded in the receptacle.

Inflexed. Turned inward.

Inflorescence. A cluster of flowers.

Internode. The area between two adjacent nodes.

Involucel. A cluster of bracteoles which subtends a secondary flower cluster.

Involucre. A circle of bracts which subtends a flower cluster.

Involute. Rolled inward.

Keel. A ridgelike process.

Laciniate. Divided into narrow, pointed divisions.

Lanceolate. Lance-shaped; broadest near base, gradually tapering to the narrower apex.

Lanceoloid. Referring to a solid object which is broadest near base, gradually tapering to the narrower apex.

Latex. Milky juice.

Leaflet. An individual unit of a compound leaf.

Legume. A dry fruit usually dehiscing along two sides at maturity.

Lemma. A scale subtending a floret in grasses.

Lenticel. Corky openings on bark of twigs and branches.

Lenticular. Lens-shaped.

Lepidote. Scaly.

Ligule. An elongated tongue-like process found on the leaf of *Isoetes* and *Selaginella;* a structure on the inner surface of the leaf of grasses and sedges at the junction of the blade and the sheath.

Linear. Elongated and uniform in width throughout.

Lip. The lowermost, often greatly modified, petal of a flower.

Lobulate. With small lobes.

Locular. Referring to the locule, or cavity of the ovary or the anther.

Locule. The cavity of an ovary or an anther.

Loculicidal. Said of a capsule which splits down the dorsal suture of each cell.

Lodicule. A small rudimentary structure at the base of some grass flowers.

Loment. A fruit divided into 1-seeded segments by transverse partitions.

Lunate. Crescent-shaped.

Lustrous. Shiny.

Lyrate. Pinnatifid, with the terminal lobe much larger than the lower ones.

Megaspore. A spore produced in a megasporangium; a spore which produces a female gametophyte.

Microspore. A spore produced in a microsporangium; a spore which produces a male gametophyte.

Moniliform. Constricted at regular intervals to resemble a string of beads.

Monoecious. Bearing both sexes in separate flowers on the same plant.

Monomorphic. Having but one form.

Mucilaginous. Slimy.

Mucro. A short, abrupt tip.

Mucronate. Possessing a short, abrupt tip.

Mucronulate. Possessing a very short, abrupt tip.

Muricate. Minutely spiny.

Nectariferous. Producing nectar.

Nigrescent. Blackish.

Node. That place on the stem from which leaves and branchlets arise.

Nutlet. A small nut.

Oblanceolate. Reverse lance-shaped; broadest at apex, gradually tapering to narrow base.

Oblong. Broadest at the middle, and tapering to both ends, but broader than elliptic.

Oblongoid. Referring to a solid object which, in side view, is nearly the same width throughout.

Obovoid. Referring to a solid object which is broadly rounded at the apex, becoming narrowed below.

Obpyramidal. Referring to an upside-down pyramid.

Obtuse. Rounded at the apex.

Ocrea. A sheathing stipule, often tubular.

Ocreola. A secondary, usually tubular, sheath.

Opaque. Incapable of being seen through.

Orbicular. Round.

Oval. Broadly elliptic.

Ovary. The lower swollen part of the pistil which produces the ovules.

Ovate. Broadly rounded at base, becoming narrowed above; broader than lanceolate.

Ovoid. Referring to a solid object which is broadly rounded at the base, becoming narrowed above.

Palea. The scale opposite the lemma which encloses the flower.
Palmate. Divided radiately, like the fingers of a hand.
Pandurate. Fiddle-shaped.
Panduriform. Fiddle-shaped.
Panicle. A type of inflorescence composed of several racemes.
Papilla. A small wart.
Papillate. Bearing small warts, or papillae.
Papillose. Bearing pimple-like processes.
Pappus. The modified calyx in the Compositae.
Papule. A pimple-like projection.
Pectinate. Pinnatifid into close, narrow segments; comb-like.
Pedicel. The stalk of a flower of an inflorescence.
Pedicellate. Bearing a pedicel.
Peduncle. The stalk of an inflorescence.
Peduncled. Provided with a peduncle.
Peltate. Attached away from the margin, in reference to a leaf.
Perennial. Living more than two years.
Perfect. Bearing both stamens and pistils in the same flower.
Perfoliate. Referring to a leaf which appears to have the stem pass through it.
Perianth. Those parts of a flower including both the calyx and corolla.
Pericarp. The ripened ovary wall.
Perigynium. A sac-like covering enclosing the achene in *Carex*.
Petal. One segment of the corolla.
Petaloid. Resembling a petal in texture and appearance.
Petiolate. Bearing a petiole, or leafstalk.
Petiole. The stalk of a leaf.
Petiolulate. Bearing a petiolule, or leaflet-stalk.
Petiolule. The stalk of a leaflet.
Pilose. Bearing soft hairs.
Pinna. A primary division of a compound blade.
Pinnate. Divided once into distinct segments.
Pinnatifid. Said of a simple leaf or leaf-part which is cleft or lobed only part way to its axis.
Pinnule. The secondary segments of a compound blade.
Pistil. The ovule-producing organ of a flower normally composed of an ovary, a style, and a stigma.
Pistillate. Bearing pistils but not stamens.
Plano-convex. Flat on the inner face, rounded on the outer face.
Plicate. Folded.

Plumose. Bearing fine hairs, like the plume of a feather.
Procumbent. Lying on the ground.
Prophyll. A bracteole.
Prophyllate. Bearing a bracteole.
Pruinose. Having a waxy covering.
Puberulent. With minute hairs.
Pubescent. Bearing some kind of hairs.
Punctate. Dotted.
Pyramidal. Shaped like a pyramid.
Pyriform. Pear-shaped.

Quadrate. Four-sided.

Raceme. A type of inflorescence where pedicellate flowers are arranged along an elongated axis.
Rachilla. The flower-bearing axis in a grass spikelet.
Rachis. A primary axis.
Receptacle. That part of the flower to which the perianth, stamens, and pistils are usually attached.
Reflexed. Turned downward.
Reniform. Kidney-shaped.
Repand. Wavy along the margin.
Repent. Creeping.
Resinous. Producing a sticky secretion, or resin.
Resupinate. Upside-down.
Reticulate. Resembling a network.
Reticulum. A network.
Retrorse. Pointing downward.
Retuse. Shallowly notched at a rounded apex.
Revolute. Rolled under from the margin.
Rhizome. An underground horizontal stem, bearing nodes, buds, and roots.
Rhombic. Becoming quadrangular.
Rosette. A cluster of leaves in a circular arrangement at the base of a plant.
Rotate. Flat and circular.
Rufescent. Reddish-brown.
Rufous. Red-brown.
Rugose. Wrinkled.
Rugulose. With small wrinkles.

Saccate. Sac-shaped.
Sagittate. Shaped like an arrowhead.

Salverform. Referring to a tubular corolla which abruptly expands into a flat limb.

Samara. An indehiscent winged fruit.

Scaberulous. Slightly rough to the touch.

Scabrous. Rough to the touch.

Scape. A leafless stalk bearing a flower or inflorescence.

Scarious. Thin and membranous.

Scurfy. Bearing scaly particles.

Secund. Borne on one side.

Sepaloid. Resembling a sepal in texture.

Septate. With dividing walls.

Septicidal. Said of a capsule which splits between the locules.

Sericeous. Silky; bearing soft, appressed hairs.

Serrate. With teeth which project forward.

Serrulate. With very small teeth which project forward.

Sessile. Without a stalk.

Seta. Bristle.

Setaceous. Bearing bristles, or setae.

Setiform. Bristle-shaped.

Setose. Bearing setae.

Silicle. A short silique.

Silique. An elongated capsule with a central partition separating the valves.

Sinuate. Wavy along the margins.

Sinus. The cleft between two lobes or teeth.

Sorus. An aggregation of sporangia.

Spadix. A fleshy axis in which flowers are embedded.

Spathe. A large sheathing bract subtending or usually enclosing an inflorescence.

Spatulate. Oblong, but with the basal end elongated.

Spicate. Bearing a spike.

Spike. A type of inflorescence where sessile flowers are arranged along an elongated axis.

Spikelet. A small spike.

Spinescent. Becoming spiny.

Spinose. Bearing spines.

Spinule. A small spine.

Spinulose. Bearing small pines.

Sporangium. That structure which produces spores.

Spore. That structure formed within a sporangium which gives rise to the gametophyte generation.

Sporocarp. A compound structure containing the sporangia in *Marsilea* and *Azolla*.

Sporophyll. A structure subtending the sporangia.

Spur. A sac-like extension of the flower.

Stamen. The pollen-producing organ of a flower composed of a filament and an anther.

Staminate. Bearing stamens but not pistils.

Staminodium. A sterile stamen.

Standard. The upper, usually enlarged, petal of a pea-shaped flower.

Stellate. Star-shaped.

Stipe. A stalk.

Stipel. A small stipe.

Stipitate. Possessing a stipe.

Stipular. Pertaining to a stipule.

Stipule. A leaf-like or scaly structure found at the point of attachment of a leaf to the stem.

Stolon. A slender, horizontal stem on the surface of the ground.

Stoloniferous. Bearing stolons.

Stomate. An opening in the epidermis of a leaf.

Stramineous. Straw-colored.

Striate. Marked with grooves.

Strigillose. With short, appressed, straight hairs.

Strigose. With appressed, straight hairs.

Strigulose. With short, appressed, straight hairs.

Style. That elongated part of the pistil between the ovary and the stigma.

Subcuneate. Nearly wedge-shaped.

Suborbicular. Nearly spherical.

Subulate. With a very short, narrow point.

Succulent. Fleshy.

Suffused. Spread throughout; flushed.

Superior. Referring to the position of the ovary when the free floral parts arise below the ovary.

Supra-axillary. Borne above the axil.

Tendril. A spiralling, coiling structure which enables a climbing plant to attach itself to a supporting body.

Terete. Round, in cross-section.

Ternate. Divided three times.

Testa. The seed coat.

Tetrad. A cluster of four.

Tomentose. Pubescent with matted wool.

Tomentulose. Finely pubescent with matted wool.

Tometum. Woolly hair.

Torulose. With small contractions.

Translucent. Partly transparent.

Trifoliolate. Divided into three leaflets.

Trigonous. Triangular in cross-section.

Truncate. Abruptly cut across.

Tuber. An underground fleshy stem formed as a storage organ at the end of a rhizome.

Tubercle. A small, wart-like process.

Tuberculate. Warty.

Turgid. Swollen to the point of bursting.

Umbel. A type of inflorescence in which the flower stalks arise from the same level.

Undulate. Wavy.

Unisexual. Bearing either stamens or pistils in one flower, but not both.

Urceolate. Urn-shaped.

Utricle. A small, one-seeded, indehiscent fruit with a thin covering.

Verrucose. Warty.

Verticil. A whorl.

Verticillate. Whorled.

Villous. With long, soft, slender, unmatted hairs.

Virgate. Wand-like.

Viscid. Sticky.

Whorl. An arrangement of three or more structures at a point on the stem.

Zygomorphic. Bilaterally symmetrical.

Index to Families and Genera

Index to Common Names